Lecture Notes in Electrical Engineering

Volume 1264

The book series *Lecture Notes in Electrical Engineering* (LNEE) publishes the latest developments in Electrical Engineering—quickly, informally and in high quality. While original research reported in proceedings and monographs has traditionally formed the core of LNEE, we also encourage authors to submit books devoted to supporting student education and professional training in the various fields and applications areas of electrical engineering. The series cover classical and emerging topics concerning:

- Communication Engineering, Information Theory and Networks
- Electronics Engineering and Microelectronics
- Signal, Image and Speech Processing
- Wireless and Mobile Communication
- Circuits and Systems
- Energy Systems, Power Electronics and Electrical Machines
- Electro-optical Engineering
- Instrumentation Engineering
- Avionics Engineering
- Control Systems
- Internet-of-Things and Cybersecurity
- Biomedical Devices, MEMS and NEMS

For general information about this book series, comments or suggestions, please contact leontina.dicecco@springer.com.

To submit a proposal or request further information, please contact the Publishing Editor in your country:

China

Jasmine Dou, Editor (jasmine.dou@springer.com)

India, Japan, Rest of Asia

Swati Meherishi, Editorial Director (Swati.Meherishi@springer.com)

Southeast Asia, Australia, New Zealand

Ramesh Nath Premnath, Editor (ramesh.premnath@springernature.com)

USA, Canada

Michael Luby, Senior Editor (michael.luby@springer.com)

All other Countries

Leontina Di Cecco, Senior Editor (leontina.dicecco@springer.com)

**** This series is indexed by EI Compendex and Scopus databases. ****

Gaurav Raj · Bhuvan Unhelker · Ankur Choudhary
Editors

Advances in Artificial Intelligence and Machine Learning

Select Proceedings of ICAAAIML 2023, Volume 1

Springer

Editors
Gaurav Raj
Department of Computer Science
and Engineering, Sharda School
of Engineering and Technology
Sharda University
Greater Noida, India

Ankur Choudhary
Accenture
Noida, India

Bhuvan Unhelker
University of South Florida
Sarasota–Manatee
Sarasota, FL, USA

ISSN 1876-1100 ISSN 1876-1119 (electronic)
Lecture Notes in Electrical Engineering
ISBN 978-981-97-9506-2 ISBN 978-981-97-9507-9 (eBook)
https://doi.org/10.1007/978-981-97-9507-9

This Springer imprint is published by the registered company Springer Nature Singapore Pte Ltd.
The registered company address is: 152 Beach Road, #21-01/04 Gateway East, Singapore 189721, Singapore

Preface

Volume 1: AIML with IoT, Blockchain, and Healthcare

In an era marked by the convergence of emerging technologies, the fusion of Artificial Intelligence and Machine Learning (AIML) with transformative fields like the Internet of Things (IoT), Blockchain, and Healthcare has ushered in a new paradigm of innovation and possibility. As we stand at the intersection of these dynamic disciplines, the potential for revolutionizing healthcare delivery, enhancing patient outcomes, and optimizing operational efficiencies is boundless.

Volume 1 of this compendium embarks on a journey through the intricate tapestry of AIML applications within the realms of IoT, Blockchain, and Healthcare. From pioneering research to real-world implementations, this volume serves as a comprehensive guide for both seasoned professionals and curious enthusiasts alike.

In these pages, readers will explore the fundamental principles underpinning AIML technologies, unravel the intricacies of IoT ecosystems, delve into the decentralized landscape of Blockchain, and witness firsthand the profound impact these innovations are making in the healthcare domain. Each chapter offers a unique blend of theoretical insights, practical methodologies, and compelling case studies, providing readers with a holistic understanding of how AIML is reshaping the future of healthcare delivery.

This volume is not merely a static compilation of knowledge; it is a dynamic testament to the ongoing evolution of AIML in tandem with IoT, Blockchain, and Healthcare. As such, it is my sincere hope that this volume will inspire readers to embark on their own explorations, spark new avenues of inquiry, and ultimately contribute to the collective endeavor of leveraging technology for the betterment of humanity.

Together, let us embark on this voyage of discovery, as we navigate the nexus of AIML, IoT, Blockchain, and Healthcare, charting a course towards a future defined by innovation, compassion, and empowerment.

Greater Noida, India
Sarasota, USA
Noida, India

Gaurav Raj
Bhuvan Unhelker
Ankur Choudhary

Contents

About the Editors

Gaurav Raj works at Sharda University, Greater Noida, India, and has done B.Tech. in CSE from UPTU, M.Tech. in SE from NIT, Allahabad, and Ph.D. in CSE from Punjab Technical University, India. His research areas are artificial intelligence, cloud computing, software engineering, and web services and security. He has more than 11 years of teaching experience. He has several research papers to his credit that have been published in reputed journals and conferences. He also has several patents published to his credit. He has vast experience in organizing faculty development programs, workshops, seminars, and conferences.

Bhuvan Unhelker is an Accomplished IT Professional and Professor of IT at the University of South Florida, the USA. He is also a Founding Consultant at Method Science and Co-Founder and Director at PlatiFi. His research areas are business analysis and requirements modeling, software engineering, big data strategies, agile processes, mobile business, and green IT. His domain experience is banking, financial, insurance, government, and telecommunications. He has designed, developed, and customized a suite of industrial courses, which are regularly delivered to business executives and IT professionals globally, including those in Australia, the USA, the UK, China, India, Sri Lanka, New Zealand, Singapore, and Malaysia. His thought leadership is reflected through multiple Cutter Executive Reports, various journals, and 21 books. He is the Winner of the Computerworld Object Developer Award (1995), Consensus IT Professional Award (2006), and IT Writer Award (2010). He has a Doctorate in the area of "Object Orientation" from the University of Technology, Sydney, in 1997. He is a Fellow of the Australian Computer Society, a Senior Member of IEEE, a Professional Scrum Master, and a Lifetime Member of the Computer Society of India.

Dr. Ankur Choudhary is an Accomplished ML Engineering Manager at Accenture, specializing in artificial intelligence and machine learning. With a Ph.D. in Computer Science and Engineering from Gautam Buddha University, he brings over 18 years of experience in technical leadership and project management. Dr. Choudhary has previously held Senior Roles at D.S. Group Ltd. and Sharda University, contributing significantly to AI research and education. He has published 50+ papers in esteemed journals and conferences, focusing on AI, image processing, and software engineering. His work has earned him multiple best paper awards, including at AICAI-2019 and DoSCI-2021. Known for his dedication to leveraging AI for transformative change, Dr. Choudhary continues to drive innovation and deliver tangible business outcomes. He remains active in academia and industry, aiming to advance the field of AI and inspire the next generation of technologists.

Leveraging Federated Explainable AI with Blockchain for Secure Healthcare Data Analysis: A Promising Approach for Improved Patient Outcomes

Swati Sharma, Vivek Tomar, and Sangeeta Arora

Abstract To address significant issues in healthcare data exchange and analysis, this research paper investigates the integration of Federated Explainable Artificial Intelligence (Fed XAI) with blockchain technology. Fed XAI is a cutting-edge method that blends Explainable AI (XAI) approaches with Federated Learning (Fed) privacy-preserving features to enable collaborative model training without jeopardising the confidentiality of sensitive patient data. Participating healthcare organisations share model updates or gradients rather than raw data, assuring data security. The ecosystem is improved using blockchain technology since it provides data integrity, transparency and immutability. The blockchain's decentralisation streamlines patient care and administrative procedures by enabling seamless data interchange and interoperability among healthcare providers. By offering concise and understandable justifications for the actions taken by AI models, explainable AI empowers healthcare providers. This encourages confidence and trust in AI-driven recommendations, enhancing clinical decision-making. Healthcare organisations are encouraged to submit data through token-based incentives, fostering collaboration and improving model precision and generalizability. The potential for combining Fed XAI and blockchain in healthcare data exchange and analysis is highlighted in this research paper. To fully realise the promise of this game-changing solution, which will ultimately improve patient care and healthcare efficiency, collaboration efforts across healthcare organisations, AI developers and blockchain experts are essential.

Keywords Federated learning · XAI · Blockchain · Machine learning · Artificial intelligence · Healthcare

---

S. Sharma (✉) · S. Arora
KIET Group of Institutions Delhi-NCR, Delhi, India
e-mail: swatish.3006@gmail.com

V. Tomar
Graphic Era (Deemed-to-be-University), Dehradun, India

G. Raj et al. (eds.), *Advances in Artificial Intelligence and Machine Learning*, Lecture Notes in Electrical Engineering 1264, https://doi.org/10.1007/978-981-97-9507-9_1

1 Introduction

Artificial intelligence (AI) and machine learning (ML) algorithms have become a fundamental cornerstone for developing cutting-edge services and applications. These technologies are now widely adopted in both the public and private sectors, playing integral roles in various aspects of daily operations. They excel in tasks such as pattern recognition, recommendation systems, natural language translation, virtual assistants, speech and text identification, predictive analytics, medical diagnosis, and more, showcasing their effectiveness as offshoots of artificial intelligence [19].

However, a crucial challenge in harnessing the power of AI and ML lies in the availability and quality of data necessary for creating precise and successful ML models. In reality, much of the data suitable for training ML models is scattered and often inaccessible. Many industries face limited access to data, and when available, the data quality may be subpar. While there is a willingness among businesses to collaborate and exchange data, numerous obstacles stand in the way. These hurdles encompass administrative, technological and protocol issues, as well as concerns regarding cost and competition within industries. All of these factors pose a significant bottleneck to realising our vision of interconnected, highly intelligent Machine Learning Models.

Recognising this challenge, the research community is increasingly alarmed. The conventional approach to training ML models typically involved collecting data at one location, cleaning it at another and conducting model training, possibly at a third location. The resulting model could then be applied elsewhere. This process relied heavily on data exchange between different parties. However, in light of evolving standards and requirements, this practice must be reevaluated to ensure data security, privacy and compliance.

In order to stop data leaks and address privacy and security issues, Google introduced a brand-new concept in 2017 called "federated learning" [5, 6, 13]. Instead of using centralised data as in the standard method, this approach builds machine learning models using dispersed data from numerous devices. The conventional method entails data owners moving their information to a centralised place or cloud, where it is combined with other information to create the models. In contrast, because the Federated approach uses data that is stored locally at each of the contributing parties' locations, all calculations, data cleaning and model creation are carried out locally. To create a machine learning model that is more comprehensive and centralised, the various models are integrated on the cloud. Each participant offers their own distinctive model in place of the data. This kind of framework utilises a framework for data partitioning, and each partition represents a portion of data samples obtained from one or more users [24]. A machine learning model for application in the healthcare industry can be successfully trained via federated learning.

Federated Learning ensures that private healthcare information is stored locally inside some institutions, eliminating the need for centralised data warehouses and reducing privacy concerns. By exchanging model updates or gradients rather than raw data, participating parties actively train an AI model while protecting the secrecy and privacy of the data.

By providing clear and understandable justification for the decisions made by the AI model, Explainable AI techniques increase the confidence of the Fed XAI system. Healthcare practitioners who comprehend the model's rationale will be better able to apply that knowledge to make confident decisions.

Parallel to this, the ecosystem is strengthened by blockchain technology, which provides immutability, transparency and data integrity. Blockchain makes sure that data is trustworthy and reliable by making sure it is impermeable and resistant to unauthorised access. Simple data sharing between healthcare providers has dramatically boosted interoperability, enhancing patient care.

Fed XAI with blockchain is a viable option for improving healthcare data analysis when considering the potential benefits in terms of better patient outcomes, increased data privacy and simplified healthcare operations. This is true despite the challenges the strategy poses, such as data quality and regulatory compliance.

2 Related Work

According to Kitchenham [4], the purpose of a systematic literature review is to broaden the scope of the existing body of knowledge and spot any gaps through in-depth investigation. These suggestions were kept in mind as the writers conducted a literature review for the current work.

An electronic record of patient or health data is necessary to deliver high-quality healthcare. The creation of a machine learning model using this information can be used to carry out a number of tasks, including the efficient distribution of medications, the provision of personalised healthcare services, and the diagnosis and prognosis of illnesses, to name a few. A decrease in pointless medical exams [15], a reduction in cost for healthcare professionals [17] and improved medical education [18] are a few examples of the benefits of using electronic records in medicine.

Machine learning techniques, particularly EMR-driven Boltzmann machines, have found valuable applications in predicting age-related muscular degeneration [9] and Type 2 Diabetes well before the onset of these conditions [11, 20]. Additionally, they have been utilised to assess the risk of suicide [9]. Various methodologies, including regression, decision trees, k-nearest neighbour and support vector machines, have been employed in these contexts. These applications have had a positive impact on stakeholders' perceptions of the healthcare system [14]. However, their effectiveness relies heavily on the availability of healthcare data, which must be easily and promptly accessible.

The sensitivity of electronic medical information has raised valid concerns regarding data privacy, security and access. Healthcare data is now distributed across various locations, including pharmacies, personal devices and beyond traditional healthcare providers such as hospitals [3, 23]. Consequently, the application of machine learning to assess healthcare data in practice can be challenging. This challenge has spurred researchers to explore innovative solutions for the swift, secure and confidential sharing of medical data among stakeholders [10, 12].

One recent solution involves the use of Unmanned Aerial Vehicles (UAVs) equipped with body sensor hives to collect individuals' health data, which is then securely stored on a Blockchain-based platform [16]. Experimental results demonstrate the practicality of this approach for safely collecting and safeguarding healthcare data against potential adversaries.

Later, federated learning was published in [7, 8, 12], and its theoretical foundations were looked at. The authors provide a useful method for learning across various devices while preserving the privacy and autonomy of each one. There have been new initiatives recently to broaden the FL framework as it is outlined in [1].

Most current attempts to practically integrate AI onto blockchain technology are white paper recommendations, with no discernible real-world applications. The methodology proposed in [2] builds a machine learning platform on SC that allows users to advertise jobs and contribute models to address existing issues. Additionally, a distributed AI computing infrastructure has been presented [25] that enables mining nodes to generate revenue by processing AI models.

3 Background of Technologies Used

3.1 Federated Learning

Federated Learning is a machine learning approach that allows multiple devices or entities to collaboratively train a shared model while keeping their data localised and decentralised. In traditional machine learning, data is typically collected, centralised and processed on a single server or data centre. However, federated learning aims to address concerns related to data privacy, security and bandwidth by enabling training on distributed data sources without sharing the raw data itself. Several advantages of federated learning involve privacy, decentralisation, security, efficiency, diversity, etc. Federated Learning has gained significant attention and interest in the healthcare domain due to its potential to address the challenges associated with sharing sensitive patient data while still benefiting from collaborative model training. Healthcare data is often highly regulated and subject to strict privacy requirements, making federated learning an attractive approach for conducting machine learning tasks without compromising patient confidentiality. Figure 1 shows the topology of federated learning.

3.2 Federated Learning with XAI

Federated Learning combined with explainable artificial intelligence (XAI) holds significant promise for healthcare applications. XAI enhances the transparency and interpretability of machine learning models, allowing healthcare professionals to

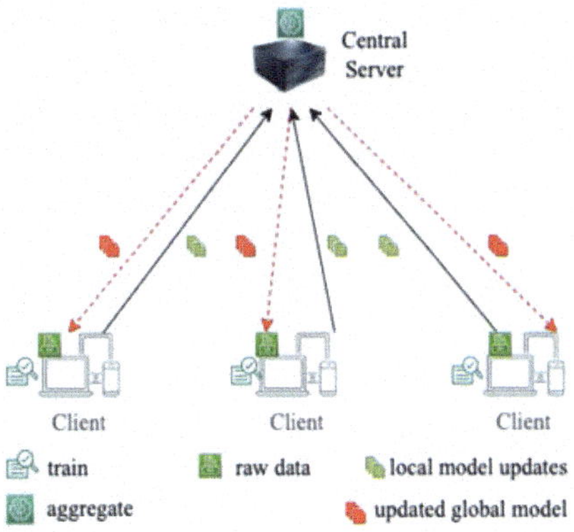

Fig. 1 Topology of traditional federated learning [22]

understand and trust the decisions made by these models. When applied in a federated learning setting, XAI can provide insights into how the model is behaving across different healthcare institutions while maintaining data privacy and security. Federated learning with XAI can benefit healthcare in several ways.

3.2.1 Interpretability

XAI techniques, such as feature importance visualisation, can help healthcare practitioners understand which factors or variables contribute to model predictions. This is crucial for clinical decision-making and helps ensure that the model's recommendations align with medical expertise.

3.2.2 Model Monitoring

XAI tools can help monitor the behaviour of federated models over time, allowing healthcare institutions to detect any unexpected shifts in model performance or biases that may arise as data distributions change.

3.2.3 Trust and Acceptance

By providing transparent explanations for model decisions, XAI helps build trust among healthcare professionals and patients, making them more likely to adopt and rely on AI-driven tools for diagnosis, treatment planning and patient care.

3.2.4 Collaborative Model Improvement

XAI can facilitate communication and collaboration among different healthcare institutions by providing intuitive explanations for model behaviour. This enables institutions to collectively identify and rectify model weaknesses. Overall, the combination of federated learning and XAI has the potential to revolutionise healthcare by providing accurate, interpretable and privacy-preserving.

3.3 Blockchain

Blockchain is a decentralized digital ledger technology that ensures secure and transparent transactions across a network of computers. It is most famously associated with cryptocurrencies like Bitcoin, allowing for peer-to-peer transactions without intermediaries. Beyond finance, blockchain finds applications in supply chain management, healthcare, identity verification, smart contracts, voting systems and more. Each transaction is recorded in a block, cryptographically linked to the previous one, creating an immutable chain of data. However, challenges like scalability, energy consumption, regulation and user experience need to be addressed as blockchain technology continues to evolve and find new ways to revolutionize industries. Some of the Applications of blockchain include Cryptocurrencies, Supply Chain Management, Healthcare, Identity Verification, Smart Contracts, Voting Systems, Financial Services, etc.

4 Proposed Methodology

In the current study, we are proposing a methodology of training the AI models on healthcare data using FED XAI with blockchain. The model uses the positive features of all the three techniques and thereby can be applied over healthcare data to provide a secured environment for the AI models to get trained. Federated explainable artificial intelligence (FED XAI) combined with blockchain technology offers a potent framework for privacy-preserving, transparent and trustworthy AI applications. FED XAI leverages federated learning and explainability techniques, enabling multiple entities to collaboratively train models while maintaining data privacy and providing human-interpretable explanations. By integrating blockchain, the transparency and security of the entire process can be elevated. Each federated model update and associated explanation can be cryptographically recorded on the blockchain, creating an immutable audit trail of model evolution and decision rationales across participating nodes. This approach enhances accountability, ensures the integrity of model updates and allows stakeholders to verify the authenticity of explanations. The combination of FED XAI and blockchain holds promise for domains like healthcare, finance and supply chain management, where privacy, interpretability and auditability are paramount (see Fig. 2).

The diagram could consist of several interconnected components and arrows to represent the flow of data and interactions. Hospitals act as data sources. Arrows extending from data sources to the federated learning network, indicating data contributions. XAI depicts a machine learning model with explainability (e.g. decision tree) within the federated learning network. Blockchain Layer is positioned beneath the federated learning network and connected to it. Arrows connecting from the blockchain layer to the federated learning network represent smart contracts that govern data sharing, access and rewards.

5 Training a FED XAI Model with Blockchain

This integration involves a collaborative process that ensures privacy, transparency and accountability. Here's how training will take place in such a model: Step 1: Data Distribution and Local Training:

- Healthcare institutions (nodes) each possess their own localized datasets.
- At each institution, XAI techniques are applied to build models that generate human-interpretable explanations for predictions.
- Local training is conducted using the institution's data and XAI-enhanced models, allowing for fine-tuning and optimization.

Step 2: Federated Learning Setup:

- A central coordinating entity manages the federated learning process.
- Model parameters or updates, such as gradients, are shared between the central entity and participating institutions. These updates are sent over secure channels, maintaining privacy.

Step 3. Aggregation with Blockchain:

- After each institution completes its local training round, model updates and associated explanations are aggregated.

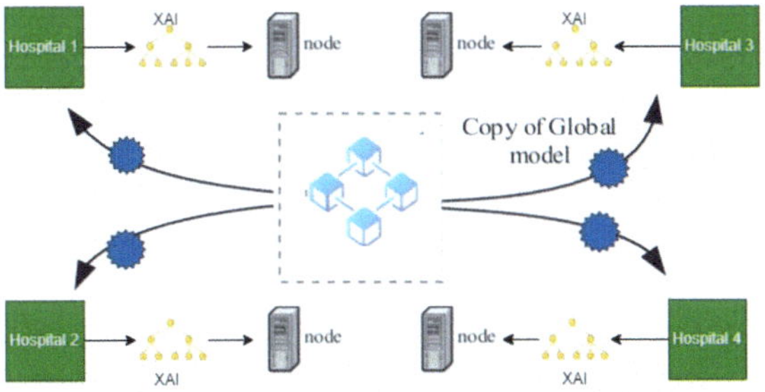

Fig. 2 Healthcare data training model using FED XAI with blockchain

- Blockchain technology is used to record each model update and its corresponding explanation in a new block.
- Cryptographic hashing links each block to the previous one, creating an immutable and tamper-proof chain of training events.

Step 4. Global Model Update:

- The central entity integrates the aggregated model updates, creating an improved global model that benefits from the collective knowledge of all participating institutions.
- This updated global model is then disseminated back to the institutions.

Step 5. Interpretability and Accountability:

- When the global model is used for making predictions, XAI-generated explanations accompany each prediction.
- These explanations are derived from the integrated model's features and the blockchain-recorded training history, providing insights into the reasoning behind predictions.

Step 6. Continuous Iteration:

- The training process continues iteratively, with local institutions periodically refining their models and contributing to the global model's advancement.
- Each iteration's model updates and explanations are added to the blockchain, maintaining a comprehensive and transparent history (see Fig. 3).

6 Incentivizing the Data Sources

The combination of FED XAI and blockchain ensures that the training process is decentralized, secure and auditable. Institutions collaborate to improve the model collectively while preserving data privacy and providing interpretable explanations for predictions. Blockchain's transparency and immutability contribute to the trustworthiness and accountability of the entire training lifecycle. Hospitals play a critical role in the healthcare ecosystem and incentivizing them to share data or participate in initiatives can lead to improved patient outcomes, research advancements and overall healthcare quality. Thus, some rewards can be provided to the participating nodes such as Financial Compensation, Research Collaboration Opportunities, Access to Advanced Medical Technologies, Clinical Trials and Drug Development, Enhanced Reputation and Recognition, etc. [21].

Rewards may be given in the form of Shapley value. We select the shapley value as the allocation method because it is frequently used in collaboration and sharing contexts. Shapley value, which can more accurately quantify the contribution of participants to the collective, is a metric based on the marginal contribution of the influence of players joining the collective in different orders on the collective

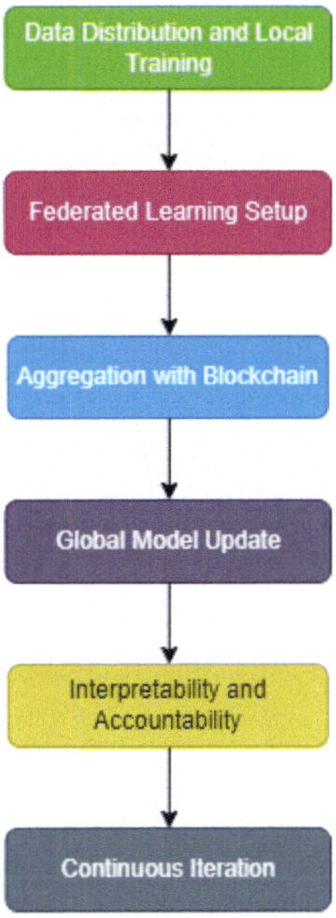

Fig. 3 Model training steps

benefits. However, there are other assessment factors besides contribution value. We offer reputation points to assess the sincerity of participants, preventing malicious individuals from sharing incorrect parameters to harm the model.

7 Conclusion and Future Scope

The use of FedXAI and blockchain technology in the healthcare sector shows enormous promise for transforming the accessibility, security and administration of data in the field. Healthcare organisations may work together to train AI models on distributed data sources using FedXAI, a cutting-edge federated learning platform, while maintaining data ownership and privacy. This decentralised strategy reduces the dangers of centralising private patient data while facilitating the creation of robust and highly accurate AI models.

The FedXAI framework enables tamper-proof, transparent and auditable data transfers by utilising blockchain technology. Smart contracts make it possible for predefined criteria to be executed automatically, improving data sharing agreements, consent management and guaranteeing that only authorised parties have access to data. In addition to providing healthcare organisations with cutting-edge AI capabilities, the synergistic combination of FedXAI and blockchain builds confidence among all stakeholders, including patients, healthcare providers, researchers and regulators.

But there are also certain difficulties with implementing FedXAI with blockchain in the healthcare industry. To fully realise the potential of this technology stack, technical challenges, scalability issues, regulatory compliance and the requirement for standardised protocols must be solved. To tackle these challenges, interdisciplinary cooperation between healthcare professionals, AI researchers, blockchain developers and policymakers is essential. Unlocking the full potential of FedXAI with blockchain in healthcare and guiding the sector towards a more resilient and patient-centric future will require continued research, development and collaboration.

References

1. Bonawitz, K., Eichner, H., Grieskamp, W., Huba, D., Ingerman, A., Ivanov, V., Kiddon, C.M., Konečný, J., Mazzocchi, S., McMahan, B., Overveldt, T.V., Petrou, D., Ramage, D., Roselander, J.: Towards federated learning at scale: system design. In: SysML 2019 (2019)
2. Chen, Z., Wang, W., Yan, X., Tian, J.: Cortex-AI on blockchain: the decentralized AI autonomous system. Cortex White Paper (2018)
3. Dubovitskaya, A., Xu, Z., Ryu, S., Schumacher, M., Wang, F.: Secure and trustable electronic medical records sharing using blockchain. In: AMIA Annual Symposium Proceedings, vol. 2017, p. 650. American Medical Informatics Association (2017)
4. Kitchenham, B.: Procedures for performing systematic reviews. Keele, UK, Keele Univ. **33**(2004), 1–26 (2004)
5. Konečnỳ, J., McMahan, H.B., Ramage, D., Richtárik, P.: Federated optimization: distributed machine learning for on-device intelligence (2016). arXiv:1610.02527
6. Konečnỳ, J., McMahan, H.B., Yu, F.X., Richtárik, P., Suresh, A.T., Bacon, D.: Federated learning: strategies for improving communication efficiency (2016). arXiv:1610.05492
7. Konecny, J., McMahan, H.B., Yu, F.X., Richtarik, P., Suresh, A.T., Bacon, D.: Federated learning: strategies for improving communication efficiency. In: NIPS Workshop on Private Multi-Party Machine Learning (2016)
8. Konečný, J., McMahan, B., Ramage, D.: Federated optimization: distributed optimization beyond the datacenter (2015). arXiv:1511.03575
9. Lee, C.S., Baughman, D.M., Lee, A.Y.: Deep learning is effective for classifying normal versus age-related macular degeneration oct images. Ophthalmol. Retin. **1**(4), 322–327 (2017)
10. Liu, D., Miller, T., Sayeed, R., Mandl, K.D.: FADL: federated-autonomous deep learning for distributed electronic health record (2018). arXiv:1811.11400
11. Mani, S., Chen, Y., Elasy, T., Clayton, W., Denny, J.: Type 2 diabetes risk forecasting from EMR data using machine learning. In: AMIA Annual Symposium Proceedings, vol. 2012, p. 606. American Medical Informatics Association (2012)
12. McMahan, B., Ramage, D.: Federated learning: collaborative machine learning without centralized training data (2017)
13. Mcmahan, H., Moore, E., Ramage, D., Com, B.: Federated learning of deep networks using model averaging Blaise Agüera y arcas (2012). arXiv:1602.05629
14. Miotto, R., Wang, F., Wang, S., Jiang, X., Dudley, J.T.: Deep learning for healthcare: review, opportunities and challenges. Brief. Bioinform. **19**(6), 1236–1246 (2018)
15. Podichetty, V., Penn, D.: The progressive roles of electronic medicine: benefits, concerns, and costs. Am. J. Med. Sci. **328**(2), 94–99 (2004)
16. Raj, J.S.: Security enhanced blockchain based unmanned aerial vehicle health monitoring system. J. IoT Soc. Mob. Anal. Cloud **3**(2), 121–131 (2021)

17. Simon, S.J., Simon, S.J.: An examination of the financial feasibility of electronic medical records (EMRs): a case study of tangible and intangible benefits. Int. J. Electron. Healthc. **2**(2), 185–200 (2006)
18. Tierney, M.J., Pageler, N.M., Kahana, M., Pantaleoni, J.L., Longhurst, C.A.: Medical education in the electronic medical record (EMR) era: benefits, challenges, and future directions. Acad. Med. **88**(6), 748–752 (2013)
19. Tomar, V., Kumar, N., Srivastava, A.R.: Single sample face recognition using deep learning: a survey. Artif. Intell. Rev., 1–49 (2023)
20. Tran, T., Nguyen, T.D., Phung, D., Venkatesh, S.: Learning vector representation of medical objects via EMR-driven nonnegative restricted Boltzmann machines (eNRBM). J. Biomed. Inform. **54**, 96–105 (2015)
21. Wang, Z., Yan, B., Dong, A.: Blockchain empowered federated learning for data sharing incentive mechanism. Procedia Comput. Sci. **202**, 348–353 (2022)
22. Wang, Z., Hu, Q.: Blockchain-based federated learning: a comprehensive survey (2021). arXiv:2110.02182
23. Yang, J.J., Li, J.Q., Niu, Y.: A hybrid solution for privacy preserving medical data sharing in the cloud environment. Futur. Gener. Comput. Syst. **43**, 74–86 (2015)
24. Yang, K., Fan, T., Chen, T., Shi, Y., Yang, Q.: A quasi-newton method based vertical federated learning framework for logistic regression (2019). arXiv:1912.00513
25. Yong, H., Lee, C., Wang, D.: Artificial intelligence computing platform driven by blockchain. Technical report, DeepBrain Chain, Singapore (2017)

Understanding the Behaviors and Detecting the Mental Health Symptoms Among Students Using Machine Learning Algorithms

Sangita Pokhrel, Swathi Ganesan, Soonleh Ling, and Nalinda Somasiri

Abstract International students studying in the UK are facing challenges in their life that can impact their mental health. To better understand and address these challenges, this research paper investigates the behaviors of the students and the potential machine learning algorithms to predict the mental health of them. A form was distributed to international students at various universities in the UK. The survey aimed to gather information on aspects like age, country of origin, sleep patterns, emotional behaviors, support from family and the university, financial pressures, experiences of loneliness, and living standards, all in an effort to gain insights into their mental well-being. The collected data has been cleaned, pre-processed, and visualized the findings such as whether male or female are affected much based on the collected data, their sleeping behavior, financial stress, and so on. Later, machine learning algorithms such as decision tree, CatBoost, random forest, and XGBoost classifiers were used to predict the mental health status of the students, and the random forest works best with 88% accuracy with the highest precision of 0.92 and F1-score of 0.82. The findings from this research have the potential to offer valuable perspectives and assistance in improving mental health services for international students in the UK. By using machine learning algorithms, this study aims to improve upon traditional methods of mental health prediction and provide a more efficient and accurate means of identifying and supporting students in need.

Keywords Emotional background · International students · Machine learning algorithms · Decision tree · CatBoost · Random forest classifier · XGBoost classifier

---

S. Pokhrel (✉) · S. Ganesan · S. Ling · N. Somasiri
York St John University, London Campus, York 14 2BA, UK
e-mail: s.pokhrel@yorksj.ac.uk

S. Ganesan
e-mail: s.ganesan@yorksj.ac.uk

S. Ling
e-mail: s.ling@yorksj.ac.uk

N. Somasiri
e-mail: n.somasiri@yorksj.ac.uk

G. Raj et al. (eds.), *Advances in Artificial Intelligence and Machine Learning*, Lecture Notes in Electrical Engineering 1264, https://doi.org/10.1007/978-981-97-9507-9_2

1 Introduction

International students in the UK come from a wide variety of countries and cultural backgrounds. They choose to study in the UK for many reasons, including the quality of education, the opportunity to learn English, and the chance to experience life in a new country. The UK has a long history of welcoming international students and has a reputation for excellence in education all over the world. In 2022, the United Kingdom boasted the world's second-largest international student population, with a staggering figure of over half a million, specifically 605,130 students. This group accounts for roughly 20% of the overall students [1]. The United Kingdom plays a crucial role in higher education and scientific research, with its foreign student community consistently showcasing outstanding academic achievements year in and year out. The universities and campuses in the UK offer courses at all levels, from undergraduate degrees to Ph.D. International students can study a wide range of subjects, including technology, engineering, public health, medicine, business, music, IT, and so on. Overall, studying in the UK can be a challenging but rewarding experience for international students, offering the opportunity to learn, grow, and make new connections in a dynamic and diverse country.

Mental health refers to a person's overall emotional and psychological well-being. Mental health is important for everyone, as it refers to the well-being of one's overall emotion and psychology. It can affect how a person thinks, feels, and behaves. Like anyone else, international students in the UK can experience issues affecting their mental health. There are several factors that can contribute to mental health problems, including adjusting to life in a new country where international students may experience homesickness, culture shock, and other challenges as they adapt to life in the UK. There may be academic pressures with high expectations and pressure to succeed academically, which can be stressful. Another factor of stress is financial stability, studying in the UK can be expensive, and may have to manage their finances carefully. And language barriers can affect their ability to socialize and participate in academic and social activities. It is important for them to be aware of the potential impact and to take necessary steps to take care of themselves. This may include seeking support from friends, family, and other trusted individuals; participating in activities that promote mental well-being; and seeking help from mental health professionals if needed.

There are several resources available to international students in the UK who are experiencing mental health problems. Many universities and colleges have counseling services that are available to students, and there are also a few organizations and charities that provide mental health support and services. It is important for students to reach out for help if they are struggling with their mental health.

2 Literature Review

In the paper [2], they have discussed a systematic review of machine learning approaches that have been used to predict issues contributing to mental health. The review searched for relevant research articles on the topic and used the PRISMA methodology in conducting the review. During the review, 30 research articles were analyzed for mental health problems (such as schizophrenia and bipolar disorder). The findings of the review are discussed, along with the challenges and limitations encountered by researchers using machine learning. The pressure on employees in the IT industry and the mental health conditions were published by the paper [3, 4]. Mental health being a measure of a person's well-being, IT workers are more vulnerable to mental disorders as a result of the increasing use of technology and constant pressure [5, 6]. The authors [7] proposed to use various machine learning algorithms to explore stress patterns in working adults based on a survey data. Machine learning algorithms were trained and tested for model accuracy and better prediction. Prediction of mental health problems in adolescence using ML techniques can be seen in [8].

This paper [9] introduces a study that aims to use artificial intelligence (AI) and machine learning to predict anxiety among university students during the COVID-19 pandemic. The study collated data through survey including general information, COVID-19 stressors, and a scale to categorize levels of anxiety. The best model performance was achieved by using AdaBoost and neural network. The methodology [10, 11] involves developing a real-time database by surveying over 190 school children between 14 and 18 years with over 26 questions to analyze their stress level. Machine Learning Algorithms were used to provide early treatment, and the highest accuracy of 88% was achieved using K Nearest Neighbors [12] which discussed the importance of early detection of mental health issues and how it can improve patient's quality of life. The study found that the Stacking technique had the highest accuracy of prediction at 81.75%. The study [13] suggests using advanced analytics and AI to create a decision support system (DSS) that can more efficiently detect and diagnose various mental disorders. The DSS uses the Network Pattern Recognition (NEPAR) algorithm to identify relevant questions and then trains machine learning models using participants' answers and historical data to predict the presence and type of mental disorder. The DSS can accurately diagnose disorders with an 89% accuracy rate using only 28 questions, which is a significant reduction in the number of questions compared to current tools, resulting in higher participation and completion rates. The study [14] provides a comprehensive analysis of Leo Bierman's random forest model proposed in the 2000s, highlighting its consistency and adaptability to sparsity, with a focus on the algorithm's rate of convergence, shedding light on its statistical properties and underlying mathematical dynamics.

In [15], the author used text analysis of social media to better understand mental health and provide early detection. The study used deep learning-based models to detect users' mental status on social media, moving away from traditional approaches. Similarly, we can analyze mental health using the text sentiment analysis from the opinion of the people who are suffering from mental issues by comparing the positive and negative feelings. For that, we can use the Vader and Roberta model to convert the sentiment into its polarity based on the word choice [16]. We can see the concept of mental health, which includes thinking, interactive, and responsive well-being in [17, 18]. It highlights that mental health can be affected due to various factors such as daily life cycle, relationships, and physical health. The article notes that stress, depression, and anxiety are conditions that can harm mental health and interfere with daily life. The study [19] aimed to address the negative impacts on people's mental health issue by using a COVID-19 Mental Health Questionnaire to study the psychological burden among Indians and using machine learning to predict the likelihood of mental health outcomes using the data from 395 Indian participants. They proposed a Psychological Disorder Prediction (PDP) tool that uses a multinomial Naive Bayes classifier to detect the onset of specific psychological disorder and classify participants into two categories: anxiety disorder and mood disorder. The study reported an accuracy of 92.15% and suggested that this automation can play a pivotal role in clinical support by identifying individuals who may need psychological help.

3 Experimental Study

Individuals who are pursuing education in the UK as international students were given the option to participate in our online survey, which was designed for data collection. In order to protect the privacy of each participant, the form was created in an anonymous manner, and therefore, no personal details such as name, email, and university were collected. Figure 1 shows the framework of the system and the work that was done starting from collecting the raw data using questionnaire survey to cleaning and pre-processing, data visualizations to calculating the performance parameters using various machine learning algorithms.

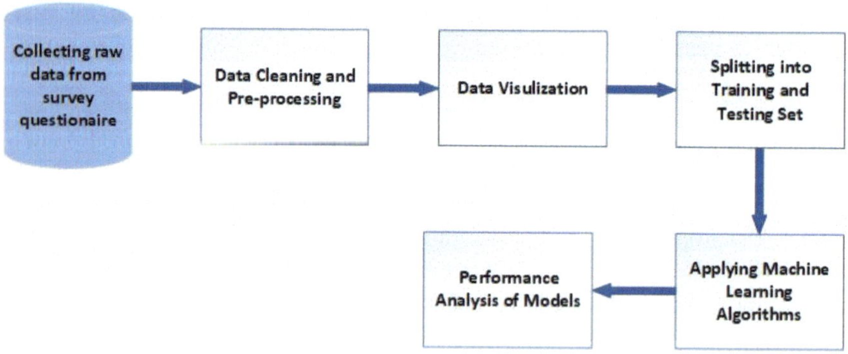

Fig. 1 Framework of the system

3.1 Data Collection

Data was collected through a survey distributed to students in various UK universities via email and a WhatsApp group between January 2023 and April 2023. The dataset contains 290 instances with 14 parameters.

3.2 Data Description

This study involved collecting real-time data from students aged 17–50, over a four-month period from undergraduate, postgraduate and Ph.D. students, using a Google form. The data includes 14 parameters including information about the students' experiences studying in the UK. The data was then compiled into an Excel sheet for further analysis. The survey contains various questions related to mental health and well-being, including age, gender, country of origin, and the following questions.

3.3 Framework of the System

The framework defines the process of applying machine learning algorithms to a dataset. The process begins with collecting data from a survey questionnaire. The data is then cleaned and pre-processed, and then visualized to gain insights into the data. The data is split for training and testing purposes. The data is trained for various machine learning algorithms and tested on new data to make predictions.

3.4 Data Cleaning and Pre-processing

Initially, a form was designed with specific objectives and multiple-choice questions to gather relevant information which is shown in Table 1. However, to ensure the integrity of the dataset, certain steps were taken. Firstly, duplicates were identified and removed, as they could skew the analysis by inflating certain responses. Secondly, null values and unrelated content were eliminated to streamline the dataset and maintain consistency. Additionally, columns that were deemed unnecessary for the analysis were deleted to focus solely on the variables of interest. Lastly, the students' opinions, captured in the form of comments in the last column, were extracted and separately utilized for text sentiment analysis. By employing these data cleaning techniques, the resulting dataset was refined and prepared for effective data visualization and subsequent analysis of the mental health of international students in the UK.

Table 1 Category and summary of the questions in the dataset

S. no.	Questions	Options
1	Level of study	Undergraduate, postgraduate, Ph.D.
2	How often did the students felt sad or depressed during their course of study in the UK?	Extremely often, very often, not so often, not at all
3	How often did the students feel lonely?	Extremely often, very often, not so often, not at all
4	Did the students received support from their university or family regarding mental health well-being during their stay in the UK?	Extremely often, very often, not so often, not at all
5	How do the students rate their work-life balance?	Excellent, good, average, poor, very poor
6	Did the students' mental health interfere with their ability to accomplish tasks?	Yes, no, not at all
7	How much stress did the students feel due to their financial situation, including university fees and living costs in the UK?	Very severe, moderate, no stress
8	How many hours did the students sleep per day?	7+, 6–7, 4–6, less than 4
9	Have the students seen a therapist recently regarding their mental health?	Yes, no
10	Overall, how would the students rate their mental health?	Excellent, good, poor, very poor
11	An optional open-ended question asking students to describe their current feelings	Feedback/feelings (optional)

3.5 Data Visualization

Based on the collected data from the survey, there are total 10 countries that the students participated who were doing their higher studies in the UK or have completed; the countries are Nepal, India, Nigeria, Sri Lanka, Pakistan, Ghana, Bangladesh, Bhutan, China, and Aruba (the Netherlands). The number of males is higher than the females with the percentage of 52.6 and 47.4% that have participated in the survey (see Fig. 2).

Among the various categories of students, postgraduate students stand out with the highest numbers, exceeding both undergraduate and Ph.D. students. When examining the age distribution, the largest cohort of students is found within the 21–30 age range, with a total of 217 individuals. Following closely behind, the 31–40 age group is represented in Table 2.

We have taken great care in crafting the questions for our survey form in order to thoroughly explore various aspects. Another factor that could potentially impact the mental well-being of students is their sleeping habits. It is evident that students face

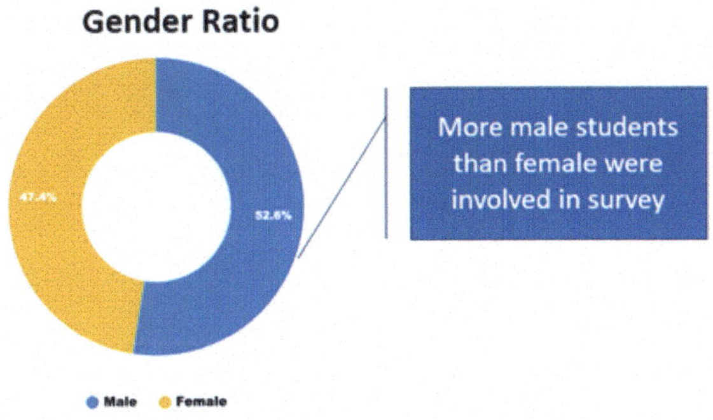

Fig. 2 Gender ratio from the participants

Table 2 Age distribution

S. no.	Age group	Count
1	11–20	10
2	21–30	217
3	31–40	55
4	41–50	6
5	51–60	1

a multitude of challenges, such as academic pressure, part-time jobs, family issues, and so on. While analyzing the distribution of sleeping hours among the students. only 18.3% of students reported getting a recommended sleep duration of 7 or more hours, indicating a relatively small proportion. The majority of students stated that they sleep for six to seven hours, while a significant number of students reported sleeping for only four to six hours per night, possibly due to various commitments and responsibilities. Those who sleep less than 4 h could be impacted in the mental health, and it can be one of the input variables while applying for the machine learning algorithms.

Given that the parameters for the three questions are identical, it is appropriate to compare them using a bar chart. Figure 3 illustrates the responses of students regarding their feelings of sadness and depression, loneliness, and whether they got support from university and family. Interestingly, the distribution of responses for the options "Very Often" and "Not at All" is nearly equal, indicating a normal distribution of mixed reactions.

The visualization of the single parameter in Fig. 4 shows the percentage of the condition of their loneliness including the count of them. There are 15.2% students who are lonely extremely often and 36.7% says very often.

The survey questionnaire included additional parameters aimed at assessing the support received by students from both the university and their families. It is worth noting that a significant majority of students indicated that they receive either minimal or infrequent support in these areas. This highlights the urgent need for universities to prioritize these aspects to assist their students effectively. Given that the data was obtained exclusively from international students, it is possible that they may

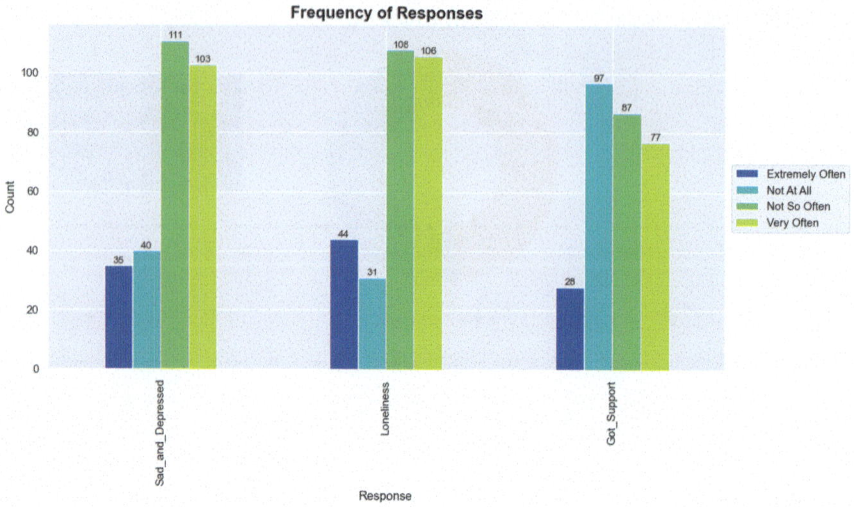

Fig. 3 Frequency of responses of three questions

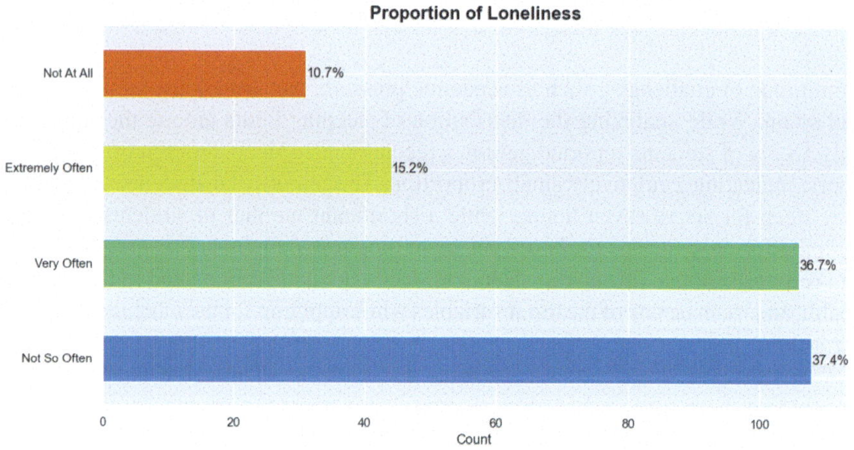

Fig. 4 Proportion of loneliness

have limited awareness of assessment patterns, research-based study requirements, plagiarism concerns, and other academic matters. Consequently, universities should consider addressing these areas to mitigate mental health issues among their students. Moreover, it is crucial to acknowledge that parental understanding plays a pivotal role. In Asian cultural backgrounds, there exist substantial differences when compared to British culture. As a result, students studying abroad may face challenges in communicating their financial stress, work-related pressures, and mental distress to their family members. This lack of communication can further contribute to their feelings of depression.

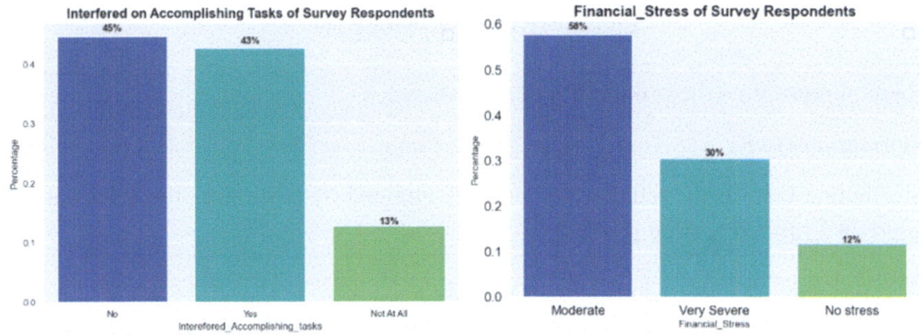

Fig. 5 Interfered accomplishing tasks and financial stress of survey respondents

Approximately 43% of participants reported that their mental health significantly interfered with their ability to complete tasks or accomplish work. However, the majority, comprising 57% of the participants, stated that their mental health did not have any impact on their productivity or task completion.

Analyzing Fig. 5, it is observed that among the survey respondents, 58% reported experiencing moderate financial stress. Additionally, 30% of participants indicated they were facing very severe financial stress, while only 12% mentioned having no stress in this regard.

3.6 *Machine Learning Algorithms*

Decision Tree

Decision tree classifiers are widely recognized and utilize machine learning algorithms in numerous fields due to their interpretability, simplicity, and capability to handle both numerical and categorical data. The decision tree algorithm is a popular choice for classification tasks as they construct a hierarchical structure of internal and leaf nodes, with each internal node representing a decision based on a specific feature and each leaf node representing a class label. The decision tree algorithm recursively partitions the feature space by selecting the most informative features to split the data, resulting in an intuitive tree-like structure that facilitates decision-making.

The decision tree classifier utilizes several formulas to guide its construction and classification process.

Entropy

Entropy is the impurity or uncertainty within a suite of samples. A high entropy indicates a more heterogeneous set of samples while a low entropy indicates a more pure or homogeneous set with a skewed distribution toward a specific class label. The formula for entropy is [20]:

$$\text{Entropy (S)} = -\Sigma p_i \log_2(p_i),$$

where p_i represents the probability of an instance.

Information Gain

Information Gain is the reduction in entropy attained by portioning the information based on a limited feature [10, 20]:

$$\text{Information Gain(S, A)} = \text{Entropy}(S) - \sum_{v \in D_A} \frac{|S_v|}{|S|} \text{Entropy}(S_v)$$

where

A is an attribute,

S is the set,

v is element of value of attribute A,

S_v is subset of S with value v.

Gini Index

The Gini Index measures the impurity of a set of samples [10]:

$$\text{Gini} = 1 - \Sigma(p_i)^2$$

Random Forest

Random forest classifier also called ensemble learning method combines multiple decision trees to make predictions. It is popular for its strength, accuracy, and capability to handle complex datasets. Some key concepts used in Random Forest are as follows.

Bagging (Bootstrap Aggregating)

Each decision tree is trained on a bootstrap model of the training data, a random model with alternative. This helps in creating diverse trees and reducing overfitting.

Feature Subset Selection

At each split in a decision tree within the Random Forest, only a subset of features is considered. This subset is randomly selected from the total set of features.

Voting or Averaging

The predictions from multiple decision trees in the random forest are combined through voting (for classification problems) or averaging (for regression problem) to make the final prediction. This aggregation process improves the overall prediction accuracy and reduces the impact of individual tree errors.

XGBoost

XGBoost (Extreme Gradient Boosting) also an ensemble learning method is an optimized execution of gradient boosting. XGBoost combines multiple weak predictive models to create accurate and robust models. The algorithm builds an ensemble of decision trees sequentially, where each subsequent tree is trained to correct the mistakes made by the previous trees [21]. The predictions of all the trees are combined to make the final prediction. XGBoost employs a gradient-based optimization approach that reduces a specific loss function, such as logistic loss for binary classification or SoftMax loss for multi-class classification:

The similarity score is $\frac{(\Sigma \text{ Residuals})^2}{\Sigma[\text{Prob} * (1 - \text{Prob})] + \lambda}$, [21] where as

$$\text{Gain} = \text{similarity (Left node)} + \text{similarity (Right node)} - \text{Similarity (Root node)}$$

CatBoost

The CatBoost Classifier is a powerful machine learning algorithm that has gained significant attention in recent years due to its outstanding performance in various classification tasks. It is a gradient boosting algorithm that incorporates innovative techniques to handle categorical features and avoid common issues such as overfitting. CatBoost supports an ordered boosting strategy, which enables it to effectively process categorical variables without the need for one-hot encoding or manual feature engineering. Moreover, it introduces a novel approach called ordered boosting, which utilizes a permutation-based method to handle categorical features' values during the training process. This technique significantly reduces the bias caused by the order of the categories. Additionally, CatBoost utilizes gradient-based optimization methods and incorporates L2 regularization to control model complexity and prevent overfitting. These strategies, combined with efficient handling of missing values and advanced tree structure optimization, make CatBoost Classifier an effective and reliable classifier for various real-world applications.

4 Results and Performance Analysis

The image shows a pie chart that represents the percentage of people with poor mental health. The pie chart shows that 20.4% of people have poor mental health, while 60.6% of people have good mental health. The remaining people have either very poor mental health or excellent mental health. This suggests that there is a need for more mental health resources and support for people who are struggling with their mental health (see Fig. 6).

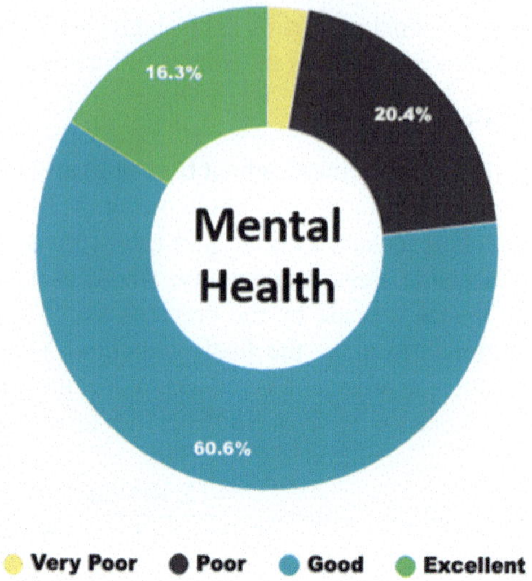

Fig. 6 Mental health condition of the participants

The bar graph in Fig. 7 represents the number of people who have seen a therapist, where the blue line represents the number of people who have seen a therapist. The trend shows that more and more people are not seeking therapy, which is a negative trend.

The word cloud shows that international students often feel lonely, isolated, and stressed when they are away from home. This is because they are missing their friends,

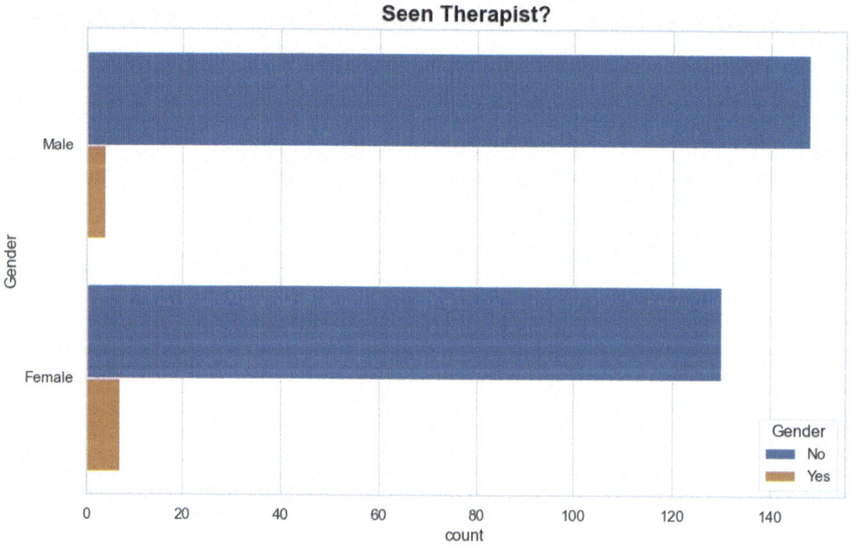

Fig. 7 Count of seeing the therapist

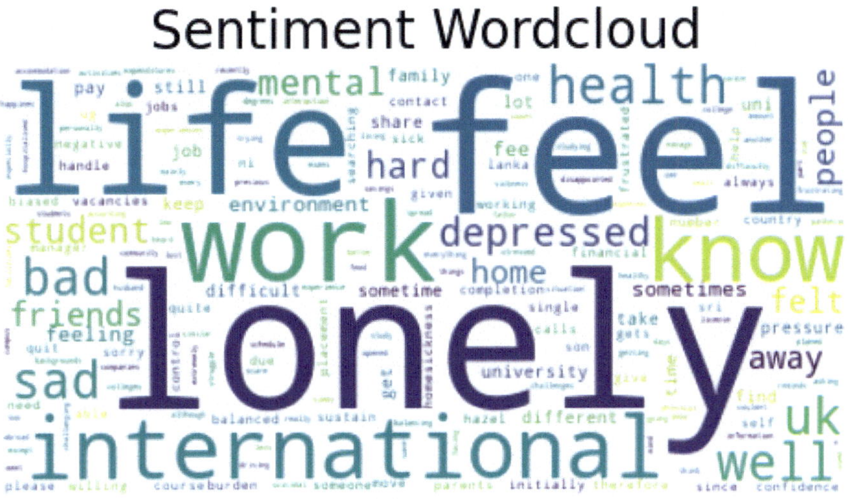

Fig. 8 Word cloud of the emotions of students

family, and familiar surroundings. They may also be feeling pressure to succeed in their studies or their new job. To support for international students who are feeling homesick or stressed could come in the form of counseling, social activities, or financial assistance (see Fig. 8).

4.1 Performance Parameters of ML Algorithms

While building a robust model, evaluating the performance metrics is vital, as this indicates the model's generalizability toward unseen data. Some of the parameters are as follows.

Confusion Matrix

It is a tabular representation of the model's outcomes, to describe the model's performance on a set of test data. A n*n matrix compares the actual target values with predicted label detailing how well the model is performing. Let's take an example of binary classification, 2*2 matrix as shown in Table 3.

Table 3 Confusion matrix

	Positive	Negative
Positive	True Positive (TP)	False Positive (FP)
Negative	False Negative (FN)	True Negative (TN)

Accuracy

Accuracy is the percentage of accurate classifications that the model achieves. It is given by the number of correct predictions over the total number of predictions across all classes:

$$\begin{aligned}\text{Accuracy} &= \text{Number of correct predictions/Total number} \\ \text{of predictions} &= (\text{TP} + \text{TN})/(\text{TP} + \text{TN} + \text{FP} + \text{FN})\end{aligned}$$

Precision

Precision, an evaluation metric, is given by the number of true positives over the total number of predictions. It gives the proportions of correct positive identification:

$$\text{Precision} = \text{TP}/(\text{TP} + \text{FP})$$

Recall

Recall is the number of correct positive results over the number of all related positive samples:

$$\text{Recall} = \text{True Positives}/(\text{True Positives} + \text{False Negatives})$$

F1-Score

The harmonical mean between precision and recall is the F1-score. The greater the F1-score, the better the performance of the model:

$$\text{F1} = 2 * 1/((1/\text{precision}) + (1/\text{recall}))$$

AUC-ROC Curve

While checking or visualizing the performance at various threshold settings of the multi-class classification problem, the Area Under Curve (AUC) Receiver Operating Characteristics (ROC) curve is used. The higher the AUC, the better the model is at predicting 0 classes as 0 and 1 classes as 1 (see Fig. 9).

In this study, we assessed the performance of four machine learning models: CatBoost, Decision Tree, Random Forest, and XGBoost, using several evaluation metrics. The models were evaluated on a specific task, and the metrics used for assessment were Precision, Recall, F1-score, and Accuracy. Among the models, Random Forest stood out with the highest Precision (0.92) and F1-score (0.84), showcasing its ability to accurately classify positive instances while maintaining a balanced trade-off between Precision and Recall. Decision Tree followed closely, achieving good Precision (0.77) and Recall (0.82) scores. XGBoost and CatBoost also performed reasonably well, with varying scores across the metrics. Furthermore,

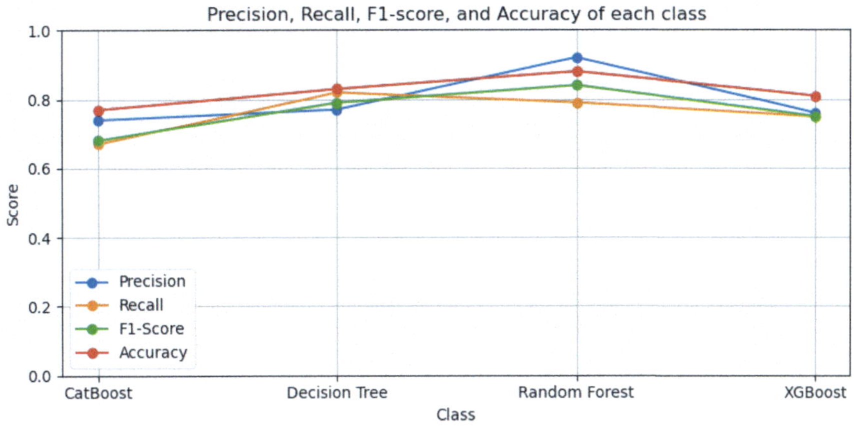

Fig. 9 Line plot comparison of evaluation matrix

Random Forest achieved the highest Accuracy (0.88), indicating its proficiency in making overall accurate predictions. Decision Tree and XGBoost also exhibited respectable Accuracy scores of 0.83 and 0.81, respectively. CatBoost had an Accuracy score of 0.77.

Receiver Operating Characteristic (ROC) curve and Area Under the Curve (AUC) are evaluation metrics used to assess the performance of a classifier when making decisions based on different classification thresholds. ROC indicates the performance of a classifier by plotting the True Positive Rate (TPR) against the False Positive Rate (FPR) where TPR is also known as sensitivity or recall and represents the proportion of positive instances correctly classified as positive and FPR is the ratio of negative instances incorrectly classified as positive. The curves in Fig. 10 show that they all are able to handle multiple classes well. The AUC for each class is relatively high, which indicates that the algorithms are able to correctly classify most of the samples in each class. XGBoost is able to classify the most difficult class to classify with the highest AUC because of its gradient boosting nature which is a type of ensemble learning algorithm that can learn from the mistakes of other algorithms, which helps them to improve their performance.

5 Conclusion and Future Work

In conclusion, this study explored the challenges faced by international students in the UK and how these challenges affect their mental well-being. Our research, which involved surveys from students across different UK universities, found that a significant portion (around 20.4%) of these students reported experiencing poor mental health. Even more concerning, only a few students are seeking therapy for their mental health issues. Using machine learning algorithms, we were able to predict

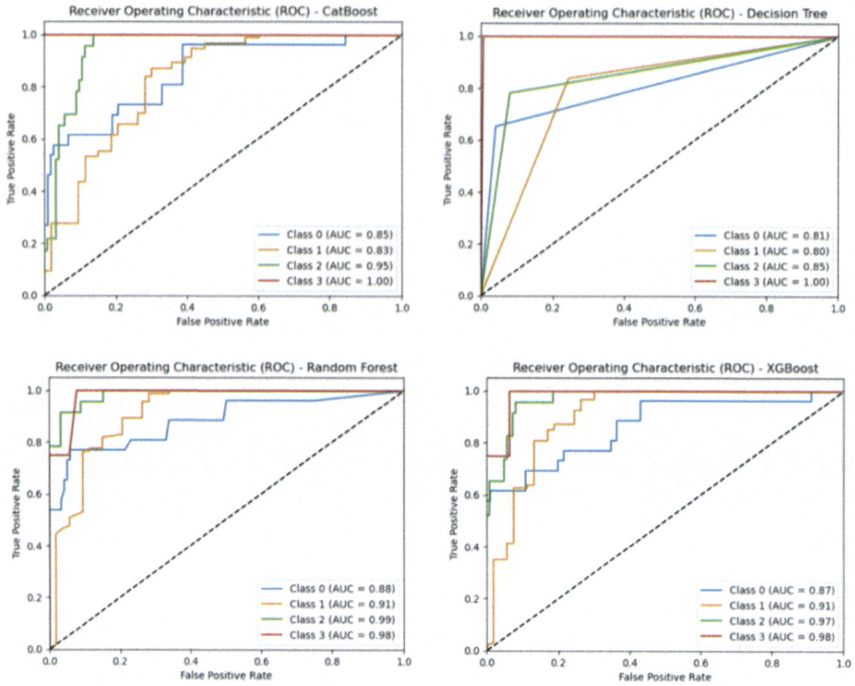

Fig. 10 ROC curve with AUC score

students' mental health status with a high level of accuracy (around 88%) by random forest, meaning we can identify those who may need help. Our findings also revealed that many international students often feel lonely, stressed, and isolated due to being far from home. To support them, the universities and parents should provide tailored assistance such as counseling, social activities, and financial support. In essence, this research provides actionable insights for universities to improve the well-being of international students, ultimately helping them succeed in their studies and enjoy a better quality of life.

In future, this work can be expanded with a larger and more diverse sample dataset, ensuring a broader representation of experiences and perspectives. Also, the questionnaire can be improved to better target specific information about the student's mental well-being and the diverse university sampling to reach out to more universities across the UK to reduce bias and ensure a broader representation of student backgrounds.

Acknowledgements This work was supported by Internal Funding from York St John University, London, United Kingdom. We extend our sincere appreciation to Dr. Soonleh Ling, the project supervisor, for his guidance, and also thanks to all the participants who were involved in the data collection.

References

1. Start your journey—Erudera. https://erudera.com/start-journey/. Accessed 20 Dec 2023
2. Chung, J., Teo, J.: Mental health prediction using machine learning: taxonomy, applications, and challenges: Appl. Comput. Intell. Soft Comput. **2022** (2022). Hindawi Limited. https://doi.org/10.1155/2022/9970363
3. Srividya, M., Mohanavalli, S., Bhalaji, N.: Behavioral modeling for mental health using machine learning algorithms. J. Med. Syst. **42**(5) (2018). https://doi.org/10.1007/s10916-018-0934-5
4. Laijawala, V., Aachaliya, A., Jatta, H., Pinjarkar, V., Somaiya, K.J.: Mental health prediction using data mining: a systematic review (2022). https://ssrn.com/abstract=3561661
5. Singh, H., Kaur, B.: Predicting student performance and its impact on mental health using machine learning (2022). https://ssrn.com/abstract=4356166
6. Fernandez, A., Salvador-Carulla, L., Choi, I., Calvo, R., Harvey, S.B., Glozier, N.: Development and validation of a prediction algorithm for the onset of common mental disorders in a working population. Aust. N. Z. J. Psychiatry **52**(1), 47–58 (2018). https://doi.org/10.1177/0004867417704506
7. Srinivasulu, U., Vivek, A., Tech, T.B., Dharun, A., Tech, B.: Machine learning techniques for stress prediction in working employees; machine learning techniques for stress prediction in working employees (2018)
8. Tate, A.E., McCabe, R.C., Larsson, H., Lundström, S., Lichtenstein, P., Kuja-Halkola, R.: Predicting mental health problems in adolescence using machine learning techniques. PLoS One **15**(4) (2020). https://doi.org/10.1371/journal.pone.0230389
9. Alharthi, H.: Predicting the level of generalized anxiety disorder of the coronavirus pandemic among college age students using artificial intelligence technology. In: Proceedings—2020 19th Distributed Computing and Applications for Business Engineering and Science, DCABES 2020, pp. 218–221. Institute of Electrical and Electronics Engineers Inc. (Oct. 2020). https://doi.org/10.1109/DCABES50732.2020.00064
10. Bisht, A., Vashisth, S., Gupta, M., Jain, E.: Stress prediction in Indian school students using machine learning. In: Proceedings of 3rd International Conference on Intelligent Engineering and Management, ICIEM 2022, pp. 770–774. Institute of Electrical and Electronics Engineers Inc. (Oct 2022). https://doi.org/10.1109/ICIEM54221.2022.9853200
11. Garriga, R., et al.: Machine learning model to predict mental health crises from electronic health records. Nat. Med. **28**(6), 1240–1248 (2020). https://doi.org/10.1038/s41591-022-01811-5
12. Vaishnavi, K., Kamath, U.N., Rao, B.A., Reddy, N.V.S.: Predicting mental health illness using machine learning algorithms. In: Journal of Physics: Conference Series. IOP Publishing Ltd. (Jan. 2022). https://doi.org/10.1088/1742-6596/2161/1/012021
13. Tutun, S., et al.: An AI-based decision support system for predicting mental health disorders. In: Information Systems Frontiers, Springer (2022). https://doi.org/10.1007/s10796-022-10282-5
14. Biau, G., Fr, G.B.: Analysis of a Random Forests Model (2012)
15. Sekulić, I.S., Strube, M.: Adapting Deep Learning Methods for Mental Health Prediction on Social Media (2020)
16. Pokhrel, S., Somasiri, N., Rebecca Jeyavadhana, C., Ganesan, S.: Web data scraping technology using TF-IDF to enhance the big data quality on sentiment analysis. In: ICDSBDA 2022: XVI. International Conference on Data Science and Big Data Analytics, pp. 1–8 (2022). https://waset.org/
17. Manvith, G.S., Reddy, G.S., Reddy, B.L.: Mental health prediction for an individual using machine learning. www.irjmets.com
18. Srividya, M., Mohanavalli, S., Bhalaji, N.: Behavioral modeling for mental health using machine learning algorithms. J. Med. Syst. **42**(5) (May 2018). https://doi.org/10.1007/s10916-018-0934-5

19. Kumar, A.: Machine learning for psychological disorder prediction in Indians during COVID-19 nationwide lockdown. Intell. Decis. Technol. **15**(1), 161–172 (2021). https://doi.org/10.3233/IDT-200061
20. Rokach, L., Maimon, O.: Decision trees. In: Data Mining and Knowledge Discovery Handbook, pp. 165–192. Springer (2006). https://doi.org/10.1007/0-387-25465-x_9
21. Chen, T., Guestrin, C.: XGBoost: A Scalable Tree Boosting System (2016). https://doi.org/10.1145/2939672.2939785

Leveraging Differently Layered Convolutional Neural Network for Speech Emotion Recognition

Prathik Kumar Gaddam, Anas Iqubal, Muqtada Hasan, and Arun Prakash Agrawal

Abstract Text stands to be one of the prominent aspects of Communication in any Technical Phenomenon. But, with Text, understanding the Emotion of a Person, which plays a vital role in decision-making for actuating a Task, isn't possible. With the Escalated Indulgence of Speech into Diverse Technological Systems, it has been necessary to formulate a modus operandi which would precisely determine the Emotion of a Person through its Speech. Different approaches have been proposed for recognizing the Emotion of a Person through Speech, but the Limitation of Modular Outlook over Speech Data seems to be Visible. In this paper, we consummated Convolutional Neural Networks and tried to understand the Impact of the Convolution Layers, i.e., 03, 04, 05, and 06—Layered CNN, over the Speech Data and thereby formulate a Potent and Efficient Solution for Speech Emotion Recognition. We utilized the well-known Open-Source Datasets such as CREMA, RAVDESS, SAVEE, and TESS, consisting of Speech Audio for various Human Emotions, and combined them into one whole Dataset. The outcome we got was phenomenal as the Accuracy surpassed the already existing methodologies, and also it gave us an Impactful Elucidation of its Real-time Applicability for varied use-cases related to Speech.

Keywords Speech Emotion Recognition (SER) · Convolutional Neural Network (CNN) · Deep Learning (DL) · Data augmentation · Natural language processing · Speech emotion features · Logistic regression

---

P. K. Gaddam
School of Engineering & Applied Science, The George Washington University, Washington, D.C, USA
e-mail: prathikkumar.gaddam@gwmail.gwu.edu

A. Iqubal
Galgotias College of Engineering and Technology, Greater Noida, India

M. Hasan
Kratikal Tech Private Limited, Noida, India

A. P. Agrawal (✉)
School of Computer Science Engineering and Technology, Bennett University, Greater Noida, India
e-mail: arunpragrawal@gmail.com

G. Raj et al. (eds.), *Advances in Artificial Intelligence and Machine Learning*, Lecture Notes in Electrical Engineering 1264, https://doi.org/10.1007/978-981-97-9507-9_3

1 Introduction

As Human Beings, we tend to understand the Speech of a Person while communicating through Emotions. Nowadays, Text being the prominent aspect of Conveying or Disseminating over our Cellular Devices has reluctantly reduced the touch of Humanness into our Lives [1]. As a result, to imbibe Speech into Conversations, a diverse modus operandi encapsulating Speech has been Actuated [2]. But, incorporating Speech, still does not solve the actual problem and thus, attempts of formulating a Speech Emotion Recognition System have been made. Different techniques for the same, have been proposed but with a plethora of Limitations, fencing the Data Creation [3], Data Unavailability [6], Feature Extraction [8], In-Efficient Modelling [10], etc. The Major Problem lies in the Dataset, as the Deep Learning Models could not take Sound as an Input, and thus, devising the Model Favourable Dataset, stands to be the Real Deal. The inculcation of Speech has been actuated into varied Sectors such as E-Commerce, Music Streaming Platforms, OTTs, and Voice Assistants [4]. All these Technologies operate on the Psychological Viewpoint of a Person and aid the Person in making Decisions [5]. This has all been done through a Person's past Choices, but now Inculcation of Speech into the above-mentioned Systems has made it more feasible for a Human Being to operate those Applications through Voice which was earlier done by Texts [7]. Thus, recognizing Emotion through Speech stands to be a Game-Changer for providing a better User-Experience to the User, for which Companies spend Millions.

In this paper, we've tried to contrive a quintessential modus operandi which focuses upon the creation of a Speech Emotion Recognizer with the utilization of Differently Layered Convolutional Neural Networks such as 03-layered CNN [6], 04-layered CNN, 05-layered CNN, and 06-Layered CNN. This study not only focuses upon creating a potent system for Speech Emotion Recognition, but also explicates upon the Impact of Layers over an Audio Dataset, which has not been done yet in any of the previous methodologies. The modus operandi employs the collection of famous Audio Datasets such as CREMA, RAVDESS [14], SAVEE, and TESS for removing the Constraint of the Data. With the help of Data Augmentation and well-defined Feature Extraction [11], our proposed outlook stands to be one of the best out there in terms of Efficacy and Model Complexity. Also, the Ultimate viewpoint, i.e., the More the Number of Layers in the Network, the More the Accuracy of the System, has also been Scrutinized. As a result, the Propounded Methodological Outlook encompasses the prevailing Modus Operandi related to Speech Emotion Recognition in Sect. 2. Section 3 demonstrates the Detailed Elucidation of the Methodology. The Experimental Outcomes have been illustrated in Sect. 4, whereas Sect. 5, induces the Conclusion and Future Scope of the System.

2 Related Work

With the Implorations focused upon Speech Emotion Recognition, we found out that there have been Major Advancements in this Domain. The Utilization of Deep Neural Networks (DNN) for Speech Emotion Recognizer over a German Database consisting of 06 Emotions namely Angry, Happy, Sad, Neutral, Bored, and Fear has been Elucidated gauging an Average Accuracy of 45% over all Emotions with a Constraint of Language focused Dataset [1].

Inculcation of Adversarial Training Methodology has also been employed for Speech Emotion Recognition with emphasis on the Regularization of the Speech Data and the Improvement over the Supervised Modus Operandi. The Average Accuracy stood to be 15–20% more efficient than Baseline DNN procuring an Average Accuracy of 65% over 04 Emotions namely Angry, Sad, Neutral, and Happy. The proposed methodology should have included more Types of Emotions to understand the Performance of the Model more explicitly [2]. Furthermore, the Proposition of Incorporation of Deep Belief Network (DBF) which extracts the Speech Emotion Features automatically has been made. While using Single DBNs the Model expelled 75.4% Average Accuracy and with Multiple DBNs the Average Accuracy stood to be at 82.6%. But with the proposed outlook, the Utilization of DBN along with a Shallow Neural Network gauged an Average Efficacy of 89.8%. The Model has been trained to Classify the 05 Class Dataset and more Data could've been included to attain more Accurate Outcomes [3].

Arabic Emotion Recognition engaging Speech has also been emphasized by utilizing Convolutional Neural Network (CNN) and Bi-Directional Long Short-Term Memory (Bi-LSTM) Models [15]. Thus, while employing CNN the Average Accuracy came out to be 85%, and while using CNN + Bi-LSTM Network, the Average Accuracy came out to be 87.2%. More Data could've been induced for more Optimum Results [4]. A study of Normalization along with CNN Network over RAVDESS Dataset for Speech Emotion Recognition has also been carried out, expelling an Accuracy of 72%. The Limitation stands to be the Data Preparation along with the Modular Outlook [5]. A unique use-case of Speech Emotion Recognition using the Internet of Things (IoT) for Healthcare Systems has also been proposed. With the utilization of 03-Layered CNN, the Model procured an Average Accuracy of 90% over a 07 Class Dataset. The only limitation that the System had was the Additional System Requirement [6]. Induction of the Mel-Frequency Cepstral Coefficient (MFCC) for Extraction of Features imposed over the LSTM Network has been Elucidated, expelling an Average Accuracy of 84.81% over Seven Emotion Datasets [7]. With an Average Accuracy of 91.6%, the Integration of Empirical Mode Decomposition (EMD) for Speech Emotion Recognition has been made with the Future Scope of Inculcation of more Voice Samples for Improved Efficacy [9]. Moreover, Deep Stride Convolutional Neural Networks (DSCNN) have also been employed for Speech Emotion Recognition over the SAVEE Dataset which comprises of 04 Emotions namely Angry, Happy, Neutral, and Sad [19]. The Average Accuracy came out to be 87.8% with the Limitation of a Less Number of Emotions along with a Less

Number of Data utilized for creating the Recognizer [13]. A Comparative Analysis between 05 Machine Learning Algorithms such as Logistic Regression, Support Vector Machine, Random Forest, XG-Boost, and K-Nearest Neighbour along with 03 Deep Learning Approaches such as Convolutional Neural Network (CNN), Multi-layer Perceptron (MLP) and Long Short-Term Memory Network has also been made. The Maximum Accuracy was obtained through Deep Neural Networks, i.e., 92% over RAVDESS Dataset [14].

Thus, in previous works, it can be seen that in each of the papers, there are some sort of limitations such as a limited number of emotions in data sets, usage of less number of layers, or combining one CNN layer to different networks/models to get a high accuracy. The number of layers and the limitation of emotions in the dataset in some of the previous papers were the main issues that arose. Therefore, through a well-versed Scrutiny of already existing Modus Operandi, we formulated our System by focusing upon the prominent aspects for Speech Emotion Recognition such as Data, Feature Extraction, and the Fine-Tuning of the Model utilized.

3 Proposed Methodology and Its Implementation

The Exegesis related to the Speech Emotion Recognition System led us to Envisage our Proposed Architecture, by inducing the Key Points required in the System Formulation, for ultimately attaining an Efficient Model with Less Complexity. Figure 1, demonstrates the Propounded Methodological Outlook.

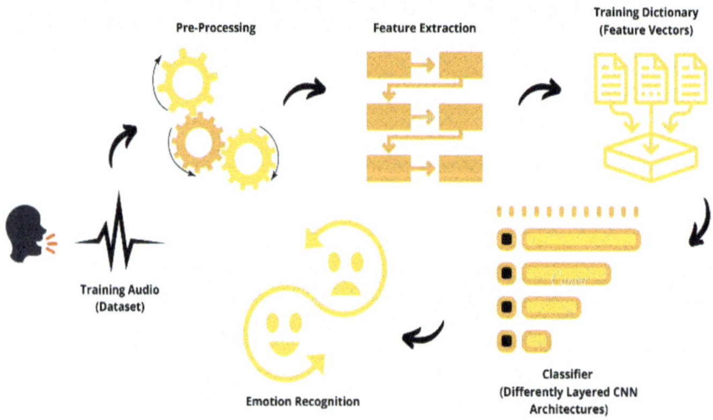

Fig. 1 Propounded system architecture

3.1 Implementation of Methodology

A. *Data Creation and Preparation*: As Data plays an Eminent Role in Evaluating a Model's Competency, we collated the widely used Datasets which are Open-Source namely CREMA, RAVEDESS [5], SAVEE, and TESS. Through this Collation our Overall Dataset Range, i.e., the Number of Audio Samples encompassing Diverse Emotions Increased [20]. As a result, We Classified the Audio Samples based on their Emotions into 07-Classes such as Disgust, Happy, Sad, Neutral, Fear, Angry, and Surprise. Figure 2, demonstrates the Count of Emotions in our Formulated Dataset.

A Plot encompassing Amplitude on the Y-Axis and Time on the X-Axis is coined as Wave-Plot which is commonly used to infer the Loudness of the Speech Sample [17], whereas A Plot encompassing Frequency on the Y-Axis and Time on the X-Axis which demonstrates Frequency Spectrum contemplating with Time is coined as Spectrogram [18]. Figure 3, represents the Wave-Plot and Spectrogram for Disgust and Happy Emotion Audio.

Similarly, Fig. 4, demonstrates the Wave-Plot and Spectrogram for Sad and Neutral Emotion Audio.

Furthermore, Fig. 5, illustrates the Wave-Plot and Spectrogram for Fear, Surprise, and Angry Emotion Audio.

To make the overall System data-driven, we also Applied Data Augmentation [6] and Converted each Audio Sample into its Noised, Stretched, Shifted, and Pitched Form. Figure 6, represents the Augmentation of a Sample Audio.

So, now we have a Dataset which comprises of 04 Different Types of Data Folders, each consisting of Unique Audio Files, which are Further Segregated into 07-Classes based on Emotion, and each Audio File has been augmented with 04 Different Form Factors.

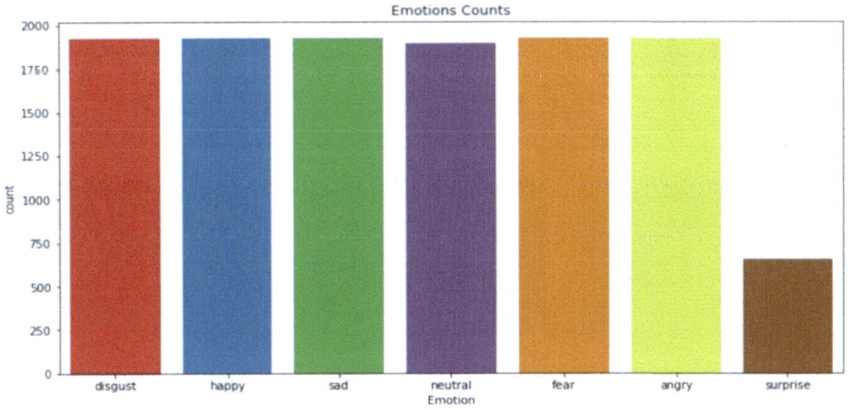

Fig. 2 Bar plot for count of data in each emotion

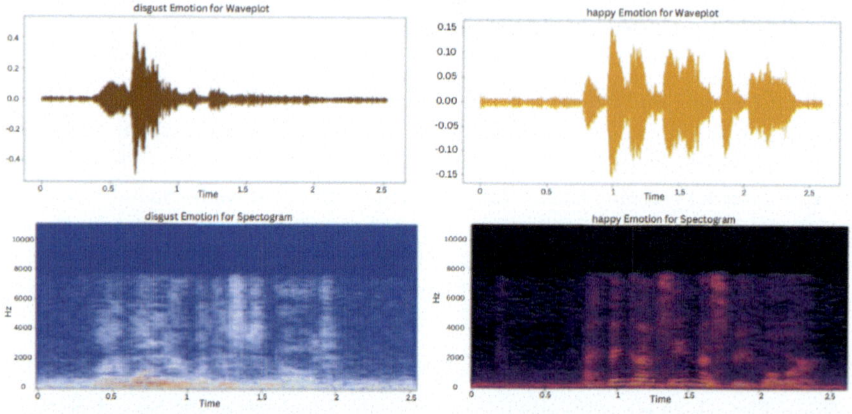

Fig. 3 Wave-plot and spectrogram plot for disgust and happy emotion audio

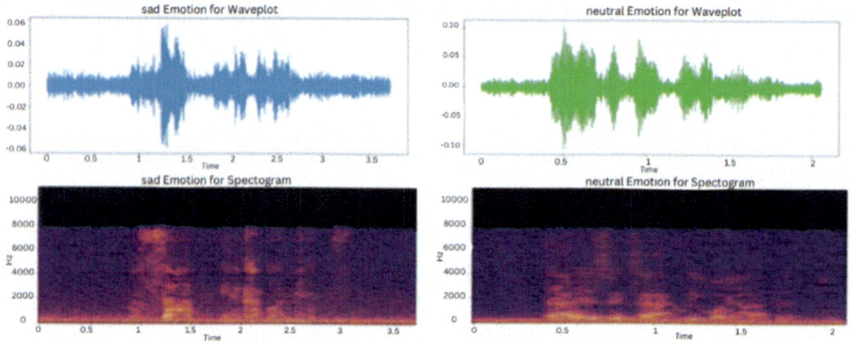

Fig. 4 Wave-plot and spectrogram plot for sad and neutral emotion audio

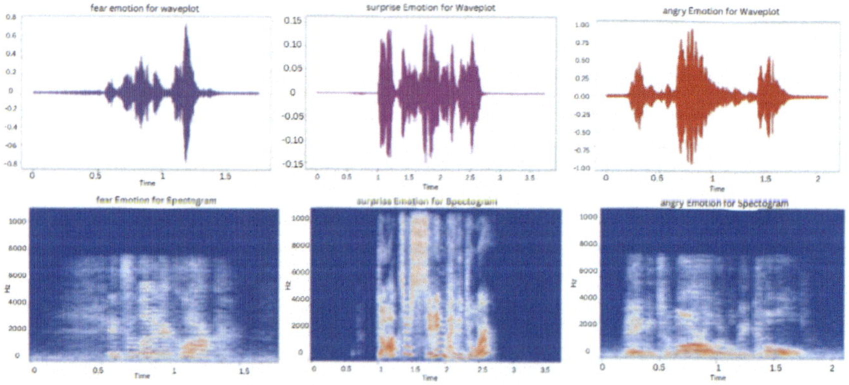

Fig. 5 Wave-plot and spectrogram plot for fear, surprise, and angry emotion audio

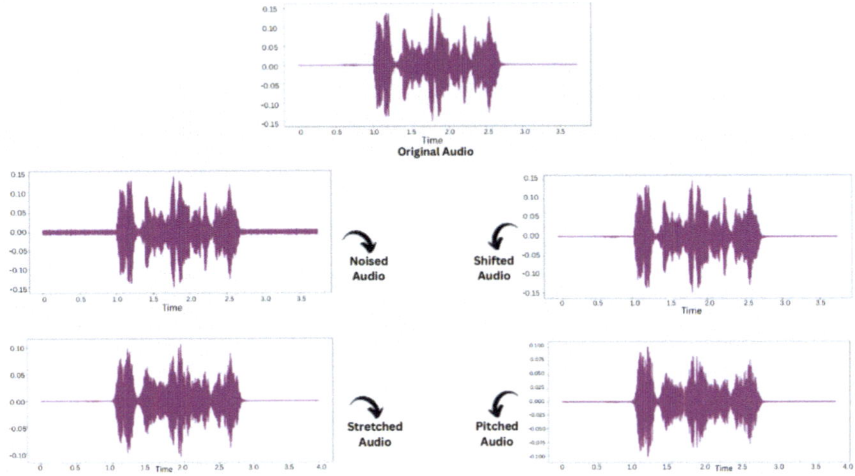

Fig. 6 Data augmentation over a sample audio file

B. *Feature Extraction*: As our Data is in Audio Format, it's not possible for our Model to Understand Audio as an Input, and therefore, Feature Extraction stands to be one of the prominent aspects of our overall Modus Operandi. Basically, a Sound Signal is a 3-D Signal in which the 03 Axes exemplifies Frequency, Amplitude, and Time [18] as demonstrated in Fig. 7.

There are Diverse Transformations which can be performed over an Audio Data, but to make our Model more Optimized, we selected the 03 Most Eminent Transformations for Feature Extraction namely, Zero Crossing Rate (ZCR), Root Mean Square (RMS) along with Mel-Frequency Cepstral Coefficients (MFCC) [15].

Zero Crossing Rate (ZCR) [18]: It basically demonstrates the Change of Sign of a Signal, with Time, i.e., whether the Commuting Signal alters from Positive to Negative, Negative to Positive, or Back. ZCR comes under Time-Domain Features.

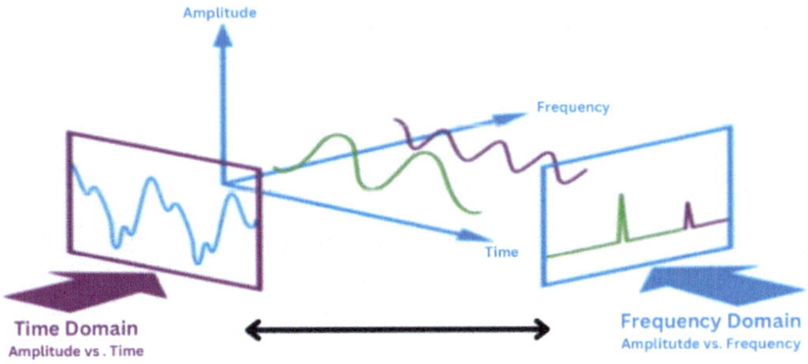

Fig. 7 3-D illustration of an audio signal [18]

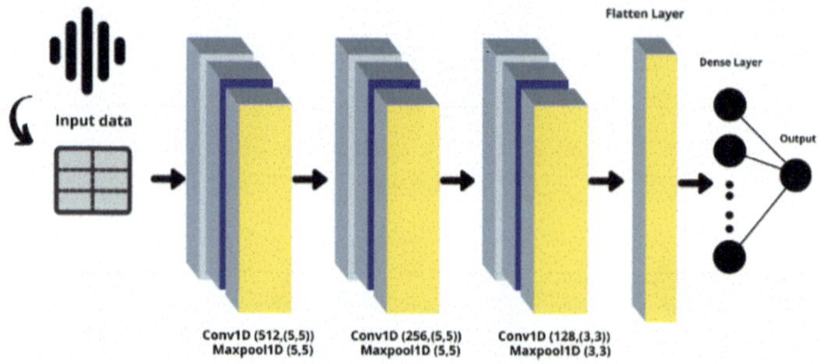

Fig. 8 Pictorial representation of CNN architecture with 03 layers

Root Mean Squared (RMS) [17]: It's basically the Average, i.e., when contemplated in the form of Noise of Sound Signal, it exacerbates the Average Power of a Continuous Wave-Plot over a longer Period of Time. RMS comes under Frequency-Domain Features.

Mel-Frequency Cepstral Coefficients (MFCC) [18]: This is one of the special Feature Extractor which is widely used when working with Audio Data. Basically, MFCCs comprises of Trivial Collection of Features, which pre-eminently delineate the comprehensive structure of a Spectral Envelope [16].

After the selection of Feature Extractors, we now consummated our Comprehensive Data consisting of 12,000 Audio Files, over these Feature Extractors, and after Processing, we stored the Normalized Values in a.csv File with Numerical Values which could then be used for Modular Tuning. Now, our Input Dataset is ready to be given as an Input to Convolutional Neural Networks (CNNs).

C. *Training the Aggregated Data*: As stated earlier, for Training, we'll be utilizing Differently Layered CNN Architectures. As our Data is in the form of Vectors, i.e., Numeric Values we'll be utilizing Conv-1D rather than Conv-2D [12]. Let's explore each Layered CNN System utilized in Detail.

3.2 Architecture with Different Numbers of CNN Layers

The Architecture of a CNN Network with 03 Layers:

Here, we Induced 03 Conv-1D Filter Blocks, i.e., (512, (5,5)), (256, (5,5)) and (128, (3,3)), which in return formulated 03-Layered CNN Architecture, collated with MaxPool-1D for Eminent Features Retrieval. After the Convolutional Layers, a Flatten Layer is introduced, followed with the Dense Layer to Procure the Output. Figure 8, depicts the Pictorial Representation of CNN Architecture with 03 Layers.

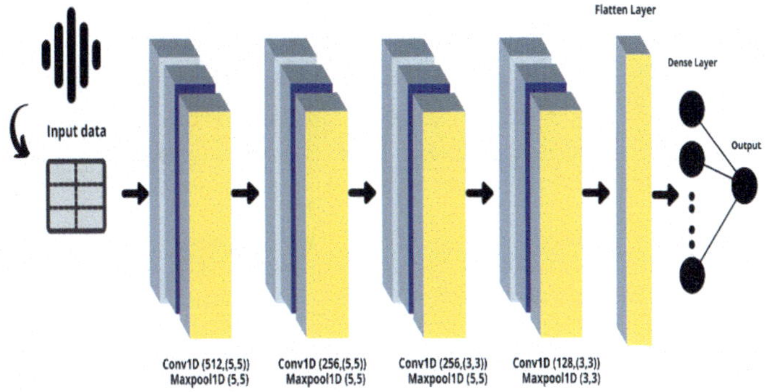

Fig. 9 Pictorial representation of CNN architecture with 04 layers

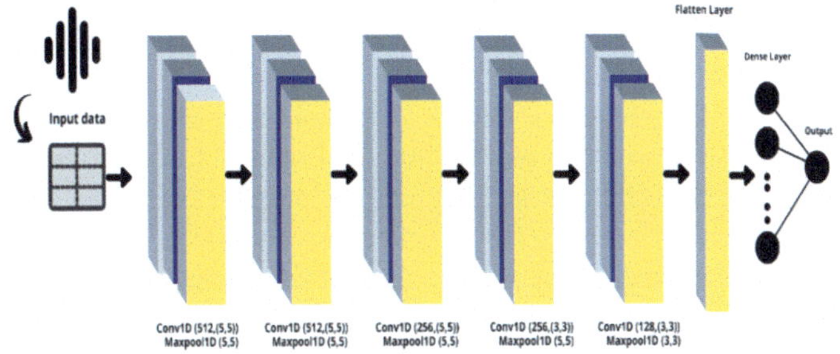

Fig. 10 Pictorial representation of CNN architecture with 05 layers

The Architecture of a CNN Network with 04 Layers:

Here, we Induced 04 Conv-1D Filter Blocks, i.e., (512, (5,5)), (256, (5,5)), (256, (3,3)) and (128, (3,3)), which in return formulated 04-Layered CNN Architecture, collated with MaxPool-1D for Eminent Features Retrieval. After the Convolutional Layers, a Flatten Layer is introduced, followed with the Dense Layer to Procure the Output. Figure 9, depicts the Pictorial Representation of CNN Architecture with 04 Layers.

The Architecture of a CNN Network with 05 Layers:

Here, we Induced 05 Conv-1D Filter Blocks, i.e., (512, (5,5)), (512, (5,5)), (256, (5,5)), (256, (3,3)) and (128, (3,3)), which in return formulated 05-Layered CNN Architecture, collated with MaxPool-1D for Eminent Features Retrieval. After the Convolutional Layers, a Flatten Layer is introduced, followed with the Dense Layer to Procure the Output. Figure 10, depicts the Pictorial Representation of CNN Architecture with 05 Layers.

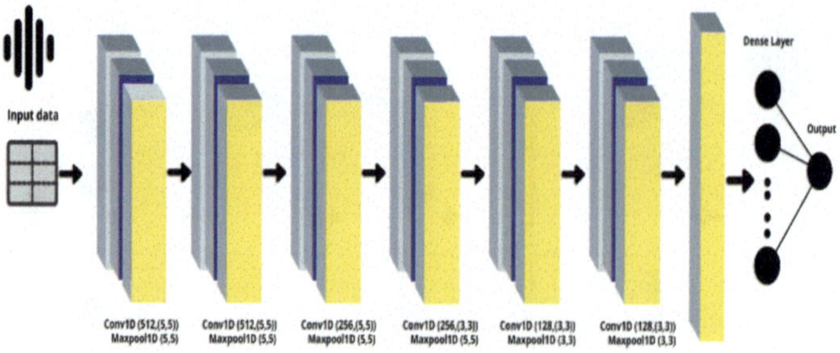

Fig. 11 Pictorial representation of CNN architecture with 06 layers

The Architecture of a CNN Network with 06 Layers:

Here, we Induced 06 Conv-1D Filter Blocks, i.e., (512, (5,5)), (512, (5,5)), (256, (5,5)), (256, (3,3)), (128, (3,3)) and (128, (3,3)), which in return formulated 06-Layered CNN Architecture, collated with MaxPool-1D for Eminent Features Retrieval. After the Convolutional Layers, a Flatten Layer is introduced, followed with the Dense Layer to Procure the Output. Figure 11, depicts the Pictorial Representation of CNN Architecture [16] with 06 Layers.

All these Differently Layered CNN Architectures have been Trained over 12,000 Audio Files over 50 Epochs with a Batch Size of 64, along with a Learning Rate of 0.00001. All these Hyperparameter Values have been Set Intentionally, as we got the Most Optimum Results with these Values. After Training, we got Exceptional Results which we have discussed in Sect. 4. These results corroborated the Potential of Convolutional Neural Networks.

4 Experimental Results and Analysis

With the Procurement of an Optimal Dataset, and Training them over the Fine-Tuned CNN Networks, exhibited Remarkable Outcomes. The 03-layered CNN Architecture gauged an Average Accuracy of 90.5%, the 04-layered CNN Architecture procured an Average Efficacy of 92%, and the 05-layered CNN Architecture obtained an Average Accuracy of 95%, whereas the 06-layered CNN Architecture expelled an Average Efficacy of 94%. All the Modular Phenomenon demonstrated Potent Results, when compared to the Existing Modus Operandi, related to Speech Emotion Recognition System. The 05-layered CNN Model performed the Best, followed by the 06-layered CNN, then the 04-layered CNN, and finally the 03-layered CNN. Figure 12, demonstrates the Graphical Result of Train and Validation Accuracy of 05-layered CNN Model. In Fig. 12, the training accuracy increases steeply with every epoch till it reaches 10 epochs approximately. After that, the graph becomes constant in accuracy with each epoch, whereas the validation accuracy also decreases but has a

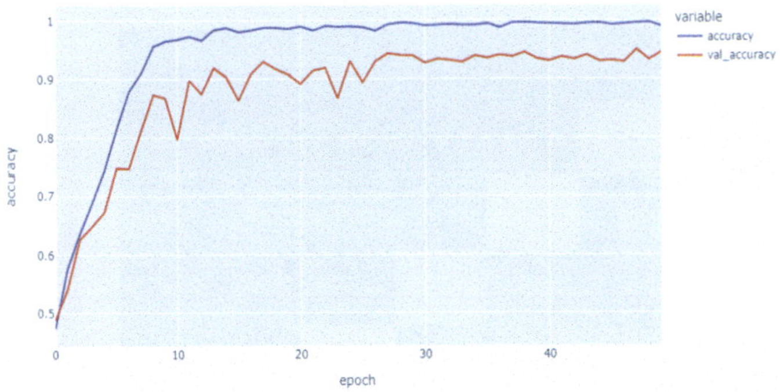

Fig. 12 Graphical illustration of train and validation accuracy of 05-layered CNN model

few spikes in it. Validation accuracy also gets constant but involves some spikes in between which shows there is fluctuation in accuracy around 95%.

Furthermore, Fig. 13, represents the Train Loss and Validation Loss of the 05-layered CNN Model. In Fig. 13, after getting into training of the CNN model, it shows an extreme decrease in the loss. It means that, after every epoch, the models get into for training, the value of training loss and the validation loss decreases significantly causing a well-trained model that produces a result with a huge accuracy of nearly 95%. As well as we can also see from the image that the training loss decreases with a smooth slope and gets consistent, whereas the validation loss takes a few spikes in between while decreasing and gets constant with a significant number of spikes in it.

Moreover, Fig. 14, depicts the Confusion Matrix, demonstrating the Overall Emotion Classification Results of the 05-layered CNN Model.

Additionally, Fig. 15, demonstrates the Output for the Individual Emotion through Precision, Recall, and F1-Score procured from the 05-layered CNN Model for the Explication of the Model's Potency.

Here, we can see that the precision and recall of every emotion involved is above 0.90 value (90%) or it can be seen that the average of each value of emotion is around 0.95 or 95%, which shows that the 05 layered CNN shows a high accuracy

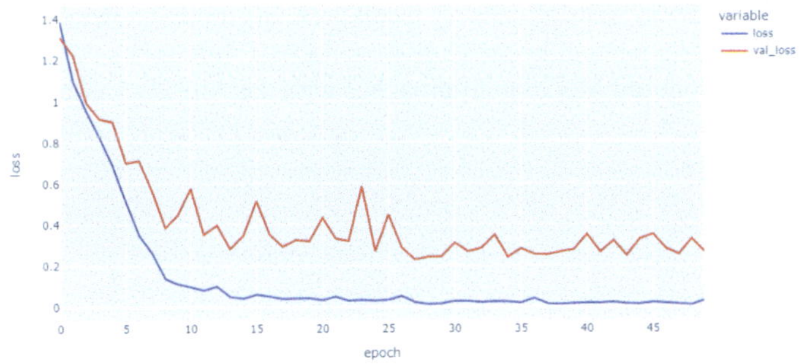

Fig. 13 Graphical illustration of train and validation loss of 05-layered CNN model

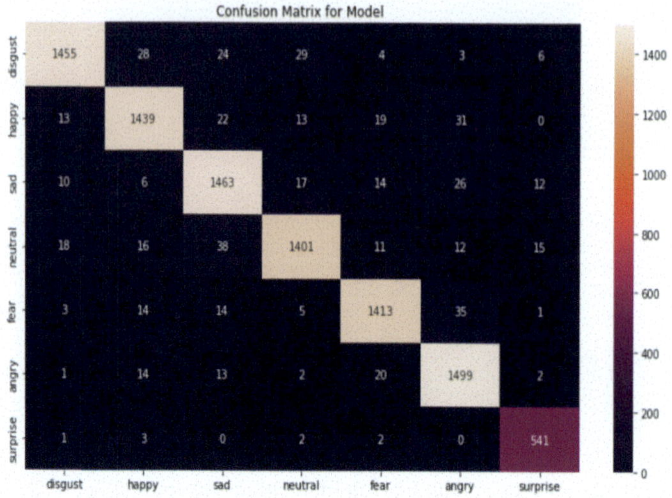

Fig. 14 Confusion matrix for 05-layered CNN model

Model Confusion Matrix

	precision	recall	f1-score	support
disgust	0.97	0.94	0.95	1549
happy	0.95	0.94	0.94	1537
sad	0.93	0.95	0.94	1548
neutral	0.95	0.93	0.94	1511
fear	0.95	0.95	0.95	1485
angry	0.93	0.97	0.95	1551
surprise	0.94	0.99	0.96	549
accuracy			0.95	9730
macro avg	0.95	0.95	0.95	9730
weighted avg	0.95	0.95	0.95	9730

Fig. 15 Explicated demonstration of confusion matrix for 05-layered CNN architecture

with 0.95 value or 96%. Also, after every recall the CNN model shows good accuracy with an average of 95 percent in each emotion. Table 1, represents the Comparative Analysis of our Proposed Modular Approaches with the Already Existing Technical Standpoints.

Table 1 Comparison between existing methodologies and proposed methodology

S. no.	Model (s) utilized	Average accuracy (%)
1	CNN + Bi-LSTM [4]	87.2
2	Empirical Mode Decomposition (EMD) [9]	91.6
3	Deep Stride Convolutional Neural Network (DSCNN) [13]	87.8
4	03-layered CNN (proposed)	90.5
5	04-layered CNN (proposed)	92
6	05-layered CNN (proposed)	95
7	06-layered CNN (proposed)	94

5 Conclusion and Future Scope

Through our Modular Explication, we found that the 05-Layered CNN Model outperformed the 06-Layered CNN Model and thus, contradicted the Ultimate viewpoint, i.e., More the Number of Layers into the Network, More the Accuracy of the System. We got to know, that the Model's Efficacy not only depends upon the Number of Hidden Layers in a Particular Network, but also the Type of Data, the Processing of the Dataset, the Feature Extraction as well as the Fine-Tuning of the Model as per the Data Requirements. Our proposed Approach, stands to be the Most Accurate and efficient one, when compared to the already existing Modus Operandi as depicted in Table 1. Thus, a 05-Layered CNN Architecture could be actuated into Real-time for Speech Emotion Recognition System for diverse use-cases. Moreover, as our Model is data-driven, more and more Data could be accumulated to make the Modular Network more Robust, with the Maximization of Efficacy. Also, more emphasis could be given on trying out different Feature Extraction Techniques as well as playing with the Hyper-Parameters of any Proposed Network. As a result, in the Domain of Speech Emotion Recognition System, the scope of Technological Advancements is Never Ending.

References

1. Abo Absa, H., Deriche, M., Mohandes, M.: A bilingual emotion recognition system using deep learning neural networks. In: 15th International Multi-Conference on Systems, Signals & Devices (SSD), pp. 1241–1245. IEEE (2018)
2. Sahu, S., Gupta, R., Sivaraman, G., Espy-Wilson, C.: Smoothing model predictions using adversarial training procedures for speech based emotion recognition. In: 2018 IEEE International Conference on Acoustics, Speech and Signal Processing (ICASSP), pp. 4934–4938. IEEE (2018)
3. Wang, J., Han, Z.: Research on speech emotion recognition technology based on deep and shallow neural network. In: 2019 Chinese Control Conference (CCC), pp. 3555–3558. IEEE (2019)

4. Hifny, Y., Ali, A.: Efficient Arabic emotion recognition using deep neural networks. In: 2019 IEEE International Conference on Acoustics, Speech and Signal Processing (ICASSP), pp. 6710–6714. IEEE (2019)
5. Sefara, T.J.: The effects of normalisation methods on speech emotion recognition. In: 2019 International Multidisciplinary Information Technology and Engineering Conference (IMITEC), pp. 1–8. IEEE (2019)
6. Tariq, Z., Shah, S.K., Lee, Y.: Speech emotion detection using IoT-based deep learning for health care. In: 2019 IEEE International Conference on Big Data (Big Data), pp. 4191–4196. IEEE (2019)
7. Kumbhar, H.S., Bhandari, S.U.: Speech emotion recognition using MFCC features and LSTM network. In: 2019 5th International Conference on Computing, Communication, Control And Automation (ICCUBEA), pp. 1–3. IEEE (2019)
8. Mustaqeem, S. M., Kwon, S.: Clustering-based speech emotion recognition by incorporating learned features and deep BiLSTM. In: IEEE Access, vol. 8, pp. 79861–79875. IEEE (2020)
9. Lin, Y.-Y., Yang, J.-Y., Kuo, C.-Y., Huang, C.-Y., Hsu, C.-Y., Liu, C.-C.: Use empirical mode decomposition and ensemble deep learning to improve the performance of emotional voice recognition. In: 2020 IEEE 2nd International Workshop on System Biology and Biomedical Systems (SBBS), pp. 1–4. IEEE (2020)
10. Priyasad, D., Fernando, T., Denman, S., Sridharan, S., Fookes, C.: Attention driven fusion for multi-modal emotion recognition. In: 2020 IEEE International Conference on Acoustics, Speech and Signal Processing (ICASSP), pp. 3227–3231. IEEE (2020)
11. Malla, S., Alsadoon, A., Bajaj, S.K.: A DFC taxonomy of Speech emotion recognition based on the convolutional neural network from the speech signal. In: 2020 5th International Conference on Innovative Technologies in Intelligent Systems and Industrial Applications (CITISIA), pp. 1–10 (2020)
12. Zayene, B., Jlassi, C., Arous, N.: 3D convolutional recurrent global neural network for speech emotion recognition. In: 2020 5th International Conference on Advanced Technologies for Signal and Image Processing (ATSIP), pp. 1–5 (2020)
13. Wani, T.M., Gunawan, T.S., Qadri, S.A.A., et al.: Speech emotion recognition using convolution neural networks and deep stride convolutional neural networks. In: 2020 6th International Conference on Wireless and Telematics (ICWT), pp. 1–6 (2020)
14. Sefara, T.J., Mokgonyane, T.B.: Emotional speaker recognition based on machine and deep learning. In: 2020 2nd International Multidisciplinary Information Technology and Engineering Conference (IMITEC), pp. 1–8 (2020)
15. Peng, Z., Li, X., Zhu, Z., Unoki, M., Dang, J., Akagi, M.: Speech emotion recognition using 3D convolutions and attention-based sliding recurrent networks with auditory front-ends. In: IEEE Access, vol. 8, pp. 16560–16572 (2020)
16. Yoon, S., Dey, S., Lee, H., Jung, K.: Attentive modality hopping mechanism for speech emotion recognition. In: 2020 IEEE International Conference on Acoustics, Speech and Signal Processing (ICASSP), pp. 3362–3366 (2020)
17. Audio signal feature extraction and clustering Homepage. https://medium.com/heuristics/audio-signal-feature-extraction-and-clustering-935319d2225. Accessed 29 Sep 2018/
18. Music feature extraction in Python Homepage. https://towardsdatascience.com/extract-features-of-music-75a3f9bc265d. Accessed 30 Dec 2018
19. Wubet, Y.A., Lian, K.-Y.: Voice conversion based augmentation and a hybrid CNN-LSTM model for improving speaker-independent keyword recognition on limited datasets. In: IEEE Access, vol. 10, pp. 89170–89180 (2022)
20. Chatterjee, R., Mazumdar, S., Sherratt, R.S., et al.: Real-time speech emotion analysis for smart home assistants. IEEE Trans. Consum. Electron. **67**(1), 68–76 (2021)

Enhancing Stock Price Prediction Accuracy Through Deep Learning Techniques: A Case Study on Nepal's Stock Market

Swathi Ganesan, Lakmali Karunarathne, Haroon Muhammed Sihan, and Nalinda Somasiri

Abstract In recent years, the emergence of Machine Learning and Deep Learning has substantially improved the precision of stock price prediction, a critical area of interest for economists and investors. This research focuses on Nepal, a developing country with high economic volatility. We use Long Short-Term Memory (LSTM), a form of Recurrent Neural Network (RNN), and data from "sharesansar" to forecast stock prices. Our LSTM model has been modified to reduce losses while maintaining constant accuracy. This enhancement not only advantages experienced traders but also allows beginner traders to engage in lower-risk trading. We use popular metrics to assess the performance of our model, such as Root Mean Square Error (RMSE), Mean Square Error (MSE), and Mean Absolute Error (MAE). These measures are used to compare the accuracy of the LSTM model to classic techniques such as Support Vector Regression (SVR) and Auto-Regressive Integrated Moving Average (ARIMA). In the case of poor countries, our research shows that LSTM outperforms SVR and ARIMA models, giving greater accuracy with reduced error rates.

Keywords Long Short-Term Memory (LSTM) · Auto-Regressive Integrated Moving Average (ARIMA) · Support Vector Regression (SVR) · Root Mean Square Error (RMSE) · Mean Square Error (MSE) · Mean Absolute Error (MAE)

S. Ganesan (✉) · L. Karunarathne · H. M. Sihan · N. Somasiri
York St John University, London EC1A 4JT, UK
e-mail: s.ganesan@yorksj.ac.uk

L. Karunarathne
e-mail: l.karunarathne@yorksj.ac.uk

H. M. Sihan
e-mail: h.sihan@yorksj.ac.uk

N. Somasiri
e-mail: n.somasiri@yorksj.ac.uk

G. Raj et al. (eds.), *Advances in Artificial Intelligence and Machine Learning*, Lecture Notes in Electrical Engineering 1264, https://doi.org/10.1007/978-981-97-9507-9_4

1 Introduction

With the advancement of machine learning, the game of prediction is engulfing the globe; in general, everyone wants to possess the ability to predict the future which can result in profit. As a result, forecasting is rising and gaining popularity in every field, including the stock market and price. To understand in simple words the stock market is a platform that connects buyers and sellers together to trade equity or stock shares of public organizations. Stock details are published to people for the buy and sell purpose by the various companies or organizations according to the rules and regulations through an authorized medium or agent. The stable stock price is unimaginable as it is non-linear and inconsistent because of the fluctuation and high-risk factors of the stock market condition. In this situation, it is also a fact that it is troublesome to extract worthwhile data from the figures. The stock market is a key figure for any economic institution. With the dependable nature of the stock exchange market with the regulation, economy, and stability the prediction of stock prices becomes more difficult in developing countries compared to developed countries. It is also proven that it is extremely hard just to predict the next day's stock price with sheer confidence and luck. To minimize the loss, algorithm models can be used to forecast which stock will go on-trend or drop by feeding the past data to the training models. There are various algorithms for predicting the next coming day's stock price but still obtaining an accurate prediction of the stock price with less error is troublesome.

SVR [1], ARIMA [2], RNN, and other traditional machine learning techniques have been used by researchers for the prediction of the stock price. The performance of these models could not provide adequate accuracy. There is always a demand for methods or techniques that can replace these and provide a prediction with accuracy even with high variations in prices. This research paper has approached LSTM as it can predict better with accuracy in comparison to other approaches. LSTM has the ability to store the cell state in long-term dependencies [3]. The main objective of this research is to focus on stock price changes and try to predict the same. So far under machine learning, various algorithms for prediction have been developed for the prediction. Many developed countries have benefited from these prediction methods. This research proposes a design to predict the stock price of the developing country Nepal where its economic status is turbulent.

2 Survey of Related Works

The research community's long-standing focus has been on stock price prediction and analysis, in order to fully comprehend the stock market and determine the best strategy for profiting from it. The stock price of a corporation, according to economic theory, reveals all information about that company. As a result, people tend to make a lot of money by investing in undervalued assets and selling them at a high-profit margin. Because the share price movement is always susceptible to uncertainty, stock price forecast is a challenging undertaking. The world is evolving around the stock with the rise and fall of the stock price. It is high time that developing countries like Nepal should change their economic flow by investing more in stocks.

Nepal's only stock exchange for the secondary market is Nepal Stock Exchange Ltd. (NEPSE) which provides a platform for investors to conduct transitions such as buying and selling shares of publicly traded companies in Nepal's various sectors. In 1976 under the industrial policy with the establishment of the Security Market Centre, the stock market of Nepal was established [4]. The trading procedure is then securitized by the Nepalese government. The purpose of establishing the securities exchange was to ease and boost the expansion of Nepal's capital market. It was the sole financial market of an institution that performed brokering, underwriting, government bonds market making, managed public offerings, and other financial services prior to its conversion to a stock exchange. Following that, in 1993, the Nepali government took the initiative to redesign the Securities Exchange Center, which was renamed Nepal Stock Exchange.

In this part, various strategies for stock price prediction, such as LSTM, SVR, and ARIMA, are investigated and analyzed. Because there is a dearth of literature available on stock price prediction in the Nepali context, this section attempts to cover as much research as it can.

In [4], the author applied SVR and Neural Networks to predict the stock price for the next day by extracting the raw data from the Nepal Stock Exchange (NEPSE) website. Mean square error (MSE), mean absolute error (MAE), root mean square error (RMSE), and Coefficient of Determination (R2) [4] were used to measure the performance instance of the two learning models namely SVR and Neural Network. The outcome of the research resulted in highlighting that SVR outperformed the Neural Network.

In this paper, a decade of data from the Tehran Stock Exchange (TSE) intraday was used to predict the coming next two months. Long short-term memory (LSTM) was compared with the autoregression integrated moving average (ARIMA) and LSTM performed better than ARIMA in terms of error accuracy, significantly. The error of prediction for the 20-day triple period in LSTM became 3.22, 2.73, and 25.76% while in ARIMA, it resulted in 9.32, 3.84, and 46.95% [5].

In "STOCK PRICE PREDICTION OF NEPAL USING LSTM" the author utilized two layers of the LSTM, the input layer, and the output layer [6]. In the beginning, the first 100 days of data were taken for training and test purposes, and the accuracy was just 30%, but the better output came after feeding the past 5 years of data to the model and the result accuracy was 80% [6]. In this way, it was conceivable to foresee the long-term stock esteem of different companies, calculate the mistake in expectation, and plot the real versus anticipated esteem in a graphical shape. The "Stock Price Prediction Using LSTM and SVR" experiment analyzes stock index data from the S&P 500, NYSE, NSE, BSE, NASDAQ, and Dow Jones Industrial Average [3]. LSTM used with Adam optimizer and sigmoid activation function is compared with the Support Vector Regression and the experiment output proved that the SVR could not perform better than the LSTM for prediction accuracy [3].

The mean absolute percentage error (MAPE) value of LSTM is less than the SVR therefore, it proves that LSTM is able to predict non-linear and high variations of price movement with adequate prediction accuracy [3].

In the [7] "Prediction of Stock Price Based on LSTM Neural Network," the study objects, three indexes of representative Chinese stocks were chosen, and the opening and closing prices, lowest and highest prices, date, and daily trading volume were among the crucial information gathered from them. Even though the LSTM neural network model had a time lag of prediction limitation, it could predict the stock prices with the attention layer. The main goal of the study was to determine the stock price trends by identifying the function of time series by examining historical stock market data and thoroughly examining its internal rules using the advanced deep learning selective memory feature of the LSTM neural network model [7]. "A Comparison of ARIMA and LSTM in Forecasting Time Series" compares the accuracy of the representative technique ARIMA and LSTM when forecasting time series data [8]. On a collection of financial data, these two techniques were put to the test, with the results revealing that deep learning-based algorithms LSTM outperformed traditional-based algorithms ARIMA. The LSTM-based method improved prediction by an average of 85% when compared to ARIMA. Furthermore, according to the study, changing the number of epochs has no impact [9]. To predict the number of COVID-19 cases in the UK, researchers developed a machine learning model using data from January 31, 2020, to March 31, 2021. The model was trained on three algorithms: SVM, RF, and LR. The RF model outperformed the other models, with a 99.47% training accuracy and a 98.26% testing accuracy. This suggests that the RF model can be used to effectively and efficiently predict new COVID-19 cases, which could help the healthcare sector take relevant control measures to stop the spread of the virus [10]. This study develops machine learning models to predict the risk of cardiovascular diseases (CVDs) in individuals. The models are trained on a dataset of patient characteristics, including demographics, medical history, lifestyle factors, and clinical indicators. Six different machine learning algorithms are used to train the models. The performance of the models is evaluated using a variety of metrics, including accuracy, precision, recall, F1 score, and AUC-ROC. The models with the best performance can be used to help healthcare professionals identify individuals at high risk of CVDs so that preventive measures can be taken to reduce their risk. The study's findings contribute to the development of more accurate and robust cardiovascular risk prediction models, which could lead to improved patient outcomes.

3 Methods

In this study, we use a strict technique to predict stock values within the framework of Nepal, a developing country with unstable economic conditions. By using the capabilities of recurrent neural networks (RNNs), particularly Long Short-Term Memory networks, we hope to improve the accuracy of stock price prediction, an area of great interest to investors and economists.

3.1 Problem Statement

Economists and investors are interested in improving the accuracy of stock price prediction, and the use of deep learning and machine learning techniques has shown promise in this regard in recent years. Still up for debate, though, is how useful these methods are in underdeveloped countries like Nepal that experience unstable economies. Finding a reliable and precise machine learning system that can anticipate stock prices more accurately than conventional techniques is the major difficulty, especially in areas with volatile markets.

The primary goals of this study are to identify the most effective machine learning algorithm for predicting stock prices in Nepal using data from the country's stock market and to evaluate the predictive accuracy of various machine learning algorithms—ARIMA, Support Vector Regression (SVR), and Long Short-Term Memory (LSTM) in particular—in the context of developing economies.

3.2 Proposed Solution

We employ a methodical approach to data analysis in our studies. First, we get unprocessed data from a certain online source. The correct information is then extracted, any missing data is cleaned up and fixed, and the data is formatted consistently, among other steps in our extensive data preparation process.

After that, we divided the data into two groups: 70% for training our models and 10% for evaluating their efficacy. Then, we use three distinct algorithms—ARIMA, SVR, and LSTM—each of which offers a special method for examining the data. We bring in the required models and technologies to do this.

Mean Square Error (MSE), Mean Absolute Error (MAE), and Root Mean Square Error (RMSE) are the three common metrics we use to evaluate the accuracy of our models. To further evaluate our model's predictions with the real data, we make visual charts (see Fig. 1).

Raw Data Collection

The study made use of 19-column data (open, close, high, and low prices) from the Sharesansar website's Nepal Stock Market Dataset on Kaggle.

Data Preprocessing

Preprocessing is done on the dataset to enhance price prediction and model performance. To get rid of duplicates, data is retrieved, combined, and cleansed. To ensure

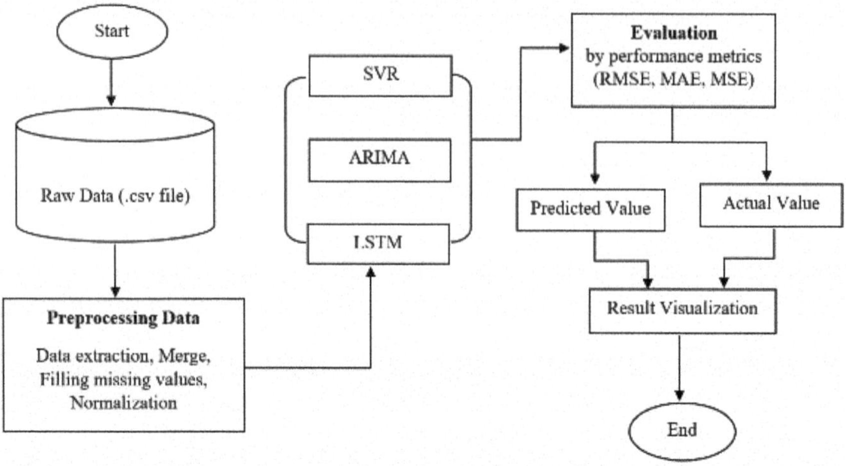

Fig. 1 High-level design of the suggested fix

quality and accuracy, the converted dataset is split into train and test sets, with 90% going towards training and 10% towards testing.

Feature Extraction

The method for examining data in a CSV file—with an emphasis on the "open" and "close" price levels—is explained in the text. As the final transaction price before to closing and a baseline for investors to track future performance, the closure price is selected as the most important reference point for training values. Newspapers employed this strategy prior to computerized trading.

Scaling

Data scaling is crucial for more accurate and effective machine learning model training, as our research has shown. For this work, we decided to use the MinMaxScaler, a useful function found in the scikit-learn library.

$$\chi_{\text{scaled}} = \frac{\chi - \chi_{\min}}{\chi_{\max} - \chi_{\min}}$$

- χ_{scaled} represents the scaled or normalized value of each data point χ.
- χ stands for the original data point.
- $\chi_{\min}$ denotes the minimum value in our dataset or the minimum value we aimed to achieve through scaling (usually 0).
- $\chi_{\max}$ signifies the maximum value in our dataset or the maximum value we intended to attain through scaling (typically 1).

By implementing this formula [11], we effectively transformed our data into a standardized range. This process ensured that the minimum value became 0, the maximum value became 1, and all other data points were proportionally adjusted between 0 and 1. Importantly, this scaling operation maintained the fundamental distribution and shape of our original data. This approach was instrumental in enabling our machine learning algorithms to operate more effectively and produce reliable outcomes.

Training Neural Network

A training dataset is utilized to train the neural network in this step. For prediction, the model employs random weights and biases. The suggested LSTM model is composed of a sequential input layer, three LSTM layers, and a dense layer with activation. Finally, a dense output layer with a linear activation function is implemented.

Output Generation

The target value and error difference are compared with the generated output of the proposed model. The backpropagation algorithm is put in use to minimize the inaccuracies and dissimilarities between the targeted and delivered output which adjusts the biases and weights of the neural network.

Analysis Metrics

To assess how well our algorithms perform, we need to check the error percentage between our predictions and the actual stock prices. To do this, we rely on three important metrics: Root Mean Square Error (RMSE), Mean Square Error (MSE), and Mean Absolute Error (MAE) for each learning algorithm we employ.

Root Mean Square Error (RMSE)

RMSE is a widely used metric for measuring the accuracy of our predictions. It essentially quantifies the differences between our predicted values and the actual values, which are often referred to as residuals. Importantly, RMSE allows us to compare how well different models perform on a specific dataset, rather than comparing models across different datasets. The RMSE formula [12] is outlined below.

$$\text{RMSE} = \sqrt{\sum_{i=1}^{n} \frac{(\hat{y}_i - y_i)^2}{n}}$$

- RMSE is the Root Mean Square Error, the metric we want to calculate.
- n is the total number of data points or observations.
- y_i represents the actual (observed) value for the i-th data point.
- $\hat{y}_i$ represents the predicted value for the i-th data point.

Mean Absolute Error (MAE)

When assessing predictive models, the simple statistic known as Mean Absolute Error (MAE) is employed. The average of the absolute discrepancies between the predicted and actual values is quantified. In contrast to several other measures, MAE does not assign greater weight to larger errors—rather, it evaluates all errors equally.

The difference between expected and actual values is called a prediction error, and it is this absolute value that is used to determine the mean absolute error (MAE) over all data points. Basically, it expresses the average deviation of our forecasts from the actual data [13].

$$\text{MAE} = \frac{1}{n}\sum_{i=1}^{n} |y_i - \hat{y}_i|^2$$

- MAE is the Mean Absolute Error, the metric we want to calculate.
- n is the total number of data points or observations.
- y_i represents the actual (observed) value for the i-th data point.
- $\hat{y}_i$ represents the predicted value for the i-th data point.

Mean Squared Error (MSE)

(MSE) is a crucial metric used to gauge the accuracy of statistical models. It quantifies the average of the squared differences between the observed values and the predicted values. In essence, it measures how far off, on average, our predictions are from the true values. If a model's predictions are perfect and contain no errors, then the MSE equals 0. However, as errors increase, the value of the MSE also rises proportionally, reflecting the increased discrepancy between predicted and actual values. The Mean Squared Error is sometimes referred to as the Mean Squared Deviation (MSD). Below is the formula [14] for MSE.

$$\text{MSE} = \frac{\sum (y_i - \hat{y}_i)^2}{n}$$

- MSE is the Mean Squared Error, the metric we want to calculate.
- n is the total number of data points or observations.
- y_i represents the actual (observed) value for the i-th data point.
- $\hat{y}_i$ represents the predicted value for the i-th data point.

3.3 Algorithm Elaboration

We will examine the algorithms we have developed in this part, using a combination of pre-existing methods and our own modifications. We've employed several fascinating techniques, such as "deep learning," which is analogous to human learning. It is a major actor in the field of data science and plays a significant role in both statistics and predictive modeling. Because deep learning enables handling, analyzing, and making sense of massive volumes of data faster and simpler, it has gained popularity among data scientists. Let's now examine a few of the algorithms we use to forecast stock prices.

Auto-Regressive Integrated Moving Average (ARIMA)

In time series analysis, ARIMA is a crucial tool for modeling time-dependent data, especially in finance. Stationarity is vital for ARIMA, and the Augmented Dickey-Fuller (ADF) test helps ensure this stationarity. It challenges the null hypothesis of non-stationarity. Additionally, the Autocorrelation Function (ACF) plays a pivotal role in quantifying data point correlations within time series. The technique involves the construction of correlation coefficient plots against various time lags, typically measured in terms of a specific number of periods or units [15]. This analytical approach adds depth to our understanding of temporal data, aiding in the development of more robust and accurate forecasting models. When compared to other machine learning techniques, the time-series algorithm has the capacity to extrapolate patterns beyond the scope of the training data [16]

Support Vector Regression (SVR)

Support Vector Regression (SVR) is a powerful supervised learning algorithm employed for predicting continuous values, applying similar principles as the well-known Support Vector Machine (SVM). In SVM, data can be categorized when a high-dimensional feature space is mapped by data, even when the data are linearly non-separable data [17]. SVR operates by finding the optimal hyperplane that best fits the given data, with the objective of minimizing prediction errors.

We employed the Support Vector Regression (SVR) algorithm as part of our analysis to compare its predictive performance with that of ARIMA and LSTM. SVR was included in our study to assess its effectiveness in modeling and predicting continuous values, alongside the other algorithms. This comparative analysis allowed us to evaluate the strengths and weaknesses of SVR in relation to ARIMA and LSTM for our specific prediction task.

Long Short-Term Memory (LSTM)

Long Short-Term Memory (LSTM) is known as a variant form of Recurrent Neural Network (RNN) which has capabilities to overcome the problem of long-term dependency of the RNN. LSTM embodies a chain-like structure that has four neural networks and different memory blocks called cells. The cell state of LSTM performs a significant role as it can perform the addition or removal of information from the cell state [18]. There are three gates known as forget gate, input gate, and output gate which manipulate the memory of the cell.

Data scaling with Scikit-learn's MinMaxScaler normalizes input data for better predictions. In LSTM, data is structured into 60 timesteps, creating a 3D array with 60 timeframes and one feature per step, allowing effective temporal pattern capture.

After defining the network architecture, it's essential to compile it with key parameters like the optimization algorithm and loss function. The choice of activation function in the output layer shapes predictions. Here, we use the Adam optimizer due to its superior performance, efficiency, and suitability for large datasets. Mean square error serves as the loss function, quantifying prediction errors.

With the network assembled, it's time to "fit" it, adjusting weights based on training data. This requires specifying input patterns (X) and matching output patterns (Y). The backpropagation algorithm trains the network, and optimization and loss functions are used as specified during model construction. Training occurs over a predefined number of epochs.

Epochs can be divided into batches, which determine how many input–output pairs the network encounters before updating weights in each epoch. Batch sizes optimize memory usage and training efficiency. In this case, we define 500 epochs and a batch size of 5.

4 Results

In the pursuit of enhancing the accuracy of our predictive modeling, we delve into the application of three distinct algorithms: ARIMA (Auto-Regressive Integrated Moving Average), SVR (Support Vector Regression), and LSTM (Long Short-Term Memory). While our primary emphasis lies on optimizing the performance of LSTM, we have also employed the other two algorithms for a comprehensive comparative analysis, evaluating key metrics such as Root Mean Square Error (RMSE), Mean Squared Error (MSE), and Mean Absolute Error (MAE).

4.1 Auto-Regressive Integrated Moving Average (ARIMA)

As we initiated our analysis with the ARIMA algorithm, Fig. 2 provides a comprehensive array of visual representations. These visuals encompass standardized residuals, a histogram overlaid with estimated density, a Normal Quantile–Quantile (Q–Q) plot, and a correlogram. These graphical depictions serve a dual purpose: they provide valuable insights into the inner workings of the algorithms under scrutiny, while also fostering a more profound comprehension of their effectiveness and appropriateness for the intricate task of forecasting stock prices.

The ARIMA model to rigorous testing using our designated test dataset, we observed a visually compelling result illustrated in Fig. 3. The predictive line is closely aligned with the actual data points, indicating a promising level of predictive accuracy.

However, a deeper analysis unveils a more nuanced picture. When we meticulously scrutinized the model's performance metrics, specifically Root Mean Square Error (RMSE), Mean Squared Error (MSE), and Mean Absolute Error (MAE), we uncovered a notable disparity. While the model exhibited a strong alignment with the test data, its performance on the training dataset revealed markedly higher error rates. This intriguing divergence prompts a closer investigation into the model's behavior,

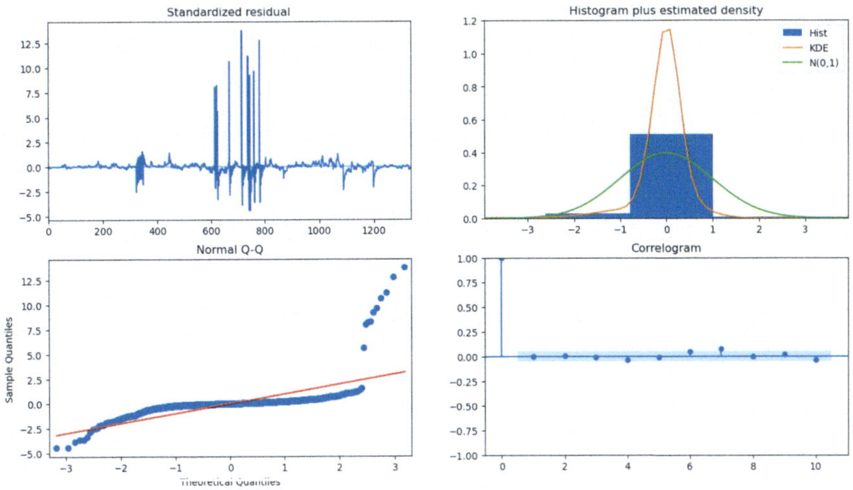

Fig. 2 ARIMA plot diagnostics

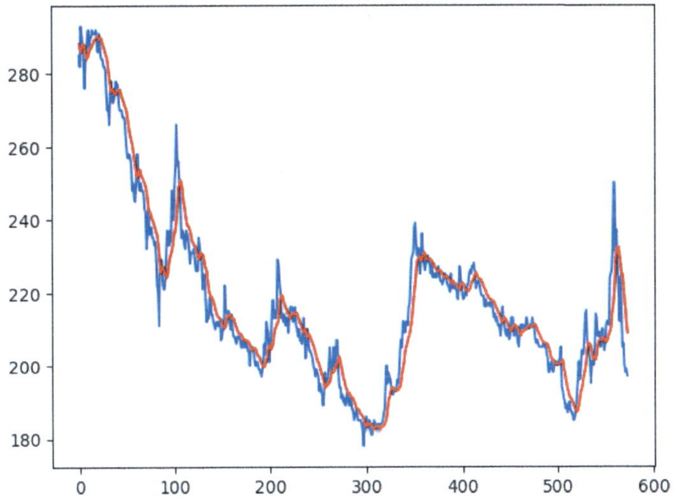

Fig. 3 Trend line forecasts against actual outcomes

potentially shedding light on the factors contributing to this observed discrepancy between training and test data performance.

Table 1 presentation of these performance metrics offers a holistic view of the model's capabilities when applied to the test dataset. While RMSE, MSE, and MAE each convey different aspects of model accuracy, together they enable a comprehensive assessment of the model's strengths and weaknesses in predicting the test data.

Table 1 Evaluate forecasts ARIMA

Evaluation metrics	Results
Test data RMSE	6.08
Test data MSE	37.00
Test data MAE	4.43

4.2 SVR (Support Vector Regression)

In our continued exploration of the Nepali stock prediction dataset, we extended our analysis to encompass the utilization of the Support Vector Regression (SVR) algorithm. This expanded examination allowed us to assess and predict critical performance metrics, namely the Root Mean Square Error (RMSE), Mean Squared Error (MSE), and Mean Absolute Error (MAE). Table 2 represents the training dataset and Table 3 represents the test dataset.

It is immediately apparent from the presented metrics that there exists a substantial disparity between the performance of our model on the training dataset and its performance on the test dataset. Notably, the RMSE for the training data is considerably higher (133.56) compared to that of the test data (18.71). Similarly, the MSE for the training data is significantly elevated (17,838.16) in contrast to the test data (350.16). These disparities underscore the importance of thorough model evaluation, revealing areas where the model's predictive accuracy may vary substantially between training and testing scenarios. The MAE values for the training dataset and test dataset are not provided, potentially limiting our ability to make direct comparisons in terms of absolute error.

This analysis serves as a critical foundation for further investigation, prompting inquiries into the factors contributing to these observed differences in model performance and offering valuable insights for model refinement and enhancement.

Figure 4 illustrates a comparison between actual and predicted close prices. While test data shows good alignment, there is a noticeable gap in the training dataset.

Table 2 Evaluation metrics RMSE, MSE, and MAE (training set SVR)

Evaluation metrics	Results
Train data RMSE	133.56
Train data MSE	17,838.16
Train data MAE	99.5

Table 3 Evaluation metrics RMSE, MSE, and MAE (testing set SVR)

Evaluation metrics	Results
Test data RMSE	18.71
Test data MSE	350.16
Test data MAE	13.84

Fig. 4 Comparison between original close price versus predicted close price SVR

Further investigation is needed to understand and address this gap for more consistent model performance.

4.3 *Long Short-Term Memory (LSTM)*

Our research centered on LSTM (Long Short-Term Memory) model optimization, and we performed a comparison analysis using ARIMA and SVR. Predicting critical performance metrics, namely Mean Absolute Error (MAE), Mean Squared Error (MSE), and Root Mean Square Error (RMSE), was our main objective. We have condensed our data into Tables 4 and 5 below to facilitate a meaningful comparison with ARIMA and SVR. These tables offer a concise summary of our research findings, facilitating a thorough evaluation of the three forecasting techniques.

Particularly when contrasted with ARIMA and SVR, these measures are essential markers of the prediction accuracy and efficacy of our LSTM model. Interestingly, LSTM performs competitively while evaluating the test data, with an RMSE of around 14.87, an MSE of roughly 221.19, and an MAE of roughly 14.06. These metrics offer insightful information about how well the model predicts the future.

Table 4 Evaluation metrics RMSE, MSE, and MAE (training set LSTM)

Evaluation metrics	Results
Train data RMSE	89.45
Train data MSE	8002.82
Train data MAE	24.37

Table 5 Evaluation metrics RMSE, MSE, and MAE (testing set LSTM)

Evaluation metrics	Results
Test data RMSE	14.87
Test data MSE	221.19
Test data MAE	14.06

Fig. 5 Comparison between original close price versus predicted close price LSTM

Though LSTM performs well in the test dataset, it's important to note that the training data has distinct properties, with a larger RMSE of around 89.46 and a matching MSE of about 8002.82. The relevance of comprehending how the model generalizes to new data in contrast to its performance on data it was trained on is shown by this disparity between training and test data.

In essence, our analysis showcases the predictive power of LSTM in the context of Nepali stock price prediction and highlights the nuances of its performance in different dataset subsets, shedding light on its potential strengths and areas for improvement.

When evaluating the performance of predictive models such as Long Short-Term Memory (LSTM), Support Vector Regression (SVR), and Auto-Regressive Integrated Moving Average (ARIMA), Fig. 5. becomes evident that LSTM emerges as the most suitable choice when considering a comprehensive analysis of multiple performance metrics. In some cases, SVR or ARIMA might still be preferred if they align better with the problem at hand or have other advantages such as simplicity or interpretability. Nevertheless, when assessing predictive models across multiple performance metrics, LSTM often stands out as the best-fit option, demonstrating its versatility and robustness in various forecasting scenarios.

5 Conclusion and Discussion

This paper presents a comparative analysis of different machine learning algorithms for predicting stock prices, focusing on ARIMA, SVR, and LSTM. The study aims to identify the most efficient machine learning algorithm for stock price prediction in Nepal, focusing on the use of data from the Nepal stock market. The research aims to identify the most useful machine learning algorithm to replace traditional methods, with close price as the key attribute. The performance metrics such as RMSE, MAE, and MSE were analyzed, and it was concluded that the LSTM neural network consistently outperforms ARIMA and SVR in terms of lower error metrics. However, using one of these models for for-profit trading seems futile. The difference between LSTM and ARIMA using error metrics is incredibly small, varying by points or decimals. Despite LSTM's superior performance when dealing with a wide range of data, the differences between the two are incredibly small, varying by points or decimals.

6 Future Work

The study on Nepal's stock price forecasting has flaws, including insufficient data and hyperparameters impacting prediction performance. A hybrid CNN-LSTM model is proposed for future research, using CNN layers for feature extraction and LSTM for sequence prediction. The focus should be on improving LSTM performance and exploring other hybrid deep learning models.

References

1. Tingyu, Z.: Short-term stock trend prediction based on improved SVR. In: 7th International Conference on Information Science and Control Engineering (ICISCE), Changsha, China, pp. 1283–1286. IEEE (2020)
2. Mahadik, A., Vaghela, D., Mhaisgawali, A.: Stock price prediction using LSTM and ARIMA. In: 2021 Second International Conference on Electronics and Sustainable Communication Systems (ICESC), Coimbatore, India, pp. 1594–1601. IEEE (2021)
3. Bathla, G.: Stock price prediction using LSTM and SVR. In: 2020 Sixth International Conference on Parallel, Distributed and Grid Computing (PDGC), Waknaghat, India, pp. 211–214. IEEE (2020)
4. Top, P.B., Tej, S.B.: Nepal stock exchange prediction using support vector regression and neural networks. In: 2018 Second International Conference on Advances in Electronics, Computers and Communications (ICAECC), Bangalore, India, pp. 1–6. IEEE (2018)
5. Abdoli, G.: Comparing the prediction accuracy of LSTM and ARIMA models for time-series with permanent fluctuation. Periódico do Núcleo de Estudos e Pesquisas sobre Gênero e DireitovCentro de Ciências Jurídicas-Universidade Federal da Paraíba, p. 9 (2020)
6. Neha, J., Pooja, B., Sadhana, K., Subarna, S.: Stock price prediction of Nepal using. In: KEC Conference, Dhapakhel, Lalitpur, Nepal (2018)
7. Dou, W.: Prediction of stock price based on LSTM neural network. In: 2019 International Conference on Artificial Intelligence and Advanced Manufacturing (AIAM), Dublin, Ireland, pp. 544–547. IEEE Computer Society (2019)
8. Sima, N.S., Tavakoli, N., Akbar, S.N.: A comparison of ARIMA and LSTM in forecasting time series. In: 2018 17th IEEE International Conference on Machine Learning and Applications (ICMLA), Orlando, FL, USA, pp. 1394–1401. IEEE (2018)
9. Ganesan, S., Somasiri, N., Jeyavadhanam, R., Karthick, G.: Improved computational efficiency of machine learning algorithm based on evaluation metrics to control the spread of coronavirus in the UK. Int. J. Comput. Syst. Eng. **17**(10), 532–537. World Academy of Science, Engineering and Technology, Open Science Index 202
10. Singarathnam, D., Ganesan, S., Pokrhel, S., Somasiri, N.: Machine learning—based predictive models for cardiovascular risk assessment in data analysis, model development, and clinical implications. Int. J. Res. Multidiscip. Res. **10**(10)
11. Chegg: How to Scale Data into the 0–1 range using Min-Max Normalization (04 May 2023). https://androidkt.com/how-to-scale-data-to-range-using-minmax-normalization/. Accessed 22 Sep 2023

12. Moody, J.: What does RMSE really mean? Towards Data Science (06 Sep 2019). https://towardsdatascience.com/what-does-rmse-really-mean-806b65f2e48e. Accessed 22 Sep 2023
13. P. M.: End-to-end introduction to evaluating Regression Models (19 Jul 2023). https://www.analyticsvidhya.com/blog/2021/10/evaluation-metric-for-regression-models/. Accessed 22 Sep 2023
14. Frost, J.: Mean squared error (MSE) (28 May 2023). https://statisticsbyjim.com/regression/mean-squared-error-mse/. Accessed 22 Sep 2023
15. Zvornicanin, E.: Baeldung (2021). https://www.baeldung.com/cs/acf-pacf-plots-arma-modeling#:~:text=The%20autocorrelation%20function%20(ACF)%20is,number%20of%20periods%20or%20units. Accessed 28 Aug 2022
16. Despot, I.: Timescale Blog (2022). https://www.timescale.com/blog/what-is-time-series-forecasting/#:~:text=In%20the%20simplest%20terms%2C%20time,specific%20point%20in%20the%20future. Accessed 29 Aug 2023
17. IBM: IBM. IBM (17 Aug 2021). https://www.ibm.com/docs/en/spss-modeler/saas?topic=models-how-svm-works. Accessed 29 Aug 2023
18. Greeks, G.F.: Greeks for Greeks (2021). https://www.geeksforgeeks.org/deep-learning-introduction-to-long-short-term-memory/. Accessed 30 Aug 2023

Decision-Level Fusion of Face and Gait Cues Using Fusion of Median Local Binary Pattern (Median-LBP) and Gabor Scale Average (GSA)-Based PCA → LDA

K. Annbuselvi, N. Santhi, and S. Sivakumar

Abstract The paper offers a methodology for recognizing humans by integrating the complementary features face, gait are extracted by fusion of Median Local Binary Pattern (Median-LBP) and Gabor Scale Average (GSA)-based Principal Component Analysis followed by Linear Discriminant Analysis (PCA → LDA) at decision level is proposed. Firstly, the face features and gait features are extracted using Median-LBP and GSA. In Median-LBP, the center pixel value in a 3×3 window is computed using every 3×3 neighborhood, encompassing the central pixel. In GSA, initially, a filter bank is defined using Gabor Wavelet with three scales $\{1, 2, 3\}$ and six angles {0, 30, 60, 90, 120, 150}. Next, a convolution operator is used for integrating the facial and gait cues with the defined filters, and the method averages six distinct angles of Gabor matrices with similar scaling into one average imaging called GSA. Then, they are enhanced via integrating Median-LBP and GSA feature vectors denoted as Median-LBP + GSA. Subsequently, lower dimensional feature vectors are obtained with the PCA → LDA algorithm. The Euclidean distance measure is utilized for obtaining the biometric system's face, gait decisions independently, then the AND/OR rule is employed to fuse the decisions at the decision level for recognition. The methodology is tested with the openly available face dataset (ORL) and gait dataset (CASIA B).

Keywords Median-LBP · GSA · PCA → LDA · Multimodal fusion at decision level

K. Annbuselvi (✉) · N. Santhi
V. V. Vanniaperumal College for Women, Virudhunagar, Tamil Nadu, India
e-mail: annbuselvi@vvvcollege.org

N. Santhi
e-mail: santhi@vvvcollege.org

S. Sivakumar
Cardamom Planter, Association College, Bodinayakanur, Tamil Nadu, India

K. Annbuselvi · N. Santhi · S. Sivakumar
Madurai Kamaraj University, Madurai, Tamil Nadu, India

G. Raj et al. (eds.), *Advances in Artificial Intelligence and Machine Learning*, Lecture Notes in Electrical Engineering 1264, https://doi.org/10.1007/978-981-97-9507-9_5

1 Introduction

Fusing face and gait biometric systems is the most explored area in recent studies due to its significance in applications related to security and human–computer interfaces [1]. Face and gait biometrics are combined primarily because their traits may be accurately extracted from low-quality pictures and are less wedged through slight adjustments on viewing angle. Hence, the combined face and gait biometric system might be adaptable to variations in the subject's distance from the camera [2, 3].

The LBP method improves the image by thresholding center pixel value among 3X3 neighboring pixels. The median value of all 3X3 neighboring pixels, which includes the centering pixel, is employed as the threshold in the Median-LBP [4, 5]. In GSA, initially, a filter bank is defined using Gabor Wavelet with 3 scales $\{1, 2, 3\}$ and 6 orientations $\{0, 30, 60, 90, 120, 150\}$. Next, a convolution operator is used for integrating the facial and gait cues with the defined filters and the method averages six distinct angles of Gabor matrices with similar scaling into single averaged imaging called GSA [6].

The Median-LBP and GSA feature enhancement methods are combined to obtain further strong features, which have resilient face features and gait features denoted as Median-LBP + GSA [7].

Subsequently, Eigen-based Fisher's features are obtained by means of PCA $\rightarrow$ LDA dimensionality reduction algorithm [8–10]. The Euclidean distance measure is utilized for obtaining the biometric system's face, gait decisions independently, then to fuse the decisions at the decision level during recognition, the AND/OR rule is used.

The paper's structure is as follows: the second part reviews relevant research, the third part offers Median-LBP, GSA, suggested methodology Median-LBP and GSA, i.e. Median-LBP + GSA, decision-level fusion of face, gait decisions using Median-LBP + GSA-based PCA $\rightarrow$ LDA. The fourth part deals with findings and discussions, and the fifth section concludes the paper.

2 Reviews of Relevant Research

Numerous multimodal systems have been developed and used commercially to date. These systems are built on a variety of algorithmic frameworks and fusion levels. In [11], coarseness, contrast, directionality, entropy, homogeneity, and energy of iris and fingerprint images have been used as features and decisions made separately and integrated at the decision level using AND and OR to recognize humans. Yet in [12], the appearance-based procedures, PCA and LDA, are combined to create face recognition system that is offered at decision-fusion level.

The facial emotion identification was put forth in [13], utilizing Gabor wavelets to preprocess human face images, considerably eliminating differences brought on by features, position, and lighting. Low-dimensional, discriminating feature vectors are extracted by PCA and LDA and used for classification. In order to improve face expression accuracy, in [14] suggested to use Gabor DCT filters in conjunction with an efficient averaging DCT filtering technique to reduce redundant features of Gabor matrices. For the automatic identification of a facial gender identification system, a dependable technique for lighting invariant concise feature extraction by Gabor filter is presented in [15].

In [16], the LBP operator is used to define more local structure information. They precisely suggested an improved LBP (ILBP) that matches each pixel comprising the central pixel through average brightness of every pixel within the 3 X 3 region [17].

Yet in [18], the Hamming LBP is proposed to improve the discrimination capabilities of LBP. They categorized non-uniform patterns based on Hamming distance rather than placing them in a single bin like LBP does. Further in [4], a method, median-based LBP is proposed to proficiently extract imperfect face and gait cue features for enhanced recognition.

Although a variety of methods for extracting features have been used to define biometrics, a new algorithm that is robust, unique, comprehensive, and easily accessible is still required.

3 Proposed Methodology

3.1 Median Local Binary Pattern (Median-LBP)

A modified version of LBP called median-LBP is used to extract more reliable feature vectors. In Median-LBP, the center pixel value in a 3×3 window is computed using all 3×3 neighborhoods, which includes the central pixel as well. The structure chart, procedure are presented in Fig. 1 and Procedure 1.

Procedure 1: Median-LBP Computation

Input: An image in grayscale.

Output: GSA Image.

Steps:

1. Select 3×3 neighborhood's including center pixel of each pixel and find the median value of selected pixels.
2. The median value is compared with the 3×3 neighborhood including centering pixel, the pixel is assigned 0 if it is smaller than the median value, and 1 otherwise.
3. Decimal equivalent of center pixel value called Median-LBP code is got by combining binary codes in clockwise direction.

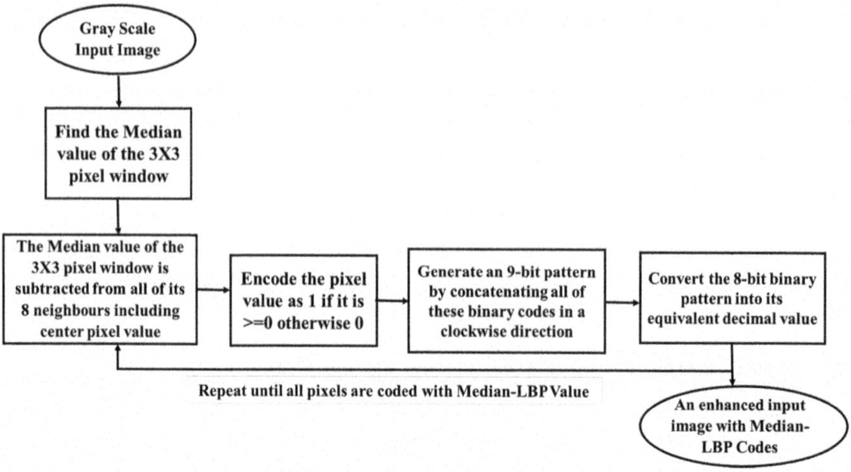

Fig. 1 Structure chart of median local binary pattern (Median-LBP)

3.2 *Gabor Scale Average (GSA)*

One of the effective methods for extracting image features is the Gabor wavelet. They can be used to extract features from images that are aligned to particular orientations. The convolution of bank of Gabor kernels with the image yields the Gabor transformation of an image. The procedure and the flow chart are presented in Fig. 2 and Procedure 2.

Procedure 2: GSA Computation

Input: An image in grayscale.

Output: Image with an enhanced input image with Median-LBP codes.

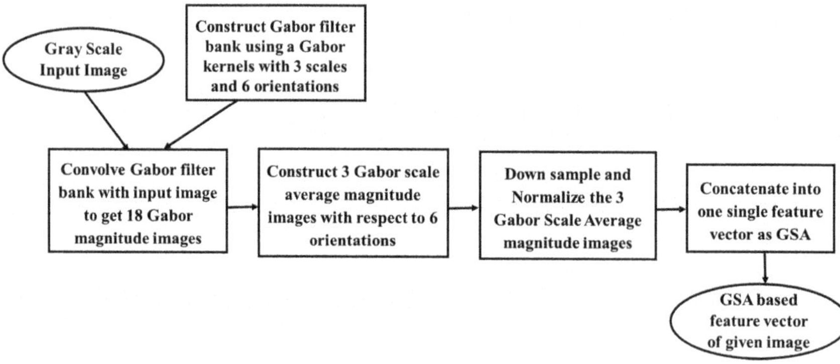

Fig. 2 Structure chart of Gabor Scale Average (GSA)

Steps:

1. Construct Gabor filter bank using Gabor kernels. The kernels are decided by the factor's scale and orientation. Here there are three scales denoted as $v = \{0,1,2\}$ and six orientations ranging between 0° and 5π/6°

$$\left(\text{i.e. } 0, \frac{\pi}{6}, \frac{2\pi}{6}, \frac{3\pi}{6}, \frac{4\pi}{6}, \frac{5\pi}{6}\right) \text{denoted as } \mu = \{0, 1, 2, 3, 4, 5\}.$$

2. Apply the convolution operator between bank of Gabor filters and the given image for obtaining 18 Gabor magnitude images represented as $G^{i}_{v,\mu,}$ where $i = 1, 2, \ldots 18$,

$$v = \{0, 1, 2\} \text{ and } \mu = \{0, 1, 2, 3, 4, 5\}$$

$$G^{i}_{v,\mu,} = \begin{pmatrix} G^{1}_{0,0}, G^{2}_{0,1,}, G^{3}_{0,2}, G^{4}_{0,3}, G^{5}_{0,4}, G^{6}_{0,5}, G^{7}_{1,0}, G^{8}_{1,1}, G^{9}_{1,2}, G^{10}_{1,3}, \\ G^{11}_{1,4}, G^{12}_{1,5}, G^{13}_{2,0}, G^{14}_{2,1,}, G^{15}_{2,2}, G^{16}_{2,3}, G^{17}_{2,4}, G^{18}_{2,5} \end{pmatrix}$$

3. For each scale, average the Gabor magnitude images with respect to six orientations separately and denoted as

$$\text{GSA}_v = \text{GSA}_0 = \text{mean}\left(G^{1}_{v,0}, G^{2}_{v,1}, G^{3}_{v,2}, G^{4}_{v,3}, G^{5}_{v,4}, G^{6}_{v,5}\right) \text{ where } v = 0$$
$$\text{GSA}_v = \text{GSA}_1 = \text{mean}\left(G^{7}_{v,0}, G^{8}_{v,1}, G^{9}_{v,2}, G^{10}_{v,3}, G^{11}_{v,4}, G^{12}_{v,5}\right) \text{ where } v = 1$$
$$\text{GSA}_v = \text{GSA}_2 = \text{mean}\left(G^{13}_{v,0}, G^{14}_{v,1}, G^{15}_{v,2}, G^{16}_{v,3}, G^{17}_{v,4}, G^{18}_{v,5}\right) \text{ where } v = 2$$

4. Finally, the normalized GSA_0, GSA_1, GSA_2 are concatenated into one single feature vector denoted as GSA.

$$\text{i.e. GSA} = [\text{GSA}_0, \text{GSA}_1, \text{GSA}_2]$$

3.3 *Fusion of Median-LBP and GSA*

In order to produce a further powerful feature vector, the suggested fusion of Median-LBP and GSA integrates, features got through Median-LBP, as shown in Sect. 3.1, and GSA, as stated in Sect. 3.2. The formula that follows is used to get the resultant feature vector through the averaged Median-LBP and GSA.

$$\text{Fusion(Median} - \text{LBP, GSA)} = \frac{\text{Median} - \text{LBP features} + \text{GSA features}}{2}$$

Procedure 3 provides the algorithm, and Fig. 3 shows the structure chart.

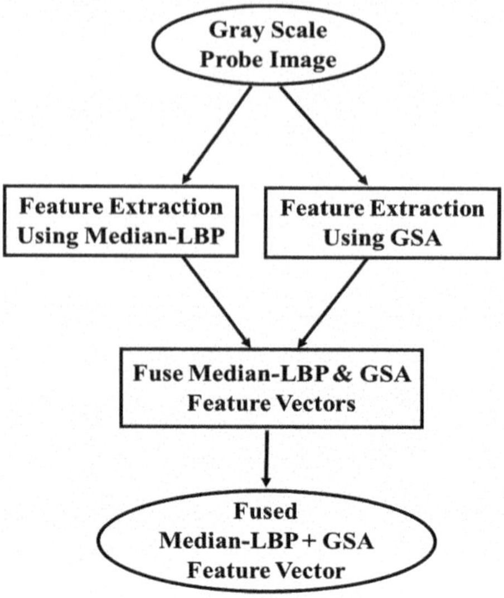

Fig. 3 Structure chart of fusion (Median-LBP, GSA)

Procedure 3. Fusion (Median-LBP, GSA)

Input: An image in grayscale.

Output: Discriminative robust feature vector by fusing Median-LBP and GSA.

Steps:

1. Construct Median-LBP features with the Median Local Binary Pattern.
2. Construct GSA features with the Gabor Scale Average.
3. To produce a more powerful and selective feature vector, combine the Median-LBP and GSA features using the above formula.

3.4 *Decision-Level Fusion of Face and Gait Cues Using Median-LBP + GSA-Based PCA → LDA*

The procedure of decision-level fusion of face and gait classifiers using fusion of Median-LBP and GSA algorithms is given in Procedure 4, and the flowcharts of fusion of face and gait Median-LBP and GSA-based biometric systems and their decision fusion are shown in Figs. 4, 5, and 6.

Procedure 4. Decision-Level Fusion of Face and Gait Cues Using Fusion (Median-LBP, GSA) based PCA → LDA

Input:

For face biometric: trained face database, face test image in gray.

For gait biometric: trained gait database, gait test image in gray.

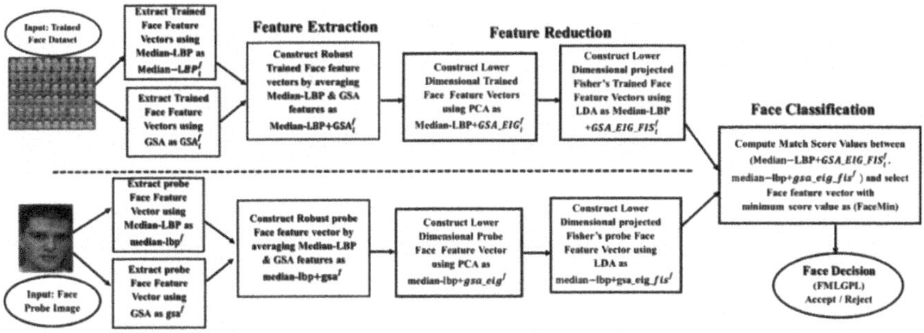

Fig. 4 Structure chart of face recognition using fusion (Median-LBP, GSA)

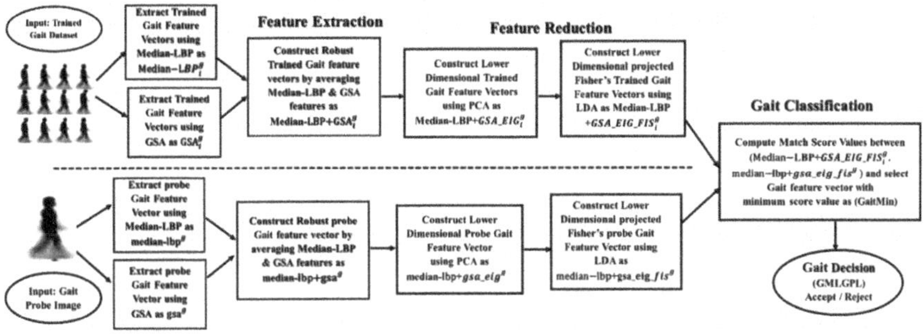

Fig. 5 Structure chart of gait recognition using fusion (Median-LBP, GSA)

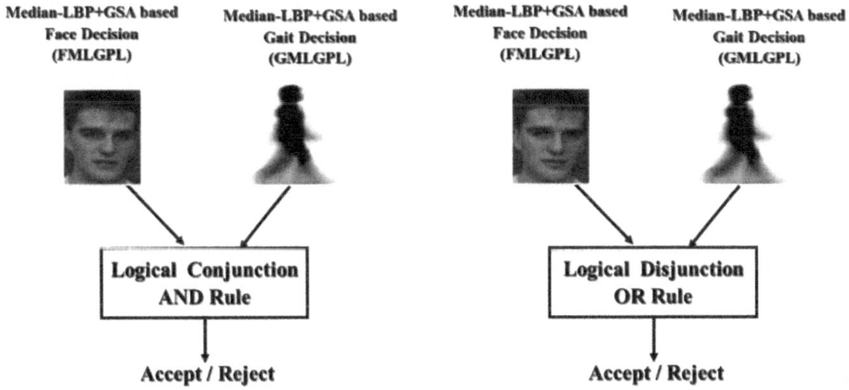

Fig. 6 Structure chart of fusion (Median-LBP, GSA)-based decision level fusion of face and gait using AND and OR fusion rules

Output:

Recognition based on Median-LBP + GSA and PCA → LDA.

Steps:

1. **Feature Extraction in Trained Face Database and Gait Database**:
   a. Extract features of face, gait for all N training images in the face and gait dataset, respectively, with the Median-LBP and GSA algorithms denoted as

$$\text{Median} - \text{LBP}_i^f \text{ and Median} - \text{LBP}_i^g \text{ and GSA}_i^f \text{ and GSA}_i^g,$$

   b. Fuse the Face Median-LBP and *GSA* features denoted as Median-LBP + GSA_i^f using the formula

$$\text{Fusion}\left(\text{Median} - \text{LBP}_i^f, \text{GSA}_i^f\right) = \frac{\left(\text{Median} - \text{LBP}_i^f + \text{GSA}_i^f\right)}{2}$$

   Similarly, fuse the Gait Median-LBP and GSA features denoted as Median-LBP + GSA_i^g using the formula

$$\text{Fusion}\left(\text{Median} - \text{LBP}_i^g, \text{GSA}_i^g\right) = \frac{\left(\text{Median} - \text{LBP}_i^g + \text{GSA}_i^g\right)}{2}$$

   c. Construct lower dimensional Median-LBP + GSA-based Eigen face and Eigen gait feature vectors using PCA denoted as

$$\text{Median} - \text{LBP} + \text{GSA\_EIG}_i^f \text{ and Median} - \text{LBP} + \text{GSA\_EIG}_i^g$$

   d. Construct lower dimensional projected Median-LBP + GSA-based Fisher's face and gait feature vectors using LDA denoted as

$$\text{Median} - \text{LBP} + \text{GSA\_EIG\_FIS}_i^f \text{ and Median} - \text{LBP} + \text{GSA\_ EIG\_ FIS}_i^g,$$

   where $i = 1, 2, \ldots, \text{N}$.

2. **Feature Extraction in Face and Gait Testing Phase**:
   a. Extract face features and gait features using Median-LBP denoted as $\text{median} - \text{lbp}^f$ and $\text{median} - \text{lbp}^g$ and GSA denoted as gsa^f and gsa^g
   b. Fuse the Face Median-LBP and GSA features denoted as $\text{median} - \text{lbp} + \text{gsa}^f$ using the formula

$$\text{Fusion}\left(\text{median} - \text{lbp}^f, \text{gsa}^f\right) = \frac{\left(\text{median} - \text{lbp}^f + \text{gsa}^f\right)}{2}$$

Similarly, fuse the Gait Median-LBP and GSA features denoted as $\text{median} - \text{lbp}^g$ using the formula

$$\text{Fusion}\left(\text{median} - \text{lbp}^g, \text{gsa}^g\right) = \frac{(\text{median} - \text{lbp}^g + \text{gsa}^g)}{2}$$

c. Construct lower dimensional Median-LBP + GSA-based Eigen face and Eigen gait feature vectors using PCA denoted as

$$\text{median - lbp} + \text{gsa\_eig}^f \text{ and median - lbp} + \text{gsa\_eig}^g$$

d. Construct lower dimensional projected Median-LBP + GSA-based Fisher's face and gait feature vectors using LDA denoted as

$$\text{median - lbp} + \text{gsa\_eig\_}\mathit{fis}^f \text{ and median - lbp} + \text{gsa\_eig\_}\mathit{fis}^g$$

3. **Classification**:

(a) Calculate Median-LBP + GSA-based face and gait matching scores between face features of training images and probe face image

$$\left(\text{Median - LBP} + \text{GSA\_EIG\_FIS}_i^f, \text{median - lbp} + \text{gsa\_eig\_}\mathit{fis}^f\right),$$

where $i = 1, 2, ..., \text{N}$ and between gait features of training images and probe gait image

$$\left(\text{Median - LBP} + \text{GSA\_EIG\_FIS}_i^g, \text{median - lbp} + \text{gsa\_eig\_}\mathit{fis}^g\right),$$

where $i = 1, 2, \ldots, \text{N}$ using Euclidean distance measure separately.

(b) From trained databases, determine the image with the least scoring, for face(Facemin), for gait(Gaitmin), so that the correlating probe image's Euclidean distance is as small as possible.

4. **Median-LBP + GSA-Based Face Recognition and Gait Recognition**:

In this face, gait recognition system using Median-LBP + GSA-based PCA $\rightarrow$ LDA, the face(FMLGPL) and gait(GMLGPL) decisions are found, by Facemin and Gaitmin.

Decision from face (FMLGPL):

$$\text{Genuine(Accept)} => \text{Facemin} <= \text{threshold}(\eta).$$
$$\text{Imposter(Reject)} => \text{Facemin} > \text{threshold}(\eta).$$

Decision from gait (GMLGPL):

Genuine(Accept) => Gaitmin <= threshold (η).

Imposter(Reject) => Gaitmin > threshold(η).

5. **Decision-Level Fusion of Face and Gait Using Median-LBP + GSA-Based PCA → LDA**:

   Rules of AND and OR:

   (a) Logical Conjunction(AND): The final fused decisions of face, gait is recognized as genuine (accepted) only if both of the decisions are true otherwise the decision is imposter (rejected).
   (b) Logical Disjunction(OR): The final fused decisions of face, gait is recognized as genuine (accepted) if either of decision is true or the decision is imposter (rejected).

4 Results and Discussion

The offered procedure is implemented with 100 images from the face database ORL [19] and 100 images from gait database CASIA B [20] with respect to 10 unique subjects with 10 images per subject. The considered images are distributed in a ratio of 80 and 20% between training and testing datasets for face and gait separately. To facilitate system usage, a database with several subdirectories for training and testing faces and gaits is automatically connected to the recognition system.

The recognition rate is used to measure each method's performance.

$$\text{Recognition Rate (in\%)} = \frac{\text{Number of subject recognized as genuine}}{\text{Total Number of subjects in the dataset}} \times 100$$

In this paper, the rates of recognition are obtained by considering the features in the database as 20, 40, 60, 80, and 100.

The recognition rates of the face classifier are displayed in Table 1 and are shown visually in Fig. 7. The gait biometric classifier results are displayed graphically in Fig. 8 and are shown in Table 2. Also Tables 3 and 4 show recognition rates of face, gait decisions combined at decision-fusion level by AND, OR rules and are displayed graphically in Figs. 9 and 10.

In accordance to the study's findings, the uni-model face biometric and gait biometric systems based on Median-LBP + GSA and PCA → LDA with 100% of the features produce recognition rates of 98% and 96.6%, respectively. On the other hand, combining the decisions of face at decision fusion with logical AND and OR conjunction and disjunction rules produces recognition rates of 99.3% and 99.4%, respectively. Additionally, it has been proposed that decision-level fusion outperforms well using disjunction (OR) fusion rules compared to conjunction (AND) rules.

Table 1 Recognition rates (in %): face classification using Median-LBP + GSA and PCA → LDA

Feature selection and reduction methods	Number of features				
	20	40	60	80	100
	Recognition rates (in %)				
Median-LBP-based PCA → LDA	94.7	95.2	95.7	96.1	96.6
GSA-based PCA → LDA	95.8	96	96.2	96.5	96.8
* Median-LBP + GSA-based PCA → LDA	96.2	96.6	97	97.5	98

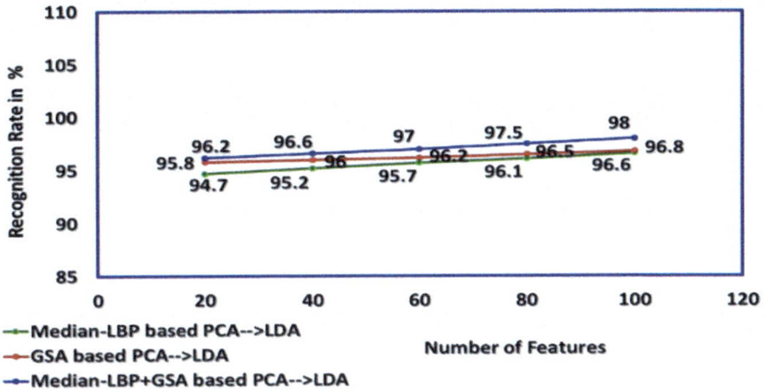

Fig. 7 Recognition rates (%): face classification using Median-LBP + GSA and PCA → LDA

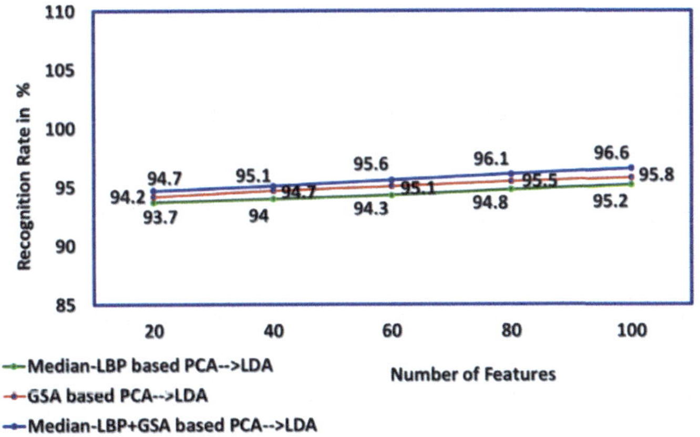

Fig. 8 Recognition rates (in %): gait classification using Median-LBP + GSA and PCA → LDA

Table 2 Recognition rates (in %): gait classification using Median-LBP + GSA and PCA → LDA

Feature selection and reduction methods	Number of features				
	20	40	60	80	100
	Recognition rates (in %)				
Median-LBP-based PCA → LDA	93.7	94	94.3	94.8	95.2
GSA-based PCA → LDA	94.2	94.7	95.1	95.5	95.8
* Median-LBP + GSA-based PCA → LDA	94.7	95.1	95.6	96.1	96.6

Table 3 Recognition rates (in %): decision fusion of face, gait biometrics using Median-LBP + GSA and PCA → LDA using logical conjunction (AND)

Feature selection and reduction methods	Number of features				
	20	40	60	80	100
	Recognition rates (in %)				
Median-LBP-based PCA → LDA	96	96.3	96.7	97.2	97.7
GSA-based PCA → LDA	96.5	97	97.4	97.8	98.2
* Median-LBP + GSA-based PCA → LDA	97	97.6	98.1	98.6	99.3

Table 4 Recognition rates (in %): decision fusion of face, gait biometrics using Median-LBP + GSA and PCA → LDA using logical disjunction (OR)

Feature selection and reduction methods	Number of features				
	20	40	60	80	100
	Recognition rates (in %)				
Median-LBP-based PCA → LDA	96.6	97	97.4	97.8	98.2
GSA-based PCA → LDA	97.3	97.7	98	98.4	98.7
* Median-LBP + GSA-based PCA → LDA	97.7	98.1	98.5	98.9	99.4

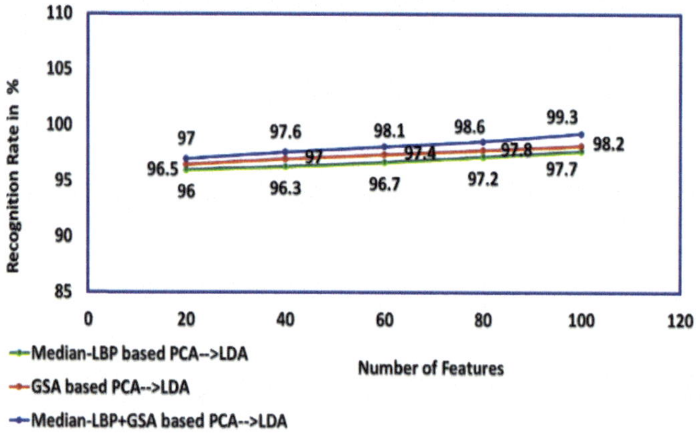

Fig. 9 Recognition rates (in %): decision fusion of face, gait biometrics using Median-LBP + GSA and PCA → LDA using logical conjunction (AND)

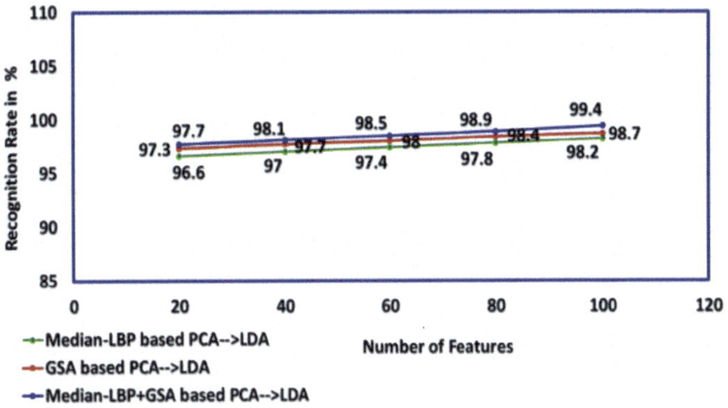

Fig. 10 Recognition rates (in %): decision fusion of face, gait biometrics using Median-LBP + GSA and PCA → LDA using logical conjunction (OR)

5 Conclusion

An efficient new procedure for recognizing face and gait cues by fusing the face features and Gait features of Median-LBP and GSA features at the decision level is provided in this paper. By integrating Median-LBP and GSA-based face features and gait features at the decision level, we can achieve 99.4% of recognition rate with logical disjunction. In future, the work might be improved by integrating three or more feature enhancement methods to produce a 100% result.

References

1. Ross, A.: An introduction to multibiometrics. In: Proceedings of the 15th European Signal Processing Conference (EUSIPCO), Poznan, Poland (2007)
2. Zhou, X., Bhanu, B., Han, J.: Human Recognition at a Distance in Video by Integrating Face Profile and Gait, Springer's Book Series Advances in Computer Vision and Pattern Recognition, pp. 533–543 (2005). https://doi.org/10.1007/11527923_55
3. Annbuselvi, K., Santhi, N.: Intelligences of fusing face and gait in multimodal biometric system: a contemporary study. Int. J. Comput. Sci. Trends Technol. (IJCST) **10**(10) (2017). ISSN: 2347-8578, 516-522
4. Annbuselvi, K., Santhi, N., Sivakumar, S.: A competent multimodal recognition using imperfect region based face and gait cues using Median-LBPF and Median-LBPG based PCA followed by LDA. Mater. Today: Proc. **62**(7), 4869–4879. Science Direct, (2022). Elsevier. https://doi.org/10.1016/j.matpr.2022.03.505
5. Annbuselvi, K., Santhi, N., Sivakumar, S.: Region based feature fusion of imperfect face and gait cues for human recognition using median-LBPF and median-LBPG based PCA followed by LDA. Int. J. Eng. Comput. Sci. **11**(1) (2022). ISSN: 2319-7242, 25483-25492
6. Santhi, N., Annbuselvi, K., Sivakumar, S.: An Efficient Gabor Scale Average (GSA) based PCA to LDA feature extraction of face and gait cues for multimodal classifier. Springer's Book Series. Advances in Intelligent Systems and Computing, pp. 153–164. ISBN 978-981-19-0474-5 (2021).https://doi.org/10.1007/978-981-19-0475-2

7. Annbuselvi, K., Santhi, N., Sivakumar, S.: Fusion of LBP and median LBP for dominant region based multimodal recognition using imperfect face and gait cues. In: 7th International Conference on Computer Vision, High Performance Computing, Smart Devices and Networks (CHSN-2022), Springer Book Series, Lecture Notes in Electrical Engineering, (Indexed by Scopus), pp 537–550 (2022). https://doi.org/10.1007/978-981-99-6690-5_41
8. Annbuselvi, K., Santhi, N.: Role of feature extraction techniques: PCA and LDA for appearance based gait recognition. Int. J. Comput. Sci. Eng. **6**(4) (2018). E-ISSN: 2347-2693, 38-44
9. Santhi, N., Annbuselvi, K.: Performance analysis of feature extraction. Int. J. Eng. Res. Comput. Sci. Eng. **5**(3) (2018). ISSN (Online) 2394-2320
10. Kumar, G., Bhatia, P.K.: A detailed review of feature extraction in image processing systems. In: 2014 Fourth International Conference on Advanced Computing & Communication Technologies (2014)
11. Garg, S.N., Vig, R., Gupta, S.: Analysis of decision level fusion in multimodal biometrics using iris and fingerprint. In: ICEEECOS-2016 (2016). ISBN: 978-1-78561-826-0. https://doi.org/10.1049/CP.2016.153
12. Marcialis, G.L.: Decision-level fusion of PCA and LDA-based face recognition algorithms. Int. J. Image Graph. (2011). https://doi.org/10.1142/S0219467806002239
13. Magesh Kumar, C., Thiyagarajan, R., Natarajan, S.P., Arulselvi, S..: Gabor features and LDA based face recognition with ANN classifier. © IEEE (2011). 978-1-4244-7926-9/11/$26.00
14. Dosodia, P., Poonia, A., Gupta, S.K., Agarwal, S.L.: New Gabor-DCT feature extraction technique for facial expression recognition. In: ICCSNT-2015 (2015)
15. Agrwa, S.L., Jhanwar, A., Goswami, K., Gupta, S.K., Vibhor: Facial gender recognition using Gabor-DCT feature extraction. J. Stat. Manag. Syst. **22** (2019)
16. Jin, H., Liu, Q., Lu, H., Tong, X.: Face detection using improved LBP under Bayesian framework. In: ICIG'04. © IEEE (2004)
17. Lin, Y., Zhang, J., Wang, S.: An Improved LBP Feature Extraction Algorithm. © Springer Nature Singapore Pvt. Ltd. (2020). https://doi.org/10.1007/978-981-15-3863-6_38
18. Yang, H., Wang, Y.: A LBP-based face recognition method with hamming distance constraint. In: ICIG'07. © IEEE (2007). https://doi.org/10.1109/ICIG.2007.144
19. The Database of Faces (ORL Face Database). In: AT&T Laboratories Cambridge (2002). http://www.cl.cam.ac.uk/research/dtg/attarchive/facedatabase.html
20. CASIA Gait Database. http://www.sinobiometrics.com

Analysis of CERN's Project Atlas Data for a Possible Relationship Between Subatomic Particles

Shailendra Narayan Singh, Farhan Asif, Duranjai Malik, and Prajawal Gupta

Abstract The European Organization for Nuclear Research also known as CERN houses the Large Hadron Collider (LHC), which plays a role in scientific exploration. It allows us to delve into the world of particles and the forces that shape our universe. Among the LHC endeavors is the ATLAS experiment, which offers a perspective on proton collisions and provides us with a wealth of valuable data. Our research focuses on utilizing machine learning models to simulate these collisions within the ATLAS experiment. This innovative approach aims to uncover hidden patterns and relationships within the realm pushing beyond the boundaries set by the Standard Model. In this paper, we present an analysis of data using machine learning techniques revealing unknown correlations and patterns among subatomic particles. Moving forward our research will involve expanding the dataset through data collection from LHC experiments refining our machine learning models and conducting investigations to validate these proposed relationships, among subatomic particles.

Keywords ATLAS · LHC · CERN · Machine learning · Neural networks

1 Introduction

The combination of machine learning and particle physics marks a phase in scientific exploration, where we delve into the fundamental building blocks of the universe. At the core of particle physics lies the Standard Model, a framework that meticulously describes nature's particles and their intricate interactions. The dominant SM background process for all-hadronic events is QCD multi-jet production [20]. While the Standard Model has proven successful in predicting outcomes it still leaves us with

S. N. Singh · F. Asif · D. Malik (✉) · P. Gupta
Amity School of Engineering and Technology Amity University Noida, Noida, India
e-mail: duranjai000@gmail.com

S. N. Singh
e-mail: snsingh36@amity.edu

G. Raj et al. (eds.), *Advances in Artificial Intelligence and Machine Learning*, Lecture Notes in Electrical Engineering 1264, https://doi.org/10.1007/978-981-97-9507-9_6

questions, such as the origins of mass, the enigmatic nature of dark matter and how to unify fundamental forces.

In this era driven by data-driven discoveries, machine learning plays a role in advancing our understanding of particle physics. Hadronic object reconstruction at the LHC is ripe for ML applications [12]. Particle accelerators like the Large Hadron Collider (LHC) generate datasets filled with insights about particles. Among these experiments, the ATLAS project stands out as a pioneer in accumulating a treasure trove of data that demands exploration.

This research endeavor diverges from physics-centric approaches. Embraces the immense potential of machine learning. Our main objectives revolve around analyzing the ATLAS dataset to uncover hidden patterns and unknown connections between subatomic particles. In addition, to this analysis, we utilize machine learning models to simulate proton collisions within the high energy environment of LHC. Deep neural network-based classifiers are used to (continuously) reweight one particle-level simulation into another and additionally use the full phase space to fit parameters within a given model [18].

Our main goal has two parts; firstly, we want to examine the data from the ATLAS experiment to uncover any connections or trends that could reveal phenomena in the world of subatomic particles. Secondly, we plan to use machine learning models to simulate proton collisions in the Large Hadron Collider (LHC) setting. By comparing these simulations with collision data, our aim is to analyze any differences and gain insights that may take us beyond what is currently understood by the Standard Model of particle physics (see Fig. 1).

The comparison between these simulations and real collision data has the potential to reveal differences and guide us towards exploring areas of physics that go beyond the limits of the Standard Model.

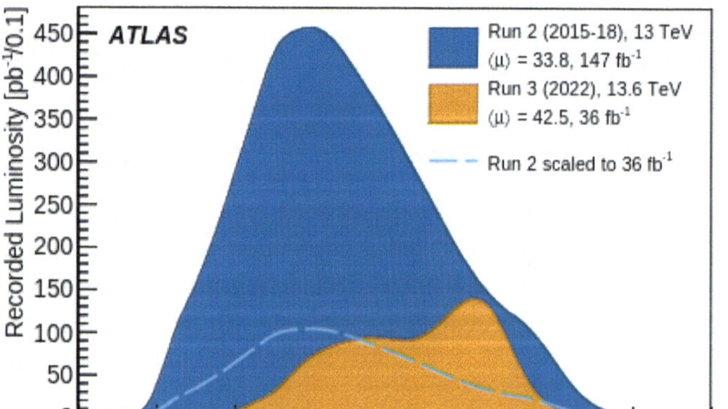

Fig. 1 Distribution of the average number of pp collisions at the ATLAS interaction point during the Run 2 and the 2022 Run 3 LHC data-taking. The dashed line depicts the Run 2 distribution normalized to the same integral as the Run 3 distribution [10]

The significance of this research is that it has the potential to uncover the connections, among particles providing insights that push the boundaries of our knowledge. It also highlights how machine learning and particle physics can work together showing that artificial intelligence can be a tool for analyzing data and identifying patterns that humans might miss (see Equation 1).

$$Gain(S, A) = Entropy(S) - \Sigma_{v \in Values(A)} \frac{|S_v|}{|S|} Entropy(S_v)$$

Equation 1 Extra

As machine learning continues to advance and transform exploration this research demonstrates the relationship between artificial intelligence and particle physics. The best AUC with a non-negligible level of noise population is achieved using Extra (Decision) Trees [15]. It represents progress in using machine learning capabilities to delve into the mysteries of the subatomic realm. In this journey, we venture into territory, guided by the belief that combining machine learning with particle physics holds great promise, for groundbreaking discoveries that go beyond traditional physics paradigms reshaping our understanding of the universe itself (see Fig. 2).

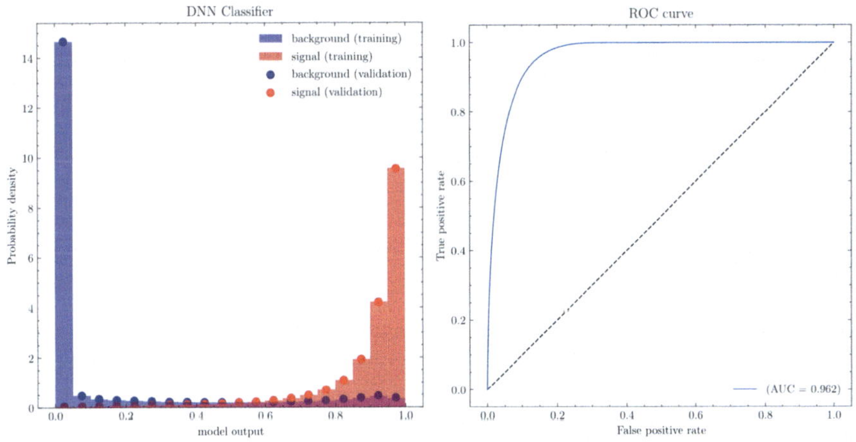

Fig. 2 Left panel: DNN output distribution for the background events in the training sample (blue shaded), for the background events in the test sample (blue dots), for the signal events in the training sample (red shaded) and for the signal events in the test sample (red dots). Right Panel: Receiver Operator Characteristics (ROC) curve for the trained model [9]

2 Literature Review

The pursuit of discovering physics is a focus in contemporary particle physics research extending beyond the boundaries of the well-established Standard Model. While the Standard Model has achieved accuracy in describing the interactions of known particles it still leaves fundamental questions about our universe unanswered. These questions involve mysteries like the origin of mass, the nature of matter and dark energy, and the intriguing dominance of matter over antimatter. In our quest for answers exploring physics becomes an avenue.

The Standard Model, which is a quantum field theory, explains how basic building blocks interact and how forces govern their existence. However, it remains a framework that cannot fully comprehend these mysteries. The search for physics offers us hope as we venture into territories to uncover solutions to these deep questions. Searches for new physics phenomena beyond those described by the Standard Model (SM) require advanced techniques to devise selections that involve a large number of variables characterizing collision events [16].

There are approaches to investigate new physics realms. In particle physics, semi-supervised machine learning is an attractive option to reduce model dependency searches beyond the Standard Model [7]. One approach involves searching for particles that have not been predicted by the Standard Model but may reveal themselves through high energy collisions with those at the Large Hadron Collider (LHC). Another method explores forces or interactions between particles that could potentially be induced by undiscovered particles.

The quest to uncover knowledge in physics is a pursuit, a never-ending journey that captivates scientists all over the world. With the LHC pushing into its lifespan, there remains an exciting possibility of unveiling groundbreaking discoveries just within reach.

In the realm of investigation driven by data, machine learning has become a tool, especially in the field of particle physics. This subset of intelligence empowers us to utilize data for a range of tasks. Within particle physics, it plays a role by providing approaches to tackle our challenges:

1. **Identifying Particles in Collisions**: Distinguishing particles in high energy collisions poses a hurdle. Machine learning comes to the rescue with its ability to recognize patterns aiding in the identification of particles based on their tracks or energy signatures. Recent studies have shown that jet identification algorithms deploying ML methods directly on the jet constituents yield significantly improved performance compared to traditional algorithms [4].
2. **Determining Particle Properties**: Extracting properties like mass, charge, and spin from the amount of data produced by particle detectors can be quite intricate. Machine learning algorithms excel at this task enabling us to extract the characteristics of particles
3. **Exploring New Physics**: One remarkable aspect of machine learning is its proficiency in uncovering patterns within data that defy the predictions made by the

Standard Model. It plays a role in our quest for physics by identifying deviations from expected behavior.

4. **Simulating Particle Collisions**: Simulating collisions between protons requires resources and effort. Machine learning models hold promise in expediting these simulations and enhancing our comprehension of collision events.
5. **Optimizing Detector Design**: Another area where machine learning shines is optimizing the design of particle detectors boosting their sensitivity and precision. Machine learning has truly revolutionized how we approach aspects within particle physics research. This optimization plays a role, in enhancing the efficiency and overall success of experiments (see Equation 2).

$$h_t = f(W_{hh} \cdot h_{t-1} + W_{xh} \cdot x_t + b_h)$$

Equation 2 Recurrent neural network

The use of machine learning in particle physics offers benefits. Designed for handling sequential data, RNNs consist of internal neural networks that process new input combined with the past processed state [6]. Neural Networks need to have almost half of the noise population to achieve even worse results than Logistic Regression [15]. It automates tasks that used to rely on involvement allowing experts to focus on complex challenges. Machine learning's ability to uncover patterns and relationships in data opens up possibilities for exploring physics and gaining deeper insights into established phenomena. Additionally, machine learning improves the efficiency of experiments by enhancing data collection and analysis making it easier to discover aspects of physics.

However, there are still challenges to overcome. Machine learning algorithms can be complex. Sometimes produce results that are difficult to interpret. If the training data are not representative of the world, biases in these algorithms can introduce errors. Incomplete knowledge of the simulation parameters translates into systematic uncertainties in the signal efficiencies [11]. Furthermore, the computational requirements of machine learning limit the complexity of problems that can be addressed.

Nevertheless, the collaboration between machine learning and particle physics holds potential. Machine learning in the top quark sector is a fast-evolving field in both the ATLAS and CMS Collaborations [17]. As machine learning continues to advance it will play a role in particle physics research. This interdisciplinary fusion between intelligence and particle physics paves the way for groundbreaking discoveries that will reshape our understanding of the universe and our place within it.

The intricate web of particles governed by the laws of the Standard Model presents a captivating storyline, within particle physics (see Fig. 3).

However, this narrative is far from being told as it remains filled with inquiries and mysterious connections, among particles. By delving into these connections, we

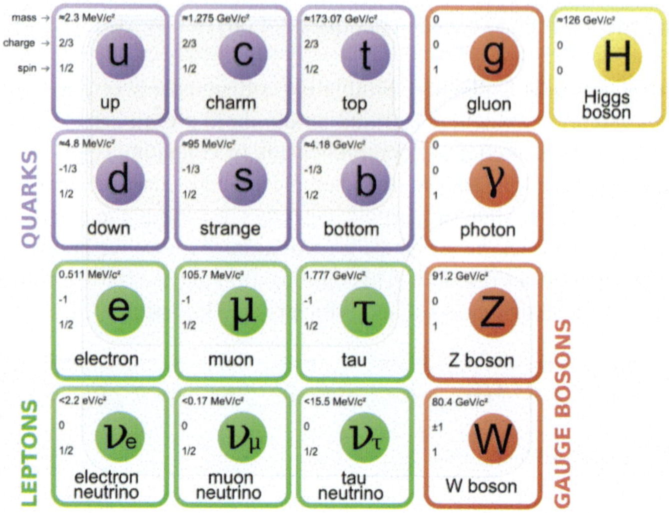

Fig. 3 This Standard Model infographic lists the names and properties (mass, charge, spin) of all known SM particles. The types of particles are color-coded. Quarks and leptons are arranged in three generations from left to right. The first row of quarks shows up-type quarks, the second down-type quarks. The shaded blobs in the background connect particles that couple to a given force mediator (gauge boson) [5]

unravel the essence of particle physics. The kinematic reconstruction of top quarks is well described by the simulation, and the resulting uncertainties are small [3]. The dominant uncertainties are those due to the descriptions of the detector material and of electron–ion recombination in the EM calorimeter [11].

Properties such as weight, electrical charge, and rotation determine the characteristics of particles. These properties govern their energy levels, interactions with particles, and angular momentum forming the basis for their behaviors and connections. The Standard Model, which encompasses the electromagnetic and gravitational forces, provides a framework to interpret these interactions. However, it is a narrative that leaves the intricacies of subatomic relationships unexplored.

Scientists employ an approach to comprehend these relationships. Through experiments, calculations, and computer simulations they decipher the connections that unite particles together. The ultimate objective is to unravel the nature of matter and understand the forces that govern the universe.

Numerous questions arise about the relationships between particles:

1. **Origin of Mass**: The Standard Model fails to shed light on how mass originates, leading researchers to explore concepts like the Higgs field and its implications.
2. **Nature of Dark Matter and Dark Energy**: The mystery surrounding matter. Which accounts for a portion of the universe's matter. Remains unresolved alongside understanding dark energy.
3. **Matter vs. Antimatter**: The Standard Model falls short in explaining why there is an abundance of matter compared to antimatter in our universe.

4. **Unification of Forces**: The unification of forces remains a goal, in the field of physics as scientists strive to develop a theory that reconciles the weak, electromagnetic, and gravitational forces.

This pursuit of knowledge is at the heart of particle physics research, which is characterized by its changing nature and the potential for new findings. By unraveling the connections between particles, we deepen our understanding of the fundamental structure of the universe and our role, within it.

Simulating proton collisions is a crucial task, in the field of particle physics research. These collisions, which are complex and unpredictable, play a role in experiments. To understand these collisions and their outcomes computer simulations are tools.

The new deployment of the ACTS software framework in ATLAS for primary vertex reconstruction results in a significant reduction of primary vertexing CPU time [10]. Various types of computer simulations are used for proton collisions:

1. **Monte Carlo Simulations**: These simulations use numbers to generate outcomes making them suitable for modeling the nature of particle interactions. Samples of Monte Carlo (MC) simulated inclusive prompt production of $J/\psi \rightarrow \mu\mu$ and $\Psi(2S) \rightarrow \mu\mu$ events were generated at leading-order (LO) accuracy using Pythia 8.186, for matrix element (ME) 4 calculations, and for the modeling of the parton shower, hadronization, and the underlying event [1]. In relevant studies, asymmetries were mostly calculated using signal only events at the parton level where background contributions were simulated by a dilution factor [9].
2. **Fluid Dynamics Simulations**: These simulations model the flow of particles during collision events. Find application, in dynamics. They help predict particle trajectories and interactions.
3. **Neural Network Simulations**: By utilizing networks, these machine learning simulations provide highly accurate predictions of collision outcomes contributing to a comprehensive understanding (see Fig. 4).

The choice of simulation type depends on requirements. We can use a parameterized machine learning method to apply a single deep neural network (DNN) algorithm for all the benchmark points taking the mass of the resonant particle as a parameter to the network [13]. Monte Carlo simulations are tools while fluid dynamics and neural network simulations cater to specialized applications.

Simulating proton collisions offers advantages:

1. **Safety and Cost Effectiveness**: Simulations allow studying events that are too risky or expensive to replicate through experiments. Deep learning methods to circumvent expensive simulations for hypothesis testing have been studied in the context of effective field theory [18].
2. **Testing**: Simulations provide a means to test and validate models and new hypotheses aiding in experimental design. Empirical results have demonstrated that DCTR is ready to be deployed for full phase-space tasks [18].

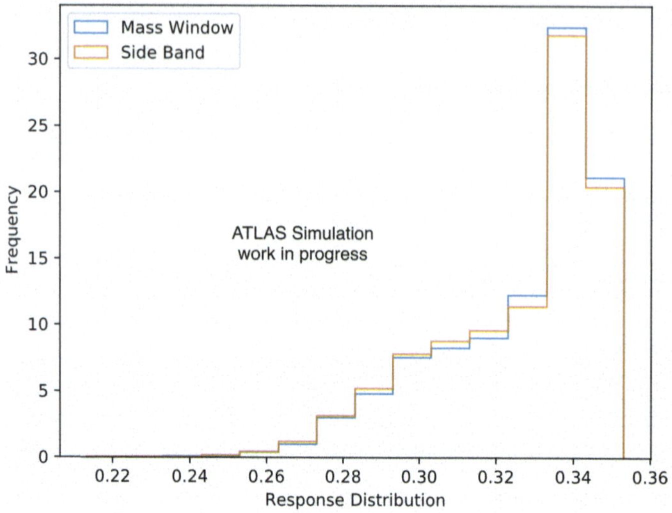

Fig. 4 Example response distribution output from DNN Toy Monte Carlo model [7]

3. **Efficiency Enhancement**: Simulations play a role in improving the efficiency of experiment design and data analysis. They help enhance data collection processes and enable researchers to make discoveries effectively.

However, there are still some obstacles that persist:

1. **Computational Intensity**: Simulations often require resources and time which limits their scope and complexity.
2. **Imperfect Models**: The simplified models used in simulations can introduce inaccuracies in the results.
3. **Representation Bias**: If the training data are not representative biases, simulation models can lead to outcomes.

Nevertheless, simulations continue to be a cornerstone of particle physics research. They provide insights into outcomes and contribute to the advancement of our understanding in this field.

The Higgs boson, a particle of importance, stands at the forefront of exploration in particle physics. This mysterious particle was discovered in 2012 at the Large Hadron Collider (LHC). Holds the key to unlocking the secrets behind mass. Its discovery confirmed the foundations of the Standard Model. Marked a milestone in our scientific journey.

However, many mysteries surrounding the Higgs boson remain:

1. **Mass**: The Higgs boson has a mass of 125 GeV/c^2, which poses a challenge for explanation.
2. **Spin**: With its spin value being 0, the Higgs boson is considered a particle. A characteristic that distinguishes it from particles within the Standard Model.

3. **Parity**: The Higgs boson possesses parity. Remains unaffected by parity transformations.
4. **Coupling**: The Higgs boson couples with all particles although the strength of these interactions may vary.

The Higgs boson is known for its nature as it quickly transforms into particles, which poses significant challenges in studying it. However, scientists persist in their pursuit of answers, to questions such as:

1. **The Origin of the Higgs Field**: Researchers remain deeply fascinated by the connection between the Higgs boson and the Higgs field.
2. **The Mystery of Mass**: The small mass of the Higgs boson continues to puzzle scientists and unraveling this mystery remains an endeavor.
3. **Exploring Dark Matter Connections**: Scientists are tantalized by the possibility of a relationship between the Higgs boson and dark matter prompting investigations. A variety of models of WIMP dark matter at the LHC involve the Higgs boson acting as a portal between the dark sector and the SM sector, either via direct Yukawa couplings to fermionic dark-matter candidates or via other mechanisms [2].
4. **Investigating Additional Higgs Bosons**: The prospect of types of Higgs bosons sparks curiosity and calls for further exploration.

The study of the Higgs boson is an ever-evolving field that advances alongside progress. As we delve deeper into understanding these particles' properties, our knowledge of the Standard Model and the fabric of our universe stands on the brink of expansion. In the Standard Model (SM) of particle physics, interactions amongst electroweak gauge bosons (γ, W, Z) are entirely determined by the non-Abelian $SU(2) \times U(1)$ structure of the electroweak sector [8]. In addition, the top quark and tt distributions are determined in the full phase space, in order to facilitate the comparison with higher-order perturbative QCD calculations [3] (see Fig. 5).

3 Analysis

The research papers provide an overview of particle physics investigations with a focus on searching for new physics phenomena, machine learning's role in these endeavors exploring intricate relationships, among subatomic particles simulating proton collisions and examining various properties associated with this mysterious particle known as the Higgs boson.

In this section of analysis, our goal is to explore these subjects in detail discussing patterns, debates, and areas of research. We will specifically examine the increasing importance of machine learning in the field of particle physics.

Search for New Physics:

The search for physics goes beyond what's currently known in the Standard Model and is a key focus in particle physics. Despite the achievements of the Standard

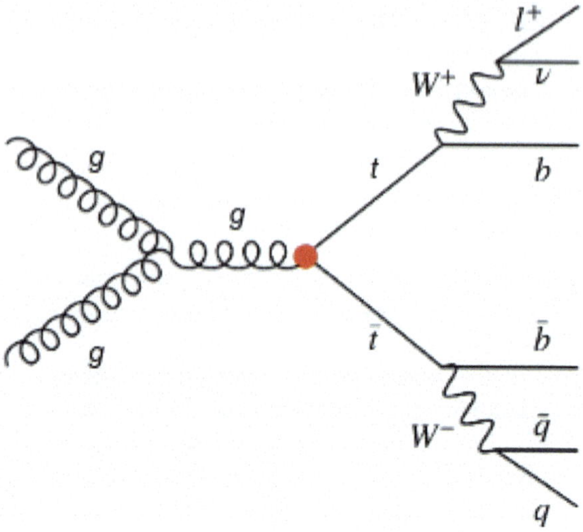

Fig. 5 Feynman diagram for the top quark pair semileptonic decay. The red dot represents the CEDM contribution [9]

Model, important questions still remain unanswered such as where mass comes from the characteristics of matter and dark energy and how to unify fundamental forces. Ongoing experiments at particle accelerators like the Large Hadron Collider (LHC) along with calculations and computer simulations have contributed to expanding our knowledge in these areas.

Current Trends:

- **Accelerator Power**: The LHC, being the particle accelerator in the world, continues to offer potential for new discoveries.
- **Experimental Advancements**: Scientists are constantly working on developing detectors and innovative experimental techniques.
- **Computational Enhancements**: The advancement of computers has greatly aided in conducting detailed simulations and analyzing data.
- **Interdisciplinary Collaboration**: Collaborations among scientists, from fields have resulted in progress.
- **International Cooperation**: The global nature of particle physics research promotes collaboration and sharing of data and expertise.

Controversial Topics:

- **Dark Matter and Dark Energy**: There is debate regarding the nature of matter and dark energy with multiple competing theories being discussed.
- **Unification of Forces**: Finding a way to unify all forces into a theory remains an unsolved challenge.
- **Quantum Gravity**: Bringing together the principles of quantum mechanics and general relativity which is commonly referred to as the quantum gravity problem presents us with difficulties.

- **Existence of Extra Dimensions**: The notion of dimensions, within the fabric of spacetime continues to be a topic of contention and debate.
- **Supersymmetry and New Particles**: Discussions surrounding the validity of supersymmetry and the existence of particles generate disputes. Should monopoles exist in the mass range accessible to a particle accelerator, they could be copiously produced at the LHC. If they were detected, the measured mass and coupling would severely restrict cosmological scenarios [11].

Open Questions in Research:

- **Origin of Mass**: One major question in the Standard Model is understanding the origin of mass, which is currently not well explained.
- **Nature of Dark Matter and Dark Energy**: Another important area of study is identifying what dark matter consists of and explaining the nature of energy.
- **Unification of Forces**: A standing challenge is achieving an unified theory that brings together all fundamental forces.
- **Quantum Gravity**: It is essential to resolve the problem of quantum gravity in order to reconcile quantum mechanics and general relativity.
- **Properties of Neutrinos**: To deepen our understanding of the universe it's critical to determine the mass and interaction characteristics of neutrinos.
- **Higgs Boson Interactions**: We still have a question regarding how the Higgs boson interacts with particles. The measured mass of the Higgs boson in the $H \rightarrow Z Z * \rightarrow 4\ell$ final state using the full Run 2 dataset is $mH = 124.99 \pm 0.18$ (stat.) $\pm$ 0.04 (syst.) = 124.99 $\pm$ 0.19 GeV [19].
- **Fate of the Universe**: The ultimate fate of the universe whether it will continue expanding or collapse remains uncertain at this point.
- **Quantum Computing**: Developing quantum computers has the potential for revolutionizing computational capabilities.
- **Harnessing Dark Energy**: Exploring and harnessing the force known as dark energy could lead to groundbreaking technologies.

Use of Machine Learning in Particle Physics:

Machine learning is becoming a tool in particle physics providing solutions to challenges and expanding research possibilities.

Trends:

- **Increasing Power of Machine Learning Algorithms**: The power of machine learning algorithms is constantly increasing, as they become more capable and sophisticated.
- **Broad Application**: Machine learning is being applied in fields such, as particle identification, anomaly detection, simulation acceleration, and data analysis.
- **Efficiency Enhancement**: Automation through machine learning is enhancing efficiency by speeding up data processing and analysis.
- **Interdisciplinary Impact**: The combination of machine learning and particle physics is leading to research approaches, with impact.

Controversies:

- **Algorithm Bias**: One ongoing concern, in the realm of machine learning algorithms, is ensuring fairness and avoiding bias.
- **Interpretability**: It can be challenging to interpret and validate results produced by machine learning models.
- **Data Quality**: Reliable outcomes rely heavily on high-quality training data and the avoidance of biases. The dominant sources of uncertainty come from the data-driven background estimates and the statistical uncertainty on data [8].

Open Research Questions:

Optimal Algorithm Selection: Determining the machine learning algorithms for specific tasks related to particle physics is still an open question.

Addressing Algorithm Bias: Developing methods to identify and address biases in machine learning models is an area of research.

Interpretable AI: Advancing techniques that enable us to interpret and explain machine learning results is another focus.

Enhancing Data Quality: Strategies aimed at improving data quality and ensuring representativeness are actively pursued.

Ethical AI

Establishing guidelines for the use of machine learning in particle physics research is an endeavor.

In conclusion, the field of particle physics continually evolves with exploration. It encompasses endeavors such as searching for physics, adopting machine learning techniques studying particles enigmatic nature simulating proton collisions, and understanding the properties of the Higgs boson. With collaboration and international cooperation shaping this field, resolving research questions and addressing controversies remain significant priorities.

The convergence of machine learning and particle physics is a field that has the power to transform research approaches and unveil revelations about the most enigmatic aspects of our universe.

4 Synthesis

The field of particle physics is vast and intricate, filled with questions that captivate physicists. While there are still mysteries to unravel, the ongoing research has already provided us with insights into the very essence of our universe.

The Standard Model and Beyond:

At the heart of our understanding of particle physics lies the Standard Model, a theory that elegantly explains the particles in our universe and the forces that govern their interactions. It has successfully accounted for a range of observations. However, it is important to note that the Standard Model is not all encompassing; it leaves gaps and unanswered questions about the nature of our cosmos.

The Dark Universe:

One of the challenges in particle physics revolves around matter and dark energy. Dark matter, a mysterious substance constituting 85% of all matter in the universe, continues to baffle us due to its composition and perplexing interactions. Additionally, we are still trying to understand the role of energy in driving the expansion of our universe.

Unifying the Forces:

Another significant challenge lies in unifying forces. The Standard Model effectively describes three out of four forces; electromagnetism, strong force, and weak force. However, finding a framework that incorporates all four forces remains a goal for physicists. However, the endeavor to combine these forces into one theory is a profound ambition and continues to be a fundamental objective of particle physics.

The Changing Landscape:

Particle physics is a field that is constantly evolving, with ongoing research reshaping our understanding of the universe. It's an exciting realm that holds the potential for groundbreaking discoveries. The knowledge gained from studying particle physics has already transformed our understanding of the cosmos. It has the possibility to drive technologies for the betterment of humanity.

Insights from the Reviewed Papers:

The research papers being examined offer a range of perspectives and insights into particle physics showcasing the nature of this field. They shed light on how various studies contribute to our comprehension of the subject making them an invaluable resource for those in delving deeper into particle physics intricacies.

Highlighted Insights from the Papers:

- The quest for physics remains an exhilarating and demanding undertaking, marked by open research questions and expected growth in this field.
- Advancements in accelerator technology, experimental techniques and computational power are enhancing efforts to explore frontiers in physics.
- Innovative theoretical models play a role in interpreting findings and pushing boundaries in new areas of physics. Approximate nearest neighbor search (K-ANNS), or more generally similarity search, is receiving growing attention in the current era of big data and large information resources [5].
- International collaboration stands as a foundation for success, in pursuing physics enabling data exchange and sharing expertise.
- The field of machine learning is bringing about a transformation, in the realm of data analysis in particle physics by uncovering concealed patterns.

- The continuous advancements in techniques to detect dark matter are providing insights into the enigmatic properties of this elusive substance.
- The study of physics, beyond the Standard Model, is an evolving field of unanswered questions and anticipated advancements.

Other Key Points from the Research Papers:

- The importance of cooperation in particle physics research emphasizes its nature.
- The ongoing need for models is to explain experimental results highlighting the interplay between theory and experimentation. An alternative method for model training is the Empirical Risk Minimization (ERM) procedure. It relies
- On the idea that if a loss function L can be defined to measure the cost of a model's decision, then the
- The objective is to minimize the expected loss over all possible observable events [14].
- The potential of machine learning is to revolutionize particle physics by improving data analysis and pattern recognition.
- The challenges involved in detecting matter and unraveling its mysteries are prominently discussed.
- Exciting possibilities on the horizon continue to captivate researchers and enthusiasts in the field of particle physics.

In summary, particle physics remains a discipline driven by a pursuit of answers to questions about our universe. The synthesis of knowledge, from sources and research papers underscores the nature of this field characterized by challenges, collaborative efforts, technological advancements, and the promise of groundbreaking discoveries that may reshape our understanding of the cosmos.

5 Conclusion

The exploration of particle physics as revealed through these research papers highlights the nature of this endeavor. Although the Standard Model of particle physics has been incredibly successful, in helping us understand the building blocks and interactions of the universe it also reminds us that there are still unexplored areas in the field of new physics.

The Exciting Quest, for New Physics:

The realm of particle physics is both thrilling and demanding. It entices researchers to explore territories where numerous opportunities for discoveries await. The changing landscape of questions from unraveling the mystery behind the origin of mass to understanding the elusive nature of dark matter and achieving a unified understanding of forces fuels the excitement in this pursuit.

The Essence of Collaboration:

International collaboration emerges as an aspect woven into the fabric of particle physics research. This spirit of cooperation is not just convenient but necessary. Particle physics requires investments and massive experimental machinery. These factors make it essential for scientists, from nations, to come together to conceive ideas, construct equipment, and conduct scientific exploration.

The Potential Revolution of Machine Learning:

An insight derived from literature review is the growing role of machine learning in particle physics. This advancing field has the potential to reshape how we approach research in particle physics. Machine learning algorithms provide us with the ability to analyze datasets obtained from particle accelerator experiments uncovering patterns that may evade human observation. This transformative capability promises to accelerate our pace of discovery and enhance our comprehension of how the workings of the universe unfold.

A Promising Future in the Field:

Particle physics is an evolving field. We can expect the Large Hadron Collider (LHC) to continue its operations for many more years. This extended mission brings with it the possibility of uncovering aspects of physics. Additionally, plans for particle accelerators are being developed, offering the potential for deeper insights into the fundamental nature of our universe.

Implications and Future Paths:

There are implications that arise from this research. Firstly, it emphasizes the importance of persisting in our pursuit of discoveries in physics. While the Standard Model has been incredibly valuable, its limitations remind us how crucial it is to keep exploring and push boundaries to potentially revolutionize our understanding of the universe.

Secondly, international collaboration plays a role again. The collaborative efforts between scientists from nations remain pillars of particle physics research. They ensure that complex experiments can be executed successfully while benefiting from expertise.

Lastly, we cannot overstate the growing potential of machine learning in reshaping particle physics research. By integrating machine learning into data analysis, we have an opportunity to shed light on patterns that were previously unseen, potentially leading to breakthroughs.

As we ponder the changing field of particle physics it's clear that there's still more to discover. While the Standard Model provides a basis in the realms of physics where we may uncover the most fascinating revelations. The search for physics driven by collaboration between nations and enhanced by machine learning continues to attract physicists, to a realm brimming with possibilities. It is here that we eagerly await the unraveling of the universe's mysteries.

References

1. ATLAS Collaboration: Studies of the Muon Momentum Calibration and Performance of the ATLAS Detector with pppp Collisions at $\sqrt{s} = 13$ TeV (2022). arXiv:2212.07338
2. Aad, G.: Search for new phenomena in events with an energetic jet and missing transverse momentum in pp collisions at. Phys. Rev. D (2021)
3. CMS Collaboration: Measurement of the differential cross section for top quark pair production in pp collisions at $\sqrt{s} = 8$ TeV (2015). arXiv:2212.07338
4. CMS Collaboration: Identification of heavy, energetic, hadronically decaying particles using machine-learning techniques. J. Instrum. (2020)
5. Tosciri, C.: Machine Learning Applications and Observation of Higgs Boson Decays into a Pair of Bottom Quarks with the ATLAS Detector. University of Oxford (2020)
6. Chiedde, N., Atlas Liquid Argon Calorimeter Group: Machine learning for real-time processing of ATLAS liquid argon calorimeter signals with FPGAs. J. Instrum. (2022)
7. Lieberman, B., Choma, J., Dahbi, S.E., Mellado, B., Ruan, X.: An investigation of over-training within semi-supervised machine learning models in the search for heavy resonances at the LHC (2021). arXiv:2109.07287
8. ATLAS Collaboration: Observation of W$\gamma\gamma$ Triboson Production in Proton-Proton Collisions at $\sqrt{s} = 13$ TeV with the ATLAS Detector (2023). arXiv:2308.03041
9. Hayreter, A., Hüdaverdi, M., Ilgın, F., Salva, S., Şimşek, E., Güyer, S.: Investigating the Violation of Charge Parity Symmetry Through Top Quark Chromo-Electric Dipole Moments by Using Machine Learning Techniques (2023). arXiv:2306.11683
10. ATLAS Collaboration: Software Performance of the ATLAS Track Reconstruction for LHC Run 3 (2023). arXiv:2308.09471
11. Apyan, A., Bensinger, J., ATLAS Collaboration: Search for magnetic monopoles and stable particles with high electric charges in s = 13 TeV pp collisions with the ATLAS detector. J. High Energy Phys.: JHEP (2023)
12. Hodkinson, B.: Improving ATLAS Hadronic Object Performance with ML/AI Algorithms (2023). arXiv:2307.01388
13. Palit, P., Shil, S.: Probing Electroweak Phase Transition in Singlet scalar extension of Standard Model at HL-LHC through bbZZ channel using parameterized machine learning (2023). arXiv: 2302.04191
14. Paganini, M.: Machine Learning Solutions for High Energy Physics: Applications to Electromagnetic Shower Generation, Flavor Tagging, and the Search for Di-Higgs Production. Yale University (2019)
15. Martínez, S.C., Salt, J., de la Hoz, S.G., Villaplana, M.: Comparison of different ML methods applied to the classification of events with ttbar in the final state at the ATLAS experiment (2020). arXiv:2006.02135
16. ATLAS Collaboration: Search for new phenomena in two-body invariant mass distributions using unsupervised machine learning for anomaly detection at $\sqrt{s} = 13$ TeV with the ATLAS detector (2023). arXiv:2307.01612
17. Keicher, P.: Machine Learning in Top Physics in the ATLAS and CMS Collaborations (2023). arXiv:2301.09534
18. Andreassen, A., Nachman, B.: Neural networks for full phase-space reweighting and parameter tuning. Phys. Rev. D (2020)
19. ATLAS Collaboration: Combined measurement of the Higgs boson mass from the $H \rightarrow \gamma\gamma$ and $H \rightarrow ZZ* \rightarrow 4\ell$ decay channels with the ATLAS detector using $s\sqrt{} = 7$, 8 and 13 TeV pp collision data (2023). arXiv:2308.04775
20. Atlas Collaboration: Search for vector-boson resonances decaying to a top quark and bottom quark in the lepton plus jets final state in pp collisions at $s\sqrt{} = 13$ TeV with the ATLAS detector (2018). arXiv:1807.10473

Machine Learning-Based Approach for Diabetic Retinopathy Detection and Grading

Shashank Goyal, Shailendra Narayan Singh, Harsh Jain, Ojasvi Jain, and Tanu Bhaskar

Abstract Diabetic retinopathy (DR) is a significant consequence of diabetes that affects the blood vessels in the retina, potentially leading to visual impairment or blindness. Traditional diagnosis entails a physical examination of fundus images, which is prone to inaccuracy. In response, researchers advocate for the use of computer vision techniques to automate DR detection. However, these approaches frequently meet difficulties, notably in reliably diagnosing early-stage DR. This research digs especially into the use of Machine Learning (ML), adopting a DenseNet-121 architecture for the analysis of DR stages. A significant dataset from the APTOS 2019 Blindness Detection competition on Kaggle serves as the basis for this exploration. The method uses patient fundus images as input, allowing the model to extract features and generate output via an activation function. With a stated accuracy of 0.9644 and a quadratic weighted kappa score of 0.9135 for DR detection, remarkable results are obtained. This highlights the revolutionary potential of AI and ML in the diagnosis and management of diabetic retinopathy. It opens the door to more accurate and efficient screening and treatment approaches.

Keywords Diabetic Retinopathy (DR) · Fundus images · Computer vision · Machine Learning (ML) · DenseNet-121 architecture

S. Goyal
Inderprastha Engineering College, AKTU, Ghaziabad, India

S. N. Singh · H. Jain · O. Jain (✉) · T. Bhaskar
Amity School of Engineering and Technology, Amity University, Noida, India
e-mail: ojasvijainfeb4@gmail.com

S. N. Singh
e-mail: snsingh36@amity.edu

G. Raj et al. (eds.), *Advances in Artificial Intelligence and Machine Learning*, Lecture Notes in Electrical Engineering 1264, https://doi.org/10.1007/978-981-97-9507-9_7

1 Introduction

Internal eye and retinal fundus image analysis has grown in relevance in biometric identification and expanded our understanding of the human visual system during the last decade. This examination is critical in the early diagnosis of numerous diseases, including diabetic retinopathy (DR), a microvascular consequence of diabetes [1]. If left untreated, DR can cause serious visual impairment and eventually blindness. Diabetes prevalence has increased globally, reaching 463 million in 2019 and is anticipated to reach 700 million by 2045, intensifying the burden of DR and posing significant difficulties to public health and the economy [2]. Recent advancements in diagnostic and therapy technology, such as optical coherence tomography and intravitreal injections, have considerably improved the early diagnosis and management of DR [3]. The incorporation of artificial intelligence algorithms for automated screening of retinal images is a significant step forward. Figure 1 offers visual representations of both a healthy retina and one affected by diabetic retinopathy (DR) [4].

Diabetic retinopathy progresses from Non-Proliferative Diabetic Retinopathy (NPDR) to Proliferative Diabetic Retinopathy (PDR) [5]. NPDR is caused by leaking from retinal blood vessels, which causes swelling and impaired vision. PDR is more severe, with significant bleeding that endangers one's vision. Early screening and regular eye exams are critical for optimal treatment. Diagnosis of DR entails recognising retinal lesions, which are described below [6]:

- Microaneurysms, which appear as minute protrusions in the blood vessels of the retina, are early markers of diabetic retinopathy. They look like little red spots and can leak fluid, blood, or lipids, resulting in vision difficulties.
- Haemorrhages, which develop when damaged blood vessels burst and leak blood into the retinal tissue. These haemorrhages appear as dot-like or flame-shaped patches, causing visual impairment in later stages.
- Soft exudates, commonly known as Cotton-Wool Spots, are white or greyish fluffy lesions that appear on the retina. They are caused by blood vessel blockages, which

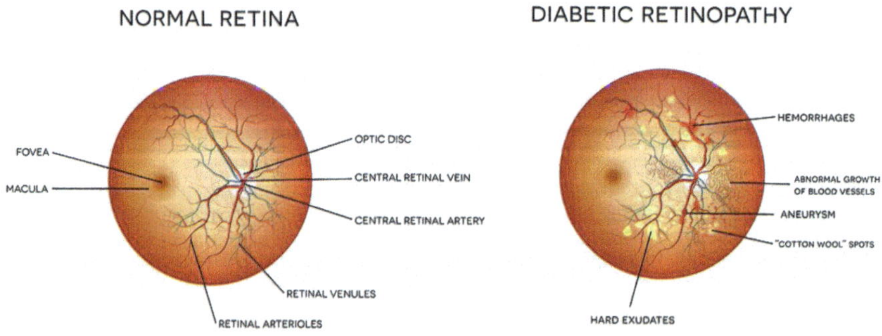

Fig. 1 Comparison between normal retina and diabetic retinopathy

Table 1 Stages of DR

Class	Name	Image
0	No DR	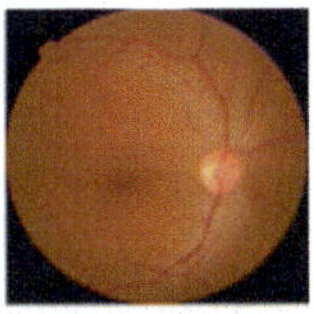
1	Mild	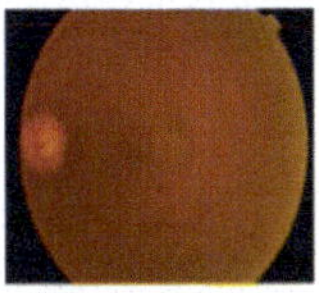
2	Moderate	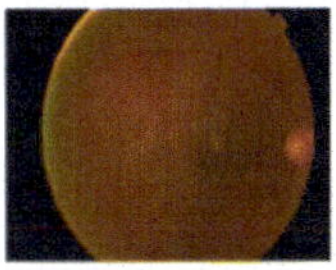
3	Severe	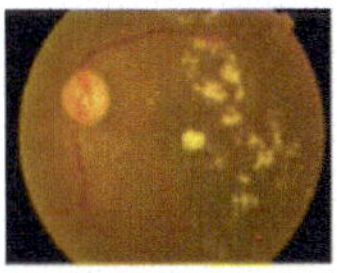
4	Proliferative DR	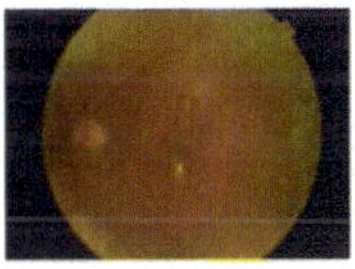

impede oxygen supply to retinal nerve fibres and indicate the existence of retinal ischemia.

- Hard exudates are yellow or white lipid deposits in the retina that have a waxy appearance. They are mostly found in the macula. They impede central vision and are common in non-proliferative diabetic retinopathy.

These retinal lesions collectively define the five stages of diabetic retinopathy: no DR, mild DR, moderate DR, severe DR, and proliferative DR [7]. Table 1 contains a brief description of these stages along with sample images.

Recent academic investigations have revealed that artificial intelligence (AI) models, particularly those based on "Deep Learning," have exhibited exceptional accuracy in exposing hidden complexity across a wide range of AI applications, with a particular emphasis on medical image analysis [8, 9]. Deep learning models may competently classify diseases and play a role in medical decision-making, ultimately improving overall patient care quality.

2 Overview of Dataset

The research makes use of a comprehensive image dataset obtained from the Aravind Eye Hospital and made available through Kaggle's APTOS 2019 Blindness Detection competition [10]. This comprehensive dataset, which is required for training models in ophthalmological image processing, includes high-resolution retinal images acquired in a variety of clinical settings. These images, obtained using a fundus camera resembling a low-powered microscope, are critical in the diagnosis procedure.

The dataset is organised into directories such as "train.csv" and "test.csv," and it contains identifiers and severity levels. It consists of 3,662 training and 1,928 testing images standardised to 224 × 224 pixels with RGB channels and divided into five DR severity classifications. The distribution of classes, as shown in Fig. 2, encompasses a variety of severity levels, increasing the dataset's relevance for the development and assessment of machine learning models dedicated to the detection and diagnosis of diabetic retinopathy.

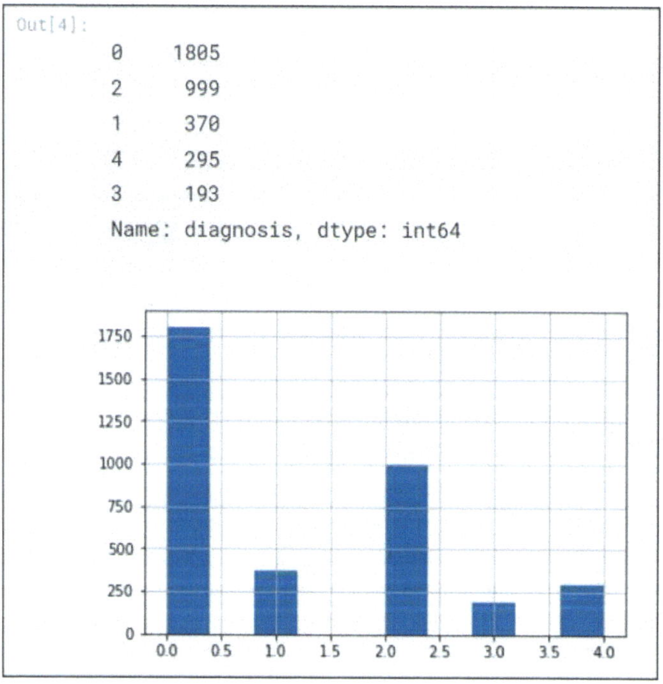

Fig. 2 Count of images in each class (train_images)

3 Proposed Methodology

3.1 DenseNet-121 Architecture

DenseNet-121 is distinguished by its dense connectivity framework. In classic CNNs, each layer takes input solely from the layer preceding it. DenseNet-121, on the other hand, connects each layer to all the following layers inside the same dense block. Because lower-level features are readily accessible to higher-level layers, this tight interconnectedness supports efficient feature reuse. As a result, dense connections considerably relieve the vanishing gradient problem and enhance gradient flow, allowing deeper networks to be trained [11]. Figure 3 provides a deeper look at the DenseNet-121 architecture, showing its layer structure and connectivity [12].

Structure of DenseNet-121. The architecture is structured with specific layers, outlined as follows [13]:

- Initial Convolution Layer: The network begins its operations with a single convolutional layer, which begins processing the input image.
- Dense Blocks: These blocks, which serve as the foundation of DenseNet-121, include densely connected convolutional layers. To prevent overfitting, the output of each layer is combined with the output of the following layers within the same block, followed by batch normalisation and ReLU activation.
- Transition Layers: The transition layers, which are strategically placed between dense blocks to control the spatial dimensions of feature maps, consist of batch normalisation, a 1×1 convolutional layer, and a 2×2 average pooling layer. The 1×1 convolution reduces the depth of the feature map and the pooling layer decreases spatial dimensions.
- Global Average Pooling: Global average pooling, which is used at the end of the network, is a step that converts the 3D feature map into a 1D vector, substantially lowering spatial dimensions while preserving feature information.

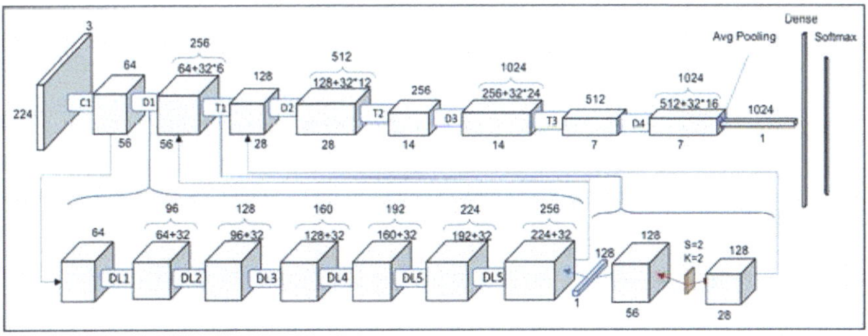

Fig. 3 A closer look at DenseNet-121 architecture

Fig. 4 ImageNet

- Fully Connected Layer: The last layer, which oversees classification, has a variable number of output neurons depending on the job, such as the classes in a classification problem.

3.2 *ImageNet*

The research involves leveraging the ImageNet dataset to pre-train, which results in improved accuracy for the DenseNet-121 model. ImageNet, founded at Stanford University by Fei-Fei Li, is a large dataset and a benchmark in the field of computer vision. It contains millions of labelled images from many categories, making it an important resource for CNN training and evaluation. The ImageNet Large Scale Visual Recognition Challenge (ILSVRC) is an annual ImageNet competition in which participants create and evaluate image classification algorithms [14]. Because of its large size and variety, the dataset is essential for testing various computer vision models. The success of deep learning models in ILSVRC competitions such as AlexNet, VGG, and ResNet has considerably increased the popularity and progress of deep neural networks in computer vision research [15]. The wide range of variation within ImageNet's labelled picture collection is depicted in Fig. 4 [16].

Advantages of Using ImageNet Pre-Trained Models for Medical Imaging. Using pre-trained models, such as DenseNet-121, which was trained on ImageNet, delivers significant benefits for medical imaging applications, notably diabetic retinopathy detection. The pre-trained model excels in feature extraction, having learned to recognise informative patterns in a wide range of images [17]. This ability is crucial in medical imaging, where visual patterns might be subtle or complicated. Furthermore, the use of pre-trained models minimises the computing cost and training time, which is very useful when dealing with large medical imaging datasets. Pre-trained models, known for their strong generalisation capabilities, show adaptability in a wide range of medical imaging applications beyond diabetic retinopathy diagnosis [18]. Models such as DenseNet-121 not only perform well on benchmark tests but also make consistent and trustworthy predictions, leading to a standardised approach to medical diagnosis.

3.3 Workflow Diagram

The comprehensive flowchart of our research is presented in Fig. 5. This flowchart is inherently explanatory, encompassing the preprocessing step, dataset information display, image shape presentation, utilisation of quadratic weighted kappa, integration of ImageNet, and concluding with the execution of epochs to obtain the desired output.

3.4 Quadratic Weighted Kappa

Quadratic Weighted Kappa (QWK), also known as Cohen's Kappa, is a statistical metric for measuring rater agreement or the dependability of a classification system. It is useful in a variety of domains, including inter-rater agreement research, machine learning, and psychology. This statistical measure is especially useful when dealing with categorical or ordinal data that lacks intrinsic ordering, which is typical with rating scales or diagnostic labels. QWK improves understanding across several fields, highlighting its importance in analysing inter-rater agreement and classification system dependability.

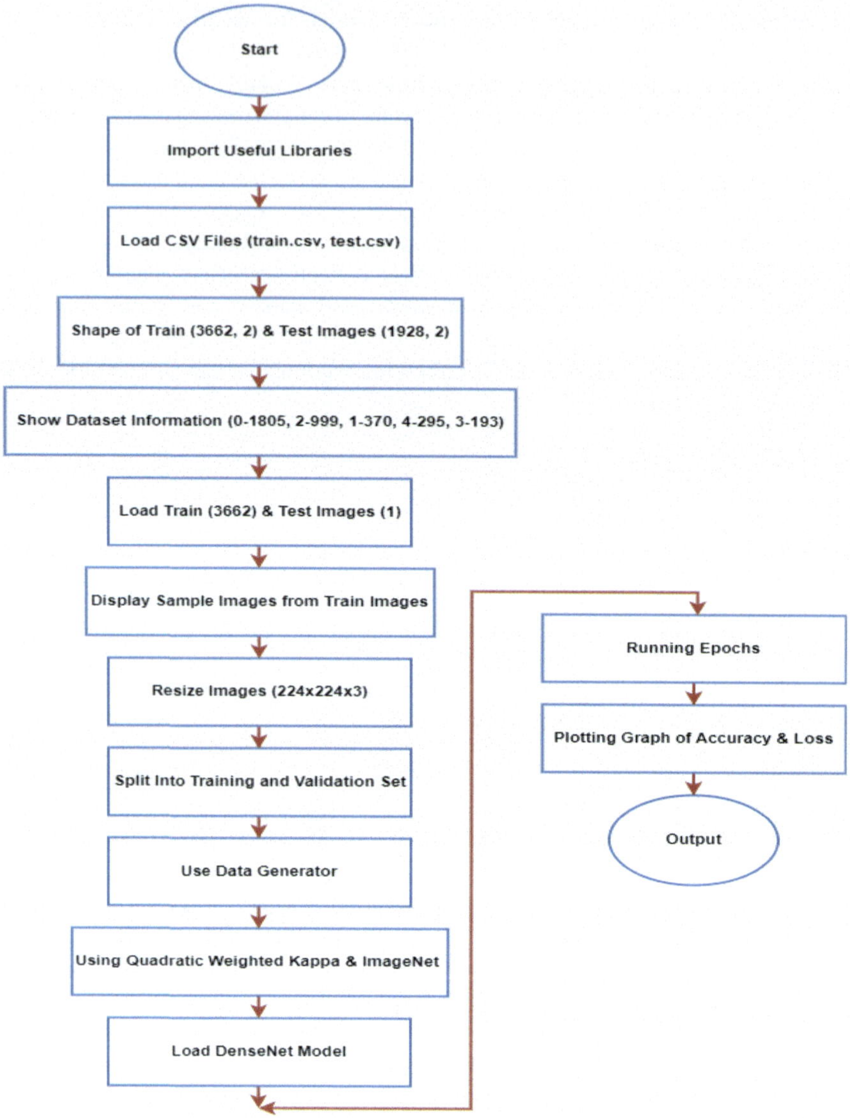

Fig. 5 Flowchart

Calculation of QWK. A Weighted Matrix w, of size NxN, is determined using Eq. (1), which considers the difference between actual and predicted values [19].

$$w_{i,j} = \frac{(i-j)^2}{(N-1)^2} \tag{1}$$

Table 2 Guidelines for interpretation

Value	Description
0.81–1.00	Almost perfect agreement
0.61–0.80	Substantial agreement
0.41–0.60	Moderate agreement
0.21–0.40	Fair agreement
0.0–0.20	Slight agreement
Less than 0	Agreement worse than chance

Table 3 Performance metrics

Architecture	Accuracy	Loss	QWK
DenseNet-121	0.9644	0.1061	0.9135

The following is the formula for calculating Quadratic Weighted Kappa, provided in Eq. (2) [19].

$$\kappa = 1 - \frac{\sum_{i,j} w_{i,j} O_{i,j}}{\sum_{i,j} w_{i,j} E_{i,j}} \tag{2}$$

where,

w = Weighted matrix, O = Histogram matrix, E = Expected matrix.

Interpretation of QWK. Refer to Table 2 for comprehensive guidelines on interpreting Quadratic Weighted Kappa [20]. It is important to note that the interpretation of Quadratic Weighted Kappa may differ depending on the specific field of application and the setting of the study. In addition, recommendations for interpretation may be included in the documentation of the individual statistical program or methodology used for computation. When analysing the results, always keep the specific conditions and types of data in mind.

4 Results and Analysis

In this section, the model's performance in identifying and grading diabetic retinopathy is thoroughly examined. The DenseNet-121 architecture is used, with pre-training on ImageNet to assist in efficient feature extraction and classification. In Table 3, a variety of performance indicators are provided to analyse its efficacy, offering a thorough grasp of the model's capabilities.

Visual representations in Figs. 6, 7, and 8 display the validation accuracy, validation loss, and validation kappa scores throughout numerous epochs to gain a deeper understanding of the model's learning trajectory and sensitivity to overfitting or underfitting.

This evaluation goes beyond testing the model's performance on known data by evaluating its generalisation capabilities on unfamiliar images. This assessment makes use of images from the test_images directory that were not part of the model's training set. Figure 9 depicts the distribution of retinopathy severity levels within these undisclosed images.

Furthermore, the research emphasises the model's versatility in assessing the severity of diabetic retinopathy across several images at the same time. Figure 10 depicts the simultaneous detection of the severity of diabetic retinopathy in a collection of numerous images. This feature is especially useful in clinical settings where ophthalmologists must review several retinal images quickly.

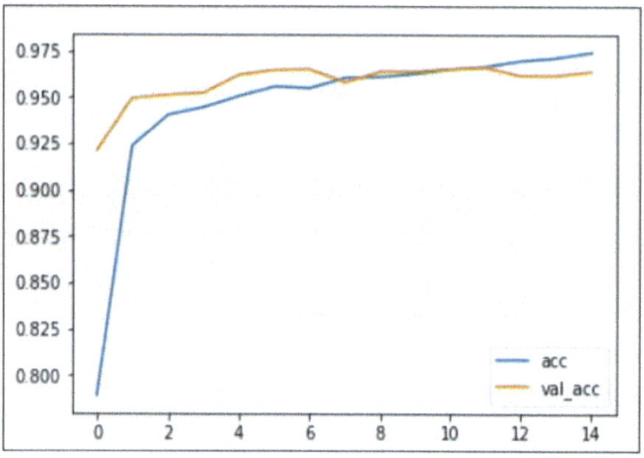

Fig. 6 Graph depicting the validation accuracy (val_acc)

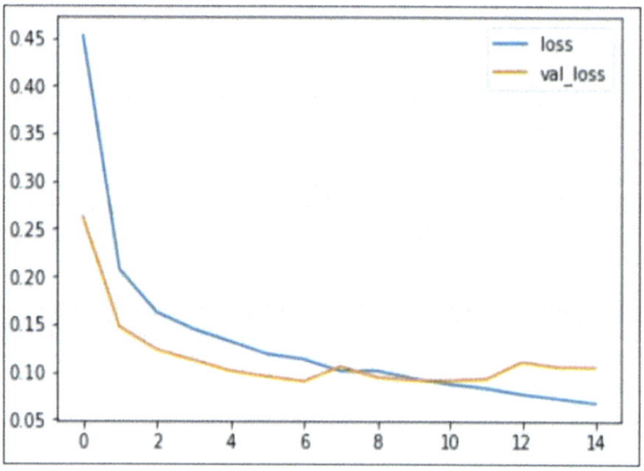

Fig. 7 Graph depicting the validation loss (val_loss)

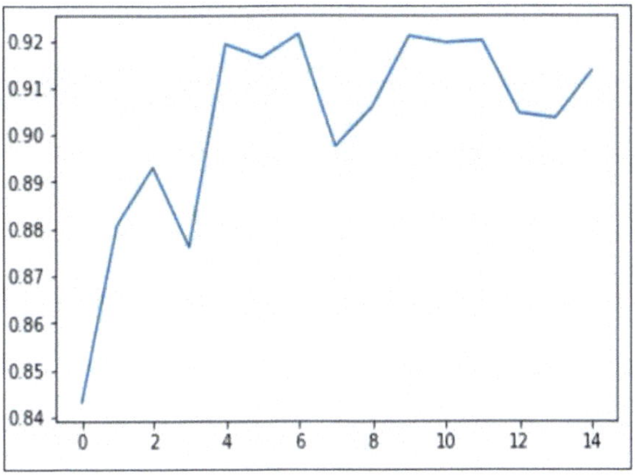

Fig. 8 Graph depicting the validation kappa scores (val_kappa)

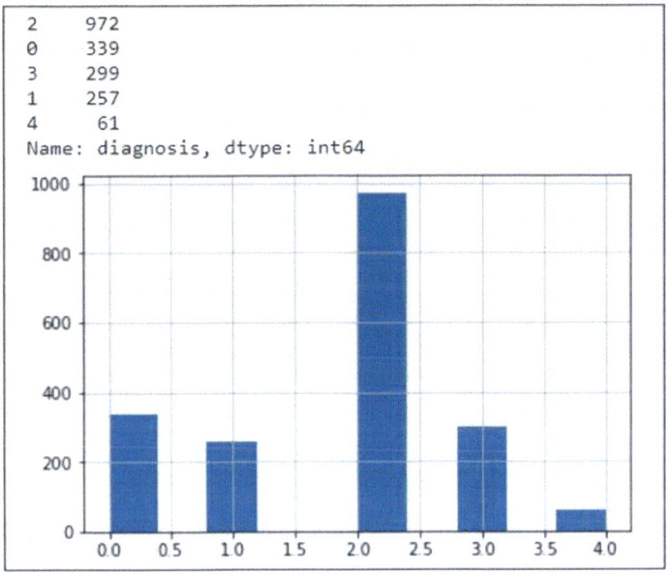

Fig. 9 Count of images in each class (test_images)

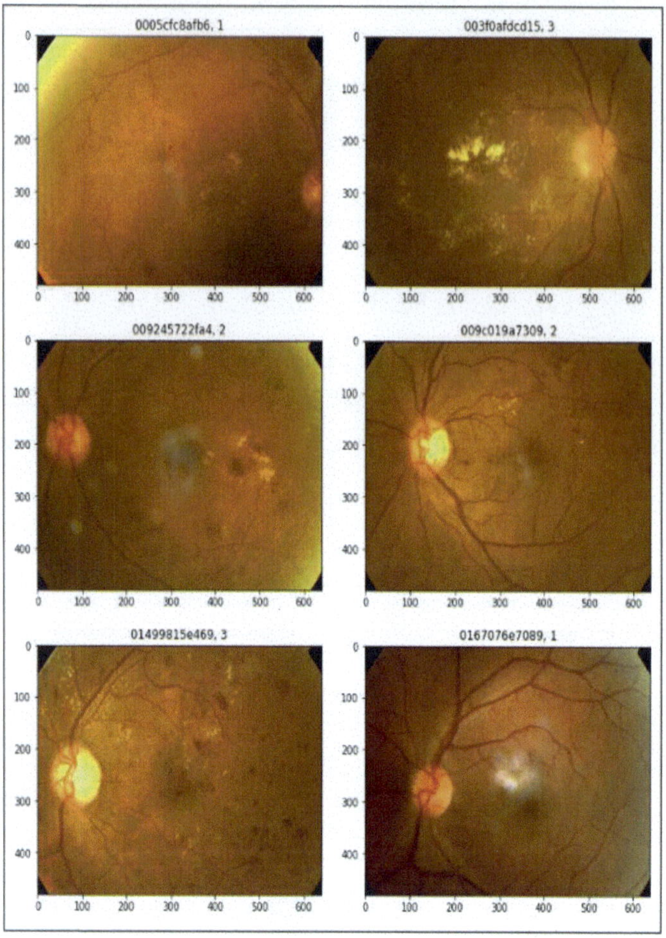

Fig. 10 DR severity detection across multiple images

5 Conclusion

According to several polls, diabetes has become an increasingly common health condition in recent years, with around a 30% chance of people having diabetic retinopathy (DR). If not discovered early, diabetic retinopathy can progress to Proliferative Diabetic Retinopathy (PDR). Manual DR diagnosis is time-consuming and difficult, necessitating the services of highly competent clinicians. Numerous computer vision algorithms for automatic DR detection and staging have been developed to overcome this difficulty. This paper offers an architecture for DR detection that uses pre-trained DenseNet-121 from ImageNet and achieves an accuracy of 0.9611 and a QWK score of 0.9135. Future research will try to enhance early-stage detection precision by training different models for each stage and using ensemble approaches.

References

1. Chetoui, M., Akhloufi, M.A., Kardouchi, M.: Diabetic retinopathy detection using machine learning and texture features. In: 2018 IEEE Canadian Conference on Electrical & Computer Engineering (CCECE), pp.1–4. IEEE, India (2018)
2. Saeedi, P., Petersohn, I., Salpea, P., Malanda, B., Karuranga, S., Unwin, N., Colagiuri, S., Guariguata, L., Motala, A.A., Ogurtsova, K., Shaw, J.E.: Global and regional diabetes prevalence estimates for 2019 and projections for 2030 and 2045: results from the international diabetes federation diabetes atlas. Diabetes Res. Clin. Pract. **157**, 107843 (2019)
3. Nawash, B., Ong, J., Driban, M., Hwang, J., Chen, J., Selvam, A., Mohan, S., Chhablani, J.: Prognostic optical coherence tomography biomarkers in neovascular age-related macular degeneration. J. Clin. Med. **12**(9), 3049 (2023)
4. Comparison Between Normal Retina and Diabetic Retinopathy. https://www.eye7.in/retina/diabetic-retinopathy/. Accessed 21 Nov 2023
5. Yang, Q.H., Zhang, Y., Zhang, X.M., Li, X.R.: Prevalence of diabetic retinopathy, proliferative diabetic retinopathy and non-proliferative diabetic retinopathy in Asian T2DM patients: a systematic review and meta-analysis. Int. J. Ophthalmol. **12**(2), 302 (2019)
6. Alyoubi, W.L., Shalash, W.M., Abulkhair, M.F.: Diabetic retinopathy detection through deep learning techniques: a review. Inform. Med. Unlocked **20**, 100377 (2020)
7. Jain, A., Jalui, A., Jasani, J., Lahoti, Y., Karani, R.: Deep learning for detection and severity classification of diabetic retinopathy. In: 2019 1st International Conference on Innovations in Information and Communication Technology (ICIICT), pp. 1–6. IEEE, India (2019)
8. Mishra, S., Hanchate, S., Saquib, Z.: Diabetic retinopathy detection using deep learning. In: 2020 International Conference on Smart Technologies in Computing, Electrical and Electronics (ICSTCEE), pp. 515–520. IEEE, India (2020)
9. Tymchenko, B., Marchenko, P., Spodarets, D.: Deep learning approach to diabetic retinopathy detection, pp. 1–9 (2020). arXiv:2003.02261
10. Kaggle Dataset. https://www.kaggle.com/competitions/aptos2019-blindness-detection/data. Accessed 23 Nov 2023
11. Liu, Q., Wu, Z., Du, Q., Xu, Y., Wei, Z.: Multiscale alternately updated clique network for hyperspectral image classification. IEEE Trans. Geosci. Remote Sens. **60**, 1–15 (2021)
12. A Closer Look at DenseNet-121 Architecture. https://towardsdatascience.com/understanding-and-visualizing-densenets-7f688092391a. Accessed 25 Nov 2023
13. Swaminathan, A., Varun, C., Kalaivani, S.: Multiple plant leaf disease classification using densenet-121 architecture. Int. J. Electr. Eng. Technol **12**, 38–57 (2021)
14. Deng, J.: A large-scale hierarchical image database. In: Proceedings of the IEEE Computer Vision and Pattern Recognition, pp. 248–255 (2009)
15. Nugraha, G.S., Darmawan, M.I., Dwiyansaputra, R.: Comparison of CNN's Architecture GoogleNet, AlexNet, VGG-16, Lenet-5, Resnet-50 in Arabic Handwriting Pattern Recognition. Kinetik: Game Technology, Information System, Computer Network, Computing, Electronics, and Control (2023)
16. ImageNet. https://cs.stanford.edu/people/karpathy/cnnembed/. Accessed 26 Nov 2023
17. Marcelino, P.: Transfer learning from pre-trained models. Towar. Data Sci. **10**, 23 (2018)
18. Saeed, F., Hussain, M., Aboalsamh, H.A.: Automatic diabetic retinopathy diagnosis using adaptive fine-tuned convolutional neural network. IEEE Access **9**, 41344–41359 (2021)
19. Understanding the Quadratic Weighted Kappa. https://www.kaggle.com/code/reighns/understanding-the-quadratic-weighted-kappa/notebook. Accessed 28 Nov 2023
20. Kempen, J.H., Ganesh, S.K., Sangwan, V.S., Rathinam, S.R.: Interobserver agreement in grading activity and site of inflammation in eyes of patients with uveitis. Am. J. Ophthalmol. **146**(6), 813–818 (2008)

Anticipating Customer Loyalty to Health Insurance Using PLS-SEM: The Role of AI Chatbots, Gamified Apps, and Customer Engagement

Dolly Mangla and Renu Aggarwal

Abstract The study examined the role of AI-based chabots and gamified apps in health insurance sector in anticipating customer loyalty by intervention role of customer engagement in NCR. Partial least squares structural equation modeling (PLS-SEM) was used to interpret the data compiled from 110 health insurance customers. The findings showed that the link between AI-based chatbots together with customer loyalty is somewhat middling by customer engagement. In the health insurance marketplace, gamified applications can help in increasing customer loyalty. Customer loyalty is immediately impacted negatively by AI-based chatbots. The findings drive health insurance companies to invest money on AI-powered chatbots in order to improve customer loyalty and engagement. Health insurance managers can also enhance customer loyalty by using gamified apps.

Keywords AI-based chatbots · Gamified health apps · Customer engagement · Customer loyalty

1 Introduction

In recent years, there has been a burgeoning concern in the implementation of AI-based chatbots in various industries. These chatbots powered by artificial intelligence also with natural language processing have proven to be highly advantageous for improving service to customers and improving communication efficiency. Being able to respond to consumer enquiries quickly and effectively is one of the most significant superiority of AI-based chatbots. Unlike human representatives who may be limited by time and resources, chatbots can engage in real-time conversations with customers, providing them with immediate solutions to their problems or answering their questions. This not only improves customer satisfaction but also helps the service center and the customer save considerable time.

D. Mangla (✉) · R. Aggarwal
Department of Management Studies, J.C. Bose UST, Faridabad, India
e-mail: dollymangla86@gmail.com

G. Raj et al. (eds.), *Advances in Artificial Intelligence and Machine Learning*, Lecture Notes in Electrical Engineering 1264, https://doi.org/10.1007/978-981-97-9507-9_8

A chatbot refers to a software program that is supported by artificial intelligence (AI) to engage in simulated conversations with users. These conversations are conducted through various platforms, such as messaging platforms, phone applications, and websites [1, 2]. Chatbots can be categorized based on various factors, including the knowledge domain they operate in, the services they offer, their goals, and the responses they generate [3]. The necessity for innovation is evident in a rising nation such as India since it is crucial for the advancement of service quality and the establishment of a globally competitive workforce [4].

1.1 Artificial Intelligence

The literature highlights the application of artificial intelligence-led software such as mobile apps, chatbots, and service robots in tourism [5], kiosks give the customer a creative service experience in the restaurant industry [6]. The inception of artificial intelligence-driven text-input chatbots [7] and voice-input chatbots [8] changed the face of the inter-linkage between banks and customers in the banking sector. AI-powered chatbot applications in e-commerce include fraud prevention and personalized shopping [9]. The most recent technology development that has influenced almost every domain of businesses is AI. Artificial intelligence-based software is also gaining attention in the insurance sector in different countries. AI-driven software also helps with efficient client communication, cost reduction, new goods and services development, and an improved competitive climate in the insurance sector [10].

Figure 1 represents the word cloud using papers related to the theme AI chatbots in service sector from Scopus database. The study highlights AI chatbots, deep learning, natural language processing, and machine learning are the most frequent used keywords used in the research of AI chatbots in service sector.

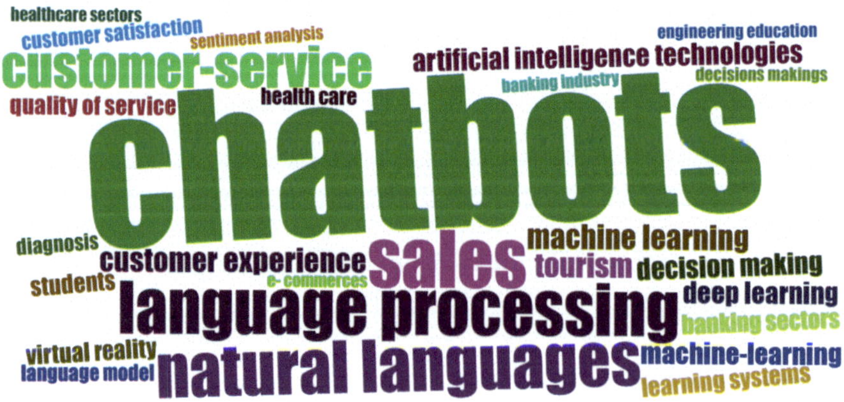

Fig. 1 Word cloud

1.2 Chatbots in the Insurance Sector

Future advancements and creativity in the insurance sector will be significantly assisted by InsurTech technologies like big data, chatbots, blockchain, biometrics, AI, and ML [11]. In order to identify their problems from the perspective of engaging in a conversation with a consumer, Nuruzzaman and Hussain investigated the methods used by existing chatbots termed Intelli Bot in creating a response. The Cornell movie script and a specially created insurance dataset served as IntelliBot's training grounds, giving it subject expertise [12]. In a blended study, Rodrguez et al. examined the factors influencing the acceptance and diffusion of chatbot technology in the insurance sector using the German insurance market [13]. Riikkinen et al. also conducted a conceptual study to characterize the link between insurance chatbots and value co-creation and generated four metaphors to depict interaction [14]. Koetter et al. researched chatbots in German insurance firms for innovation, digitalization, and creative business experiences [15]. Based on the extant literature, it is evident that there is a dearth of research articles investigating the implications of gamified applications and artificial intelligence (AI) chatbots on fostering consumer loyalty within the domain of health insurance. The purpose of this article is to investigate the intention for adopting AI-based chatbots and gamified apps in the insurance sector for the purpose of predicting consumer loyalty.

2 Methods and Materials

The instrument used in this study to collect data was a survey questionnaire. Responses based on a Likert scale were used to measure the variables, and there were five categories: strongly disagree (score 1) to strongly agree (score 7) and requested to rank their scale of agreement or disagreement with a few object-related propositions. Judgmental sampling was used to choose the sample and implement PLS-SEM technique. Participants were Indians living in the Faridabad city who had purchased health insurance. Since 2018, there has been an increase in the utilization of artificial intelligence chatbots and gamified applications within the service business. From a total of 150 questionnaires distributed, 110 valid responses were received. On the basis of judgment sampling, an online questionnaire was sent out to all customers. The questionnaire has four sections. The main goal of this section is to gather fundamental the customer demographic information. The second section is devoted to accumulating information on the willingness of customers to use AI chatbots. Collecting data pertaining to gamified health apps is the focus of the third section. The loyalty and engagement of customers are analyzed in the fourth section. We selected four items from [16] to gauge the insurance industry's readiness for using AI chatbots. We used four items to measure the adoption intention towards gamified Apps [17]. There are four items that have been utilized to assess customer engagement in the past by researchers [18, 19]. Customers' loyalty in the health insurance industry was also examined across five items [20]. A total of 22 questions in all were formulated, and SPSS was used to do an exploratory factor analysis. Four factors (constructs) were created from the choice of 17 out of the 28 items whose factor loading was over 0.70 (see Table 1).

With the intent of analytical analysis, the current investigation adopts partial least squares structural equation modeling (PLS-SEM) [21]. Furthermore, PLS-SEM is the appropriate technique for this investigation because it is experimental in nature. It is advised that researchers consider two parts of the measuring model when attempting to demonstrate convergent validity: the outer loadings of the items and the average

Table 1 Scale used in research

Construct		Items	References
Customer engagement	CE1	Using health insurance, I am interested to learn more about health insurance plans	Hollebeek et al. [19] and Kosiba et al. [18]
	CE2	I am very positive about my health insurance company	
	CE3	I am very active in using the services of my health insurance plan	
	CE4	I feel good when I am able to use my health insurance at the time of need	
AI chatbots	AI1	I feel that chatbots are competent	Pillai and Sivathanu [16]
	AI2	I feel that chatbots for insurance companies are knowledgeable	
	AI3	I feel that chatbots for insurance companies are sensible	
	AI4	Chatbots have the ability to respond to my specific need quickly	
Customer loyalty	LOY1	I would recommend my health insurance plans to friends and family	Fernandes and Moreira [20]
	LOY2	I will buy my health insurance plan again	
	LOY3	I will buy health insurance even if everyone in family is healthy	
	LOY4	I am faithful to my health insurance provider	
	LOY5	I am a loyal customer to my health insurance provider	
Gamified apps	GA1	I feel that Gamified health app facilitates social interaction with friends and family	Eisingerich et al. [17]
	GA2	Gamified health app provides me claim related information as soon as possible	
	GA3	Gamified health app updates me about latest progress that I have made in the app	
	GA4	Gamified health app offers me health check-up appointments	

variance extracted (AVE) [22]. Cross-loading and the Fornell-Larcker criterion were also proposed as ways to establish discriminant validity [22]. Heterotrait-Monotrait analysis was also proposed as a criterion to measure discriminant validity [23]. In order to evaluate the structural model, we will calculate the path coefficients as well as the R2 value. As a result, we will evaluate the measurement along with structural models using all the above criteria.

3 Theoretical Framework and Research Hypothesis

In the present study, customer adoption intention of chatbots as a technology in the health insurance sector was measured using the TAM components of perceived intelligence and perceived usefulness [24]. Gamified health app adoption intent was also measured (see Fig. 2).

Based on the aforementioned information, this research work intends to examine and assess the validity of the following hypotheses:

H1: AI chatbots positively influence the customer engagement towards health insurance sector.

H2: AI chatbots positively influence the customer loyalty towards health insurance sector.

H3: Customer engagement positively influences the customer loyalty towards health insurance sector.

H4: Gamified apps positively influence the customer loyalty towards health insurance sector.

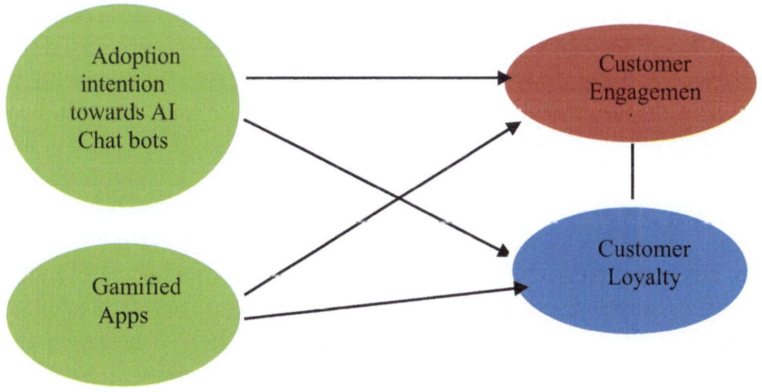

Fig. 2 Our proposed model

H5: Customer engagement positively mediates the relationship in AI Chatbots and customer loyalty towards health insurance sector.

4 Results and Discussion

4.1 Demographics

We first examine the demographics of customers of health insurance using AI-based chat bots. The study included 58% of the men and 42% of the women customers. The study also highlights the level of chatbots awareness among customers. 23% of customers are highly aware of chatbots, 61% of medium awareness, and 16% of low awareness.

4.2 Measurement Model Results

The outer loading of each item can be used as a gauge of its reliability. If the outside loadings for each item are larger than 40 but less than 0.7 but the AVE, composite reliability, and Cronbach's alpha values in PLS-SEM show a green color. Therefore, we will keep the items.

Table 2 shows that with the exception of items CE_1, CE_2, GA_1, and LOY_2, all items are reliable and fulfill the standards laid down. PLS-SEM shows Cronbach's alpha, composite reliability, and average absolute verifiability values in green color. This means that we are not eliminating anything from the structure of the construct. Furthermore, the average variance extracted (AVE) is the standard metric for verifying convergent validity, and it is described as the global mean value of the squared loadings of the items related to the construct. If the AVE is larger than 0.5, then the construct explains more than 50% of the variation among its components [22]. Cronbach's alpha and composite dependability are both over 0.7, and AVE values are also above 0.5, as indicated in Table 2. This proves that the constructs have convergent validity.

The Fornell-Larcker criterion and the Heterotrait-Monotrait ratio should be analyzed to determine discriminant validity. Table 3 shows that the present study fulfills the Fornell-Larcker criterion, which requires that each variable's square root of the AVE be greater than the correlation of the latent variables. When comparing cross-loadings, it is preferable for each indicator's loading to be larger than the loadings of the variables they represent.

A value for the Heterotrait-Monotrait ratio (HTMT) less than 0.85, as shown in Table 4, should be verified. The HTMT requirement is fulfilled, which proves the discriminant validity.

Table 2 Measurement model results

Constructs	Items	Outer Loadings	AVE	Composite Relibility (rho_c)	Cronbach's alpha
Customer Engagement	CE_1	0.623	0.525	0.814	0.706
	CE_2	0.66			
	CE_3	0.817			
	CE_4	0.78			
Gamified Health Apps	GA_1	0.674	0.598	0.855	0.778
	GA_2	0.86			
	GA_3	0.789			
	GA_4	0.759			
Customer Loyalty	LOY_1	0.806	0.592	0.879	0.838
	LOY_2	0.681			
	LOY_3	0.742			
	LOY_4	0.808			
	LOY_5	0.804			
Adoption intention to use AI Chatbots	AI_1	0.574	0.572	0.84	0.764
	AI_2	0.83			
	AI_3	0.799			
	AI_4	0.794			

Table 3 Fornell-Larcker criterion results

	AI	CE	GA	LOY
AI	0.758			
CE	0.303	0.725		
GA	0.145	0.183	0.776	
LOY	0.047	0.253	0.289	0.776

Table 4 Heterotrait-Monotrait ratio (HTMT) results

	AI	CE	GA	LOY
AI				
CE	0.35			
GA	0.201	0.24		
LOY	0.156	0.296	0.332	

4.3 Structural Model Assessment

One way to evaluate the model ability is to measure the degree of difference found in the dependent variables. Essential metrics like the R square value and the path coefficients are taken into consideration while analyzing the structural model [22]. These measures provide valuable insights into the extent to which the model is able to describe the variations observed in the dependent variables. By analyzing the R-square value, one can determine the scale of the total variation in the dependent variables that is marked for by the model. Additionally, the path coefficients reveal the strength and direction of the relationships among the variables in the model. As displayed in Fig. 3, the model has R-square value of 0.123 for loyalty (shows significant variance) and 0.092 for customer engagement which indicates that greater the R-square value the better. The findings of this investigation provide evidence to support the relationships between the independent and dependent variables.

The path analysis results, as illustrated in Fig. 3 and Table 5, indicate that four out of five hypotheses were supported, highlighting the significant paths between the variables. This implies that the independent variables have a sound impact on the dependent variable, customer engagement. The p-values obtained further confirm the significance of these relationships. These results provide valuable observation into understanding the factors that influence customer engagement. H2 ($B = -0.058$, $p < 0.05$) describes the path between Adoption motive towards AI chatbots and customer loyalty; indicating that the Adoption intention towards AI chatbots negatively influence customer loyalty in health insurance sector. H1 ($B = 0.303$, $p < 0.05$) shows the path among adoption intention towards AI chatbots as well as

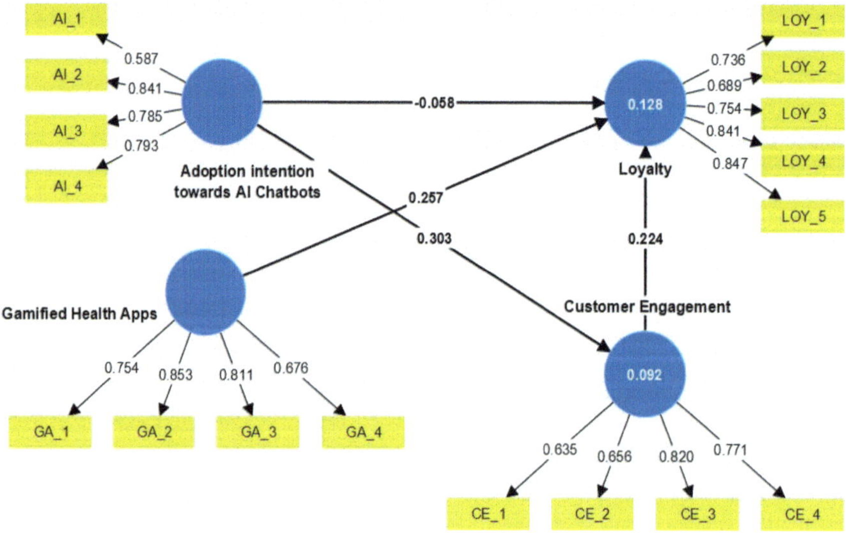

Fig. 3 Path analysis results

Table 5 Hypothesis test results

Path	Hypothesis	P value	Path coefficient	Remarks
AI > CE	H1	0	0.303	Supported
AI > Loyalty	H2	*0.339*	*−0.058*	Not supported
CE > Loyalty	H3	0.018	0.224	Supported
GA > Loyalty	H4	0.001	0.257	Supported
AI > CE > Loyalty	H5	0.045	0.068	Supported

customer engagement; representing that the adoption intention towards AI chatbots leverages the customer engagement.

H3 ($B = 0.224$, $p < 0.05$) demonstrates the path between customer engagement and customer loyalty; revealing that customer engagement positively influences the customer loyalty. H4 ($B = 0.257$, $p < 0.05$) depicts the path enclosed by gamified health apps and Loyalty; indicating that gamified health apps are significantly affecting the loyalty. H5 ($B = 0.068$, $p < 0.05$). The findings recommend that the mediating effect of client engagement plays a role in impacting purchaser loyalty. Specifically, the indirect effects of AI-based chatbots on customer loyalty through CE were found to be significant.

These results highlight the importance of cultivating positive customer engagement through AI chatbots in order to enhance customer loyalty. By focusing on providing opportunities for customer engagement, health insurance companies can foster higher levels of engagement and ultimately, greater customer loyalty.

5 Conclusion

This study examined the customer loyalty among the customers by adopting AI-based chatbots and the gamified apps. The Partial Least Squares Structural Equation Modelling (PLS-SEM) approach was used to inspect a sample of 110 customers of health insurance residing in the city Faridabad. The Partial Least Square-Structural Equation Modelling method has gained significant recognition as a robust analytical approach in the field of research.

This method is particularly well-suited for exploratory studies that aim to establish new structural relationships among multiple variables. The PLS-SEM approach involves a two-step procedure, namely, the measurement model assessment and the structural model assessment. The results show that while consumer loyalty is unaffected by AI-based chatbots, customer engagement and gamified health apps only have a favorable impact on customer trust.

This study adds to the system of the factors influencing customer engagement in the insurance sector by demonstrating the direct link between loyalty, the perceived intelligence also with functionality of AI-based chatbots. By analyzing the mediating function of CE, this study also helps to close a gap in the literature on the insurance industry. In particular, it was discovered that CE had a very small mediating role in the relationship between customer loyalty, perceived intelligence, and usefulness of AI-based chatbots.

Managerial Implications

The study focuses on chatbots that use artificial intelligence and gamified health applications and provides companies with valuable insights into the requirements of customers by examining the factors that influence their loyalty. The results may provide some insight into the psychological composition of this group, given the majority of people in the study's host city of Faridabad are middle class.

Limitations/Future Research

The study was conducted in just one location, Faridabad. Drawing from the limits of this work, some recommendations for future research areas can be made. First, the perceived usefulness and perceived intelligence TAM variables are used in this investigation without any further additions. Therefore, more research should concentrate on identifying additional variables that could affect consumers' acceptance of AI chatbots. Moreover, since the study's focus is on health insurance customers, its findings might not apply to other insurance kinds. Subsequent investigations may examine the impact of demographic factors within the suggested framework. This would offer a deeper comprehension of the ways in which age, gender, marital status, and level of education affect people's decisions about health insurance coverage. Furthermore, further factors like customer experience that may be crucial in predicting consumer engagement and loyalty may be added to the model in subsequent studies.

References

1. Ranoliya, B.R., Raghuwanshi, N., Singh, S.: Chatbot for university related FAQs. In: International Conference on Advances in Computing, Communications and Informatics (ICACCI). IEEE (2017)
2. Shevat, A.: Designing Bots: Creating Conversational Experiences. Reilly Media, Inc. (2017)
3. Daniela, L., Lytras, M.D.: Themed issue on enhanced educational experience in virtual and augmented reality. Virtual Reality **24**(3), 1–3 (2019)
4. Sandu, R., Gide, E., Karim, S.: The impact of innovative strategies to influence the adoption of cloud based service success in Indian small and medium enterprises (SMEs). Int. J. Arts Sci. **10**(2), 403–413 (2017)
5. Solakis, K., Katsoni, V., Mahmoud, A.B., Grigoriou, N.: Factors affecting value co-creation through artificial intelligence in tourism: a general literature review. J. Tour. Futur. (ahead-of-print) (2022)
6. Blöcher, K., Alt, R.: AI and robotics in the European restaurant sector: assessing potentials for process innovation in a high-contact service industry. Electron. Mark. **31**(3), 529–551 (2021)
7. Suhel, S.F., Shukla, V.K., Vyas, S., Mishra, V.P.: Conversation to automation in banking through chatbot using artificial machine intelligence language. In: 8th International Conference on Reliability, Infocom Technologies and Optimization (ICRITO), pp. 611–618 (Jun. 2020)

8. Kim, S., Goh, J., Jun, S.: The use of voice input to induce human communication with banking chatbots. In: ACM/IEEE International Conference on Human-Robot Interaction, pp. 151–152 (Mar. 2018)
9. Erem Ceylan, I.: The effects of artificial intelligence on the insurance sector: emergence, applications, challenges, and opportunities. In: The Impact of Artificial Intelligence on Governance, Economics and Finance, vol. 2, pp. 225–241 (2022)
10. Biswal, A.: AI Applications: Top 14 Artificial Intelligence Applications in 2023. Simplilearn.com. https://www.simplilearn.com/tutorials/Top. 14 Artificial Intelligence (AI) Applications in 2023. Assessed 13 Sept 2022
11. Volosovych, S., Zelenitsa, I., Szymla, W., Mamchur, R.: Transformation of insurance technologies in the context of a pandemic. Insur. Mark. Co. **12**(1) (2021)
12. Nuruzzaman, M., Hussain, O.K.: IntelliBot: a dialogue-based chatbot for the insurance industry. Knowl.-Based Syst. **196** (2020)
13. Rodríguez, D., Werth, O., Schönborn, S., Breitner, M.H.: A mixed methods analysis of the adoption and diffusion of Chatbot Technology in the German insurance sector. In: Twenty-fifth Americas Conference on Information Systems, Cancun (2019)
14. Riikkinen, M., Saarijärvi, H., Sarlin, P., Lähteenmäki, I.: Using artificial intelligence to create value in insurance. Int. J. Bank Mark. (2018)
15. Koetter, F., Blohm, M., Kochanowski, M., Goetzer, J., Graziotin, D., Wagner, S.: Motivations, classification and model trial of conversational agents for insurance companies (2018). arXiv: 1812.07339
16. Pillai, R., Sivathanu, B.: Adoption of AI-based chatbots for hospitality and tourism. Int. J. Contemp. Hosp. Manag. **32**(10), 3199–3226 (2020)
17. Eisingerich, A., Marchand, B., Fritze, A., Dong, L.: Hook vs. hope: how to enhance customer engagement through gamification. Int. J. Res. Mark. **36**(2), 200–215 (2019)
18. Kosiba, J.P., Boateng, H., Okoe, A.F., Hinson, R.: Trust and customer engagement in the banking sector in Ghana. Serv. Ind. J. **40**(13–14), 960–973 (2020)
19. Hollebeek, L.D., Glynn, M.S., Brodie, R.J.: Consumer brand engagement in social media: Conceptualization, scale development and validation. J. Interact. Mark. **28**(2), 149–165 (2014)
20. Fernandes, T., Moreira, M.: Consumer brand engagement, satisfaction and brand loyalty: a comparative study between functional and emotional brand relationships. J. Prod. Brand. Manag. **28**(2), 274–286 (2019)
21. Ringle, C.M.: Partial least squares structural equation modelling (PLS-SEM) using SmartPLS 3. In: Computational Data Analysis and Numerical Methods VII WCDANM, Portugal (2015)
22. Hair, J.F., Hult, G.T., Ringle, M.C., Sarstedt, M.: A Primer on Partial Least Squares Structural Equation Modeling (PLS-SEM). Sage Publications (2014)
23. Henseler, J., Ringle, C.M., Sarstedt, M.: A new criterion for assessing discriminant validity in variance-based structural equation modeling. J. Acad. Mark. Sci. **43**(1), 115–135 (2015)
24. Davis, F.D.: Perceived usefulness, perceived ease of use, and user acceptance of information technology. MIS Q. **13**(3), 319–340 (1989)

Enhancing Sensitivity Control for Improved Gesture Recognition in Unconstrained Environments

Jay Jain, Shyam Bhuva, Sahil Bhagat, Deepkumar Patel, Kriti Srivastava, and Pranit Bari

Abstract The mouse activities are carried out by the virtual mouse system through the utilization of hand motions. Over time, people have developed a greater awareness of the surfaces they come into contact with. Mouse systems, which are extensively utilized in a variety of contexts including workplaces, educational institutions, and other environments, are among the objects that are handled the most frequently. Individuals frequently experience difficulties in controlling their gadgets when they are at a greater distance from them or when their hands are dirty, which hinders their ability to use the mouse and keyboard. In this way, the motion mouse is brought into play as a factor. It is necessary to suggest a solution for situations in which physical contact is difficult to achieve and mouse operations can only be carried out through finger motions. It is not the objective to replace the mouse systems that are now in use. In order to ensure that a system is user-friendly, it is essential that it be able to accommodate the specific hand movements and gestures of the user, while simultaneously providing the user with feedback to improve their understanding of the system and their ability to utilize it. In order to accomplish this, it may be necessary to employ machine learning algorithms that are able to learn from the interactions that the user has with the system and gradually improve the system's accuracy and responsiveness.

Keywords YOLO · Sensitivity control · Computer vision · Machine learning · Gesture detection

J. Jain (✉) · S. Bhuva · S. Bhagat · D. Patel · K. Srivastava · P. Bari
Dwarkadas J, Sanghvi College of Engineering, Mumbai, Maharashtra, India
e-mail: jay4emails@gmail.com

K. Srivastava
e-mail: kriti.srivastava@djsce.ac.in

P. Bari
e-mail: pranit.bari@djsce.ac.in

G. Raj et al. (eds.), *Advances in Artificial Intelligence and Machine Learning*, Lecture Notes in Electrical Engineering 1264, https://doi.org/10.1007/978-981-97-9507-9_9

1 Introduction

1.1 Motivation

The goal of research is to change the way people and computers interact by using virtual hand gestures. By combining transfer learning and computer vision technologies, this is trying to make a system that is easy to use and can be used for a wide range of tasks, such as creating an immersive virtual reality experience or helping people who have trouble with mobility. This advanced system uses YOLO, OpenCV, and Python to let you precisely control a mouse from 4 m away and in a range of lighting conditions. With this practical implementation, it can be used in everyday situations like education and gaming, helping to solve problems like reducing latency. Making user experiences better across all fields, will help make computing easier to use, more fun, and more a part of everyday life.

1.2 Problem Statement

A lot of research shows that mouse systems that use virtual hand motion are a big step forward in how user interfaces are made [1]. With these gadgets, people can use their natural hand movements to control cursors in virtual worlds. These days, algorithms are so good that they can move, click, and scroll on your screen just like you would with your hands. This new idea streamlines the user experience and makes it easier for people to interact with computers. Smart devices with digital displays are becoming more common, and more often they can be controlled remotely from a user interface [2]. Virtual hand motion-based mouse systems can work with a lot of different devices, so standard Smart TV remote controls might become obsolete in the future [3]. Users of the Xbox and PlayStation can now use a virtual mouse instead of a regular controller, which makes a lot of new games possible [4]. This virtual mouse might also be useful for people who are used to using physical mice on laptops and desktops since they won't have to replace the mouse as often [5]. The main goal of this work is to make a mouse that only responds to hand movements [6]. Users may be more productive if they can do more than one thing at once with gestures instead of switching between input devices. A virtual mouse is a very useful and flexible alternative to real mice that lets you use your computer without touching anything. It's great for making solutions that use gestures and you can use it from afar. As scientists learn more about this technology and make progress on it, the chances of a more natural and cohesive computer experience are growing [7].

1.3 Organization of the Paper

The writers looked at 32 sources and 20 papers to write the literature survey. The papers were about different models and tools used for estimating distances. These papers show various methods for figuring out distances, such as famous object detection models like YOLO and SSD, as well as long-range detection methods that use special hardware. The literature review gives an in-depth look at the most recent approaches to figuring out distances between things and points out the pros and cons of each one. This knowledge can help researchers find gaps in their current work and come up with new ways to accurately estimate distances.

The project's goal is to build a virtual hand-based mouse system that can be used on computers, smartphones, and game platforms. Instead of a real mouse, the system will use advanced computer vision and machine learning techniques to correctly track and understand the user's hand movements in a virtual area. The project is about gesture-detecting software that will be used by the system to pick up on the user's hand movements and translate them into acts on the screen. The YOLOv5 model will be retrained and changed to fit the needs of the new user. This will help it better classify movements.

Bright lighting settings, followed by mid-lighting conditions and low lighting conditions, are associated with the greatest average confidence levels. Several variables are considered in the analysis, including lighting intensity at 2 m for both bright and low-lighting conditions, lighting intensity at 2 m for mid-lighting conditions, lighting intensity at 4 m for both bright and mid-lighting conditions, and lighting intensity at more than 4 m for bright lighting conditions.

2 Literature Review

2.1 Related Work

Much research has been done on the Virtual Hand Gesture-based mouse that can be used to perform basic operations of a mouse-like scroll, drag, and click. In Begum's research [8], computer vision was used to implement deep learning to prevent the COVID-19 spread. The MediaPipe framework is utilized for hand tracking and gesture detection, and the OpenCV library is used for computer vision. The algorithm tracks and recognizes the hand motions and hand tips using machine learning methods.

The hand interaction system by Krishnamoorthi [9] has four steps: detection, processing, tracking, and recognition. Skin color segmentation is used in many hand detection methods. The color space must be carefully considered to mimic skin tone. Color spaces include RGB, normalized RGB, HSV, YCrCb, YUV, and others. PyTorch is the native framework for 2 YOLOv5 to speed up training for 3D Model-Based Detection.

The Ranawat's collection [10] contained 216 photos. Also, these photographs were divided into five categories. To initiate the relevant mouse clicks and cursor movements, PyAutoGUI modules were used. OpenCV is used to capture the image, image preprocessing is performed, hand detection is performed, gesture recognition is performed, CNN (Convolution Neural Network) is performed, Max-pooling layers, Dropout, Fully Connected layers with activation function ReLU, softmax activation is performed, and Adam optimizer is used as optimizer with 98.47% testing accuracy.

Using YOLOv5 and several datasets, Karthi's work [11] explores object-location methods and provides a brief analysis of each dataset. Since obtaining high accuracy, recent research has focused on the best detection method and dataset. The object detection challenge is explained simply in this review using dataset features and YOLOv5's efficacy. After this assessment, the framework may be chosen swiftly and effectively based on the issue statement and data type. YOLOv5 can recognize and detect various elements in targeted photos with a certain level of accuracy and supports rigorous training and testing.

A virtual mouse that can replace real input devices including mice, keyboards, and remote controls is proposed by Shibly's study [12] utilizing gesture detection. By working with computers that lack physical input devices, this method improves mouse cursor functionality. Using a basic camera, it captures hand motions. Using OpenCV, a computer vision library, the system is implemented in Python. This technology's application in low light is limited by poor illumination. Its new approach to human–computer interaction notwithstanding, and the system's usability and functionality were not tested in the research. The system's functionality, accuracy, and user satisfaction in practical situations may all be assessed in subsequent studies.

Using Python and the OpenCV package, the authors created the gesture recognition system, MediaPipe in Swamy's proposed system [13]. Over 90% of user commands were recognized by the system, demonstrating its efficacy. In order to get user input, the authors also polled users, the findings indicated that the method was well-liked and thought to be beneficial for enhancing accessibility. In general, the study presents an optimistic perspective on assistive technology for those with physical limitations.

An OpenCV program has been utilized in Guha's paper [14] for media piping. People with impairments will find it easier and simpler to do their tasks with the aid of this technology. Furthermore, it is clear from the major limitations during the pandemic that contactless technology is important in today's society, and prototype offers one of the greatest ways to do it.

You Only Look Once (YOLO)-based kitchen hand detection algorithms are investigated in Van Staden's study [15] to improve human–machine interaction in culinary contexts. Three YOLO-based hand motion detection models—YOLOv2, YOLOv3, and Tiny YOLOv3—are compared by the authors. For the trials, a special collection of hands from around the home was employed. By virtue of its mean average 3 precision (mAP) of 94.1% and processing speed of 31 frames per second, YOLOv3 outperformed the alternative models in both accuracy and speed. The authors further examined the potential applications of hand detection technology in assisting those with disabilities to use kitchen equipment.

Utilizing two cameras in position, Shajideen's research [16] aims to enhance human–computer interaction systems via the use of hand gestures in three-dimensional space. A mapping is performed to the screen coordinate system based on an estimation of the hand-pointing action. Additionally, they finish the operation of the virtual mouse with additional hand motions. By using hand gestures to indicate where an item or folder is located on the screen and for performing further activities like menu selection.

Subsequent research endeavors may explore techniques to mitigate the disparity in the distance and enhance the accuracy of hand gesture recognition across different distances. The sensitivity gap refers to a situation where the cursor responds or moves more rapidly than the user intends, leading to user frustration and a sub-optimal user experience. To enhance the accuracy and promptness of feedback, it is recommended that forthcoming research should focus on augmenting the sensitivity of hand gesture detection systems.

2.2 Research Gap

Some of the provided research publications [17, 18] depend on specialized hardware, such as the Kinect sensor, to get depth pictures for hand gesture identification. As such, there are hardware constraints that must be taken into account. However, it is possible that some users may not have access to or possess the ability to utilize such devices. Subsequent research could explore hardware options that are more accessible and cost-effective, yet still possess the ability to attain a high degree of precision.

Generalizing the findings of the provided research studies to other locations or situations may be challenging because they were evaluated under restricted circumstances and specialized datasets. Subsequent research endeavors could focus on developing hand gesture recognition systems that exhibit precision across diverse environments and situations in constrained environments paper [19] can solve this using the OpenCV library.

Hand gesture recognition algorithms may encounter difficulty distinguishing between different movements when the user is positioned at varying distances from the camera or sensor as related work with the help of SSD (Single Shot Multi-box Detector) as portrayed in the paper [20]. Subsequent research endeavors may explore techniques to mitigate the disparity in the distance and enhance the accuracy of hand gesture recognition across different distances. The sensitivity gap refers to a situation where the cursor responds or moves more rapidly than the user intends, leading to user frustration and a sub-optimal user experience. To enhance the accuracy and promptness of feedback, it is recommended that forthcoming research should focus on augmenting the sensitivity of hand gesture detection systems.

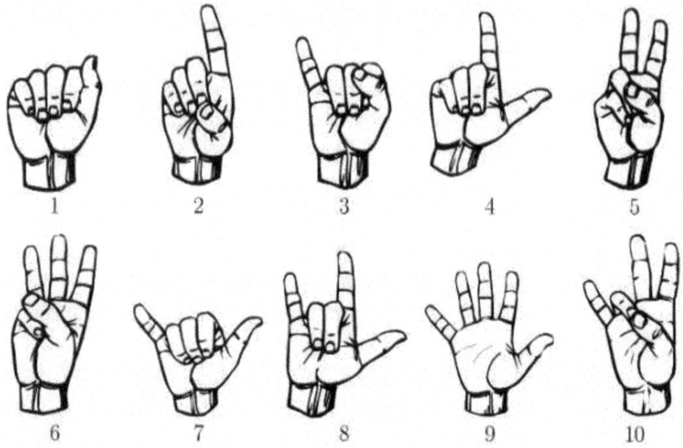

Fig. 1 Proposed hand gestures

3 Methodology

3.1 Dataset Description

The proposed system's methodology involves training and testing using a dataset of 3900 photos that includes 10 distinct hand motions as in Fig. 1. The dataset is divided into a training set of 3500 photographs and a testing set of 400 images, which comprises images of hands in varied postures, lighting situations, and backdrops, to guarantee the system's resilience to diverse real-world settings.

3.2 Gesture Recognition

The YoloV5 object detection algorithm is used for hand identification and recognition. YoloV5 is a cutting-edge item recognition algorithm that is very effective and well-known for its accuracy. Since it uses a single neural network to simultaneously conduct object detection and classification, it is especially well suited for real-time applications. The suggested method intends to achieve high accuracy and efficiency in detecting and recognizing hand gestures by using the capabilities of YoloV5.

3.3 Model Architecture

Technology in Fig. 2 allows users to control the mouse using hand movements captured by the laptop's webcam, going beyond the quick access choices for hand

and mouse gestures offered on many computers. Users may perform basic actions like moving the mouse cursor, selecting and deselecting items with left clicks, and more by utilizing solely hand gestures. Once the gesture is recognized, the system retrieves the bounding box coordinates of the hand and converts them into numerical values that can be interpreted by the PyAutoGUI Python module. PyAutoGUI is a Python module that enables cross-platform control of the mouse and keyboard. The cursor on the user's screen responds to the hand movements, and each gesture performs a specific operation. For example, gesture 1 moves the cursor, while gesture 2 executes a right-click operation.

A sensitivity option was added to the system, as illustrated in Fig. 3 to control various hand movement speed rates. The system may adjust the cursor movement speed using the sensitivity option to the user's preferences and age. With this setting, the system functioned flawlessly for users of all ages and hand movement intensities. By changing the sensitivity parameter in real-time, the user may fine-tune the system to match their specific requirements. The system is more user-friendly and the functionality improves the user experience.

The sensitivity parameter denoted as s controls the sensitivity of the system. The hand position in pixels is represented by the coordinates (x_h, y_h) in the pixel space. Similarly, the mouse position in pixels is denoted by the coordinates (x_m, y_m). It can then use these variables to define the relationship between the sensitivity parameter and the movement of the hand and mouse as follows: The movement of the mouse is proportional to the movement of the hand:

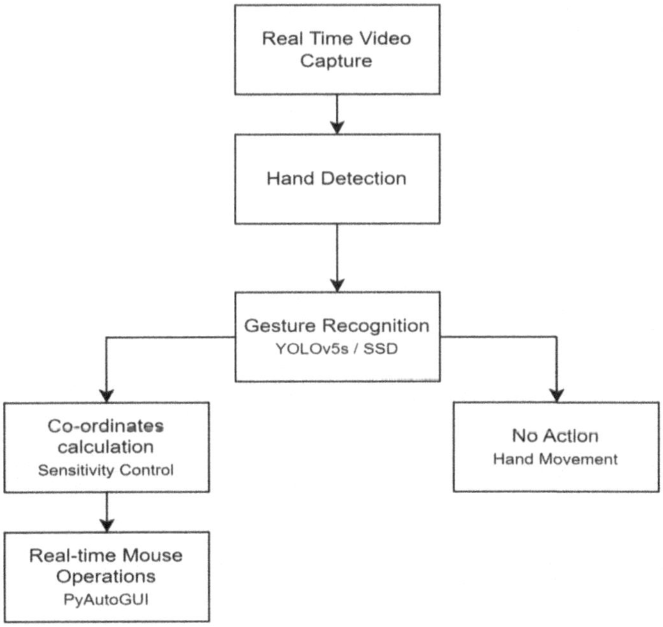

Fig. 2 Proposed system design

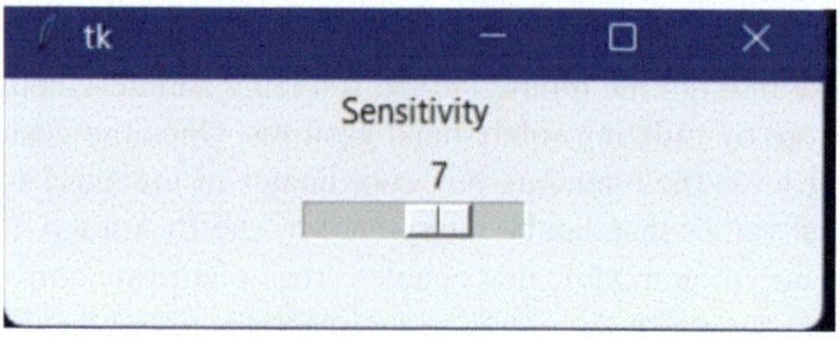

Fig. 3 Sensitivity control for user input

$$(x_m, y_m) = s * (x_h, y_h) \tag{1}$$

The sensitivity parameter scales the movement of the hand and mouse in Eq. (1).

$$(x_m, y_m) = s * k_m * (x_h, y_h) \tag{2}$$

$$(x_h, y_h) = s * k_h * (dx, dy) \tag{3}$$

where k_m and k_h are scaling factors for the mouse and hand movements, respectively, and (dx, dy) is the change in hand position.

It can further simplify this formula by expressing it in terms of the sensitivity parameter s and the change in hand position in Eqs. (2) and (3) (dx, dy):

$$(x_m, y_m) = s^2 * k_m * (x_h, y_h) \tag{4}$$

$$(x_h, y_h) = s^2 * k_h * (dx, dy) \tag{5}$$

These formulas show how the sensitivity parameter affects the movement of the hand and mouse in a linear way. The scaling factors k_m and k_h can be used to adjust the movement of the mouse and hand independently as in Eqs. (4) and (5).

Each bounding box for gesture classification has a confidence value associated with it. The confidence is calculated by YOLOV5s as follows:

$$PrPr(\text{object}) * \text{IoU}_{\text{pred}} \tag{6}$$

The model makes it possible to recognize hand movements. Employ the deep learning model YOLOv5, which excels at object recognition, to recognize these gestures. With the help of a model, it's able to create a straightforward and trustworthy hand gesture detector that a human user could use to communicate control data to a computer system.

The aforementioned gadget is made to identify certain hand movements and associate them with particular tasks. In contrast to Hand-2, which is used for right-clicking, Hand-1 is used to scroll. Hand-5 is used for double-clicking, whereas Hand-6 is used for single-clicking. The Hand-7 device facilitates cursor movement.

The remaining hand gestures for additional purposes may also be configured and customized by users to improve their experience. This feature gives customers the freedom to customize the functionality of the gadget to their own needs, potentially boosting productivity and user happiness.

4 Results and Evaluation

4.1 Experimental Results

Across a variety of discrete distances, including 1, 2, and 3 m for short-range gestures and 5, 6, and 7 m for long-range gestures. The frequency of individual hand is mentioned in Fig. 4.

Different test cases, such as light and distance, were used to get the results. Light was put into three groups: low, medium, and bright, and distance ranged from 1 to 4 m.

Figure 5 illustrates that the system can identify up-hand movements from more than 4 m away with a confidence level of 0.85 for gesture hand-1in bright conditions and from 4 m away with a confidence level of 0.91 for gesture hand-2 in bright-lighting conditions.

Figure 6 illustrates that the system can identify up-hand movements from less than 4 m for gesture hand-9 away with a confidence level of 0.91 and 0.90 in low-lighting conditions.

The line in Fig. 7 shows that the highest average trust levels are found in bright-lighting conditions, followed by mid-lighting conditions and low-lighting conditions. The study looks at a lot of different factors, such as lighting intensity at 2 m for both

Fig. 4 Data description

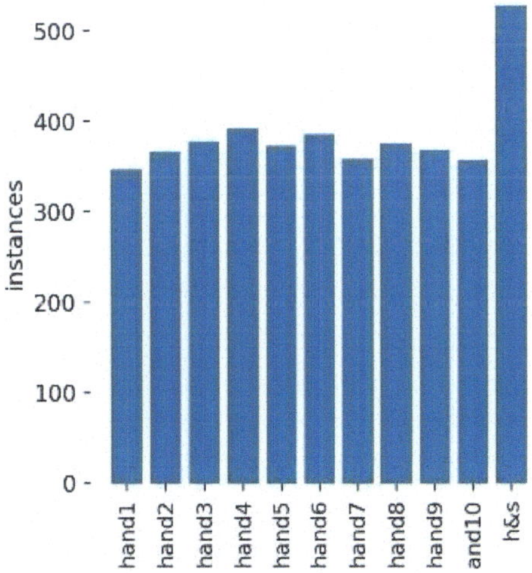

Fig. 5 Illustration of hand gesture recognition beyond and at 4 m in bright-lighting conditions

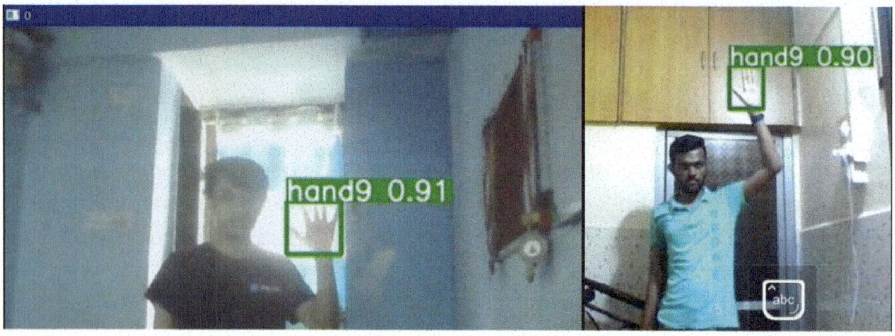

Fig. 6 Illustration of hand gesture from less than 4 m away in low lighting conditions

bright and low lighting, lighting intensity at 2 m for mid-lighting, lighting intensity at 4 m for both bright and mid-lighting, and lighting intensity at more than 4 m for bright lighting.

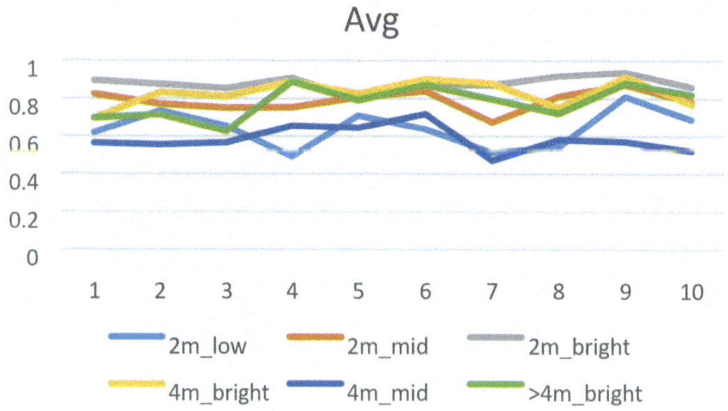

Fig. 7 Confidence scores average graph

5 Conclusion

The project uses Python, YOLO, OpenCV, a transfer learning algorithm, and a diverse lighting-conditioned dataset to control a mouse from a maximum distance of 4 m away. There are plans for games, smart boards in the classroom, and more. The mouse cursors are moved and chosen by hand movements. It finds motions, gets hand-bounding box coordinates, and turns them into numbers that the PyAutoGUI Python module can understand. This module sets the cursor movement and right-click gestures so that the mouse and keyboard can be used on different platforms. With 3900 pictures of 10 hand movements, it trains and tests. To make sure the resistance works in real life, the dataset has 3500 training photos and 400 testing photos with different lighting, backgrounds, and poses.

Transfer learning and computer vision are used in the virtual hand gesture-based mouse technology to make a new computer interface. The system learns from how people use it and can track and understand hand gestures for virtual reality, games, and helping people move around. Hand-cursor latency is being kept as low as possible for real-world accuracy in a range of lighting conditions. This new idea could make user displays easier to understand and more useful. Future endeavors may work on improving accuracy and responsiveness in changing real-life situations. With the help of new sensors and machine learning techniques, the system can adapt better to lighting and other environmental factors, which means it can be used in more industrial settings. Researchers may explore adaptive algorithms for dynamic sensitivity control.

References

1. Xu, C., et al.: Robust 3D hand detection from a single RGB-D image in unconstrained environments. Sensors **20**(21), 6360 (2020)
2. Tran, D.-S., et al.: Real-time virtual mouse system using RGB-D images and fingertip detection. Multimed. Tools Appl. **80**, 10473–10490 (2021)
3. Fu, Y., et al.: Research on long-distance hand recognition based on depth information. In: Journal of Physics: Conference Series, vol. 1187, no. 4, p. 042108. IOP Publishing (Apr. 2019). https://doi.org/10.1088/1742-6596/1187/4/042108
4. Sarkar, S., et al.: A vision base application for virtual mouse interface using hand gesture. Int. J. Innov. Sci. Res. Technol., 2456 216 (2021)
5. Zhou, L., et al.: Long-range hand gesture recognition via attention-based SSD network. In: 2021 IEEE International Conference on Robotics and Automation (ICRA). IEEE (2021)
6. Jung, H.-K., Choi, G.-S.: Improved yolov5: Efficient object detection using drone images under various conditions. Appl. Sci. **12**(14), 7255 (2022)
7. Shriram, S., et al.: Deep learning-based real-time AI virtual mouse system using computer vision to avoid COVID-19 spread. J. Healthc. Eng. **2021** (2021)
8. Begum, S.S., et al.: Design of a virtual mouse for hand gesture recognition and characterization. In: International Conference on Information and Management Engineering. Springer Nature Singapore, Singapore (2022)

9. Krishnamoorthi, M., et al.: Virtual mouse using YOLO. In: 2022 International Conference on Innovative Computing, Intelligent Communication and Smart Electrical Systems (ICSES). IEEE (2022)
10. Ranawat, M., et al.: Hand gesture recognition based virtual mouse events. In: 2021 2nd International Conference for Emerging Technology (INCET). IEEE (2021)
11. Karthi, M., et al.: Evolution of yolo-v5 algorithm for object detection: automated detection of library books and performance validation of dataset. In: 2021 International Conference on Innovative Computing, Intelligent Communication and Smart Electrical Systems (ICSES). IEEE (2021)
12. Shibly, K.H., et al: Design and development of hand gesture based virtual mouse. In: 2019 1st International Conference on Advances in Science, Engineering and Robotics Technology (ICASERT). IEEE (2019)
13. Swamy, T.J., et al.: Voice and gesture based virtual desktop assistant for physically challenged people. In: 2022 6th International Conference on Trends in Electronics and Informatics (ICOEI). IEEE (2022)
14. Guha, J., et al.: AI virtual mouse using hand gesture recognition. In: Int. J. Res. Appl. Sci. Eng. Technol. (IJRASET) **10**(4) 3070–3076 (Apr. 2022). https://doi.org/10.22214/ijraset.2022.41981
15. van Staden, J., Brown, D.: An evaluation of YOLO-based algorithms for hand detection in the kitchen. In: 2021 International Conference on Artificial Intelligence, Big Data, Computing and Data Communication Systems (icABCD). IEEE (2021)
16. Shajideen, S.M.S., Preetha, V.H.: Hand gestures-virtual mouse for human computer interaction. In: 2018 International Conference on Smart Systems and Inventive Technology (ICSSIT). IEEE (2018)
17. Mazhar, O., et al.: Towards real-time physical human-robot interaction using skeleton information and hand gestures. In: 2018 IEEE/RSJ International Conference on Intelligent Robots and Systems (IROS). IEEE (2018)
18. Chareonsuk, W., et al.: Face and eyes mouse for ALS patients. In: 2016 Fifth ICT International Student Project Conference (ICT-ISPC). IEEE (2016)
19. Gupta, B., Saini, S.: Virtual mouse using hand gestures with python and OpenCV. In: 2018 3rd International Conference on Computing, Communication, Control and Automation (ICCUBEA), pp. 1–6. IEEE (2018)
20. Yi, C., et al.: Long-range hand gesture recognition with joint SSD network. In: 2018 IEEE International Conference on Robotics and Biomimetics (ROBIO). IEEE (2018)

Health Risk Predictor Analysis to Assess the Impact of Heavy Alcohol Consumption

P. G. Sunitha Hiremath, Vivek V. Pais, and Neha Tarannum Pendari

Abstract The study aims to explore the factors that influence risk alcohol drinking, their relationship and impact on health using epidemiological data. We collected data of 2972 male patients from Shri Dharmasthala Manjunatheshwara De-Addiction and Research Centre, Ujire, Karnataka State, India. The data was in raw format, which was preprocessed and cleaned for further analysis. Techniques such as statistical data description measures and simple correlation analysis were used for analysis. From our analysis, there were around 73.2% of patients in the age range of 26–46 years, 34.8% had suffered with at least one adverse childhood experience (ACE) and 27% with at least one problematic behaviour in their childhood/adolescence. Around 77.6% of patients were in very high-risk level of alcohol consumption. A positive correlation was observed between alcohol risk and any one type of ACE and childhood/adolescence problematic behaviour, whereas a negative correlation between the quantity of alcohol consumption and age of alcohol initiation was exhibited. It is also observed that maximum patients who were in the very high risk or high-risk level have developed more psychiatric problem than physiological problems.

Keywords Data preprocessing · Binning · Co-relation analysis · Physiological health · Alcohol risk factors

P. G. Sunitha Hiremath (✉) · N. T. Pendari
KLE Technological University, Hubballi, Karnataka, India
e-mail: pgshiremath@kletech.ac.in

N. T. Pendari
e-mail: neha_ip@kletech.ac.in

V. V. Pais
SDM Deaddiction and Research Centre, Ujire, India
e-mail: vivekvpais@gmail.com

G. Raj et al. (eds.), *Advances in Artificial Intelligence and Machine Learning*, Lecture Notes in Electrical Engineering 1264, https://doi.org/10.1007/978-981-97-9507-9_10

1 Introduction

Problematic alcohol consumption or harmful drinking pattern by the people is one of the social issue faced in India. This kind of drinking pattern leads to certain physiological and/or psychological disorders, which in turn affects the individual's family and society [1]. In India, the percentage of people who consume alcohol is increasing at a faster pace as it is easily accessible legal drug. According to a Government Report, about 14.6% of population (16 crore persons) consume alcohol in India [2]. The uncontrolled harmful drinking pattern leads to a chronic disease called Alcohol Use Disorder (AUD). People who suffer with AUD take treatment in hospitals or rehabilitation centres. There are many factors that influence the harmful drinking pattern of individuals. This study focuses on investigating the factors that are responsible for causing harmful drinking patterns in Indian context. To carry out this work detailed information about alcohol dependent patients was required. We collected 2972 alcoholic patient's data who took treatment at Shri Dharmasthala Manjunatheshwara De-addiction (SDM) and Research Centre, Ujire, a place in Karnataka state in India. This centre runs deaddiction camp for alcohol-dependent patients and collects data by taking their consent. Each camp serves around 60–70 patients. We collected 44 camps patients (male) data that were conducted from 16-04-2019 to 09-03-2020.

There is a well-designed tool (Intake Proforma) that acts a case sheet for every patient, was used to collect their data. Trained counsellors collect patient's data under five sections, namely (i) sociodemographic data, stressors, drinking pattern, and withdrawal symptoms (ii) medical comorbidities and past treatment history, (iii) family history, temperament, and personality, (iv) causes for relapse/factors responsible for continuation, and (v) adverse childhood experiences (ACEs), problematic childhood/adolescence behaviour, and education history. Our main objective is to analyse the patients' data based on Alcohol Age of Initiation (AAI) below 18 years and later age groups in relation to drinking pattern, risk calculation, health comorbidities and adverse childhood experiences, and problematic childhood/adolescence behaviour.

Alcohol consumption beyond specific daily limits is called risk drinking. World Health Organization has defined gender-specific risk drinking levels based on number of grams of ethanol consumed per day [3]. It classified the risk levels as Low (male = 1–40 g, female = 1–20 g), Medium (male = 41–60 g, female = 21–40 g), High (male = 61–100 g, female = 41–60 g), and Very High Risk (male >101 g, female >61 g). The risk level is given by the product of average quantity of alcohol beverage consumption, its percentage of ethanol content and alcohol conversion factor (1 ml ethanol = 0.79 g) [4]. For example, if gender is male, average quantity of consumption is 250 ml, brand of alcohol is Old Tavern, type of drink is Whiskey, Ethanol content in that drink is 40% then the Alcohol consumption can be calculated as follows: $250*0.4*0.79 = 79$ g, which indicates the risk level as High.

Section 2 discusses the state of the art to explore the adverse childhood experience & problematic childhood/adolescence behaviour on risk level, effect of AAI and duration of consumption on risk level, analysis of effect of risk level on health based on AAI. In Sect. 3, we discuss about the detailed analysis of alcoholic patients' data with respect to the above-said factors followed by conclusion in Sect. 4.

2 Related Work

Section 2.1 outlines some of the works carried on adverse childhood effects on alcoholism, the impacts of alcoholism on psychiatric health and physiological health are discussed in Sects. 2.2 and 2.3, respectively.

2.1 *Relationship Between ACE and Alcohol Initiation*

In [5], authors focused on finding the relationship between the co-occurrence of ACEs (ACE Score) and alcohol initiation at age less than 14, 15–17 and 18–20 years. It was discovered that there was a strong relationship between ACE score and alcohol initiation at less than 14 and 15–17 years but for 18–20 years, it was moderate. They used ACEs like abuse and neglect in childhood, children growing up in problematic household environment, alcohol use in adolescence and adulthood. In [6], authors carried out a study on 1200 adolescents by applying Multiple Correspondence Analysis on their data and found that family and social factors were associated with early initiation of alcohol. Authors of [7] stated that parent's substance use and their lack of control on their children behaviour are the causes for early initiation of alcohol. In some cases, overprotection by parents also will affect, this was shown in the study carried out by authors in [8], which showed that parents' overprotection of their children has led to adolescent alcohol use.

2.2 *Alcohol Effect on Psychiatric Health*

Many works have been carried out to investigate the impact of alcohol initiation at early ages on health during later ages. The author in [5] discussed about various health issues and social harms caused due to heavy drinking. The authors in [9] discussed the results of previous prospective studies that were finding the causal relationships between alcohol usage by adolescents and their psychological changes and cognitive impairments. It was shown that heavy drinking during adolescence leads to neurotoxicity and accompanying cognitive deficits. The authors in [10, 11] showed that heavy alcohol consumption is highly correlated to depressive symptoms and weakly to anxiety symptoms. Increased consumption of alcohol in adolescents will result in to depressive symptoms [12].

2.3 Alcohol Effect on Physiological Health

Authors in [13] found that alcohol consumption is inversely proportional to the risk of diabetes in men and women but frequent consumption of alcohol (average weekly consumption) has shown an association with lower risk of diabetes. In a study carried in, authors found that there was no relationship between alcohol consumption and risk of type 2 diabetes in elderly and middle-aged men, but there was an increase in the risk of type 2 diabetes if the average alcohol consumption is greater than 2 g/ day or greater than 7 times/week. The study carried out by [14] found a stronger relationship between heavy alcohol intake in early adulthood and increased risk of type 2 diabetes. But later reduction in alcohol intake did not show reduction in risk level. The authors of [15] found that later AAI and short drinking duration showed decreased risk on type 2 diabetes, where early AAI and longer duration of drinking did not show decreased risk. Heavy alcohol consumption may adversely affect the lungs. This was investigated by authors [16] and found a strong correlation between heavy alcohol consumption and pulmonary infections and acute respiratory distress syndrome (ARDS).

No intensive work is carried out considering the alcohol rehabilitation patients' data to investigate the impact of ACEs, problematic childhood/adolescence behaviour on age of alcohol initiation and also the impact of age of alcohol initiation on health. We worked on the rehabilitation patient's data to investigate the effects of above-said parameters on age of alcohol initiation and alcohol effects on health.

3 Predictor Variables Analysis

This section identifies and discusses three prominent factors that lead to heavy alcohol consumption which in turn affects the health of individuals.

3.1 Data Description

In this data, 1.74% of patient's records which had multiple missing field values were eliminated [17]. The numerical field values which were missing were filled with statistical measure mean/median. The unwanted text that was tampered with the numerical values was removed. One hot encoding technique is applied to extract new features for the attributes with multiple values.

In this study, we investigated the effect of ACE, problematic childhood/ adolescence behaviour and AAI and duration of consumption on alcohol risk level.

Table 1 Patient's risk in each ACE category

ACE category	#Total patients	#Patients in VHR	#Patients in HR
Poverty/severe debts	832	649	140
Early parental loss	156	116	28
Extramarital affairs of parents	35	30	5
Broken home/single parenting	174	140	27
Sexual abuse	26	24	2
Violence	164	133	24

3.2 Adverse Childhood Experience Effect on Risk Level

Even though the risk level is based on the consumption of number of grams of ethanol per day, there are various other factors that influence the amount of ethanol consumption by the patients. One of the factors that leads to alcoholism is adverse childhood experiences (ACE) and problematic behaviour in childhood and/or adolescence [18]. The ACEs are the unpleasant events experienced by the people in their childhood [19, 20]. The tool collects the childhood/adolescence history with respect to ACE factors such as poverty/severe debts, early parental loss, extramarital affairs of parents, broken home/single parenting, sexual abuse and violence. The behavioural problems identified in the childhood/adolescence include running away from home, frequent physical fights and violence, destruction of others property, stealing, scholastic backwardness, experimenting with drugs, alcohol, gambling, and sexual issues.

This study aimed to find the relationship between ACE factors to the risk levels of patients. It was found that some patients had more than one adverse experience. Table 1 shows the number of patients in each of ACE categories and their risk level based on the quantity of alcohol consumption. Around 34.8% of the total (2972) patients had suffered with ACEs, out of which 27.9% suffered with poverty/severe debt, followed by 5–6% in early parental loss, broken home or single parenting and violence, 1% in sexual abuse and extramarital affairs of parents. It is observed that around 75–80% of the patients who suffered with any kind of ACE have developed very high risk level of drinking.

3.3 Analysis of Problematic Childhood/Adolescence Behaviour on Risk Level

The adverse childhood experiences may lead to some behavioural problems in childhood and/or adolescence. This section discusses which of the problematic childhood/adolescence behaviour of the patients had an effect on their alcohol consumption risk level. Table 2 shows different category of problematic behaviour of patients,

Table 2 Problematic behaviour category and risk level

Problematic behaviour category	#Total patients	#Patients in VHR	#Patients in HR
Running away from home	200	170	26
Scholastic backwardness	288	227	48
Alcohol consumption	182	152	23
Destruction to others property	68	55	10
Experimenting with drugs	40	33	4
Stealing	21	17	1

number of patients in each category and their observed risk level. It's observed that around 27% of the total (2972) patients had problematic behaviour in their childhood/adolescence, out of which running away from home is (6.7%), scholastic backwardness is (9.6%), alcohol consumption is (6%), destruction to others' property is (2.2%), Experiment with drugs is (1.3%) and stealing is (0.7%). Whichever the problematic behaviour the patients had in their childhood, now currently maximum of them have developed very high or high risk level of alcohol consumption. This study found that individuals who suffer with any type of ACE or any type of problematic behaviour, there is a higher probability of them developing risk level of alcohol consumption in their later life.

3.4 Effect of AAI and Duration of Consumption on Risk Level

This section discusses the effect of age, AAI on quantity of alcohol consumption, and risk level. Age bins are created to know the number of patients in each age group. Figure 1 shows the visualization of the number of patients in each age bin with mean age of 38.23 (±9.67). It's observed that 73.2% of the patients lie in the age range of 26–46 years and the remaining 26.8% of patients lie outside this range.

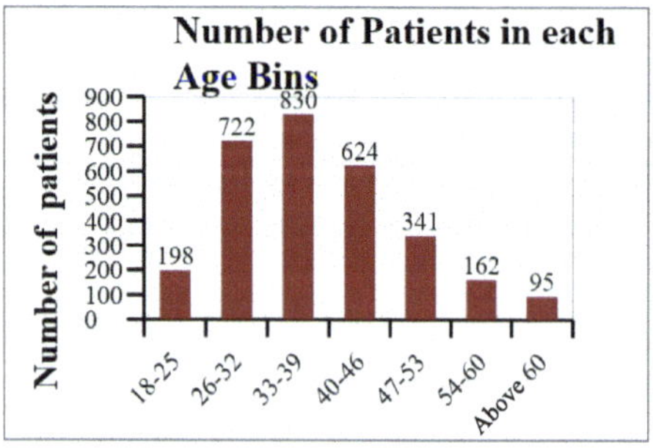

Fig. 1 Patients in each age category

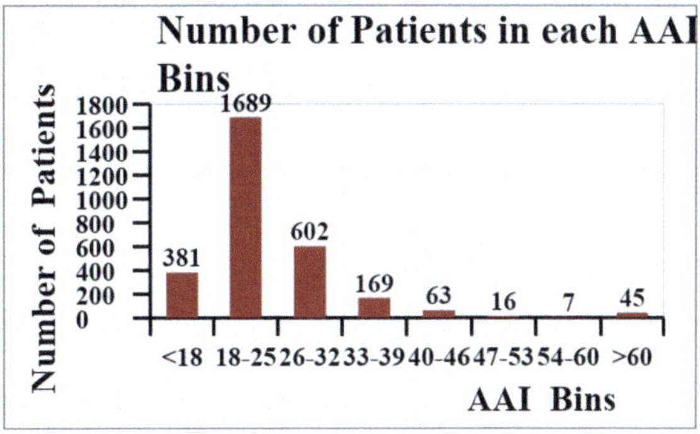

Fig. 2 Patients in each AAI category

To investigate the effect of age of alcohol initiation of these patients (age at initiation AAI) and their duration of alcohol consumption on the risk of alcohol consumption, AAI age bins are created. The number of patients in each AAI category is shown in Fig. 2 with a mean AAI of 25 (±14.85). It is observed that 56.8% of patients are in the AAI age group of (18–25), 20.2% in (26–32) AAI age group. This indicates that maximum patients (77%) started consuming alcohol in the age range of (18–32) years.

The risk level for each of the patient is computed as described in the previous section. It was found that 2307 (77.6%) patients lie in the very high risk level and 513 (17.26%) in the high risk level. Remaining patients (5.14%) lie in the other risk levels. Patients have many reasons to start and to continue the consumption of alcohol [21]. The possibilities of reasons that the tool collected about the patients are: To try as Someone in family or friends were using, To feel better and confident or Happy, To avoid problems and sadness, and other reasons. The possibilities for continued use were Liked effect and wanted more of it, Friends Forced, Gave Confidence, Craving, Felt relaxed and reduced physical exertion and Other. The data were pre-processed using a one hot encoding method to create a new feature that takes binary input (1/0) for each of the reason to start and continued use. Patients gave more than one reason but maximum patients (74.5%) gave the reason as 'to try as someone in family or friends were using', 28% as 'to feel better and confident or happy', 15% 'to avoid problems and sadness' and 2% gave other reasons. One of the reasons to start may be due to the adverse childhood experiences which we discussed in the previous section. We did not look into the correlation between reasons to start with the ACE factors instead analysed which ACE factor was predominantly seen in the data and its relationship to the risk level.

The AAI and duration of alcohol consumption will have an impact on the quantity of alcohol consumption [22]. The quantity of consumption of alcohol is analysed with respect to the age at initiation (AAI) of <18 years age, current age, and duration of consumption. There were 381 patients whose AAI was less than 18 years of age. The patients were categorized into seven age bins as per the present age as shown in Table 3. The number of patients in each age bin was 56, 110, 101, 52, 36, 15, and 11. The attribute values shown in Table 3 are their mean ± the standard deviation.

Table 3 Alcohol consumption analysis

Age groups	Duration of consumption (years)	Average quantity of alcohol consumption (mL)
18–25	6.7 ± (3.15)	550 ± (541)
26–32	14.1 ± (5.91)	534.7 ± (348.3)
33–39	18.89 ± (5.61)	579.6 ± (354)
40–46	23.4 ± (7.9)	595.19 ± (382)
47–53	30.25 ± (9.99)	474.7 ± (319.4)
54–60	36.6 ± (13.8)	447 ± (274.4)
>60	49.7 ± (7.93)	315 ± (90)

It is observed that as the duration of alcohol consumption is increased, the average quantity of alcohol consumption is also increased starting from age 18–25 to age bin 40–46. For the later ages from 47 to 53 onwards, even though duration of consumption is increased, the average quantity of alcohol consumption has decreased. The correlation obtained between age and quantity of consumption was −0.0331, which indicates a weak negative correlation. The possible reasons are slower processing of alcohol at older age and higher Blood alcohol concentration as the amount of water in the body reduces with age.

There is an increased physical or mental health risk associated with heavier alcohol consumption [5]. In the next section, we discuss about the effects of problematic drinking pattern on health of the patients who initiated alcohol consumption in all age groups. The authors in [17] have carried an extensive analysis on factors affecting alcohol risk level based on epidemiological data.

3.5 Analysis of the Effect of AAI-Based Risk Level on Health

It is observed from the data that 45.12% of the total patients suffered with health issues, out of which 11.92% were reported chronic health issues, 27.6% with psychiatric complications and 5.5% with both. We classified patients into six categories of AAI. For each AAI category, number of patients in each of the risk levels and for each of the risk levels, the percentage of patients suffering with chronic health and psychiatric type issues are listed as shown in Table 4. It is observed that the number of patients increased from <18 AAI group up to (26–32) AAI group. From the next AAI group (33–39) onwards, the number of patients were decreased. In each of the AAI group, maximum patients are in the VHR and HR risk level of alcohol consumption. In each of the AAI group, maximum patients have suffered with psychiatric issues than chronic health issues. The results in Table 4 show that the effect of risk on health is seen more in AAI group of (18–25) and (26–32) but decreased in the later AAI age groups. And also, the risk influence was more on psychiatric problems when compared to the chronic health problems. We did not investigate the reasons why psychiatric problems are more when compared to chronic health issues, which is beyond the scope of our work.

Table 4 Effect of risk level on health in each AAI group

AAI	RL	#Patients	Chronic health issue (% patients type)	Psychiatry issue (% patients type)
<18	VHR	309	16.82% (Diabetes)	39.63% (Confusion)
	HR	62	8.06% (Respiratory problem)	25.80% (Hallucinations)
	MR	4	–	50% (Paranoid ideas)
	LR	6		–
18–25	VHR	1369	15.92% (Diabetes)	34.69% (Confusion)
	HR	269	12.63% (Diabetes)	30.48% (Confusion)
	MR	10	10% (Bronchitis)	40% (Aggressive outbursts)
	LR	30	20% (Diabetes)	33.33% (Depression)
26–32	VHR	463	20.95% (Diabetes)	28.29% (Confusion)
	HR	110	14.54% (Diabetes)	22.72% (Confusion)
	MR	10	40% (Bronchitis)	20% (Hallucinations)
	LR	30	17.39% (Others)	43.47% (Confusion)
33–39	VHR	112	27.67% (Diabetes)	31.25% (Hallucinations)
	HR	46	21.73% (Diabetes)	43.47% (Depression)
	MR	2	40% (Bronchitis)	20% (Hallucinations)
	LR	9	44.44% (Bronchitis)	33.33% (Depression)
40–46	VHR	41	14.63% (Diabetes)	19.51% (Confusion)
	HR	19	5.26% (Cardiac problem)	10.52% (Confusion)
	MR	–	–	–
	LR	3	33.33% (Diabetes)	–
47–53	VHR	10	20% (Diabetes)	–
	HR	3	–	–
	MR	–	–	–
	LR	3	–	–
54–60	VHR	1	–	–
	HR	4	20% (Diabetes)	–
	MR	–	–	–
	LR	2	–	–
>60	VHR	–	–	–
	HR	–	–	–
	MR	3	66.66% (Diabetes)	100% (Confusion)
	LR	40	7.5% (Diabetes)	10% (Confusion)

4 Conclusions and Future Work

In this paper, we discussed about the outcomes of the epidemiological data analysis that was carried out considering 2972 Indian male patients who were suffering with harmful drinking problems. We explored the data with respect to Adverse Childhood Experiences and Problematic Childhood/Adolescence Behaviour effect on Risk Level and the effect of AAI-based Risk Level on Health.

It was discovered that poverty/severe debts, early parental loss, single parenting, violence were the prominent ACE factors that initiated alcohol consumption in the patients. The patients who had alcohol initiation from <18 years of age to 32 years have developed very high risk level or high risk level of alcohol consumption. These patients had more psychiatric problems than physiological problems but we could not explore the reasons, which is beyond the scope of our work. The prominently observed psychiatric problems are confusion, hallucinations, aggressive outbursts and depression whereas diabetes, bronchitis are the prominently seen chronic health problems. It is also observed that as the duration of alcohol consumption is increased, the average quantity of alcohol consumption increased starting from 18 to 46 age. For the later age onwards, even though duration of consumption is increased, the average quantity of alcohol consumption has decreased. As a future work, we plan to develop solution for assessing the status of quality of life of alcohol dependant persons by selecting the prominent features and doctor's notes from the case sheet.

Acknowledgements We thank Dr. D. Veerendra Heggade President "SDM De-addiction and Research Centre", Ujire, Karnataka State, India, for permitting us to carry out research that has helped us to assess the impact of heavy alcohol consumption on Health and provide insights to public regarding Health Risks.

References

1. Eashwar, V.M.A., Umadevi, R., Gopalakrishnan, S.: Alcohol consumption in India—an epidemiological review. J. Fam. Med. Prim. Care **9**(1), 49–55 (2020). https://doi.org/10.4103/jfmpc.jfmpc_873_19
2. Ambekar, A., Agrawal, A., Rao, R., Mishra, A.K., Khandelwal, S.K., Chadda, R.K.: On behalf of the group of investigators for the National Survey on Extent and Pattern of Substance Use in India. Magnitude of Substance Use in India. Ministry of Social Justice and Empowerment, Government of India, New Delhi (2019)
3. Burkhardt, G., Adorjan, K., Kambeitz, J., Kambeitz-Ilankovic, L., Falkai, P., Eyer, F., Koller, G., Pogarell, O., Koutsouleris, N., Dwyer, D.B.: A machine learning approach to risk assessment for alcohol withdrawal syndrome. Eur. Neuropsychopharmacol. **35**, 61–70 (2020)
4. IRIS homepage. https://iris.who.int/publications. International guide for monitoring alcohol consumption and related harm. World Health Organization. Accessed 12 Dec 2023
5. Dube, S.R., et al.: Adverse childhood experiences and the association with ever using alcohol and initiating alcohol use during adolescence. J. Adolesc. Health **38**(4), 444-e1 (2006)
6. Trujillo, C.A., et al.: An examination of the association between early initiation of substance use and interrelated multilevel risk and protective factors among adolescents. PloS One **14**(12), e0225384 (2019). https://doi.org/10.1371/journal.pone.0225384

7. Maggs, J.L., et al.: Very early drinking: event history models predicting alcohol use initiation from age 4 to 11 years. Addict. Behav. **89**, 121–127 (2019). https://doi.org/10.1016/j.addbeh.2018.09.030
8. Visser, L., et al.: Do child's psychosocial functioning, and parent and family characteristics predict early alcohol use? The TRAILS Study. Eur. J. Public Health **25**(1), 38–43 (2015). https://doi.org/10.1093/eurpub/cku072
9. Peeters, M., et al.: Psychological changes and cognitive impairments in adolescent heavy drinkers. Alcohol Alcohol. (Oxford, Oxfordshire) **49**(2), 182–186 (2014). https://doi.org/10.1093/alcalc/agt162
10. Johannessen, E.L., Andersson, H.W., Bjørngaard, J.H., Pape, K.: Anxiety and depression symptoms and alcohol use among adolescents—a cross sectional study of Norwegian secondary school students. BMC Public Health **17**, 494 (2017). https://doi.org/10.1186/s12889-017-4389-2
11. Skogen, J.C., Sivertsen, B., Lundervold, A.J., Stormark, K.M., Jakobsen, R., Hysing, M.: Alcohol and drug use among adolescents: and the co-occurrence of mental health problems. Ung@hordaland, a population-based study. BMJ Open **4**(9), e005357 2014
12. Homman, L., Perra, O., Higgins, K., O'Neill, F., O'Neill, F.: The longitudinal relationship of alcohol problems and depressive symptoms and the impact of externalising symptoms: findings from the Belfast Youth Developmental Study. Soc. Psychiatry Psychiatr. Epidemiol. **54** (2019). https://doi.org/10.1007/s00127-019-01696-9
13. Holst, C., et al.: Alcohol drinking patterns and risk of diabetes: a cohort study of 70,551 men and women from the general Danish population. Diabetologia **60**(10), 1941–1950 (2017). https://doi.org/10.1007/s00125-017-4359-3
14. Han, T., et al.: Eighteen-year alcohol consumption trajectories and their association with risk of type 2 diabetes and its related factors: the China Health and Nutrition Survey. Diabetologia **62**(6), 970–980 (2019). https://doi.org/10.1007/s00125-019-4851-z
15. Li, H., et al.: The association between age at ınitiation of alcohol consumption and type 2 diabetes mellitus: a cohort study of 0.5 million persons in China. Am. J. Epidemiol. **189**(12), 1478–1491 (2020). https://doi.org/10.1093/aje/kwaa119
16. Simet, S.M., Sisson, J.H.: Alcohol's effects on lung health and ımmunity. Alcohol Res.: Curr. Rev. **37**(2), 199–208 (2015)
17. Sunitha Hiremath, P.G., Nigam, S., Kartik, K., Sharan, S., Suraj, G., Pais, V.V.: Explorative analysis of factors ınfluencing alcohol risk level based on epidemiological data and prediction using machine learning techniques. In: Proceedings of the 2nd International Conference on Cognitive and Intelligent Computing Cognitive Science and Technology, pp. 383–393 (2023)
18. Dube, S.R., et al.: Adverse childhood experiences and personal alcohol abuse as an adult. Addict. Behav. **27**(5), 713–725 (2002)
19. Damodaran, D.K.: The unveiled Indian picture of adverse childhood experiences.pdf. IAHRW Int. J. Soc. Sci. Rev., n. pag. Print (2018)
20. Kramer, J.R., et al.: Multiple-domain predictors of problematic alcohol use in young adulthood. J. Stud. Alcohol Drugs **69**(5), 649–659 (2008). https://doi.org/10.15288/jsad.2008.69.649
21. Poikolainen, K., et al.: Predictors of alcohol intake and heavy drinking in early adulthood: a 5-year follow-up of 15–19-year-old Finnish adolescents. Alcohol Alcohol. (Oxford, Oxfordshire) **36**(1), 85–88 (2001). https://doi.org/10.1093/alcalc/36.1.85
22. Rehm, J.: The risks associated with alcohol use and alcoholism. Alcohol Res. Health: J. Natl. Inst. Alcohol Abus. Alcohol. **34**(2), 135–143 (2011)

Genomic Insights into Infertility Using Neural Network

Payal Gupta, Garima Bhardwaj, Shivani Dubey, Tarun Tayal, Abhishek Sengupta, and Priyanka Narad

Abstract About 48.5 million couples i.e., one in four couples experience infertility worldwide, which is notably prevalent in India. This study investigates the multifaceted relationships between metabolic diseases like hyperthyroidism, diabetes type 2, recurrent implantation failure (RIF) and reproductive health. Understanding that these disorders have genetic roots, we used co-expression analysis to discover significant genes and pathways associated with reproductive health. This gives us novel insight into the molecular cause of infertility. We systematically processed and examined the pattern of gene expression for hyperthyroidism (GSE178996), RIF (GSE26168) and diabetes (GSE16532) using a publicly available database from the National Center for Biotechnology Information Gene Expression Omnibus (NCBI-GEO). In each type of disease, differential expression analysis found several genes that were both up-regulated and down-regulated genes. To understand the differences and similarities between these diseases, co-expression networks were built, with a focus on identifying key hub genes crucial in controlling diseases. Our research analysed probable disease-associated genes ABCC6 and SLC2A14 and elucidated their functions in diseases. GO analysis, revealed enriched biological pathways and processes important for reproductive health. Lastly, we used the Feed-Forward Neural Network (FFNN) model with dropout and L2 regularization for text clarification, which distinguished between the investigated conditions with an overall accuracy of 96.43%. In conclusion, the complex landscape of genes connected to reproductive health and their relationship with diseases are revealed by this thorough study. These results create the basis for more investigation that could lead to reproductive health-related therapeutics and diagnostic approaches.

Keywords Co-expression analysis · Diabetes type 2 · Hyperthyroidism · Recurrent Implantation Failure (RIF) · Infertility · Metabolic disorders

P. Gupta · G. Bhardwaj · S. Dubey · T. Tayal · A. Sengupta
Amity Institute of Biotechnology, Amity University Uttar Pradesh, Noida, UP 201301, India

P. Narad (✉)
Division of Development Research, Indian Council of Medical Research, Ansari Nagar, New Delhi, India
e-mail: priyanka.narad@gmail.com

G. Raj et al. (eds.), *Advances in Artificial Intelligence and Machine Learning*, Lecture Notes in Electrical Engineering 1264, https://doi.org/10.1007/978-981-97-9507-9_11

1 Introduction

Infertility being a physical and emotionally exhausting issue affects 60–80 million couples worldwide and approximately 15–20 million couples in India [1, 2]. There can be various other factors that may cause infertility in females, while hyperthyroidism, diabetes mellitus type 2, and repeated implantation failure (RIF) are more focused in the underlying research [3]. Various genes are involved in causing these diseases. The relationship between genes and their expression patterns has gained importance in analyzing the disease-causing roots [4]. Co-expression analysis can find important genes and pathways involved in infertility providing meaningful insights into molecular processes [5]. Couples who are trying to conceive come across various diseases which include hypothyroidism, diabetes, and recurrent implantation failure. These diseases also have a wide impact on the women's health along with prolonged infertility.

Fertility in women can be significantly impacted by hyperthyroidism, an endocrine disease defined by an overactive thyroid gland and excessive thyroid hormone production [6]. These thyroid hormones, thyroxine (T4) and triiodothyronine (T3) are responsible for controlling the body's metabolism and an enormous number of physiological functions that are involved in reproduction. In the presence of hyperthyroidism, the delicate hormonal balance which is essential for healthy conception and pregnancy might be disturbed. Diabetes mellitus type 2 is another reason that impacts female fertility, a condition marked by insulin resistance and high blood sugar levels. Type 2 diabetes mellitus has a very complicated and complex interaction with female infertility. For reproduction, the hormonal balance required can be disrupted by fluctuating insulin levels. Because of such hormonal imbalance, women would not be able to conceive naturally due to irregular menstrual periods and anovulation (lack of ovulation) [7]. Additionally, Recurrent implantation failure (RIF), a novel pathophysiological state characterized by frequent failures to achieve pregnancy after embryo transfer (ET), is also a major cause of infertility.

Potential connections between hyperthyroidism, diabetes mellitus type 2, and RIF can be found by combining information from co-expression analysis of these diseases [8, 9]. The links between genes performing various biological functions may be examined using network-based analysis of gene expression profiles. Consequently, a brief explanation of each network is essential to understand the underlying mechanism causing a disease [10, 11].

In this study, we have analysed transcriptomics dataset of hyperthyroidism, diabetes type 2 and recurrent implantation failure (RIF) by processing the expression profiles [12]. In the underlying project, we have used Limma, affy, and many more packages [13–15]. The differentially expressed genes were determined followed by functional enrichment analysis and network analysis. The network analysis was used to identify key players also called hub genes and proteins in the network which have a significant impact on the regulation of different biological processes. Lastly, a Feed Forward Neural Network (FFNN) model was developed to conduct test clarification on diabetes, hyperthyroidism, RIF, and diabetes. The model had dropout and

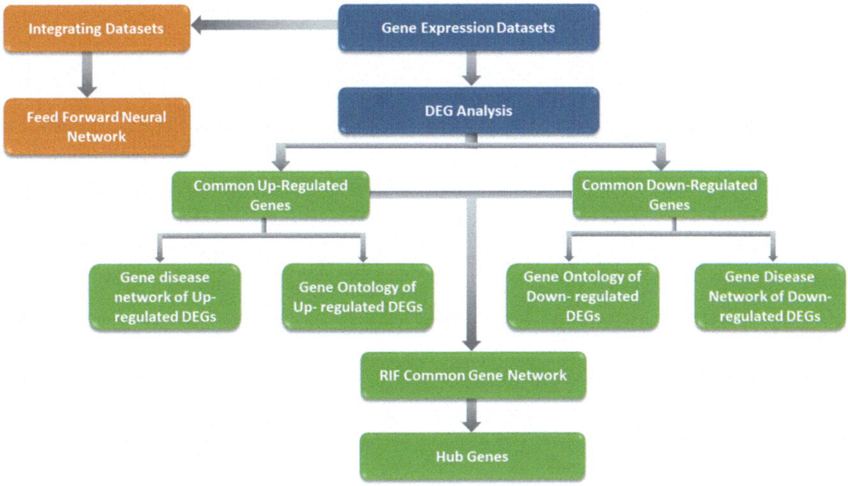

Fig. 1 This figure shows the integrated bioinformatics analytical approach used in this work

L2 regularization added to it [16, 17]. This machine learning model offers a few ways to distinguish between and classify these circumstances using textual inputs. We explored each component of methodology in depth in the following sections, presenting thorough findings that show the complicated interactions between RIF, diabetes hyperthyroidism, and reproductive health while also showcasing the potential of machine learning in diagnostic applications. The Integrated bioinformatics approach is shown as a workflow in Fig. 1.

2 Methodology

2.1 Data Collection

The data was collected from the National Center for Biotechnology Information Gene Expression Omnibus database (NCBI-GEO) https://www.ncbi.nlm.nih.gov/geo/ to study the impact of reproductive health and its relationship with other metabolic diseases [18]. We selected 3 different independent microarray datasets with accession IDs as GSE178996 for hyperthyroidism [19], GSE26168 for diabetes [20] and GSE16532 for Recurrent implantation failure (RIF) [21]. The datasets that met the following inclusion criterion were included in the study: (1) Gene expression profiling by array or high throughput sequencing; (2) The sample is of coding RNAs; (3) Diseased and corresponding control samples were contained in one experiment; (4) The sample size is at least six, with three patients in each group (5) Raw data were available.

2.2 Data Preprocessing

The gene expression data was normalized. Data was pre-processed using R studio software. Datasets were annotated using "hgu133plus2" and "org.Hs.eg.db" R Bioconductor packages depending on the platform. The probes without any annotation were removed. The values of multiple probe IDs mapping to the same gene were averaged. The code was run under the R environment version 4.0.3.

2.3 DEG Identification

To identify differentially expressed genes (DEGs), analysis was performed using R packages—limma for microarray datasets. The differentially expressed genes were identified by applying a threshold of p-value < 0.05 and $|\log2$ fold change (FC)$| > 0.5$. DEGs were ranked in accordance with p-value and $|\log2$ FC$|$. To determine the co-expressed genes a venn diagram was generated for downregulated and upregulated genes using a venn tool https://bioinformatics.psb.ugent.be/webtools/Venn/ tool [22].

2.4 Gene Disease Network Analysis (GDNA)

GDN is a bioinformatics approach used to uncover and understand the relationship between genes and diseases. We used a network construction method to find the common differentially expressed genes and to build a co-expression network. In this method, every gene is represented by a node, and the edges signify strong co-expression connections between genes.

The gene-disease association network was created and visualized using Cytoscape v 3.8.2 [23, 24]. This analysis can provide insights into the genetic basis of diseases, potential therapeutic targets, and disease mechanisms [25]. To perform GDN, we used the STRING-DB Cytoscape App https://apps.cytoscape.org/apps/stringapp. It is a powerful tool for integrating PPI data from the STRING database directly into Cytoscape, a popular network visualization and analysis platform. This allows us to create and analyze PPI network, visualize functional associations, and perform network-based analyses with cytoscape using STRING data [26, 27]. The genes which were common in all three diseases were selected as hub genes.

2.5 Gene Ontology Analysis

Gene ontology is a powerful approach for functional genomics research and is widely used in fields such as genetics, genomics, and molecular biology to elucidate the biological context of gene sets and prioritize candidate genes for further investigation. To understand the functional significance of a set of genes or proteins, Enrichr tool https://maayanlab.cloud/Enrichr/ was used, which is user-friendly tool to understand the biological pathways and processes associated with specific gene list [28, 29].

2.6 *Feed-Forward Neural Networks (FFNNs)*

A Feed Forward Neural Network is a type of artificial neural network where the connections between the nodes don't create a cycle because the input is analysed solely in a single direction, the feed-forward model is considered the most basic type of neural network [30]. However, information may transit through several hidden nodes, it continually proceeds forward and never reverses. The feed-forward neural network can be reduced to a single-layer perceptron when it is simplified. As inputs enter the layer, they are multiplied by weights. The weighted input values are then joined together to yield the sum. If the total of the values exceeds a specific threshold, which is normally set at zero, the output result is usually 1, whereas if it drops beyond the threshold, the result is generally -1. The single-layer perceptron is frequently utilised for classification as a feed-forward neural network model [30, 31].

In this paper, we proposed a neural network model with dropout and L2 regularization for text classification in which the algorithm was trained on the dataset of Diabetes, Hyperthyroidism and RIF. The code included the import of necessary Python libraries [16, 32]. To solve class imbalance, a tool for dealing with unbalanced datasets, such as a Random Under-sampler imported from SMOTE, was used [33]. A dataset was loaded into a Pandas DataFrame, which was then divided into feature (X) and target variable (y) components. For numerical representation, label encoding was applied to the target variable. A 70/30 split was used for training and testing, followed by Random Undersampling for class balance and StandardScaler for feature scaling [34, 35]. A TensorFlow Keras neural network model with input, hidden, and output layers was built using dropout and L2 regularisation. Model training entailed compiling using the Adam optimizer and halting early based on validation loss for overfitting prevention [36, 37]. It analyses the model on test data after training, decodes predictions, and returns accuracy and a classification summary with measures such as precision and recall.

3 Results

3.1 GEO Datasets

Datasets for Hyperthyroidism with accession ID GSE178996 contains 6 samples in total, out of which 3 were identified as control samples and 3 as case samples. Similarly, datasets for RIF with ID GSE16532 holds a total of 9 samples, out of which 5 were identified as control samples and 4 as case samples. Diabetes mellitus type 2 with ID GSE26168 holds total 17 samples out of which were identified as 8 control samples and as 9 case samples. Detailed information regarding the datasets is shown in Table 1.

3.2 DEG Identification and Analysis

Analysis and identification of DEG was done using R Studio. Hyperthyroidism dataset GSE178996 was performed resulting in the identification of 587 DEGs meeting the conditions, including 267 upregulated DEGs and 320 downregulated DEGs. Volcano plot of DEG in diabetes, hyperthyroidism, and RIF are shown in Fig. 2. Similarly, analysis of Diabetes mellitus type 2 dataset GSE26168 was performed resulting in the identification of 1734 DEGs meeting the conditions, including 789 upregulated DEGs and 945 downregulated DEGs. Also, analysis of RIF dataset GSE16532 was performed resulting in the identification of 5661 DEGs meeting the conditions, including 1924 upregulated DEGs and 3737 downregulated DEGs (see Figs. 3 and 4).

3.3 Gene Disease Network Analysis (GDNA)

GDNA was performed for both up-regulated and down-regulated genes. Analysis of Fig. 5b shows that gene SLC2A14 was found to be the potentially disease-associated downregulated gene out of all the down-regulated genes from hyperthyroidism, RIF and Diabetes mellitus type 2 diseases. Similarly, Fig. 5a shows that the gene ABCC6 was found to be the potential disease-associated in up-regulated gene.

3.4 Gene Ontology Analysis

Gene ontology analysis revealed that the top molecular function, biological process, cellular component, and pathway for upregulated genes were insulin-like growth factor receptor binding, regulation of T cell proliferation, secretory granule lumen,

Table 1 Detailed information about the datasets

Disease name	GEO number	Tissue name	Control samples	Case samples	Significant genes	Up-regulated genes	Down-regulated genes
Diabetes	GSE26168	Pancreas, liver, adipose and skeletal muscle tissue	8	9	1734	789	945
Hyperthyroidism	GSE178996	Skeletal muscle tissue	3	3	587	267	320
RIF	GSE16532	Endometrium	5	4	5661	1924	3737

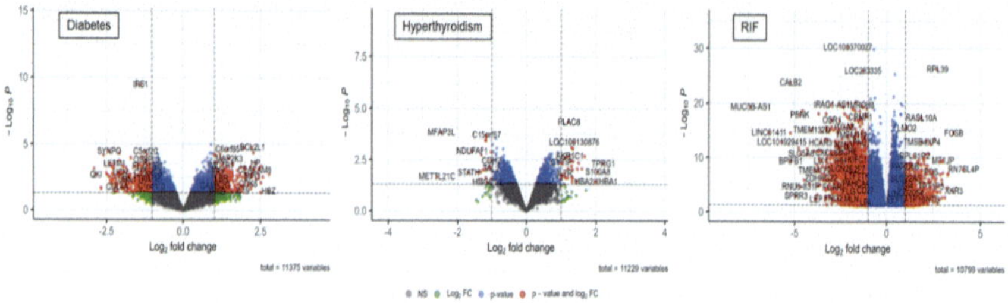

Fig. 2 Volcano plot of Differentially expressed genes in diabetes, hyperthyroidism, and RIF

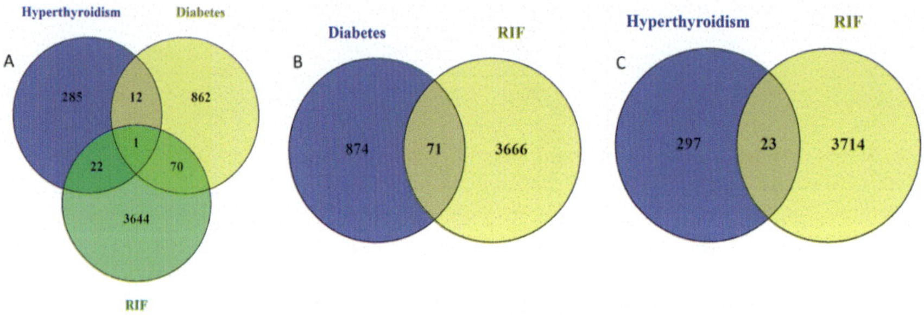

Fig. 3 Up-regulated common genes among Hyperthyroidism, Diabetes, and RIF. **a** Hyperthyroidism and Diabetes, **b** Hyperthyroidism and RIF, and **c** Diabetes and RIF

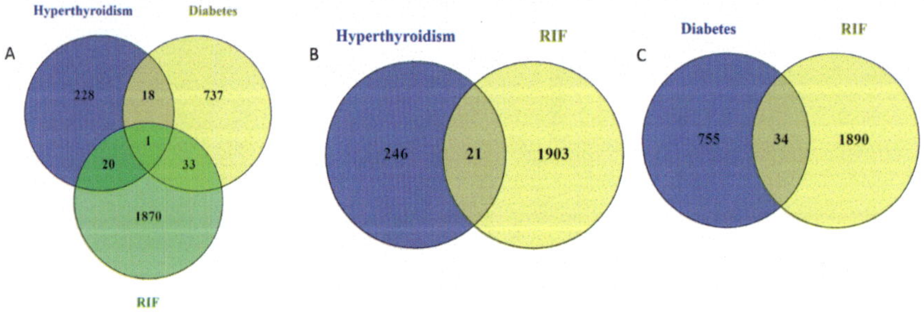

Fig. 4 Down-regulated common genes among Hyperthyroidism, Diabetes and RIF. **a** Hyperthyroidism and Diabetes, **b** Diabetes and RIF, and **c** Hyperthyroidism and RIF

and glucagon signaling pathway. The results are depicted in Fig. 6 showing enriched terms for input gene set based on −log10(p-value). The top molecular function, biological process, cellular component, and pathway for downregulated genes were voltage gated monoatomic cation channel activity, regulation of cation activity, voltage gated potassium channel complex and morphine addiction. Figure 7 shows the top 10 enriched terms, along with their corresponding p-values. Coloured bars

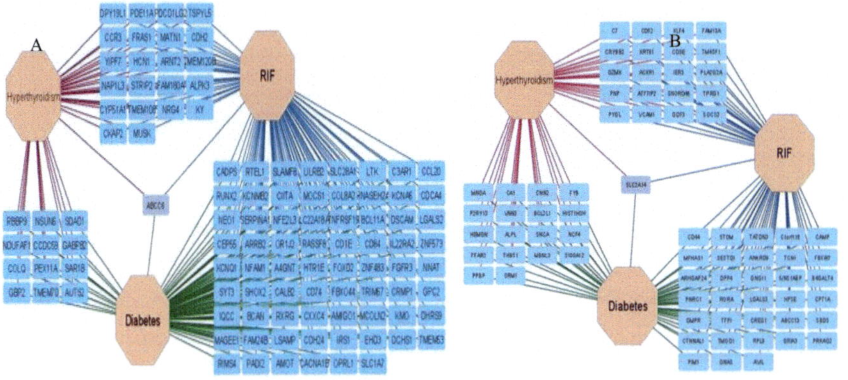

Fig. 5 Gene Disease Network model showing the PPI network between the common DEGs of down-regulated and up-regulated genes. **a** ABCC6 is found to be the potentially disease associated down-regulated gene. **b** SLC2A14 is found to be the potentially disease associated up-regulated gene

correspond to terms with significant p-values (<0.05). An asterisk (*) next to a p-value indicates the term also has a significant adjusted p-value (<0.05).

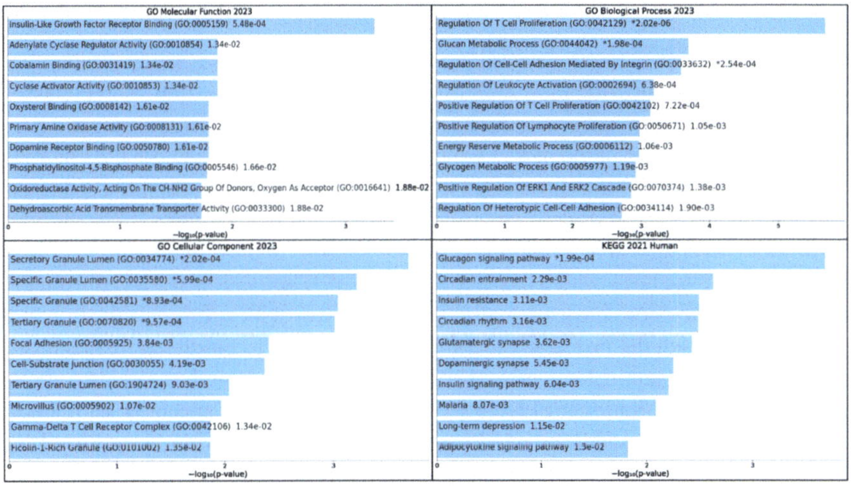

Fig. 6 GO of Up-regulated common genes Bar chart of top enriched terms from the GO_Cellular_Component_2023 gene set library. The top 10 enriched terms for the input gene set are displayed based on the −log10(p-value), with the actual p-value shown next to each

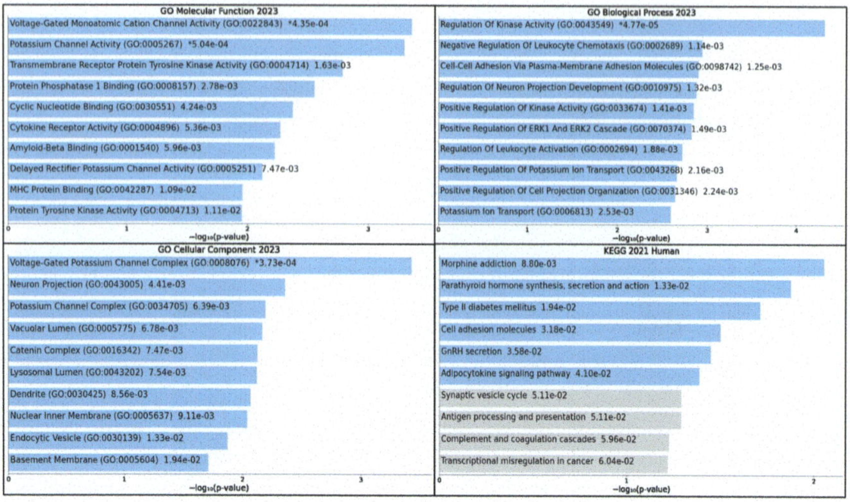

Fig. 7 GO of down-regulated genes. The bar chart shows the top 10 enriched terms in the chosen library, along with their corresponding p-values. Coloured bars correspond to terms with significant p-values (<0.05). An asterisk (*) next to a p-value indicates the

3.5 Clustering Using Feed-Forward Neural Network Analysis

In this paper, we presented a neural network model which is a machine learning algorithm for text categorization with dropout and L2 regularisation. Three classes of text documents were used to train the model: diabetes, hyperthyroidism, and RIF. There were four layers in the model which included the input layer, text documents served as the input for this layer. Dense layer with ReLU activation and 64 units in which the supplied text was used to teach this layer new features. 50% dropout rate dropout layer, this layer randomly removed neurons to prevent overfitting during training. The dense layer which has 32 units and ReLU activation, retrieved from the layer before it has more intricate features. Output layer with softmax activation and three units in which the probable outcome of each class was predicted by this layer.

Feed-Forward Neural Networks (FFNNs) are powerful models but they come with their set of limitations. FFNNs are prone to overfitting specially with small training data. In this model, to avoid overfitting, L2 regularisation was utilised for the weights of the first two layers that were dense shown in Fig. 8. The sparse categorical crossentropy loss function and the Adam optimizer were used to train the model. Early halting was used to keep an eye on the validity loss and terminate training when it stopped getting better.

The categorization model produced excellent results, with 96.43% overall accuracy shown in Fig. 9. With a flawless precision and recall score of 1.00, it displayed remarkable performance in differentiating Diabetes patients. Also, with a precision of 1.00 and a recall of 0.95, it performed effectively in detecting RIF cases. The model performed well in the recall, with a value of 1.00, despite having a precision

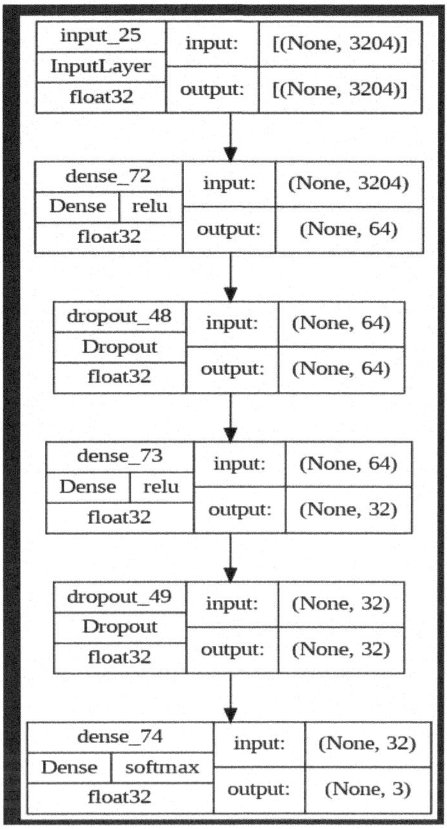

Fig. 8 This figure shows the architecture of a neural network model for text classification with dropout and L2 regularization

of 0.67 for detecting hyperthyroidism. A strong overall performance is indicated by the macro-average F1-score of 0.92, which makes this model a good option for additional study and applications in medical diagnostics.

Considering a neural network model that was trained on the datasets, the graph displays the model loss and the model accuracy [38]. The validation split is 0.3, and the model is trained for 100 epochs. As training advances, the model loss, which represents the negative log-likelihood of the correct class label, declines, while model accuracy, or the percentage of properly predicted labels, increases [39]. This pattern denotes effective learning.

Notably, the validation loss is projected to be greater than the training loss due to unseen data. It is a reasonable performance estimate [40]. The model exhibits robust generalisation to new data with a validation accuracy of 96.4%. In summary, the neural network accurately predicts class labels in text documents with high accuracy.

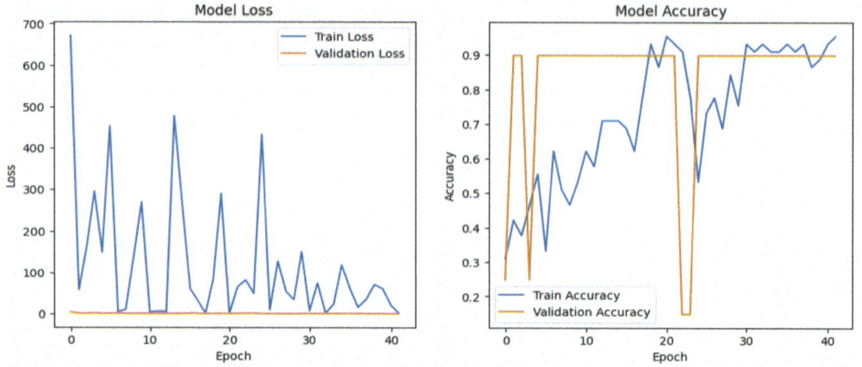

Fig. 9 This figure shows the accuracy and loss of a neural network model trained on a text classification task. The model achieves high accuracy (96.4%) on the validation data, indicating that it can generalize well to unseen data

4 Discussion

The investigation of the complex interactions between metabolic conditions such as hyperthyroidism, type 2 diabetes, and recurrent implantation failure (RIF) and their effects on reproductive health is the focus of the study described here. Co-expression analysis, Gene Disease Network Analysis (GDNA), Gene Ontology analysis, and a Feed-Forward Neural Network (FFNN) model with dropout and L2 regularization are some of the many approaches used in this study [5, 41]. The findings provide significant insight into the molecular processes that underpin such diseases and how they relate to reproductive health [42, 43].

Co-expression analysis is an effective method for revealing gene interactions and identifying important genes associated with specific diseases [44]. In this research, gene expression patterns in hyperthyroidism, diabetes, and RIF were examined using microarray datasets from the NCBI-GEO database. A critical first step is the identification of differentially expressed genes (DEGs) in each condition. This research showed that each disease has a considerable amount of upregulated and downregulated genes, indicating to the disorders' complicated genetic roots [45, 46].

To investigate the interactions between the genes within every disease type, co-expression networks were built [47]. Hub genes, which are crucial in the control of disease, can be identified through these networks. Particularly fascinating is the discovery of common differentially expressed genes among the three illnesses. The common molecular pathways that may lead to problems with reproductive health in people with these diseases can be identified with the use of this understanding [48].

GDNA is a useful method for understanding out the complex relationship between genes and illnesses. In this work, protein–protein interaction data were combined, and functional relationships were shown using the STRING-DB Cytoscape App. The discovery of two genes (SLC2A14 and ABCC6) that may be associated with disease provides a basis for future investigation into the genes' functions in hyperthyroidism, diabetes, and RIF [49, 50]. ABCC6 (ATP Binding Cassette Subfamily C Member 6) is a protein coding gene responsible for actively extruding physiological compounds, and xenobiotics from cells and SLC2A14 (Solute Carrier Family 2 Member 14) is a hexose transporter that can mediate the transport of glucose and dehydroascorbate across the cell membrane. It may be possible to develop new treatment targets by better understanding how these genes operate in the setting of these disorders.

Analysis of the gene ontology provides information on the functional importance of gene sets. The top molecular function, biological process, cellular component, and pathway for upregulated genes were insulin-like growth factor receptor binding, regulation of T cell proliferation, secretory granule lumen, and glucagon signaling pathway. For downregulated genes, the top molecular function, biological process, cellular component, and pathway were voltage gated monoatomic cation channel activity, regulation of cation activity, voltage gated potassium channel complex and morphine addiction. Researchers can better comprehend the biological context of discovered genes by discovering enriched pathways and biological processes. Future research into the specific mechanisms behind reproductive diseases in the context of hyperthyroidism, diabetes, and RIF can be guided by the results of this analysis.

For identifying articles related to the three disorders, machine learning techniques were used, more especially the FFNN model with dropout and L2 regularization [51]. The model's ability to distinguish between the circumstances showed excellent accuracy [52]. The model appears to be successful in detecting diabetes and RIF, but there is potential for improvement in recognizing instances of hyperthyroidism, according to its high accuracy and recognized scores. The performance of the FFNN model overall indicates its potential application in medical diagnostics [53, 54].

The complex network of genetic linkages that underlie hyperthyroidism, diabetes, and RIF and their potential effects on reproductive health are clarified by this study. The discovery of common genes and pathways creates the basis for more study. New treatment strategies and diagnostic instruments may become available as a result of a better understanding of the molecular processes behind these disorders. This work is an important step toward understanding the genetic intricacies of metabolic disorders and how they relate to reproductive health. The research provides a comprehensive understanding of these disorders by combining several analytical perspectives and sets the path for more research that may ultimately enhance the diagnosis and care of those who experience problems with their reproductive health.

5 Conclusions

In conclusion, our study investigated the complicated relationships between reproductive health and metabolic disorders, with an emphasis on hyperthyroidism, type 2 diabetes, and recurrent implantation failure (RIF). We discovered major genes and pathways linked to reproductive health using co-expression analysis, offering new insights into the molecular reasons for infertility. Our study began with a careful selection of datasets from the NCBI-GEO database, with strict experiment type and sample criteria. Following that, we discovered 587 DEGs for hyperthyroidism, 1734 for type 2 diabetes mellitus and 5661 for RIF, offering critical genetic insights. Thereafter, Gene Disease Network Analysis (GDNA) highlighted significant gene-disease connections, with ABCC6 and SLC2A14 identified as probable disease-related genes in hyperthyroidism, RIF, and type 2 diabetes. The Gene Ontology analysis performed by Enrichr revealed the roles of DEGs in enriched pathways and biological processes. In addition, a neural network model with dropout and L2 regularization for text classification performed admirably. The model has a 96.43% overall accuracy, flawless precision and recall for diabetes, and good performance in recognising RIF cases. Despite a somewhat lower precision for hyperthyroidism, the model's overall performance, as evidenced by a macro-average F1-score of 0.92, places it as a viable medical diagnostic tool. The model displayed good learning during training, with reduced loss and rising accuracy. Importantly, it demonstrated robust generalisation to previously encountered data, with a validation accuracy of 96.4%. Overall, our study effortlessly merged bioinformatics analysis and machine learning, giving useful insights into the genetic aspects of reproductive health and metabolic illnesses while demonstrating a sophisticated neural network model for high-accuracy text classification. Further experimental investigation could specify the roles of the genes elucidated in the study and assess their potential as therapeutic targets. In-depth exploration of the identified genes could provide their specific roles in reproductive processes, providing a solid basis for the development of diagnostic intervention. Moreover, the developed neural network model can be scaled for larger population and used for prediction purposes.

Ethics Approval and Consent to Participate Not applicable.

Acknowledgements We would like to acknowledge Dr. Ashok K. Chauhan, Founder, and President, Amity University, Uttar Pradesh for providing us the opportunity to conduct research. We would also like to thank Centre for Computational Biology and Bioinformatics, Amity Institute of Biotechnology, Amity University for providing us with necessary resources.

Competing Interests No potential conflict of interest was reported by the authors.

Funding Information This research did not receive any specific grant from funding agencies in the public, commercial, or not-for-profit sectors.

Author Contributions Conceptualization: Payal Gupta and Priyanka Narad.

Data Curation: Garima Bhardwaj, Shivani Dubey and Tarun Tayal.

Writing—Original Draft Preparation: Payal Gupta.

Revision and critical appraisal: Priyanka Narad and Abhishek Sengupta.

References

1. Mascarenhas, M.N., Flaxman, S.R., Boerma, T., Vanderpoel, S., Stevens, G.A.: National, regional, and global trends in infertility prevalence since 1990: a systematic analysis of 277 health surveys. PLoS Med. **9**(12) (2012). https://doi.org/10.1371/JOURNAL.PMED.1001356
2. Inhorn, M.C., Patrizio, P.: Infertility around the globe: new thinking on gender, reproductive technologies and global movements in the 21st century. Hum. Reprod. UpdateReprod. Update **21**(4), 411–426 (2015). https://doi.org/10.1093/HUMUPD/DMV016
3. JCDR—Female infertility, Invitro Fertilisation, Assisted Reproductive Technology, Clomiphene Citrate, Intracytoplasmic Sperm Injection, Faith healing, Ayurveda, Pumsavana karma. (n.d.). https://www.jcdr.net/article_fulltext.asp?id=2486. Accessed 30 Sep 2023
4. Yatsenko, S.A., Rajkovic, A.: Genetics of human female infertility. Biol. Reprod.Reprod. **101**(3), 549 (2019). https://doi.org/10.1093/BIOLRE/IOZ084
5. Hossain, M.A., Al Amin, M., Hasan, M.I., Sohel, M., Ahammed, M.A., Mahmud, S.M.H., Rahman, M.R., Rahman, M.H.: Bioinformatics and system biology approaches to identify molecular pathogenesis of polycystic ovarian syndrome, type 2 diabetes, obesity, and cardiovascular disease that are linked to the progression of female infertility. Inform. Med. Unlocked **30**, 100960 (2022). https://doi.org/10.1016/J.IMU.2022.100960
6. Mintziori, G., Kita, M., Duntas, L., Goulis, D.G.: Consequences of hyperthyroidism in male and female fertility: pathophysiology and current management. J. Endocrinol. Invest.Endocrinol. Invest. **39**(8), 849–853 (2016). https://doi.org/10.1007/S40618-016-0452-6
7. Condorelli, R.A., la Vignera, S., Mongioì, L.M., Alamo, A., Calogero, A.E.: Diabetes mellitus and infertility: different pathophysiological effects in Type 1 and Type 2 on sperm function. Front. Endocrinol.Endocrinol. **9**(May), 25 (2018). https://doi.org/10.3389/FENDO.2018.00268
8. Rinehart, J.: Recurrent implantation failure: definition. J. Assist. Reprod. Genet.Reprod. Genet. **24**(7), 284 (2007). https://doi.org/10.1007/S10815-007-9147-4
9. Cha, J., Sun, X., Dey, S.K.: Mechanisms of implantation: strategies for successful pregnancy. Nat. Med. **18**(12), 1754 (2012). https://doi.org/10.1038/NM.3012
10. Ma, J., Gao, W., Li, D.: Recurrent implantation failure: a comprehensive summary from etiology to treatment. Front. Endocrinol. **13** (2022). https://doi.org/10.3389/FENDO.2022.1061766
11. Wang, C.: The relationship between Type 2 diabetes mellitus and related thyroid diseases. J. Diabetes Res. (2013). https://doi.org/10.1155/2013/390534
12. Olex, A.L., Turkett, W.H., Fetrow, J.S., Loeser, R.F.: Integration of gene expression data with network-based analysis to identify signaling and metabolic pathways regulated during the development of osteoarthritis. Gene **542**(1), 38 (2014). https://doi.org/10.1016/J.GENE.2014.03.022
13. Topol, E.J.: High-performance medicine: the convergence of human and artificial intelligence. Nat. Med. **25**(1), 44–56 (2019). https://doi.org/10.1038/s41591-018-0300-7
14. Gentleman, R.C., Carey, V.J., Bates, D.M., Bolstad, B., Dettling, M., Dudoit, S., Ellis, B., Gautier, L., Ge, Y., Gentry, J., Hornik, K., Hothorn, T., Huber, W., Iacus, S., Irizarry, R., Leisch, F., Li, C., Maechler, M., Rossini, A. J., Zhang, J., et al.: Bioconductor: open software development for computational biology and bioinformatics. Genome Biol. **5**(10), 1–16 (2004). https://doi.org/10.1186/GB 2004 5 10-R80
15. Ritchie, M.E., Phipson, B., Wu, D., Hu, Y., Law, C.W., Shi, W., Smyth, G.K.: Limma powers differential expression analyses for RNA-sequencing and microarray studies. Nucleic Acids Res. **43**(7), e47–e47 (2015). https://doi.org/10.1093/NAR/GKV007
16. Srivastava, N., Hinton, G., Krizhevsky, A., Salakhutdinov, R.: Dropout: a simple way to prevent neural networks from overfitting. J. Mach. Learn. Res. **15**, 1929–1958 (2014)
17. (PDF) Natural Language Processing (Almost) from Scratch. (n.d.). https://www.researchgate.net/publication/266201822_Natural_Language_Processing_Almost_from_Scratch. Accessed 30 Sep 2023
18. Home—GEO—NCBI (n.d.). https://www.ncbi.nlm.nih.gov/geo/. Accessed 30 Sep 2023

19. Major, E., Győry, F., Horváth, D., Keller, I., Tamás, I., Uray, K., Fülöp, P., Lontay, B.: Smoothelin-like protein 1 regulates development and metabolic transformation of skeletal muscle in hyperthyroidism. Front. Endocrinol.Endocrinol. **12**, 751488 (2021). https://doi.org/10.3389/fendo.2021.751488
20. Karolina, D.S., Armugam, A., Tavintharan, S., Wong, M.T.K., Lim, S.C., Sum, C.F., Jeyaseelan, K.: MicroRNA 144 impairs insulin signaling by inhibiting the expression of insulin receptor substrate 1 in type 2 diabetes mellitus. PloS One **6**(8) (2011). https://doi.org/10.1371/JOURNAL.PONE.0022839
21. Altmäe, S., Martínez-Conejero, J.A., Salumets, A., Simón, C., Horcajadas, J.A., Stavreus-Evers, A.: Endometrial gene expression analysis at the time of embryo implantation in women with unexplained infertility. Mol. Hum. Reprod.Reprod. **16**(3), 178–187 (2010). https://doi.org/10.1093/MOLEHR/GAP102
22. Draw Venn Diagram. (n.d.). https://bioinformatics.psb.ugent.be/webtools/Venn/. Accessed 30 Sep 2023
23. Shannon, P., Markiel, A., Ozier, O., Baliga, N.S., Wang, J.T., Ramage, D., Amin, N., Schwikowski, B., Ideker, T.: Cytoscape: a software environment for integrated models of biomolecular interaction networks. Genome Res. **13**(11), 2498–2504 (2003). https://doi.org/10.1101/GR.1239303
24. Otasek, D., Morris, J.H., Bouças, J., Pico, A.R., Demchak, B.: Cytoscape automation: empowering workflow-based network analysis. Genome Biol. **20**(1) (2019). https://doi.org/10.1186/S13059-019-1758-4
25. Barabási, A.L., Gulbahce, N., Loscalzo, J.: Network medicine: a network-based approach to human disease. Nat. Rev. Genet. **12**(1), 56–68 (2010). https://doi.org/10.1038/nrg2918
26. Cytoscape App Store—stringApp. (n.d.). https://apps.cytoscape.org/apps/stringapp. Accessed 30 Sep 2023
27. Szklarczyk, D., Gable, A.L., Lyon, D., Junge, A., Wyder, S., Huerta-Cepas, J., Simonovic, M., Doncheva, N.T., Morris, J.H., Bork, P., Jensen, L.J., von Mering, C.: STRING v11: protein-protein association networks with increased coverage, supporting functional discovery in genome-wide experimental datasets. Nucleic Acids Res. **47**(D1), D607–D613 (2019). https://doi.org/10.1093/NAR/GKY1131
28. Ashburner, M., Ball, C.A., Blake, J.A., Botstein, D., Butler, H., Cherry, J.M., Davis, A.P., Dolinski, K., Dwight, S.S., Eppig, J.T., Harris, M.A., Hill, D.P., Issel-Tarver, L., Kasarskis, A., Lewis, S., Matese, J.C., Richardson, J.E., Ringwald, M., Rubin, G.M., Sherlock, G.: Gene ontology: tool for the unification of biology. Nat. Genet. **25**(1), 25–29 (2000). https://doi.org/10.1038/75556
29. Kuleshov, M.V., Jones, M.R., Rouillard, A.D., Fernandez, N.F., Duan, Q., Wang, Z., Koplev, S., Jenkins, S.L., Jagodnik, K.M., Lachmann, A., McDermott, M.G., Monteiro, C.D., Gundersen, G.W., Maayan, A.: Enrichr: a comprehensive gene set enrichment analysis web server 2016 update. Nucl. Acids Res. **44**(W1), W90–W97 (2016). https://doi.org/10.1093/NAR/GKW377
30. Feed Forward Neural Network Definition | DeepAI. (n.d.). https://deepai.org/machine-learning-glossary-and-terms/feed-forward-neural-network. Accessed 30 Sep 2023
31. Understanding Feed Forward Neural Networks in Deep Learning. (n.d.). https://www.turing.com/kb/mathematical-formulation-of-feed-forward-neural-network. Accessed 30 Sep 2023
32. Classical regularization methods. Appl. Math. Sci. (Switzerland) **160**, 7–48 (2005). https://doi.org/10.1007/0-387-27132-5_2/COVER
33. Sebastiani, F.: Machine learning in automated text categorization. ACM Comput. Surv.Comput. Surv. **34**(1), 1–47 (2001). https://doi.org/10.1145/505282.505283
34. Chawla, N.V., Bowyer, K.W., Hall, L.O., Kegelmeyer, W.P.: SMOTE: synthetic minority over-sampling technique. J. Artif. Intell. Res. **16**, 321–357 (2002). https://doi.org/10.1613/JAIR.953
35. Deep Learning with Python, Second Edition—Francois Chollet—Google Books (n.d.). https://books.google.co.in/books?hl=en&lr=&id=mjVKEAAAQBAJ&oi=fnd&pg=PR9&dq=Chollet+F.+Keras.+2015.&ots=AfgZwG_AVj&sig=bVk3hUPInDzlaY6wG_kC_--WN8s#v=onepage&q&f=false. Accessed 30 Sep 2023

36. Kingma, D.P., Ba, J.L.: Adam: a method for stochastic optimization. In: 3rd International Conference on Learning Representations, ICLR 2015—Conference Track Proceedings (2014). https://arxiv.org/abs/1412.6980v9
37. (PDF) Early Stopping—But When? (n.d.). https://www.researchgate.net/publication/2874749_Early_Stopping_-_But_When. Accessed 30 Sep 2023
38. [1811.12808] Model Evaluation, Model Selection, and Algorithm Selection in Machine Learning. (n.d.). https://ar5iv.labs.arxiv.org/html/1811.12808. Accessed 30 Sep 2023
39. Deep Learning. (n.d.). https://www.deeplearningbook.org/. Accessed 30 Sep 2023
40. Zhang, J.: Optimization Theory and Machine Learning Practice: Mind the Gap (2016)
41. Wu, B., Xi, S.: Bioinformatics analysis of differentially expressed genes and pathways in the development of cervical cancer. BMC Cancer **21**(1), 1–15 (2021). https://doi.org/10.1186/S12885-021-08412-4/TABLES/3
42. Ranea, J.A.G., Perkins, J., Chagoyen, M., Díaz-Santiago, E., Pazos, F.: Network-based methods for approaching human pathologies from a phenotypic point of view. Genes **13**(6), 1081 (2022). https://doi.org/10.3390/GENES13061081
43. Ashley, E.A.: The precision medicine initiative: a new national effort. JAMA **313**(21), 2119–2120 (2015). https://doi.org/10.1001/JAMA.2015.3595
44. Long, T., Liu, Z., Zhou, X., Yu, S., Tian, H., Bao, Y.: Identification of differentially expressed genes and enriched pathways in lung cancer using bioinformatics analysis. Mol. Med. Rep. **19**(3), 2029 (2019). https://doi.org/10.3892/MMR.2019.9878
45. Wu, R., Wu, C., Zhu, B., Li, J., Zhao, W.: Screening and validation of potential markers associated with uterine corpus endometrial carcinoma and polycystic ovary syndrome based on bioinformatics methods. Front. Mol. Biosci.Biosci. **10**, 1192313 (2023). https://doi.org/10.3389/FMOLB.2023.1192313/BIBTEX
46. Serin, E.A.R., Nijveen, H., Hilhorst, H.W.M., Ligterink, W.: Learning from co-expression networks: Possibilities and challenges. Front. Plant Sci. **7**(APR2016), 185898 (2016). https://doi.org/10.3389/FPLS.2016.00444/BIBTEX
47. Deng, L., Yang, H., Cao, J., Xu, B., Shen, F., Cai, W., Gan, X., Feng, J., Chen, Z., Guo, M., Wei, F.: Prediction of genetic factors of hyperthyroidism based on gene interaction network. Gene interaction network. Front. Cell Dev. Biol. **9**, 700355 (2021). https://doi.org/10.3389/fcell.2021.700355
48. Váradi, A., Szabó, Z., Pomozi, V., de Boussac, H., Fülöp, K., Arányi, T.: ABCC6 as a target in Pseudoxanthoma Elasticum (n.d.). www.ncbi.nlm.nih.gov/lovd/home.php?select_db=ABCC6. Accessed 30 Sep 2023
49. Shen, L., Zhang, Y., Chen, W., Yin, X.: The application of artificial intelligence in predicting embryo transfer outcome of recurrent implantation failure. Front. Physiol. **13**, 885661 (2022). https://doi.org/10.3389/FPHYS.2022.885661/BIBTEX
50. Jha, R., Bhattacharjee, V., Mustafi, A.: Increasing the prediction accuracy for thyroid disease: a step towards better health for society. Wireless Pers. Commun.Commun. **122**(2), 1921–1938 (2022). https://doi.org/10.1007/S11277-021-08974-3/FIGURES/7
51. Bukhari, M. M., Alkhamees, B. F., Hussain, S., Gumaei, A., Assiri, A., & Ullah, S. S. An Improved Artificial Neural Network Model for Effective Diabetes Prediction. Complexity, 2021. (2021). https://doi.org/10.1155/2021/5525271
52. Chang, C.M., Lin, T.K., Chang, C.W.: Applications of neural network models for structural health monitoring based on derived modal properties. Measurement **129**, 457–470 (2018). https://doi.org/10.1016/J.MEASUREMENT.2018.07.051
53. Reza, M.S., Harun-Or-roshid, M., Islam, M.A., Hossen, M.A., Hossain, M.T., Feng, S., Xi, W., Mollah, M.N.H., Wei, Y.: Bioinformatics screening of potential biomarkers from mRNA expression profiles to discover drug targets and agents for cervical cancer. Int. J. Mol. Sci. **23**(7), 3968 (2022). https://doi.org/10.3390/IJMS23073968/S1
54. Narad, P., Kumar, A., Chakraborty, A., Patni, P., Sengupta, A., Wadhwa, G., Upadhyaya, K.C.: Transcription factor information system (TFIS): a tool for detection of transcription factor binding sites. Interdiscip. Sci.: Comput. Life Sci. **9**, 378–391 (2017). https://doi.org/10.1007/s12539-016-0168-5

Classification Analysis of Motor Imagery EEG Signals Using Bidirectional LSTM Model

R. Helen, T. Thenmozhi, S. Mythili, N. Raghavan, and S. Griharan

Abstract In the field of Brain–Computer Interface (BCI), identification of Motor Imagery (MI) EEG signals is a challenging task. People with neuromuscular impairments can use the MI-BCI system to perform their day-to-day activities on their own. In this work, MI-EEG signals are obtained from BCI Dataset Iva, and spatial features (CSP) are helpful to identify the distinct characteristics of MI signal. The motor imagery actions are further classified by Bidirectional LSTM (BiLSTM) Neural Network with maximum mean accuracy of 98.36%. An enhanced LSTM model called the BiLSTM's ability to handle the complex temporal dynamics of EEG signals showcases the effectiveness of this robust approach. This BiLSTM with spatial features is compared with Random Forest, XGBoost, and LSTM models and outperforms by 4.7%. This approach mainly focuses on minimizing the validation loss and contributes to a reduced misclassification rate. This research highlights the

R. Helen
Department of Biomedical Engineering, Saveetha Engineering College, Chennai, India
e-mail: helenraj2804@gmail.com; helenr@saveetha.ac.in

T. Thenmozhi (✉)
Department of Artificial Intelligence and Data Science, Velammal College of Engineering and Technology, Madurai, India
e-mail: tth@vcet.ac.in

S. Mythili
Department of Biomedical Engineering, PSNA College of Engineering and Technology, Dindigul, India
e-mail: smbme@psnacet.edu.in

N. Raghavan · S. Griharan
Department of Electrical and Electronics Engineering, Thiagarajar College of Engineering, Madurai, India
e-mail: nraghavannarayanasamy@gmail.com

S. Griharan
e-mail: griharan2001@gmail.com

G. Raj et al. (eds.), *Advances in Artificial Intelligence and Machine Learning*, Lecture Notes in Electrical Engineering 1264, https://doi.org/10.1007/978-981-97-9507-9_12

potential of deep learning algorithms by means of improved accuracy and reliability of BCI systems. These implications made a significant impact on the advancement of future BCI systems and could potentially lead to the development of more sophisticated assistive devices for motor impaired patients.

Keywords BiLSTM · EEG · Deep learning · Motor imagery · BCI

Abbreviations

ANN	Artificial Neural Network
BCI	Brain–Computer Interface
BiLSTM	Bidirectional LSTM
CCS	Correlation-Based Channel Selection
CNN	Convolutional Neural Networks
CSP	Common Spatial Pattern
CWT	Continuous Wavelet Transform
DNN	Deep Neural Network
EEG	ElectroEncephalogram
HT	Hilbert Transform
ICA	Independent Component Analysis
LSTM	Long Short-Term Memory
RCSP	Regularized Common Spatial Pattern
XGBoost	Extreme Gradient Boosting

1 Introduction

In recent times, BCI has gained popularity due to its direct channel of communication between the human brain and an application interface. Electroencephalography (EEG) is a commonly used technique for recording brain activity and has been used in BCI to detect and classify various brain states, including motor imagery signals [1]. The UN Statistical Commission estimates that 2.21% of Indians have been affected by arm/leg amputations [2]. Prosthetic devices are operated under human thoughts to improve the quality of life of these impaired persons [3]. Motor Imagery is perfectly helpful for paralyzed persons. It involves mental stimulation of movement without using physical movement. Nowadays, the non-invasive EEG-based BCIs serve for the persons with mobility impairments to operate prosthetic devices by utilizing their brain signals [4]. The most valuable EEG-based BCI application is to classify different types of hand movements, such as grasping, pinching, and pointing.

EEG, a non-stationary time series, contains complex and noisy signals, making it difficult to process, and analyzed using conventional signal processing techniques.

Hence, Motor imagery EEG features are extracted by statistical, frequency domain and spatial domains [5]. In order to decrease the dimensional size and improve the efficiency, feature selection methods are introduced [6, 7]. Additionally, optimal EEG channels are selected through traditional methods such as filter, embedded, and wrapper methods [8]. Finally, these motor signals are discriminated by machine learning models such as SVM, LDA, and KNN [9].

Initially, the MI signal is differentiated by four kinds of regularized Common Spatial Pattern (RCSP) methods with LDA classifier and validated by 17 subjects from BCI competition III and IV datasets [10]. But complex non-linear EEG signals were incompatible with LDA classifier. Further, investigation of MI classification is continued with Bayesian Spatio-Spectral Filter Optimization (BSSFO) by competing six different CSPs, and outperforms with mean accuracy of 75.46% for BCI dataset IVa [11]. The research is extended by dividing MI signal into nine frequency subbands and extracting RCSP features correspondingly. PCA in conjunction with the SVM classifier reduces this feature vector and separates the MI features. This technique is examined by BCI Dataset IVa and achieves an accuracy of 82.69% [12]. The Misclassification rate and calibration time of MI signal are reduced by Symmetric Positive Definite matrix (SPD). In conjunction with SPD, motor organs are classified by Riemann minimum distance to mean approach. Hence, the effectiveness of this method is improved by 86.13 and 92.22% accuracy from BCI Datasets IVa and IIIa [13]. Later, the MI movements are identified by Event Related Patterns (ERP), and Hilbert Transform (HT) is computed for these alpha and beta patterns. These MI movements are detected by two classifiers and validated by two BCI databases. SVM classifier discriminates 86.11 and 75.56% by logistic regression classifier [14]. But, the dynamic nature of EEG signal cannot be handled by SVM.

Selecting appropriate weights in Artificial Neural Networks (ANN) is a Challenging Task [15].

However, deep learning models [16–18] improve the classification of MI signals without extracting, and selecting features. Though conventional methods like Deep Neural Networks (DNN) [19] greatly help in detecting motor movements, increased dimensional size and computational complexity affect the performance rate. Since long short-term memory (LSTM) neural network captures the temporal dependencies of the unprocessed EEG signals while preserving some of the spatial information, BiLSTM reduces the time-consumption while building the model, and it also reduces the dependence on professional expertise. The unique information processing mechanism of this neural network characterizes spatio-temporal dynamics in time series signal. Deep representation of this neural network can be fused and transformed by fully connected layer. Hence, in this research, a novel BiLSTM model with spatial features is implemented for decoding MI movements.

Recently, decoding of MI from EEG signal through unsupervised autoencoder neural network with SVM classifier achieves 95.33% for dataset IIIa [20]. A transformer encoder-based neural network with an attention mechanism is proposed for identification of MI movements [21]. Nowadays, Artificial Intelligence techniques have been developed in various fields [22, 23].

2 Materials and Methods

The BCI system comprises the following stages for classifying two MI tasks which are implemented using deep learning models as depicted in Fig. 1.

2.1 Signal Acquisition

The MI EEG signal is obtained from BCI-IVa dataset [24]. In this dataset, five male individuals were chosen to perform motor imagery movements. EEG data from 118 channels were recorded for 3.5 s to distinguish between the foot and right hand. The period of relaxation is about 0.5 s for every trial. Each subject has undergone 280 trials with a sampling rate of 100 Hz. Figure 2 depicts the time slot diagram for each trial. Table 1 demonstrates the division of trials conducted on five subjects.

2.2 Data Preprocessing

In BCI architecture, data preprocessing plays a vital role for removing unwanted frequencies, rejecting bad channels and eliminating the artifacts and noise from EEG signal. Hence, Alpha (8–13 Hz) and Beta (14–30 Hz) frequencies are extracted

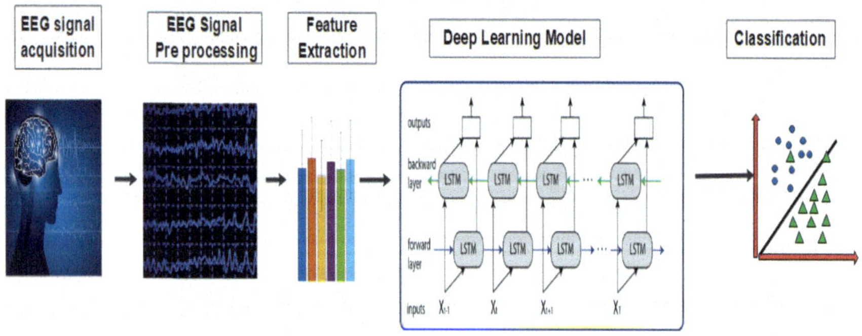

Fig. 1 Structure of BCI using deep learning model

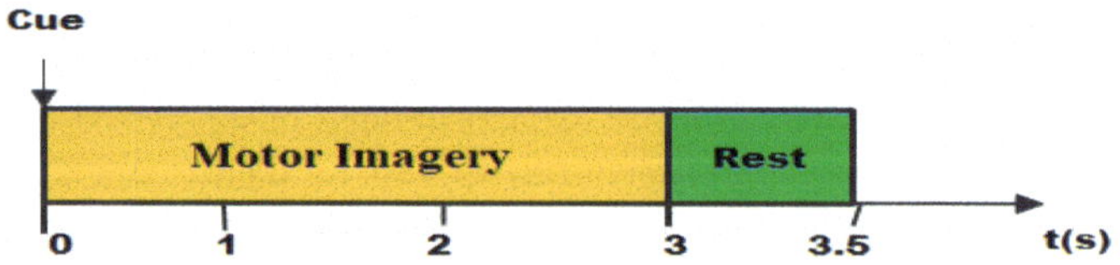

Fig. 2 Time slot diagram of dataset IVa

Table 1 Number of trials conducted on five subjects in BCI dataset IVa

Subjects	Training data	Testing data
1(aa)	168	112
2(al)	224	56
3(av)	84	196
4(aw)	56	224
5(ay)	28	252

using a fifth order IIR Butterworth band-pass filter. This filtering mainly helps for perfect classification of motor imagery signals.

2.3 Feature Extraction

To enhance the efficacy of predicting motor imagery EEG signal, a unique, informative set of features needs to be preserved from the preprocessed signal. Hence, feature extraction extracts predominant features from a raw EEG signal to maintain consistent classification performance. This process helps to minimize the noise from the raw EEG data. As a result, a model can be constructed with minimum machine effort thereby boosting the model's speed. Here spatial EEG features are extracted which is explained below.

Common Spatial Pattern (CSP)

This feature is extensively applied in the field of BCI to extract spatial features. For a non-stationary EEG data, CSP performs as a spatial filter to distinguish between the left and right feet. This spatial filter increases the variance of preprocessed EEG signals corresponding to one target class while decreasing the variance of the other target class. For every trial, preprocessed EEG signal 'F' is represented by T × U dimensional vector, where T and U denote the number of EEG channels and time samples recorded on individual channel. Then, the spatial covariance matrix for two target classes (left, right) is calculated by using two preprocessed EEG matrices G_l and G_r:

$$R_l = \frac{G_l G_l\prime}{trace(G_l G_l\prime)}; R_r = \frac{G_r G_r\prime}{trace(G_r G_r\prime)} \quad (1)$$

G_l' is the transpose of G_l, and the aggregation of the diagonal entries present in the matrix gives the trace of a matrix. The composite spatial covariance matrix R_s is computed by adding mean covariance matrices R_r and R_l which is given in Eq. (2):

$$R_s = R_l + R_r = E_i \lambda_s E_i' \quad (2)$$

From Eq. (2), R_s can also be described with the help of eigenvectors (E_i) and their diagonal matrix (λ_s). The whitening transformation matrix (M_s) is calculated using E_i and λ_s which is shown in Eq. (3):

$$M_s = \sqrt{\lambda_s{}^{-1}} E_i' \tag{3}$$

M_s transforms covariance matrix with H_l *and* H_r.

$$H_l = M_s R_l M_s' \tag{4}$$

$$H_r = M_s R_r M_s' \tag{5}$$

Both H_l *and* H_r have same eigenvectors and a unit matrix is formed by adding their corresponding eigenvalues which is described below:

$$H_l = V \lambda_{sl} V' \tag{6}$$

$$H_r = V \lambda_{sr} V' \tag{7}$$

$$\lambda_{sl} + \lambda_{sr} = I \tag{8}$$

The eigenvectors of H_l and H_r have smaller and larger eigenvalues and vice versa. The projection matrix M_m (spatial filter) is calculated as

$$M_m = M_s V' \tag{9}$$

After obtaining the spatial filter, EEG input data matrix B is projected over the spatial vector pattern. Hence, product of the spatial filter and input EEG signal produces CSP projected data. The column vectors of $Wt_m{}^{-1}$ are considered as spatial vectors. Finally, the logarithmic variance of the projected signal is utilized as a common spatial pattern feature which is described below:

$$Y_m = M_m B;\ B = M_m{}^{-1} Y_m \tag{10}$$

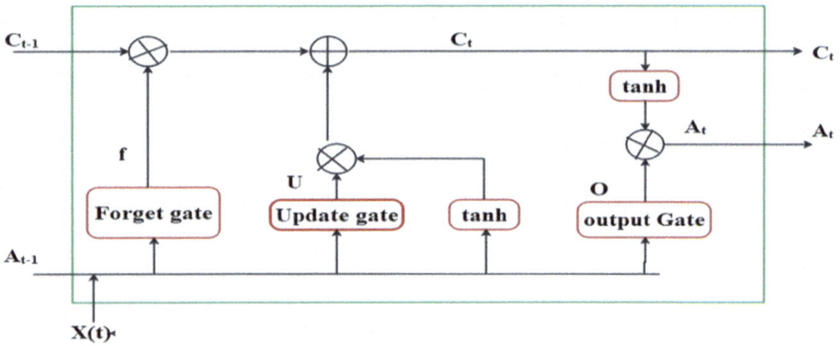

Fig. 3 Structure of LSTM model

2.4 *Classification by Deep Learning Model*

Model Architecture

LSTM models are the subdivision of RNN models, which is predominantly used in EEG signal recordings to analyze the temporal dynamics of the signal. The LSTM performs by utilizing input (X), forget (f) and output (O) gates; the input (X) gate controls the storage capacity of new cell state, the forget (f) gate eliminates the unwanted information from the cell, and the output (O) gate controls the amount of cell state linked to the upcoming layers present in the model. The LSTM can retain information about features for a long time and enable backpropagation over numerous limited non-linearities, which lessens the likelihood of the vanishing gradient. This is since they can preserve memories from earlier activations rather than completely replacing them, like a vanilla RNN. The performance of these models can vary depending on the particular task and dataset. Based on above merits, LSTM model which is demonstrated in Fig. 3 is perfectly suitable for detecting motor imagery data.

These gates utilize sigmoidal activation function to obtain values within 0 and 1, where "1" denotes an "open" condition, while "0" indicates that the gate is "closed" in which the model does not transmit any information. The weight matrix and bias are modified only once during the training process. The following expressions change the characteristics of a LSTM unit at time 't':

$$\lambda_f = \sigma(Wi_f.[b_{t-1}, x_t] + c_f \tag{11}$$

$$\tilde{d_t} = tanh(Wi_c.[b_{t-1}, x_t] + c_c \tag{12}$$

$$\lambda_u = \sigma(Wi_u.[b_{t-1}, x_t] + c_u \tag{13}$$

$$d_t = \lambda_u * \tilde{d_t} + \lambda_f * \tilde{d_{t-1}} \tag{14}$$

$$\lambda_o = \sigma(Wi_o.[b_{t-1}, x_t] + c_o \tag{15}$$

$$b_t = \lambda_o * tanh\ tanh\ d_t \tag{16}$$

$$\widehat{U_t} = softmax(b_t) \tag{17}$$

where d, σ, b represent memory cell, sigmoid activation function, activation value respectively. The biases c_f, c_u, c_o are included to the corresponding forget, update. and output gates. A candidate memory cell ($\tilde{d_t}$) with tanh activation function is used to decide and update the present memory cell with the input x_t. Finally, b_t begins the cycle of the next time step.

BiLSTM

In general, the LSTM model receives information from the past input observations, but fails to utilize the upcoming input observations. But, in the BiLSTM model, two LSTMs act to transfer the information both in the forward as well as in the backward direction and predicts the output value.

In this research, a BiLSTM model is developed to detect the motor imagery EEG signals which is depicted in Fig. 4. This EEG signal passes through the first LSTM unit of the hidden layer in the forward direction $\overrightarrow{G_t}$. It also continues to perform in the backward direction $\overrightarrow{G_t}$. These combined LSTMs linked with the output layer in turn predict the output classes. In this work, the dropout on the BiLSTM layers is used to regularize the proposed approach, and the dropout ratio is set to 0.2. The encoding layers have 128 LSTM units. The full connected layer is linked with the hidden layer through sigmoidal activation function. The classification accuracy of the movements of the motor organs is determined by the cross-entropy loss. In addition, Adam optimizer is employed in this research which runs faster and occupies less memory. In particular, the combined momentum approach and adaptive learning rate greatly help to converge quickly which in turn minimizes the loss.

3 Proposed Method

In this work, EEG signal is acquired from BCI dataset IVa to validate the performance of proposed BiLSTM model. In this dataset, five healthy subjects are involved to imagine the movements of sensory motor organs. These EEG signals are preprocessed by filtering alpha and beta frequencies using a Butterworth band-pass filter. The preprocessed signal is then passed into proposed BiLSTM model. The flowchart

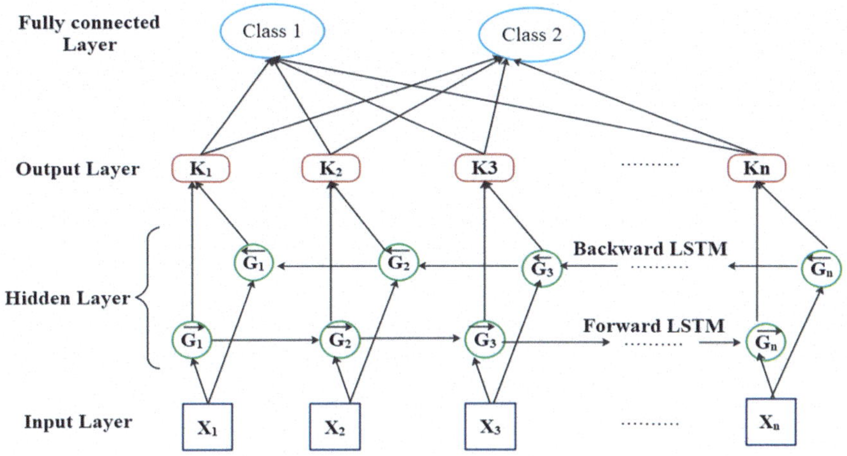

Fig. 4 Structure of BiLSTM model

of proposed approach is demonstrated in Fig. 5. In this model, a sequential bidirectional layer of 128 hidden LSTM units followed by a dense output layer with a softmax activation function is implemented. This method involves several steps, such as splitting data, data preprocessing, model selection, model training, and testing followed by evaluating the potency of the model.

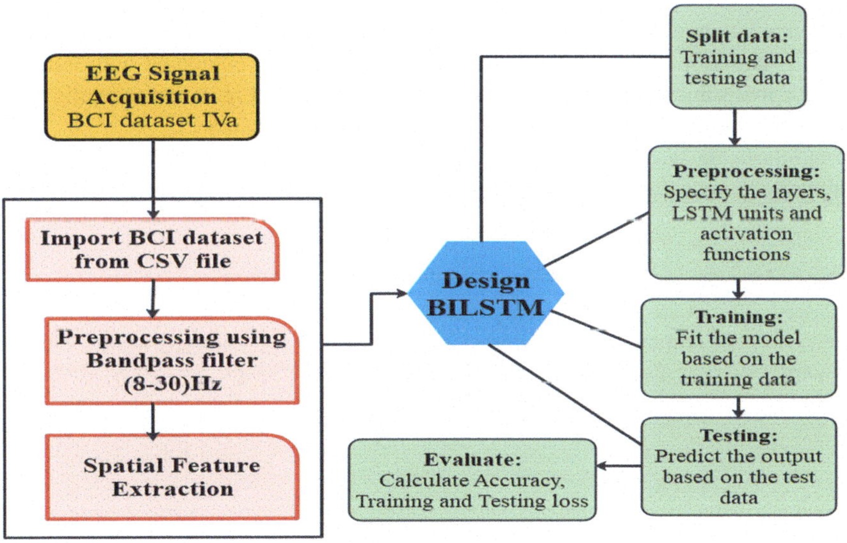

Fig. 5 Proposed BiLSTM model combined with spatial features

Table 2 Parameters of BiLSTM model

Layer	Output shape	Parameters
Bidirectional	(None,1,256)	252,928
Dense	(None,1,2)	514

To enhance the efficacy of the proposed model, common spatial features are extracted and the above procedural steps are executed. 280 motor imagery trials across 118 channels are fed into BiLSTM model (totally 253,442 parameters) which is indicated in Table 2.

The EEG samples are trained and tested with 100 epochs. The cross-entropy loss is minimized by Adam optimizer and testing accuracy is improved. Ten-fold cross-validation is implemented to justify the efficiency of the model.

4 Results and Discussion

This work is implemented in Python 3.7 version using TensorFlow V2.4.0. The model's effectiveness is assessed by classification accuracy. From the results, it is clearly shown that BiLSTM model with spatial features is effective when compared with BiLSTM model with raw EEG signal. This proposed (BiLSTM with spatial feature) model achieves highest accuracy of 98.36% which is 7% increase when compared with raw EEG signal as indicated in Table 3.

The spatial EEG features are classified through Random Forest, XGBoost, and LSTM models. The performance of BiLSTM model outperforms by 4.7% increase in accuracy than other machine learning models as indicated in Table 4.

The validation loss is computed for 100 epochs and is compared with training loss for five subjects in the given dataset. It is clear from Fig. 6 that the validation loss is minimum than the training loss for every individual subject.

Similarly, the testing accuracy is computed for 100 epochs and is compared with training accuracy for five subjects in the given dataset. It is clearly evident from

Table 3 Accuracy details of BiLSTM model with and without features

Subjects	Classification accuracy (%)	
	BiLSTM with raw EEG signal	BiLSTM with spatial features
aa	98.21	100
al	100	100
av	96.42	98.97
aw	85.71	96.42
ay	78.5	96.42
Mean	91.76	98.36

Table 4 Accuracy details of BiLSTM with spatial features and other machine learning models with spatial features

Subjects	Classification accuracy (%)			
	Random forest	XGBoost	LSTM	BiLSTM
aa	98.21	91.07	96.42	100
al	99.1	98.21	98.2	100
av	93.75	78.06	93.75	98.97
aw	95.32	93.75	95.53	96.42
ay	92.87	90.47	90.21	96.42
Mean	95.85	90.31	94.82	98.36

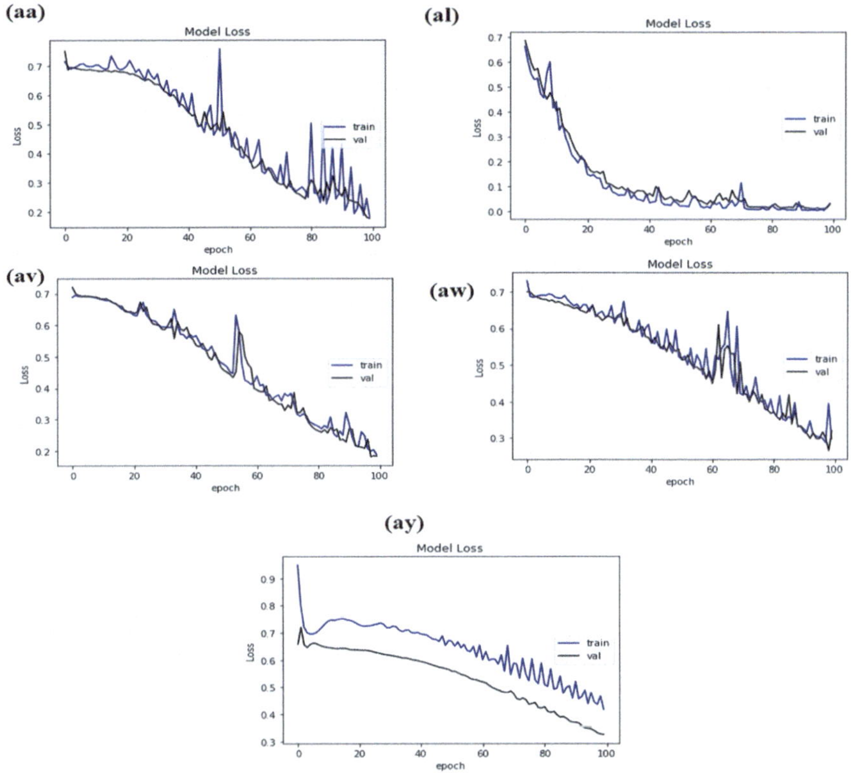

Fig. 6 Comparison of training and validation losses over five subjects

Fig. 7 that the testing accuracy achieves highest accuracy than the training accuracy for every individual subject.

The generic structure of this BiLSTM network has the ability to remember and process previous complex values for a long time. Hence, the in-built memory cells

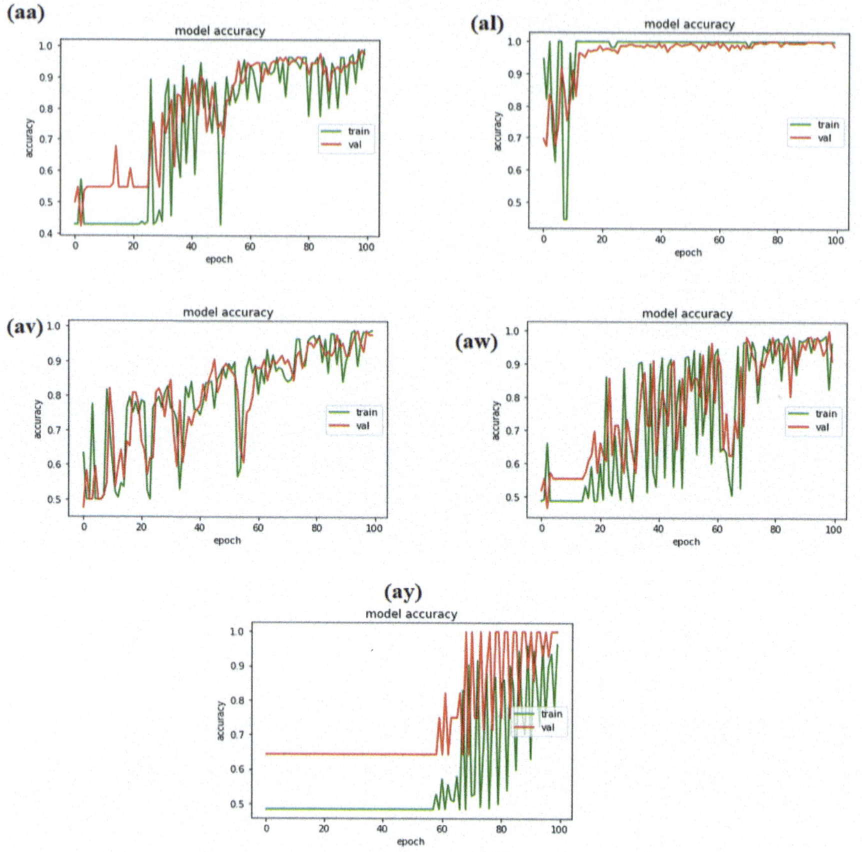

Fig. 7 Comparison of training and validation accuracy over five subjects

mainly help for preserving previous outcomes and eliminate redundant information for perfect MI movement discrimination. Therefore, in this dataset, the classification performance of BiLSTM with CSP features outperforms other existing literary methods which is illustrated in Table 5. Small Motor Imagery training dataset is utilized in this research work. Large number of training MI samples enhances the performance and reliability of BCI system. In addition to spatial feature, time–frequency and frequency domain MI features greatly help in perfect discrimination of MI signals. Decomposition of frequency bands using transform domain will reduce the noise and artifacts from EEG signal. Researchers are investigating on the integration of emerging technologies like AI-driven algorithms and neuromorphic computing for improved BCI systems.

Table 5 Details of proposed work with other literary methods on IVa dataset

Literary method	Year	Approach	Accuracy (%)					
			aa	al	av	aw	ay	Mean
Lu et al. [25]	2010	Regularized CSP	76.8	98.2	74.5	92.9	77.0	83.90
Zhang et al. [26]	2013	Z-score with LDA classifier	77.7	100	68.4	99.6	59.9	81.10
Park et al. [12]	2017	SBR	86.61	98.21	63.78	89.05	77.78	82.69
Ortiz-Echeverri et al. [27]	2019	ICA-CWT features with CNN	97.81	94.47	94.78	94.06	92.18	94.66
Jin et al. [28]	2019	CCS with CSP	80.7	96.8	70.4	92.9	92.1	86.6
Proposed method	2023	BiLSTM with CSP	100	100	98.97	96.42	96.42	98.36

5 Conclusion

This research work provides valuable insights into the performance of BiLSTM model in classifying hand movements using EEG data. BiLSTM outperforms the LSTM model by analyzing the signal in both forward and backward directions. BiLSTM with spatial features achieves maximum accuracy of 98.36% with minimum validation loss. The proposed method surpasses other machine learning and ensemble models. This accurate and effective method will be helpful for analyzing EEG data related to hand movements. The robotic arm could be developed using MI-EEG signals which could have significant implications for the identification and rehabilitation of neurological disorders. The potential expansion of BCI applications beyond motor imagery, such as cognitive assessment, mental health monitoring, and communication aid, can be discussed based on this robust approach.

References

1. Pfurtscheller, G., Neuper, C.: Motor imagery and direct brain-computer communication. Proc. IEEE **89**(7), 1123–1134 (2001)
2. United Nations Statistical Commission. Department of Economic and Social Affairs/Statistics Division. https://unstats.un.org/unsd/demographic-social/sconcerns/disability/statistics/#/countries
3. Wolpaw, J.R., Birbaumer, N., McFarland, D.J., Pfurtscheller, G., Vaughan, T.M.: Brain-computer interfaces for communication and control. Clin. Neurophysiol. **113**(6), 767–791 (2002)
4. Hwang, H.J., Kim, S., Choi, S., Im, C.H.: EEG-based brain-computer interfaces: a thorough literature survey. Int. J. Hum.-Comput. Interact. **29**(12), 814–826 (2013)
5. Herman, P., Prasad, G., McGinnity, T.M., Coyle, D.: Comparative analysis of spectral approaches to feature extraction for EEG-based motor imagery classification. IEEE Trans.

Neural Syst. Rehabil. Eng. **16**(4), 317–326(2008). https://doi.org/10.1109/TNSRE.2008.926694
6. Thenmozhi, T., Helen, R.: Feature selection using extreme gradient boosting Bayesian optimization to upgrade the classification performance of motor imagery signals for BCI. J. Neurosci. Methods **366**, 109425 (2022)
7. Salazar-Varas, R., Gutiérrez, D.: An optimized feature selection and classification method for using electroencephalographic coherence in brain–computer interfaces. Biomed. Signal Process. Control. pp. 11–18 (2015). https://doi.org/10.1016/j.bspc.2014.11.001
8. Arvaneh, M., Guan, C., Ang, K.K., Quek, C.: Optimizing the channel selection and classification accuracy in EEG-based BCI. IEEE Trans. Biomed. Eng. **58**(6), 1865–1873 (2011)
9. Lotte, F., Congedo, M., Lécuyer, A., Lamarche, F., Arnaldi, B.: A review of classification algorithms for EEG-based brain–computer interfaces. J. Neural Eng. **4**(2), R1 (2007)
10. Lotte, F., Guan, C.: Regularizing common spatial patterns to improve BCI designs: unified theory and new algorithms. IEEE Trans. Biomed. Eng. **58**(2), 355–362 (2010). https://doi.org/10.1109/TBME.2010.2082539
11. Suk, H.I., Lee, S.W.: A novel Bayesian framework for discriminative feature extraction in brain-computer interfaces. IEEE Trans. Pattern Anal. Mach. Intell. **35**(2), 286–299 (2012)
12. Park, S.H., Lee, S.G.: Small sample setting and frequency band selection problem solving using subband regularized common spatial pattern, IEEE Sens. J. **17**(10), 2977–2983 (2017)
13. Singh, A., Lal, S., Guesgen, H.W.: Reduce calibration time in motor imagery using spatially regularized symmetric positives-definite matrices based classification. Sensors **19**(2), 379 (2019)
14. Bagh, N., Reddy, M.R.: Hilbert transform-based event-related patterns for motor imagery brain computer interface. Biomed. Signal Process. Control **62**, 102020 (2020)
15. Mansoor, A., Usman, M.W., Jamil, N., Naeem, M.A.: Deep learning algorithm for brain-computer interface. Sci. Program. **2020**, 1–12 (2020)
16. Zhang, R., Zong, Q., Dou, L., Zhao, X., Tang, Y., Li, Z.: Hybrid deep neural network using transfer learning for EEG motor imagery decoding. Biomed. Signal Process. Control **63**, 102144 (2021)
17. Craik, A., He, Y., Contreras-Vidal, J.L.: Deep learning for electroencephalogram (EEG) classification tasks: a review. J. Neural Eng. **16**(3), 031001 (2019)
18. Wang, P., Jiang, A., Liu, X., Shang, J., Zhang, L.: LSTM-based EEG classification in motor imagery tasks. IEEE Trans. Neural Syst. Rehabil. Eng. **26**(11), 2086–2095 (2018)
19. Tiwari, S., Goel, S., Bhardwaj, A.: MIDNN-a classification approach for the EEG based motor imagery tasks using deep neural network. Appl. Intell. 1–20 (2021). https://doi.org/10.1007/s10489-021-02622-w
20. Phadikar, S., Sinha, N., Ghosh, R.: Unsupervised feature extraction with autoencoders for EEG based multiclass motor imagery BCI. Expert Syst. Appl. **213**, 118901 (2023)
21. Hameed, A., Fourati, R., Ammar, B., Ksibi, A., Alluhaidan, A.S., Ayed, M.B., Khleaf, H.K.: Temporal–spatial transformer based motor imagery classification for BCI using independent component analysis. Biomed. Signal Process. Control **87**, 105359 (2024)
22. Yadav, D., Sharma, A., Ahmad, S., Chandra, U.: Political sentiment analysis: case study of haryana using machine learning. In: Mobile Radio Communications and 5G Networks: Proceedings of Second MRCN 2021, pp. 479–499 (2022)
23. Khan, M.A., Ahmad, S.: An emergence of AI in data mining and KDD: ANN its strength & weakness. Int. J. Recent. Technol. Eng. (IJRTE). **9**(1), 380–383 (2020)
24. BCI competition datasets http://www.bbci.de/competition/iii
25. Lu, H., Eng, H.L., Guan, C., Plataniotis, K.N., Venetsanopoulos, A.N.: Regularized common spatial pattern with aggregation for EEG classification in small-sample setting. IEEE Trans. Biomed. Eng. **57**(12), 2936–2946 (2010)
26. Zhang, R., Xu, P., Guo, L., Zhang, Y., Li, P., Yao, D.: Z-score linear discriminant analysis for EEG based brain-computer interfaces. PLoS ONE **8**(9), e74433 (2013)
27. Ortiz-Echeverri, C.J., Salazar-Colores, S., Rodríguez-Reséndiz, J., Gómez-Loenzo, R.A.: A new approach for motor imagery classification based on sorted blind source separation, continuous wavelet transform, and convolutional neural network. Sens. **19**(20), 4541 (2019)

28. Jin, J., Miao, Y., Daly, I., Zuo, C., Hu, D., Cichocki, A.: Correlation-based channel selection and regularized feature optimization for MI-based BCI. Neural Netw. **118**, 262–270 (2019)

RideGuard: Empowering Riders with Intelligent Accident Detection and Aid

Sudhanshu Kumar, Himanshu Yadav, Ashutosh Kumar, and Mandeep Kaur

Abstract With the increase in use of two-wheeler vehicles, the traffic accidents have been reported at an alarming rate each year. Reasons such as not wearing a helmet while riding a two-wheeler vehicle such as bikes or riding the bike while being intoxicated with alcohol are worth mentioning as one of the primary causes of road accidents. Therefore, this paper is initiated with an aim to provide safety to riders. This work is a complete accident detection system that is based on a smart helmet that can identify whether or not the rider is wearing a helmet, determine whether or not the rider is intoxicated, monitor the rider's co-ordinates via real-time position tracking, and recognize unexpected violent jerks to determine whether or not an accident has occurred to notify the rider's relatives and friends via a smartphone application. The helmet comprises Arduino Nano microcontroller, IR sensor, MQ-3 Alcohol Sensor, ADXL-335 accelerometer, Wi-Fi, and a battery module.

Keywords Smart helmet · Internet of things (IoT) · SOS alerts · Alcohol detection android · Sensors · Arduino Nano

1 Introduction

Accidents on the road have become a big source of fear in our everyday lives. Many people are killed in road accidents as a result of the enormous population, with an official rate of 60 deaths per 10,000 motor vehicles. Many bike riders are expected to perish in road accidents every day, and because there is little information about the incidents, those riders are not liable to be saved as the event gets notified to the concerned authorities after the calamity had occurred. However, as the population expands, so does the number of motorcycles on the roads and highways. The most important and essential item that a rider must wear in order to avoid any potential road accidents is a helmet. It protects the motorcyclist's head from a deadly strike caused by an accident, provides bike riders with guaranteed protection, and ensures 100

S. Kumar · H. Yadav · A. Kumar · M. Kaur (✉)
Department of CSE, Sharda University, Greater Noida, India
e-mail: drmandeepkaur10@gmail.com

G. Raj et al. (eds.), *Advances in Artificial Intelligence and Machine Learning*, Lecture Notes in Electrical Engineering 1264, https://doi.org/10.1007/978-981-97-9507-9_13

percent user reliability. It has been discovered that the annual fatality rate from street accidents is high. If emergency medical workers had received mishap information and gotten on the scene sooner, several lives may have been spared. Moreover, half of all avenue visitors are killed by pedestrians, motorcyclists, and cyclists.

According to the data published by Ministry of Road Transport and Highways [1] in November 2022, fatality rate has maintained a consistent figure despite some reduction in road accidents and injuries over the years. Even after various laws and policies enforced by the government, accidents occur on a consistent figure basis. Table 1 depicts a vivid description of the figures (Fig. 1).

After going through the details of the figures of accidents, fatalities, and injuries, it is very essential to understand the involvement of mode of transport in accidents and fatality rate. Figure 2 visualizes the involvement of different means of transport that are involved in road accidents and result in fatality as a consequence [2].

Figure 3 concludes that helmet and trauma injury, and drink and drive are the major reasons behind the cause of fatalities for two-wheelers accidents. According to the WHO, proper helmet wearing can result in a 42% decrease in the risk of fatal injuries and a 69% reduction in the risk of brain injuries. Hence, analysis of these data concludes that there is a need to develop a technology that can contribute in deliberately cutting down accident cases and fatalities. So, the development of a must-wear helmet with smartness equipped by the use of IoT is the need of the hour.

Table 1 Accident, fatalities, and injury trends in India from 2016 to 2021 [1]

Year	Accidents	Fatalities	Persons injured
2016	4,80,652	1,50,785	4,94,624
2017	4,64,910	1,47,913	4,70,975
2018	4,67,044	1,51,417	4,69,418
2019	4,49,002	1,51,113	4,51,361
2020	3,66,138	1,31,714	3,48,279
2021	4,12,432	1,53,972	3,84,448

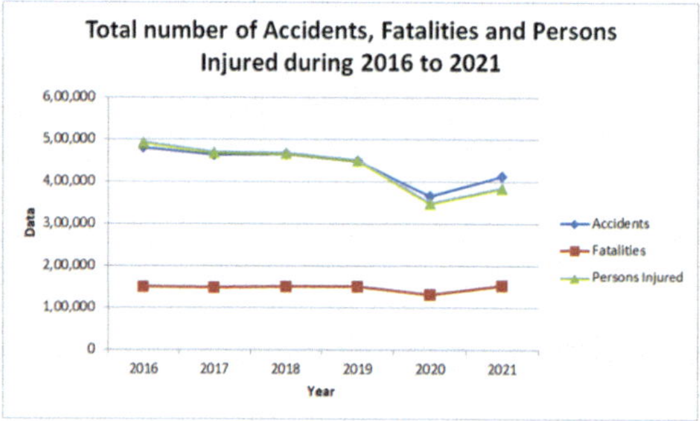

Fig. 1 Comparative analysis of accident trends

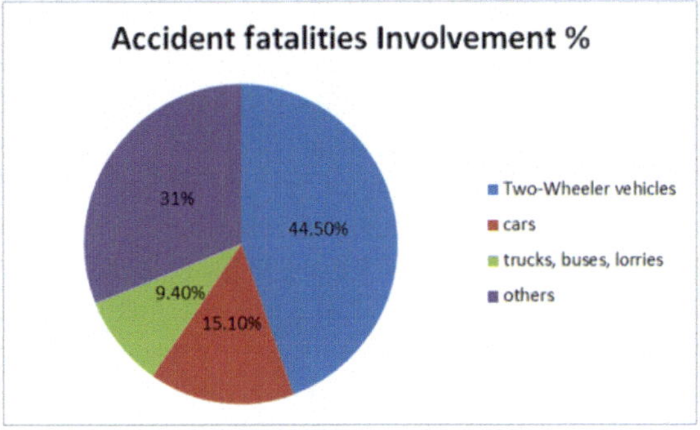

Fig. 2 Accident fatalities' involvement

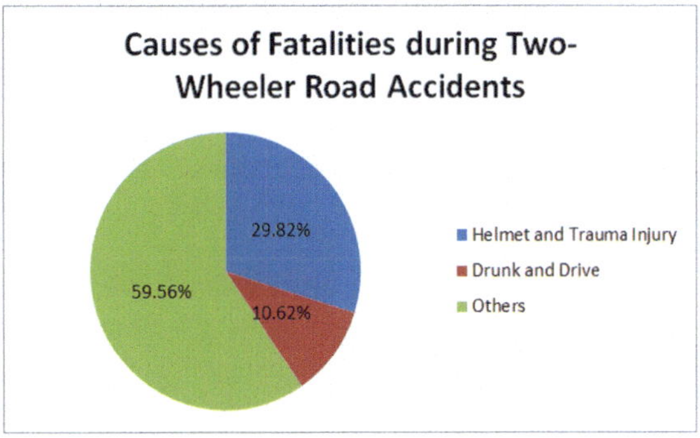

Fig. 3 Causes of fatalities

2 Literature Review

Today, the only work on safety helmets that is visible is that which is sold in the market. The DOT certification is used for helmet accreditation. Although these helmets do not give any electrical assistance, the odds of lives being saved are increased. In the recent past, there have been start-ups attempting to develop helmets that are a step ahead of the current generation helmets.

According to Kulkarni et al. [3], an ultrasonic sensor is used to inform the rider about the bumpers and potholes coming in way for the rider, which can bring awareness and help in avoiding accidents. In case of an accident, the smart helmet sends the co-ordinates of the rider's position using GSM and MEMS technologies.

According to Gupta et al. [4], a collision sensor is responsible for detecting an accident or if the driver is intoxicated with alcohol. ATMega32u4 with AI Thinker A7 GSM-GPS module is used in the helmet for GPS location monitoring, collision detection, and sending SOS help messages. The Raspberry Pi black-box will help in the investigation of the cause of the accident. The helmet also has an alcohol sensor that detects the intoxication level of the rider. The film of the course of accident will be saved for further access. It will contribute in making biking safer and convenient.

Varma et al. [5] proposed a system in which the touch sensor is used to determine whether the rider is wearing the helmet or not. The proposed system is also equipped with an alcohol sensor that detects the intoxication level of the rider. The system will not let the bike start if the rider is found to be intoxicated or not wearing the helmet. During the course of the accident, the state of the motorcycle is identified and notified. The GPS module present in the helmet will then get triggered, and the location co-ordinates will be sent to the authorized numbers and local hospitals. This feature of the helmet makes it a very safer and efficient tool for the riders in case of an emergency.

According to Bhattacharya et al. [6], the proposed helmet is a real-time monitoring system that detects the biker's acceleration and deceleration. The real-time monitoring system also signals the upcoming turns, traffic signs, and external environment updates so that driver can make perfect decisions and can avoid road accidents. The serial communication between the helmet unit and bike unit is carried out using RF receiver and transmitter.

According to Alim et al. [7], nRF24L01 RF module is employed in the smart helmet to process the signal communication between the helmet unit and the motorcycle unit. The modules are employed both on the transmitter as well as receiver side.

A study on detecting the cause of an accident is done by [8] developing a mechanism which consists of sensors and accelerometer that collects data and sends it to the API. The API measures the change in x, y, z values and calculates the probability of cause of accident. The accident detection system is built on SVM model which takes in data from the sensors as input and trains the model.

A study is carried out by [9] where visual proximity warnings are provided to the rider with the help of camera module as it detects nearby vehicles' driving behavior and alerts the rider in case of chances of collision. A camera module is deployed which takes in input images, and ML model predicts the chance of a collision which is conveyed to the user via Bluetooth module.

Sathish et al. [10] presented a smart bike system capable of detecting both bicycle and gasoline theft. It also shows the quantity of fuel in the tank and alerts you when it is low. It also detects alcohol through the helmet and does not allow the bike to start until the helmet is properly fitted. Despite these features, the model was not entirely reliant on the helmet because sensors such as GPS and GSM, which might help in monitoring and detecting the rider, were installed on the bike rather than the helmet. Furthermore, no provision was created for notification of accident discovery.

Eldho et al. [11] created a jacket cum helmet to increase wearer safety. By using Peltier's module, it enabled temperature modification of the helmet and jacket, allowing the user to avoid feeling stuffy and uncomfortable.

Sivagnam et al. [12] designed a helmet to promote comfort and ease to riders while riding. The helmet had a Bluetooth speaker controller, a speaker, and a microphone, allowing the rider to listen to music and communicate with the pillion rider without turning around. It also has an airflow chamber for the comfort of the rider and is powered by solar energy. However, the helmet does not detect accidents or track the rider's location, nor does it detect whether or not the rider is wearing the helmet.

While much work has been done in the literature [3–12], none of them has a compact system that can indicate whether or not the helmet is worn, track the rider's location, detect sudden severe jerks to identify that an accident has occurred, indicate whether or not the rider is drunk, and also alert the riders' relatives and friends via a mobile application. As a result, the goal of our effort here is to create such a comprehensive accident detection system.

3 Proposed Model

This section describes the design and implementation of a Smart Helmet that focuses on alcohol and accident detection. The project is built using Arduino Nano microcontroller. The fundamental block diagram of the Helmet unit is shown in Fig. 4.

Figure 4 describes the structural architecture of the project in which hardware equipment like sensors, accelerometer, camera module, and Wi-Fi module are used. A brief description of the used hardware components is stated below.

Arduino Nano Microcontroller: Arduino Nano is a compact and versatile microcontroller board based on the ATmega328P processor. It is a smaller version of the Arduino Uno, designed for projects with space constraints shown in Fig. 5.

Sharp Infrared Sensor: An IR (Infrared) sensor is a device that detects infrared radiation emitted or reflected by objects. When an object is in proximity to the sensor, it reflects or emits IR radiation, which is then detected by the receiver shown in Fig. 6.

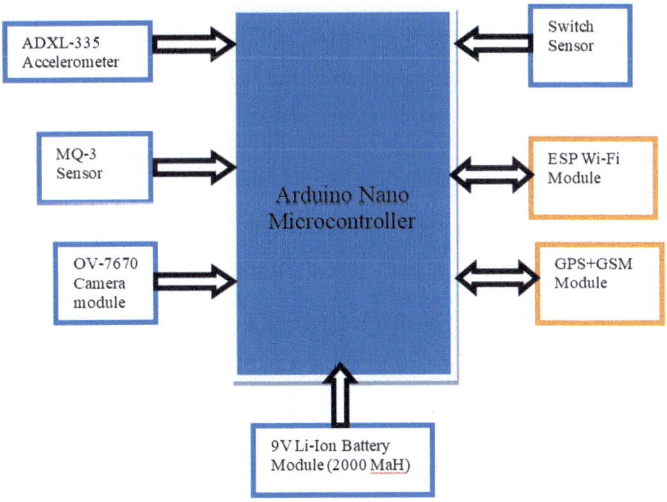

Fig. 4 Structural architecture of smart helmet

Fig. 5 Arduino nano [13]

Fig. 6 Infrared sensor [14]

MQ3 Sensor: The MQ-3 sensor is a type of gas sensor designed to detect alcohol vapor concentration in the air. The MQ-3 sensor module typically includes a sensitive element composed of SnO2 (tin dioxide) that reacts to alcohol gases shown in Fig. 7.

ADXL-335 Accelerometer: In the context of a smart helmet, an accelerometer plays a crucial role in detecting accidents or sudden impacts. The accelerometer can sense abrupt changes in velocity or deceleration, which are indicative of a collision or accident shown in Fig. 8.

ESP Wi-Fi module: A Wi-Fi module is a hardware component that enables devices to connect to a wireless local area network (WLAN) and communicate with other devices or access the Internet. It typically includes both the hardware and software necessary for wireless communication using the Wi-Fi protocol. Wi-Fi modules are commonly used in various electronic devices to provide wireless connectivity shown in Fig. 9.

OV7670 Camera Module: The OV7670 is a popular camera module designed by Omni Vision Technologies. It is commonly used in various electronic projects and applications due to its compact size, low power consumption, and reasonable image quality shown in Fig. 10.

Fig. 7 MQ3 gas sensor [15]

Fig. 8 ADXL-335 accelerometer [16]

Fig. 9 ESP Wi-Fi module [17]

Fig. 10 OV7670 camera module [18]

4 Methodology

The development of a groundbreaking new design represents a significant leap forward in enhancing the compactness of smart helmets, surpassing all previously proposed solutions [19, 20]. This innovative design provides separation between the microcontroller unit, battery backup, and rest of the components. This design is formulated because the microcontroller unit as well as battery module consumes a lot of space inside the helmet which degrades the user wearing experience [21]. Figure 11 shows the developed design of the helmet where a separate IoT box is incorporated for storing microcontroller and battery supply and doesn't interfere with the internal space of the helmet. This breakthrough in compactness sets a new standard for smart helmet designs, paving the way for safer and more accessible smart headwear across various industries and applications.

The Helmet provides the following functionalities: wear detection, alcohol level detection, and accident detection and observing the riding behavior of the rider, also recording the cause of the accident. The services offered to family members or the ones that have been subscribed to the helmet include message alerts and live location navigation to user in case of accident detection [22]. The IR sensor present in the helmet senses the proximity that the rider has worn the helmet or not. After the helmet worn is detected, the MQ-3 sensor detects the alcohol level by sensing the breath of the rider; if found drunk, then the family members are informed about the state of the rider with the exact location and navigation [23]. In case of sober state, the helmet behaves just like a safety gear. Accident is detected by ADXL-335 Accelerometer which measures the sudden change in orientation of the helmet; the quality of the

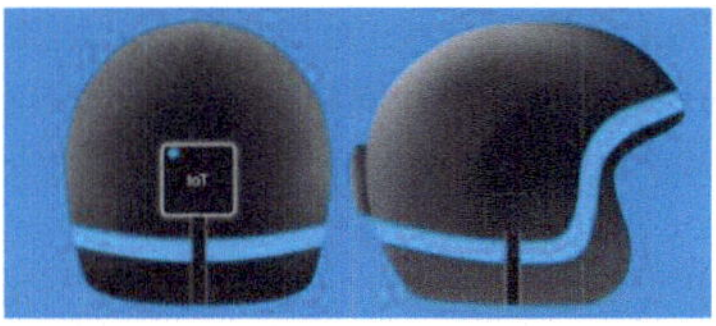

Fig. 11 Design of helmet

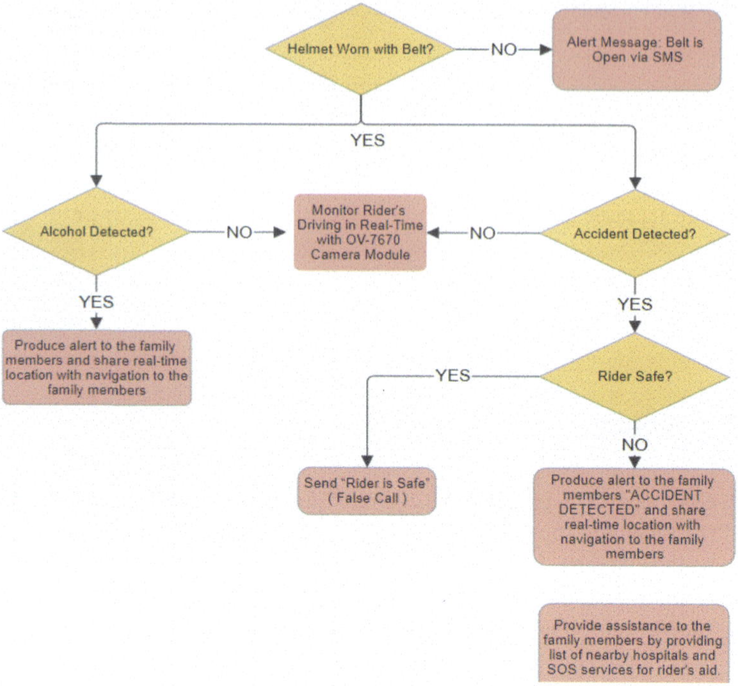

Fig. 12 Workflow of helmet module

sensor makes it capable of ignoring the small impacts that reduces the chance of false alerts. If the sensor detects accident then an SOS message will be triggered to the relatives after false call counter cancelation time limit of 60 s expires. With the SOS alert, the family members are also informed about the accident spot location, and navigation to the spot is provided. The OV-7670 camera module installed in the helmet records the cause of accident which can be further retrieved. The camera module also records the riding behavior of the rider so that constructive feedbacks can be provided to the rider on improving the driving skills of the rider to avoid accidents. Figure 12 provides the workflow chart of the smart helmet project.

5 Result Analysis

The smart helmet is developed and tested for various conditions to evaluate its operational efficiency. The first test was done for the compactness of the circuit. Figure 13 displays the circuit that has been fixed inside the helmet so that they do not contact the head of the driver. The microcontroller and battery backup are placed at the side of the helmet inside an IoT box which solves the problem of comfort for driver [24].

The size and weight of the proposed helmet is the same just like any other helmet, and it is very comfortable to wear.

The smart helmet holds the capability of getting connected with user application and performs data communication through the ESP Wi-Fi module. When the helmet is connected with mobile's Wi-Fi network, the helmet unit passes the real-time data containing accelerometer co-ordinate readings and alcohol level as shown in Fig. 14, based on which alerts and actions are triggered. This data is flashed on the user interface screen of the application. Figure 15 shows the user interface of the mobile application where green button demonstrates that the helmet is worn and the Wi-Fi communication is established.

The helmet generates alerts in three scenarios: helmet belt open, alcohol threshold level breached, and accident detected as shown in Fig. 16. In the scenario, when accident occurs and the impact is measured by the ADXL-335 accelerometer, a timer of 60 s is provided to the riders so that they can counter the false accident calls generated by the helmet. As shown in Fig. 17, the rider is provided with two swipe functionalities; if the helmet has produced a false call then the rider can swipe

Fig. 13 Circuit of the prototype of helmet

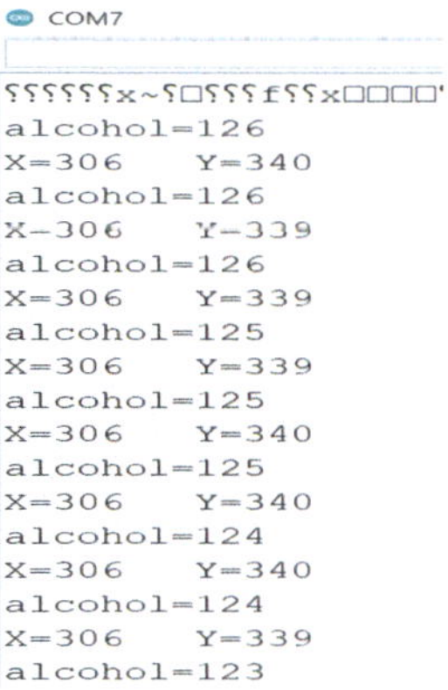

Fig. 14 Real-time data generated by helmet

Fig. 15 User interface of application

for cancel which will not produce any alerts and if the rider is severely injured and requires immediate medical aid then he can swipe for call emergency; this swipe feature will also activate automatically if there's no response from the rider within the time limit.

During the course of an accident, the bike rider's family members will be notified that an accident has occurred with a Google Map link to navigate the exact location of the rider as demonstrated in Fig. 18. This functionality of the helmet will help the family members to reach the accident spot quickly and provide the necessary help to the rider, thus, eventually saving the life of the rider.

The smart helmet undergoes testing to assess its accuracy in accident detection, involving an evaluation of 50 real-world cases. The evaluation included four key scenarios:

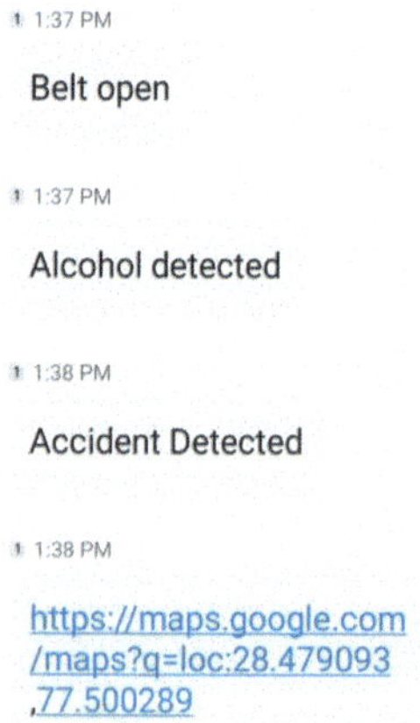

Fig. 16 Alert messages generated by helmet

Fig. 17 False call counter functionality

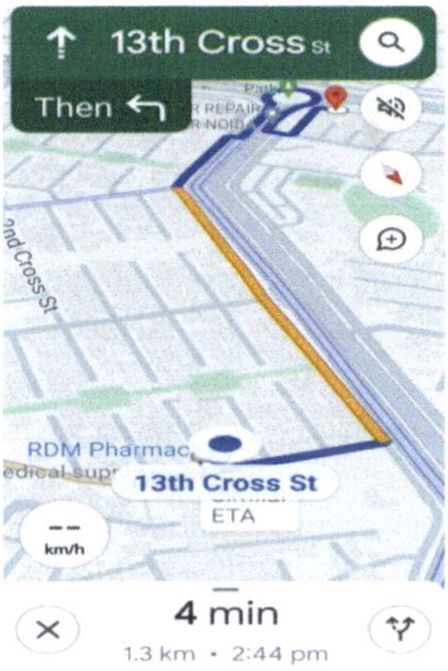

Fig. 18 Navigation mechanism to the accident spot

- True-Positive (TP): A case where the rider suffers from accident and helmet correctly generates alert for the same.
- False-Positive (FP): A case where the rider is safe but the helmet generates an alert for the accident.
- True-Negative (TN): A case where the rider is safe and there is no alert from helmet.
- False-Negative (FN): A case where the rider suffers from accident but the helmet doesn't produce any alert in response.

This testing is done to understand the helmet's capabilities to distinguish between the regular and emergency conditions. The result of the testing is demonstrated in Table 2.

Table 2 Results for unit testing of accident detection

Scenario	TP	FP	FN	TN	Accuracy
Accident detection	20	6	2	22	84%

The helmet's performance in various conditions was rigorously tested during this testing procedure to enhance its dependability and decrease the likelihood of false calls, thereby boosting its usefulness in delivering immediate help during accidents. The developed helmet produced an accuracy score of 84% which is quite remarkable in these scenarios.

6 Conclusion

The proposed Smart Helmet is an intelligent device which will aid more secured bike riding. According to the numbers in Table 1, the death rate for two-wheeler riders is high. Because smart helmets are so effective at preventing fatalities, they have become a necessity. A smart helmet would not only offer the necessary assistance, but would also prevent the accident from occurring in the first place. RideGuard achieved an overall accuracy of 84%, demonstrating its ability to discern between accident and non-accident scenarios. The project's success in minimizing false positives ensures that the helmet does not generate unnecessary alerts, contributing to its practicality and user acceptance. Moreover, the system's ability to effectively detect true positives, coupled with a low rate of false negatives, underscores its reliability in critical situations, laying the foundation for timely and life-saving interventions. This accuracy rate demonstrates the durability of the developed algorithms as well as the dependability of the sensor fusion methods used in the helmet. The Smart Helmet project is a promising addition to the field of smart safety systems, demonstrating how the Internet of Things might be used to lessen the effects of traffic accidents. The project not only meets the present demand for emergency response but also establishes the foundation for the future development of intelligent, life-saving solutions in the field of road safety by utilizing technology for quick accident detection and prompt assistance.

7 Future Scope

The project should aim to understand the potential users' interest, preferences, and acceptance of the Smart Helmet with Accident Detection and SOS System using IoT. It assesses how well the target users are likely to adopt and embrace the product. To achieve user acceptance, the following USPs should be provided through this project:

Minimizing false call alerts: Future research should be done on minimizing the false calls generated by the helmet to improve the accident detection accuracy.

Interactive user interface: User interface provided through the mobile application should be intuitive and easy to use. Controls for activating the SOS systems must be straightforward and accessible during emergencies.

Training and support: users would benefit from training or tutorials on using the Smart Helmet and its features effectively.

References

1. https://morth.nic.in/sites/default/files/RA_2021
2. https://scroll.in/article/981989/in-india-six-two-wheeler-riders-die-every-hour-in-road-accidents
3. Kulkarni, S.A., Sowmya, C., Subhalakshmi, P., Tejashwini, S., Sanusha, V., Amitha, S., Jha, V.: Design and development of smart helmet to avoid road hazards using IoT. In: 2020 IEEE International Symposium on Sustainable Energy, Signal Processing and Cyber Security (iSSSC), pp. 1−6 (2020)
4. Gupta, S., Sharma, K., Salvekar, N., Gajra, A.: Implementation of alcohol and collision sensors in a smart helmet. In: 2019 International Conference on Nascent Technologies in Engineering (ICNTE), pp. 1–5 (2019)
5. Varma, V.B.: Smart Helmet for Accident Avoidance
6. Bhattacharya, S., Jha, H., Nanda, R.P.: Application of IoT and artificial intelligence in road safety. In: 2022 Interdisciplinary Research in Technology and Management (IRTM), pp. 1–6 (2022)
7. Alim, M., Ahmad, S., Dorabati, M.N., Hassoun, I.: Design & implementation of IoT based smart helmet for road accident detection. In: 2020 11th IEEE Annual Information Technology, Electronics and Mobile Communication Conference (IEMCON), pp. 576–581. IEEE (2020)
8. Tapadar, S., Ray, S., Saha, H.N., Saha, A.K., Karlose, R.: Accident and alcohol detection in bluetooth enabled smart helmets for motorbikes. In: 2018 IEEE 8th Annual Computing and Communication Workshop and Conference (CCWC), pp. 584–590. IEEE (2018)
9. Kim, Y., Choi, Y.: Smart helmet-based proximity warning system to improve occupational safety on the road using image sensor and artificial intelligence. Int. J. Environ. Res. Public Health. **19**(MDPI), 16312 (2022)
10. Sathish, E., Manikandan, G., Bhuvaneswari, G.: WITHDRAWN: Design and Development of Multi Controlled Smart Bike (2022)
11. Eldho, A.S.T., Paul, J.T., John, K., Jose, J., Paul, B.: Design and analysis of a smart-attachment to jacket and helmet used by two-wheeler riders using Peltier-module. Mater. Today: Proc. **42**, 862–869 (2021)
12. T.A.H., Sivagnam, S.S.H., Gsangaya, K.R., Sultan, M.T.H., Hua, L.S.: Smart motorcycle helmet for enhanced Rider's comfort and safety. Mater. Today. Proc. (2021)
13. https://m.indiamart.com/proddetail/arduino-nano-10181545333.html
14. https://forum.arduino.cc/t/distance-with-ir-sensor-fc-51/341717
15. https://robokits.co.in/sensors/gas-sensors/mq-3-alcohol-ethanol-gas-sensor-module
16. https://components101.com/tags/vibration-sensor
17. https://linuxhint.com/interface-bluetooth-module-arduino-uno/
18. https://m.indiamart.com/proddetail/ov7670-640x480-vga-cmos-camera-module-2850018803997.html
19. Chandran, S., Chandrasekar, S., Elizabeth, N.E.: Konnect: an internet of things (IoT) based smart helmet for accident detection and notification. In: 2016 IEEE Annual India Conference (INDICON), pp. 1−4. IEEE (2016)

20. Cheng, C.A., Chiu, H.W.: An artificial neural network model for the evaluation of carotid artery stenting prognosis using a national-wide database. In: 2017 39th Annual International Conference of the IEEE Engineering in Medicine and Biology Society (EMBC), pp. 2566–2569 (2017)
21. Ehsanul, A.M., Ahmad, S., Dorabati, M.N., Hassoun, I.: Design & Implementation of IoT Based Smart Helmet for Road Accident Detection. In: 2020 11th IEEE Annual Information Technology, Electronics and Mobile Communication Conference (IEMCON), pp. 0576−0581 (2020)
22. Pathak, P.: IoT based smart helmet with motorbike unit for enhanced safety. In: 2020 2nd International Conference on Advances in Computing, Communication Control and Networking (ICACCCN), pp. 528–534. IEEE (2020)
23. Sinha, S., Teli, E., Tasnin, W.: An IoT-based automated smart helmet. In: Sustainable Communication Networks and Application: Proceedings of ICSCN 2021 (2023)
24. Rahman, M.A., Ahsanuzzaman, S., Rahman, I., Ahmed, T., Ahsan, A.: IoT based smart helmet and accident identification system. In: 2020 IEEE Region 10 Symposium (TENSYMP), pp. 14−17 (2020)

Statistical Feature Based Traffic Classification and Quality of Service Assessment in Software Defined Networks

Abhishek Kumar Gaur, Deepak Kumar Sharma, and Amarjit Malhotra

Abstract In current scenarios, a variety of network applications are being developed for meeting the growing business needs for Industries. These applications are complex in nature, use new protocols and port numbers from the range of the dynamic ports. For Quality of Service (QoS) assessment of these applications, traffic classification is the first and most important step. Each network application has distinct characteristics and its generated traffic also implicitly follows these characteristics. The main focus is on collecting the set of those traffic statistical features which are most suitable to correctly represent the applications under consideration for the classification task. In this paper, focus is on traffic classification in the Software Defined Networks (SDN). This paper has a proposed traffic classification and QoS assessment architecture. This proposed architecture is used for classifying the incoming traffic at the network edge devices. For classification purpose, six different machine learning algorithms have been used. Various type of traffic have been generated by using Distributed-Internet Traffic Generator. Further, the performance of these classifiers have been evaluated using the confusion matrix. A comparative performance study is done in the last. This work is based on the traffic statistical feature analysis and collected statistical features are evaluated for finding their relative importance in the task of classification. The proposed architecture works in the existing SDN environment and requires low computation, as only the important features have been passed and analyzed for the traffic classification.

Keywords Feature extraction · Decision tree · Support vector machine · Ternary content addressable memory · Deep packet inspection · Openflow protocol · Feature selection · Flow pattern

---

A. K. Gaur (✉) · A. Malhotra
Department of Information Technology, Netaji Shubhas University of Technology, Delhi, India
e-mail: abhishekergaur@gmail.com

A. Malhotra
e-mail: amarjit.Malhotra@nsut.ac.in

D. K. Sharma
Department of Information Technology, Indra Gandhi Delhi Technical University, Delhi, India

G. Raj et al. (eds.), *Advances in Artificial Intelligence and Machine Learning*, Lecture Notes in Electrical Engineering 1264, https://doi.org/10.1007/978-981-97-9507-9_14

1 Introduction

In the present age of digitization, all kinds of services ranging from communication, payment system, entertainment, banking, and many more, all are being digitized. Industrial Internet of Things (IIoT) has added a huge number of devices to the current network and as a result generated traffic volume has grown exponentially, hence the network requires the generated traffic to be classified. Online availability of these services have created new business opportunities and all-time availability of services. Internet is the backbone, for making these services available and accessible to everyone, everywhere on a variety of devices. There are various type of internet-based services and the traffic generated by these services has its own network resources requirements. Some services require a minimum guaranteed bandwidth, some require less jitter or are time-sensitive, and some require that the packet drops should be less than a given threshold, while some require the maximum throughput. From a network administrator's viewpoint, both the QoS and optimal network resource utilization are important. QoS parameters help in ensuring Quality of Experience (QoE) for users, and optimal resource utilization. Hence meeting QoS in a network preserves the network resources from wastage and enables the network to meet the demands of a comparatively more number of traffic requests without introducing additional resources.

Internet traffic can be classified based on its sensitivity to network resource provisioning. Following this idea, in order to find optimal Internet traffic classification approaches for improving the QoS, a rich and complete set of Internet traffic is required. Various solutions to the problem of traffic classification have already been proposed in different ways. Some of the solutions are based on classifying the traffic at the network edge router, while others are based on finding the solutions in the rest of the network (core network). Further, these solutions either work on the inspection of the generated packets header fields [1, 2] or patterns derived from statistics of the flows. Many researchers have taken traffic flow individually and generated the unique identification of these flows on the basis of source/destination IP address, or source/destination port. Each of these uniquely identified flows are then classified on the basis of the transport layer protocol, port numbers, and other useful fields of the packets. There are a variety of network applications that uses the traditional protocols as well as newly developed protocols. Hence there is a continuous requirement to update these packet content inspection-based approaches.

Since port-based traffic classification is unable to deal with Virtual Private Network (VPN) traffic, encrypted traffic and traffic on the dynamic ports (i.e. these port are not originally allocated for this kind of service). Many previous works have focused on discovering application specific traffic signatures and these signatures were generated on the basis of packet content. For getting the packet content of a selected flow, a manual or automatic reverse engineering of all involved applications/protocols needed to be done. Doing reverse engineering puts the constraints on the scalability of this approach, and makes it very time-consuming when it needs to deal with the applications that are newly introduced or developed for the current network.

Signature-based approach is very limited in term of its usefulness, when previously generated signatures are used on new developed network.

Predicting the network services which are generating the traffic can be done using the information of transport layer protocol and port number used in transmission. AT present various applications run on dynamic ports. Since these ports may change every time the applications start communicating or working. So generated traffic of these applications cannot be the right candidate for the port based traffic classification. Network traffic may be encrypted or unencrypted due to various security concerns. Decoding the packet's contents is not possible and as a result traffic classification based on the deep inspection of the packet contents may not be possible every time.

Reverse engineering-based approaches can't work accurately for newly developed applications and for variations in message formats for existing applications. Development of the new enhanced versions of the same applications may use different messages formats or protocols e.g. for SMTP servers EHLO and HELO messages, and for RTCP version 1 and 2, etc. Each application's traffic has statistical differences from other application flows. These statistical differences features are derived from flows, then these features are passed to the appropriate machine learning or deep learning algorithms, and flow classification can be achieved [3–5]. Because of a huge number of network applications and new innovations in the networking domain, port-based classification is unsuitable to deal with encrypted traffic [6, 7]. While Deep packet inspection (DPI) [8, 9] consumes too much time for finding a hidden pattern in traffic flows and the existing patterns become frequently invalidated [10].

In this paper, traffic classification has been done in SDN environment. Traffic data is generated by Distributed Internet Traffic Generator (D-ITG). Traffic statistics are being collected in forward and backward i.e. both directions. RYU-controller is used as the remote SDN controller. The main contributions of this work are listed below:

- Generation of Traffic dataset along 23 features.
- Traffic statistic features evaluation.
- Selection of Important features.
- Generated data set preprocessing and cleaning
- Training and testing of the selected machine learning classifiers.
- Performance evaluation and comparative study of the Classifiers

Present work is organized into 6 sections. In 2 section, previous research works related to feature extraction, traffic classification, and cluster approaches have been discussed. In Sect. 3, Openflow protocol is discussed as the SDN enabler. The proposed model of traffic classification and QoS assessment is discussed. In Sect. 4, Experiment environment setup, traffic generation and classifier selected are been covered. In Sect. 5, experimental results and findings are discussed and Sect. 6 covers the conclusion from the above work's output and future work.

2 Related Work

Measuring and achieving a consistent network QoS is very important for applications such as online gaming, video on demand and VoIP, etc. Sometimes QoS is misinterpreted as a violation of net neutrality principles. Network-level QoS may be measured in an active or passive manner. Real-time applications [11] uses TCP and UDP protocols and their measurement matrix is designed on throughput, latency, jitter, and packet loss. In active probing [12, 13], flow latency, jitter, and packet loss can be measured. In [14] K-Nearest Neighbor (KNN) and Linear Discriminant Analysis (LDA) have been used for traffic classification. In [15–17] Bayesian classifiers and Decision Trees (DT) approaches have been used classification of traffic generated by HTTP, FTP, P2P, games, video or audio streaming applications. In [18] a passively traffic is replayed and a packet parser was used for continuously extracting packet header's information, and traffic temporal information. Collected data is converted into a labeled dataset by the network profile trainer tool and then training of the classifier [19] has been done on this data. Once the model is trained, the same model is used for upcoming traffic classification. In [5] classification of the traffic is performed on the basis of packet payload contents. The developed system is capable of traffic detection with 97% precision, for applications that may be tunneled, using the random port and may be newly developed [20]. There may be some complex applications that may require various network performance parameters to be met at the same time. RFC791 [21] has provided a way of specifying these constraints in the generated flow packets. RFC791 has used the Type of Service (ToS) field (1 byte) of the IPv4 header for this task. Based on the IP precedence field value, RFC791 has 7 traffic classes, named Routine, Priority, Immediate, Flash, Flash Override, Critical, Internetwork control, and Network control. RFC791 also has the provision to represent sub-classes default, monetary cost, reliability, throughput, and delay of these 7 traffic classes. For specifying the network application's QoS in the generated packets, there are 2 other standards RFC3168 [22], RFC2475 [23], and RFC2597 [24]. RFC3168 has a provision of assigning a DSCP code based on traffic aggregate behavior, for classifying it at the boundaries of the network, further rest of the network, deals with the packets as per their DSCP code. RFC2475 has defined various DSCP codes and class selectors. RFC 2597 has taken the queuing requirement of the traffic and packet drop probability (high, medium, or low) into consideration. For denoting the lowest latency, least bandwidth requirements DSCP code also has an expedited forwarding class.

Traffic classification is essentially the first step for traffic engineering and QoS. In [25] a scheme named TSCRNN has been proposed for encrypted traffic. TSCRNN have used CNN for extracting optimal spatiotemporal feature set for traffic classification, for traffic encryption and for the rapid growth of the network, with the increase in the encryption standards and huge volume of traffic, real-time traffic classification is becoming harder than before [26]. Classification has been port-based, payload-based or statistics-based methods [27]. In [28] deep learning methods have been used for encrypted traffic identification. In [29] a set of packet-level statistical features have been selected and a semi-supervised learning framework in developed for classify the encrypted traffic. In [30] a logistic regression model has been used for traffic classification on features like TLS flow's meta-information, packet length, and inter-arrival times. Similarly, In [31] KNN classifier has been used on a set of 111 flow features. In [32] DT classifiers has been used on the features that do not include IP, port, etc. features to obtain classification accuracy of up to 90%. In [33] a hybrid machine learning, deep learning-based approach and mutli-level feature extraction have been used for classifying encrypted traffic in 4 classes.

Supervised classification algorithms are used only for handling labeled traffic flow data, and are not for handling unknown applications with missing labeled data. In [34] unsupervised learning algorithm (K-Means) has been used for grouping the traffic based on its statistical properties and packets' payload. This approach is suitable to group the traffic flows into a small number of application-based clusters. In [35] neural networks have been utilized for extracting flow features and training traffic classifiers. In [36] an unsupervised learning framework was developed for classifier training. Generated traffic data for real-time applications is not always correctly labeled. In [37] a Deep Packet Inspection (DPI) and semi-supervised learning-based architecture is developed for flow classification. In [38] various Machine learning algorithms have been utilized for VPN traffic classification and their aim is to improve the traffic classification accuracy using Ensemble learning-based Classifiers. They have used Bagging Decision Tree and Gradient Boosting algorithms for getting classification performance improvement up to 93%, in place of using KNN, Multi-layer Perceptron model (MLP), and DT classifiers. Traffic classification helps in allocating higher bandwidth to Video on-demand, Voice over Internet Protocol (VoIP) applications [39]. In [40] an iterative tuning Support Vector Machine (SVM) has been used. For fast training of classification algorithm and application traffic classification, statistical features of traffic were utilized. In [41] a combined approach of density clustering learning algorithms and SVM has been used to minimize the size of the training dataset and achieve improved traffic classification. In [42] deep learning (DL) algorithms have been utilized for the real-time application traffic classification. In [43] it has been compared and concluded that for traffic classification and intrusion detection, DL algorithms perform better than machine learning-based approaches. The performance and QoS on a network can be improved if the significance of classifying network flow is understood. In [44] a DPI based technique have been applied for network traffic classification. In [45] a semi-supervised learning-based traffic has been used for classification for large traffic classification. In [46] a multi-layer perceptron neural network is used for building a model for doing VPN and the non-VPN traffic classification.

2.1 Motivation

Traditional network does not have the central global view (i.e. network resources, resource's state, utilization pattern etc.) of the network. These networks are neither flexible nor programmable. This puts a lot of challenges in the task of traffic classification. Work presented in this paper is motivated by SDN's centralized global view and control over the network resources. Since SDN is flexible and programmable, it gives the network administrator control over the network configuration. Traffic classification in SDN is supposed to be more accurate and fast. Here an assumption has been made, that statistical features of the traffic data are more unbiased and accurate. Hence the classifiers' accuracy is supposed to be high in SDN. So, motivations for this work are listed below.

- Utilizing SDN's central view of the network.
- Collecting unbiased traffic features.
- Getting more accurate traffic classification.
- Getting more accurate QoS assessment of applications.

3 OpenFlow Protocol

OpenFlow [47] standard is at the core of the development of the SDN. SDN's data plane devices i.e. switches or routers have flow-table which are stored in the Ternary Content Addressable Memory (TCAM) of these devices. These flow tables have the flow rules (i.e. forwarding rules) to forward or manipulate the traffic flow. Initially, these flow tables are empty and at the time of initialization of the network, these tables are populated with the flow-miss entries. When a new flow is being introduced to the network and there are no matching rules to handle that flow's packet at the data plane device which is directly connected to the flow generating source. For the first packet of this flow, a Packet-In-Request is created and forwarded by the concerned data plane device to the controller, which in turn is been handled by the controller. Now controller populates the flow table with the suitable flow entry that can satisfy the QoS of the flow generating application in an optimal possible manner. For generating the flow rule at the controller or for forwarding the traffic in the SDN, matching of some selected field of the packet is done to the flow entries that are already present in the concerned data plane device's flow table.

3.1 Proposed Framework for Traffic Classification and QoS Assessment

At the time of generation of the new traffics at the source hosts, switches connected to these hosts send the first generated un-buffered packet of these flows to the controller. This packet is analyzed by the QoS assessment module, either on the basis of the port number or the deep inspection of the IP header is done on the basis of the ToS field (i.e. DSCP code). Deep Inspection of the packet can also consider the packet header fields and payloads. But this way of doing classification is computationally costly and consumes a lot of resources. For efficient traffic classification, classification based on the statistical analysis of the traffic flow is the best choice in terms of computation resources requirement. For traffic classification, various flow statistics features of the applications' traffic are extracted, collected, and analyzed by applying machine learning and deep learning algorithms. These features include flow duration, packet flow rate, byte flow rate, inter-packet arrival time, max active time, min active time, etc. There are various applications such as Browsing, Chat, Email, VoIP, Streaming, File Transfer, P2P, VPN VoIP, and VPN File Transfer which can be classified/Identified on the basis of these statistics features of the traffic.

Developed framework is for flow classification and QoS assessment in SDN environment in Fig. 1. This framework uses the benefits of SDN, such as flexibility, global view of the network, centralized control on the network resources, and ease of deployment of the network policies. This framework also utilizes the Open flow architecture and protocol. In this framework, when an application starts generating a flow, generated packets are inserted into the network by the edge device (router/modem). This device checks its TCAM for finding the suitable flow entry to handle the entering traffic flow. If the edge device is unable to find any matching flow entry to handle the entering traffic, the first packet of the flow is forwarded to the controller and network statistics monitoring module by using the selected OpenFlow protocol version. The Monitoring module collects the traffic features on the basis of decoding the Open flow header fields or the flow statistics features. Although for extracting the features for packet-level inspection, only decoding of the first packet is sufficient. While for extracting statistical flow features, the controller periodically sends a statistical request message to the network edge devices. Controller collects or records these responses received from edge devices. These collected responses are used for extracting the flow statistics. After these flow features collection, these features are passed to the suitable classifiers for flow classification. Once the flow classification is done, application QoS concerns are addressed by SDN.

4 Experiment Environment

Present work is performed in the mininet emulation environment. Here, we have used the RYU remote controller and mininet topology given in (Fig. 2). We have generated traffic for 5 types DNS, Ping, Interactive Gaming, VoIP and Telnet respectively. This traffic is generated by the D-ITG [48] traffic generator. For traffic generation using D-ITG one host is selected as the client and other host is selected as the server. Traffic data collection has been performed at the traffic source end. Traffic dataset has 30 min traffic of each category. For the classification of the generated traffic, we have utilized 6 machine learning algorithms: Decision Tree(DT) [49], Linear Regression(LR) [50], Random Forest(RF) [51], Naïve Bayes(NB) [52], Multilayer perceptron (MP) and Logistic Model Tree (LMT) [53]. Traffic generated by the D-ITG generated is collected in both forward and backward directions. From this

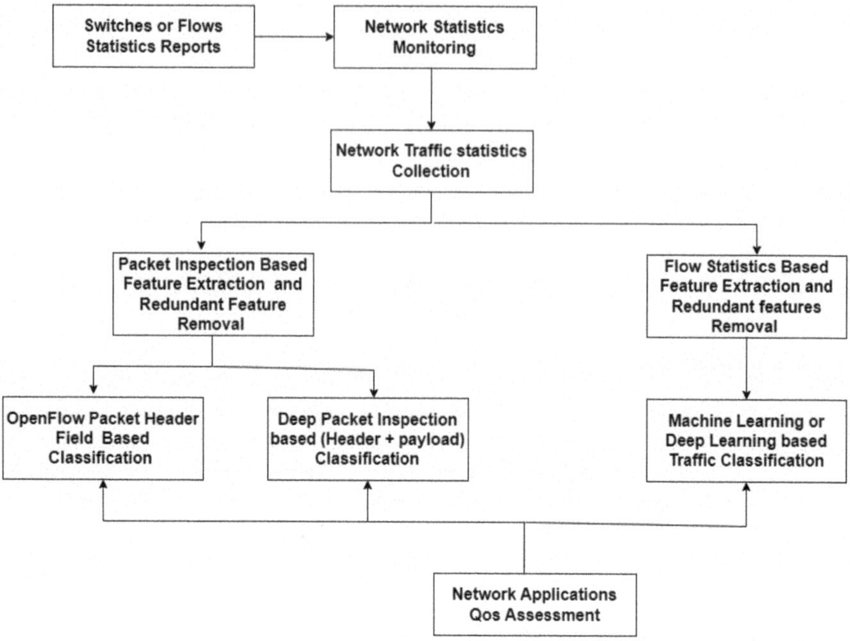

Fig. 1 Traffic classification and applications QoS assessment

collected traffic, 16 most relevant traffic statistics features are extracted. Since all the features are not of importance, so we evaluate these features with help of Random Forest Machine Learning algorithm. We select the n most important features out of the set of the given features based on the evaluated values. In last, the database of the selected features is given as input for the training of the above models and then subset of this database is used for the testing purpose.

5 Experiment Results and Result Analysis

In this paper, we have considered the 5 different traffic classes; DNS, Telnet, VoIP, Ping, and Interactive traffic. Generated traffic is collected and reported by the Open-Virtual switches to RYU framework repeatedly after a certain time interval. This Traffic has 16 statistical features for traffic moving in forward and backward directions. These features are evaluated and important features are been selected. Further models are been trained and test over the reduced feature database. Overfitting of the classification model is been handled by using the K-fold cross-validation (in our case $k = 10$). For checking and getting the Importance features for classification, we have used the Random Forest Algorithm and evaluated features' importance. After

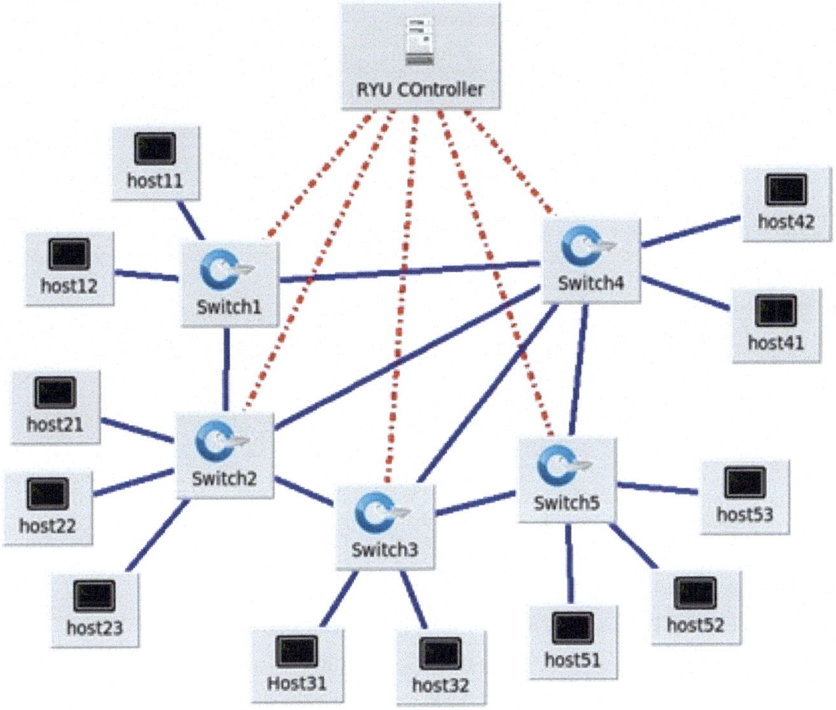

Fig. 2 Network topology for experiment

this, a suitable subset of the important traffics feature is been selected and given as input for the classification module.

Here we have done SDN traffic statistics-based classification. We have separately trained and tested the Decision tree, Random Forest, Linear Regression, Naive Bayes, Multilayer Perceptron models on the generated dataset. We have used the 70% of the dataset for training and the remaining dataset for testing purposes. We have used the confusion matrix for the purpose of performance evaluation of these models. Performance evaluation of each model in terms of the True Positive Rate (TPR), False Positive Rate (FPR), Precision, Recall, F-measure and ROC Area is given below in Table 1, 2, 3, 4, 5 and 6. Decision Tree's TPR is the highest in all the models for DNS traffic classification, for telnet traffic Logistic model, Random Forest, Multilayer perceptron, and linear Regression have the highest TPR. Similarly for Voice traffic Random forest and Logistic model, for Ping traffic Linear Regression and Random forest Classifier, and Interactive Gaming traffic Random Forest show the highest TPR. The result shows that LMT, DT, and RF are the most suitable classifier. The result shows that LMT, DT, and RF are the most suitable classifier based on their true positive rate. While Naïve Bayes Classifier performs the worst in the traffic classification.

Table 1 Decision tree based classifier performance

TP rate	FP rate	Precision	Recall	F-measure	ROC area	Class
0.999	0	0.998	0.999	0.999	0.999	DNS
0.997	0	0.998	0.997	0.998	0.998	TELNET
0.996	0	1	0.996	0.998	0.998	VOICE
0.999	0.001	0.998	0.999	0.999	0.999	PING
0.998	0.001	0.995	0.998	0.997	0.999	INTERACTIVE GAMING

Table 2 Random forest based classifier performance

TP rate	FP Rrte	Precision	Recall	F-measure	ROC area	Class
0.992	0	1	0.992	0.996	0.999	DNS
1	0	1	1	1	1	TELNET
0.997	0	1	0.997	0.999	0.999	VOICE
1	0	1	1	1	1	PING
1	0.003	0.989	1	0.994	1	INTERACTIVE GAMING

Table 3 Naïve bayes based classifier performance

TP rate	FP rate	Precision	Recall	F-measure	ROC area	Class
0.994	0.082	0.73	0.994	0.842	0.994	DNS
0.551	0.002	0.984	0.551	0.706	0.997	TELNET
0.991	0	1	0.991	0.996	0.998	VOICE
0.993	0	1	0.993	0.996	1	PING
0.646	0.099	0.596	0.646	0.62	0.942	INTERACTIVE GAMING

Table 4 Logistic regression based classifier performance

TP rate	FP rate	Precision	Recall	F-measure	ROC area	Class
0.997	0.003	0.989	0.997	0.993	0.999	DNS
1	0	1	1	1	1	TELNET
0.994	0	1	0.994	0.997	0.998	VOICE
1	0	1	1	1	1	PING
0.992	0.001	0.994	0.992	0.993	1	INTERACTIVE GAMING

Table 5 Multilayer perceptron based classifier's performance

TP rate	FP rate	Precision	Recall	F-measure	ROC area	Class
0.91	0.009	0.956	0.91	0.932	0.978	DNS
1	0.001	0.994	1	0.997	1	TELNET
0.936	0.001	0.994	0.936	0.964	0.966	VOICE
0.996	0.021	0.947	0.996	0.971	0.989	PING
0.975	0.01	0.956	0.975	0.966	0.997	INTERACTIVE GAMING

Table 6 Logistic model tree based classifier performance

TP rate	FP rate	Precision	Recall	F-measure	ROC area	Class
0.997	0	1	0.997	0.999	0.998	DNS
1	0.001	0.994	1	0.997	1	TELNET
0.997	0	1	0.997	0.999	0.998	VOICE
0.998	0	1	0.998	0.999	0.999	PING
0.994	0.002	0.992	0.994	0.993	1	INTERACTIVE GAMING

In the present paper, statistical features are collected in the mininet emulated SDN and Ryu controller is used for the statistical feature collection out of the packets from the router which is introduce the traffic to the current SDN. Collected data is the labeled data and same dataset have been used for separately training of the selected classification models (Decision tree, Random Forest, Linear Regression, Naive Bayes, Multilayer Perceptron models). From the experimental result shown in Tables 1, 2, 3, 4, 5 and 6, it is clear that all other selected models except the NB model show value of F-Measure greater than 0.99 for all type of traffic classes. While NB model can only classify the voice and ping traffic. This poor performance of the Naïve Bayes classification is because of the selected statistical attributes of the traffic classic are not independent to each other. While this issue does not affects rest other classifiers performance. From this study, it can be clearly stated that traffic classification based on the statistical features of the traffic is efficient in term of the network resource utilization and classification accuracy. From the comparative analysis of the results, it can be noted that classification based on the statistical features of the traffic is high sensitive to feature extraction step and independence among the selected features.

6 Conclusion and Future Work

Statistical feature-based traffic classification in the SDN environment is more accurate. This classification can be easily done at the edge of the network, from where traffic is entering into the network. These classifications helps in correctly recognizing the QoS of the traffic generating services. On the basis of this QoS assessment, optimal traffic reshaping, network resources utilization, and traffic engineering sub-modules can be automatically triggered to achieve better QoE. Above discussed work is only limited to covering the traffic classification and QoS assessment of the simulated traffic of mentioned traffic classes. The presented framework first requires offline training and testing of the classification model, once the training is done, model can also be used for Real-time traffic classification. Study shows that or the above traffic classification LMT, DT, and RF model are the best choices. Naïve Bayes classifier's performance is very poor for the classification of the above traffic dataset. Once the traffic is classified, different DSCP code values are attached to packets of the flows to denote the traffic class. Future work will be to extend this work for real-time network traffic assessment, designing policy for meeting QoS requirements of real-time network applications and minimizing the statistical feature based traffic classifier module's training time.

References

1. Bujlow, T., Carela-Español, V., Barlet-Ros, P.: Independent comparison of popular DPI tools for traffic classification. Comput. Netw. Netw. **76**, 75–89 (2015)
2. Antonello, R., Fernandes, S., Kamienski, C.A., Sadok, D.: Deep packet inspection tools and techniques in commodity platforms challenges and trends. J. Netw. Comput. Appl.Netw. Comput. Appl. **35**(6), 1863–1878 (2012)
3. Bocchi, E., et al.: Magma network behavior classifier for malware traffic. Comput. Netw. Netw. **109**, 142–156 (2016)
4. Muelas, D., et al.: Dictyogram a statistical approach for the definition and visualization of network flow categories. In: 11th International Conference on Network and Service Management (CNSM), IEEE (2015)
5. Tongaonkar, A., et al.: Towards self-adaptive network traffic classification. Comput. Commun. Commun. **56**, 35–46 (2015)
6. Qi, Y., et al.: Packet classification algorithms: from theory to practice. In: IEEE INFOCOM 2009. IEEE (2009)
7. Madhukar, A., et al.: A longitudinal study of P2P traffic classification. In: 14th IEEE International Symposium on Modeling, Analysis, and Simulation. IEEE (2006)
8. Sherry, J., et al.: Blindbox: deep packet inspection over encrypted traffic. In: Proceedings of the 2015 ACM Conference on Special Interest Group on Data Communication. ACM SIGCOMM (2015)
9. Bonfiglio, D., et al.: Revealing skype traffic: when randomness plays with you. In: Proceedings of the 2007 Conference on Applications, Technologies, Architectures, and Protocols for Computer Communications. ACM SIGCOMM (2007)
10. Yu, S., et al.: Predicted packet padding for anonymous web browsing against traffic analysis attacks. IEEE Trans. Inf. Forensics Secur. **7**(4), 1381–1393 (2012)
11. Beuran, R., et al.: Network quality of service measurement system for application requirements evaluation. Simul. Ser. **35**(4), 380–387 (2003)
12. Gummadi, K., Saroiu, S., Gribble, S.: King: estimating latency between arbitrary internet end hosts. In: Proceedings of the 2nd ACM SIGCOMM Workshop on Internet measurement. ACM SIGCOMM (2002)
13. Kwon, M. et al.: Use of network latency profiling and redundancy for cloud server selection. In: 2014 IEEE 7th International Conference on Cloud Computing. IEEE (2014)
14. Roughan, M., et al.: Class-of-service mapping for QoS: a statistical signature-based approach to IP traffic classification. In: Proceedings of the 4th ACM SIGCOMM Conference on Internet measurement, Association for Computing Machinery, New York, NY, United States (2004)
15. Zander, S., Nguyen, T., Armitage, G.: Automated traffic classification and application identification using machine learning. In: IEEE Conference on Local Computer Networks 30th Anniversary (LCN'05) l. IEEE (2005)
16. But, J., Branch, P., Le, T.: Rapid identification of BitTorrent traffic. In: IEEE Local Computer Network Conference. IEEE (2010)
17. Branch, P., Heyde, A., Armitage, G.: Rapid identification of skype traffic flows. In: Proceedings of the 18th International Workshop on Network and Operating Systems Support for Digital Audio and Video, Association for Computing Machinery, New York, NY, United States (2009)
18. Middleton, S., Modafferi, S.: Scalable classification of QoS for real-time interactive applications from IP traffic measurements. Comput. Netw. Netw. **107**, 121–132 (2016)

19. Hall, M., et al.: The WEKA data mining software: an update. ACMSIGKDD Explor. Newsl. **11**(1), 10–18 (2009)
20. Dewan, R., et al.: IoT-based energy efficient and longer lifetime compression approach for healthcare applications. In: Wiley Online Library (2023). https://doi.org/10.1002/ett.4843
21. PASTEL, J.: Internet Protocol Specification. RFC791 (1981)
22. Ramakrishnan, K., Floyd, S., Black, D.: RFC3168: The Addition of Explicit Congestion Notification (ECN) to IP (2001)
23. Blake, S. et al.: RFC2475: An Architecture for Differentiated Service (1998)
24. Heinanen, J. et al.: RFC2597: Assured Forwarding PHB Group (1999)
25. Lin, K., Xu, X., Gao, H.: TSCRNN: a novel classification scheme of encrypted traffic based on flow spatiotemporal features for efficient management of IIoT. Comput. Netw. Netw. **190**, 107974 (2021)
26. Rezaei, S., Liu, X.: Deep learning for encrypted traffic classification: an overview. IEEE Commun. Mag. **57**(5), 76–81 (2019
27. Ducange, P., et al.: A novel approach for internet traffic classification based on multi-objective evolutionary fuzzy classifiers. In: 2017 IEEE International Conference on Fuzzy Systems (FUZZ-IEEE). IEEE (2017)
28. Aceto, G., et al.: Toward effective mobile encrypted traffic classification through deep learning. Neuro Comput. **409**, 306–315 (2020)
29. Liu, H., Wang, Z., Wang, Y.: Semi-supervised encrypted traffic classification using composite features set. J. Networks **7**(8), 1195 (2012)
30. Anderson, B., Paul, S., McGrew, D.: Deciphering malware's use of TLS (without decryption). J. Comput. Virol. Hacking Tech. **14**(3), 195–211 (2018)
31. Yamansavascilar, B., et al.: Application identification via network traffic classification. In: 2017 International Conference on Computing, Networking and Communications (ICNC). IEEE (2017)
32. Alshammari, R., Zincir-Heywood, A.N.: Can encrypted traffic be identified without port numbers, IP addresses and payload inspection? Comput. Netw. Netw. **55**(6), 1326–1350 (2011)
33. Tong, V., et al.: A novel QUIC traffic classifier based on convolutional neural networks. In: 2018 IEEE Global Communications Conference (GLOBECOM). IEEE (2018)
34. Zhang, J., et al.: Unsupervised traffic classification using flow statistical properties and IP packet payload. J. Comput. Syst. Sci.Comput. Syst. Sci. **79**(5), 573–585 (2013)
35. Parsaei, M.R., et al.: Network traffic classification using machine learning techniques over software defined networks. Int. J. Adv. Comput. Sci. Appl.Comput. Sci. Appl. **8**(7), 220–222 (2017)
36. McGregor, A., et al.: Flow clustering using machine learning techniques. In: Barakat, C., Pratt, I. (eds.) Passive and Active Network Measurement PAM 2004, pp. 205–214. Springer, Berlin Heidelberg (2004)
37. Yu, C., et al.: QoS-aware traffic classification architecture using machine learning and deep packet inspection in SDNs. Procedia Comput. Sci. **131**, 1209–1216 (2018)
38. Afuwape, A.A., et al.: Performance evaluation of secured network traffic classification using a machine learning approach. Comput. Stand. Interfaces **78**, 103545 (2021)
39. Nguyen, T.T.T, Armitage, G.: A survey of techniques for internet traffic classification using machine learning. IEEE Commun. Surv. Tutor. **10**(4), 56–76 (2008)
40. Hong, Y., et al.: Iterative tuning support vector machine for network traffic classification. In: 2015 IFIP/IEEE International Symposium on Integrated Network Management (IM). IEEE (2015)
41. Yong, S., et al.: An improved cluster analysis algorithm using for network traffic flow. In: 10th International Conference on Computer Science and Education (ICCSE). IEEE (2015)
42. Hinton, G.E., Osindero, S., The, Y.: A fast learning algorithm for deep belief nets. Neural Comput.Comput. **18**(7), 1527–1554 (2006)
43. Dong, B., Wang, X.: Comparison deep learning method to traditional methods using for network intrusion detection. In: 8th IEEE International Conference on Communication Software and Networks (ICCSN). IEEE (2016)

44. Shafiq, M., et al.: Network traffic classification techniques and comparative analysis using machine learning algorithms. In: 2nd IEEE International Conference on Computer and Communications (ICCC). IEEE (2016)
45. Pacheco, F., et al.: Towards the deployment of machine learning solutions in network traffic classification: a systematic survey. IEEE Commun. Surv. Tutor. **21**(2), 1988–2014 (2018)
46. Miller, S., Curran, K., Lunney, T.: Multilayer perceptron neural network for detection of encrypted VPN network traffic. In: IEEE International Conference on Cyber Situational Awareness, Data Analytics and Assessment (Cyber SA). IEEE (2018)
47. McGeer, R.: A safe, efficient update protocol for open flow networks. In: Proceedings of the First Workshop on Hot Topics in Software Defined Networks, Association for Computing Machinery, New York, NY, United States (2012)
48. Botta, A., et al.: COMICS (COMputer for Interaction and CommunicationS) Group Department of Electrical Engineering and Information Technologies University of Napoli. Federico II, D-ITG 2.8.1. https://traffic.comics.unina.it/software/ITG/manual/. Accessed 28 October 2013
49. Kamiński, B., et al.: A framework for sensitivity analysis of decision trees. Cent.L Eur. J. Oper. Res. **26**(1), 135–159. https://doi.org/10.1007/s10100-017-0479-6. PMC 5767274. PMID 29375266
50. Tolles, J., Meurer, W.J.: Logistic regression relating patient characteristics to outcomes. JAMA. **316**(5), 533–4 (2016). https://doi.org/10.1001/jama.2016.7653. ISSN 0098–7484. OCLC 6823603312. PMID 27483067
51. Ho, T.K.: Random decision forests. In: Proceedings of 3rd International Conference on Document Analysis and Recognition 1. IEEE (1995)
52. Joyce, J.: Bayes' theorem. In: Zalta, Edward, N. (ed.), The Stanford Encyclopedia of Philosophy (Spring 2019 ed.). Metaphysics Research Lab, Stanford University, retrieved 2020–01–17
53. Landwehr, N., Hall, M., Frank, E.: Logistic model trees (PDF). Mach. Learn. **59**, 161 (2005). https://doi.org/10.1007/s10994-005-0466-3

An AI and ML-Enhanced Approach with IoT for Pothole Detection to Ensure Pedestrian Safety During Flooding

Pancham Singh, Mrignainy Kansal, Purvika Panwar, Manas Mishra, Kartikeya Patel, and Avinash Kumar Sharma

Abstract Road safety and flood management are major concerns, especially in water-prone areas where pedestrian safety is at risk due to hidden potholes and road hazards to address this issue, a real-time pothole detection and reporting system is proposed, aiming to enhance pedestrian safety during flooding. The system offers many benefits, such as improved road safety, reduced accidents and injuries due to potholes, real-time monitoring, and reporting of road conditions to enable priority action. The application uses the Map API for real-time pothole location tracking and uses the YOLOv8 module, an advanced object detection technique, to identify potholes from images taken by users. Known pothole locations are stored in a database that can be accessed by real-time users and government officials. User feedback and app usage statistics are included in the dataset to evaluate the efficiency of the application and the accuracy of the YOLOv8 module. Ethical considerations such as user data privacy and informed consent are considered to ensure responsible data handling. The proposed machine's contribution to society is tremendous, with capacity benefits such as stepped-forward street protection, decreased vehicle damage and repair prices, extended performance in avenue protection, better travel experiences, and effective environmental impacts from reduced fuel intake and emissions. The machine's destiny scope consists of incorporating additional features like actual-time traffic statistics, street closures, and protecting greater avenue hazards

P. Singh (✉) · M. Kansal · M. Mishra · K. Patel
Department of Information Technology, Ajay Kumar Garg Engineering College, Ghaziabad, Uttar Pradesh, India
e-mail: pancham.akg@gmail.com

M. Mishra
e-mail: manas2013011@akgec.ac.in

K. Patel
e-mail: kartikeya2013082@akgec.ac.in

P. Panwar
Department of Computer Science and Engineering, Ajay Kumar Garg Engineering College, Ghaziabad, Uttar Pradesh, India

A. K. Sharma
School of Engineering and Technology, Sharda University, Greater Noida, India

G. Raj et al. (eds.), *Advances in Artificial Intelligence and Machine Learning*, Lecture Notes in Electrical Engineering 1264, https://doi.org/10.1007/978-981-97-9507-9_15

inclusive of debris and flooding. Machine mastering and AI technology may be in addition employed to enhance pothole detection precision. Overall, this research paper provides a comprehensive method for tackling road safety issues in the course of floods through a value-effective and green Real-time Pothole Detection and Notification System. The ability effect of this machine on pedestrian protection and avenue protection, at the side of destiny improvements, gives promising possibilities for a smarter and safer transportation network.

Keywords Yolov8 · Real-time monitoring · Map API · Pedestrian safety · Potholes detection · AI and ML

1 Introduction

Road safety is a major concern in our modern lives, and there is an urgent need to address hazards that can pose a danger to drivers and pedestrians. The invitation in aquatic doomsday fields is high, while hidden potholes and potholes on the road can lead to serious accidents, so the safety of the killing areas is sufficient for Khazars to minimize this and increase the safety of the execution [1]. There is a key objective of this program is to achieve the impact of expenditure by experts by providing solutions that help detect potholes and other road hazards in real-time thus, the system is aimed at reducing accidents, and vehicle damage, and enabling informed decisions in timely road maintenance and repair, all contributing to a safe and smooth travel experience for drivers and pedestrians.

The proposed system leverages current technologies, which include Kotlin, an Android app development programming language, and a map API, to enable real-time tracking of pothole locations. The YOLOv8 module, a brand-new object detection approach, is included in the system to correctly discover potholes from snapshots taken by way of customers. The identified pothole locations are then stored in a centralized database available to both actual-time customers and government officers, ensuring seamless tracking and reporting [2]. The ability societal impact of this research is far-accomplishing. By improving avenue protection and lowering accidents as a result of potholes, the proposed system can save lives and reduce accidents. Furthermore, it can lead to sizable cost financial savings for each person and government, as vehicle harm and associated repair expenses are minimized. The green actual-time tracking and reporting of avenue conditions will empower the street government to prioritize protection and restore paintings efficiently, optimizing the utilization of available assets.

2 Related Work

In rock identification systems: a comprehensive review [3] proposed a rock detection system using hardware components and cloud-based storage. The system uses an ATmega328P micro-controller, an HC-SR04 ultrasonic sensor for telemetry and a NEO 6 m GPS module for location coordinates The HC-05 Bluetooth module is used to transfer data from the micro-controller to the Android application. Collected data, including rock depth and location, is stored in the Firebase cloud for real-time access. The Android application provides a user-friendly interface and shows pothole locations on a map, enabling drivers to take precautions and authorities to better plan road maintenance. Future developments include improving GPS accuracy and integrating additional sensors for better accuracy in pothole detection.

It [4] offers an alternative approach to the issue of potholes in road surfaces. Authors propose a system that uses visualization and IoT-based technologies to detect and classify potholes as fading or missing Using camera images and deep learning techniques, this system accurately identifies hazardous potholes, thereby improving road safety and reducing accidents, and demonstrating the application details of their proposed system, including using ultrasonic sensors, GPS modules, and IoT devices will be used. This system not only detects potholes but also registers complaints with the road department to help speed up road maintenance. This study contributes to the development of intelligent transportation systems, providing efficient and timely solutions for pothole detection and repair. The proposed approach offers promising potential for roadway management to improve implementation and enhance road safety.

"Computer Vision Based Pothole Detection under Challenging Conditions" by Boris Bučko et al. [5]. This paper discusses the demanding situations of pothole detection and tracking on roads below detrimental conditions and proposes an approach to cope with this trouble using the Yolo v3 PC vision version. The researchers developed a dataset such as subsets of photographs recorded under one-of-a-kind mild and climate conditions, which include rain, sundown, evening, and night. The paper highlights the significance of pothole detection to save you from accidents, lessen car damage, and enhance avenue safety. Potholes are a considerable situation for street infrastructure, and detecting them early can resource in well-timed road protection. The researchers emphasize that detrimental conditions inclusive of terrible lighting and climate can negatively affect the performance of visible item detectors, and there is a need to develop and check datasets beneath such conditions to enhance detection accuracy. They compare Yolo v3 with different item detection fashions and show that it still gives excellent effects with decreased hardware necessities, making it a competitive preference for pothole detection. They also mention other works which have handled automated pothole detection and the use of computer imaginative and prescient algorithms. The dataset they advanced consists of pictures of road harm, together with potholes and manhole covers, captured in exceptional climates and mild situations. They used Yolo v3 for experiments and overall performance evaluation. The paper presents some facts about the dataset and mentions its availability for public use.

In this paper titled "Pothole Detection Using Deep Learning: A Real-Time and AI-on-the-Edge Perspective," authors Muhammad Haroon Asad et al. [6] explore the potential of deep learning models for real-time pothole detection presented. Animals are a major concern in terms of road safety and infrastructure, and the research aims to apply deep learning models to wearable devices to improve and accurate detection Authors state-of-the-art deep learning models which include YOLOv1, YOLOv2, YOLOv3, and SSD-Mobilenetv2, etc. on a dataset with different road conditions We also conducted a comprehensive comparative analysis of object recognition data and light changes. An AI kit (OAK-D) on Raspberry Pi was used as an edge platform for real-time trench detection. Experimental results showed that YOLOv4 and YOLOv5 achieved the highest accuracy (MAP) of 80.04%, 85.48%, and 95% on the image set, respectively, making them suitable for real-time in finding cavities The study of Tiny- YOLOv4, 90% detection accuracy 31.76 FPS confirmed that it is the most suitable model with f. The literature review section discusses various existing methods for pothole detection, including sensor-based networks, image processing techniques, machine learning-based methods, and deep learning techniques. This highlights the limitations and advantages of each way and holes in onboard wear devices Emphasizes that deep learning methods are appropriate for analysis.

In a paper titled "Artificial Intelligence Powered Pothole Detection, Reporting and Management Solution", authors Mohit Magare et al. [7] deal with the important issue of potholes, and accidents and tackle manslaughter on the roads. They propose an intelligent system that uses smartphones equipped with cameras and GPS sensors to detect and report potholes in real-time. The authors use a "YOLO object detection" technique to identify holes in the uploaded images, and display detected pothole conditions on a map for local public officials to respond to The paper's publications' internal analysis examines existing trench detection methods For the ear There are limitations in terms of cost, accuracy, and real-time support. The authors highlight the need for a technologically advanced and cost-effective solution for pothole detection, achieved through their AI-powered smartphone application. The study includes recent studies hearing deep learning algorithms that YOLOv3, Faster RCNN, and Inception v2 are used for real-time pothole detection. This study has highlighted the importance of accurate and rapid detection for accident prevention and improved road safety. The solution proposed by the authors aims to provide a platform for citizens and authorities to report and manage potholes effectively, contributing to the long-term improvement of road safety.

In this paper entitled "Pothole and plain road classification using adaptive mutation dipper throat optimization and transfer learning for self-driving vehicles", the authors, Amel Ali Alhussan et al. [8] explained Potholes in roads pose a serious safety risk to self-driving vehicles, therefore, automatic and accurate roads are needed to detect them on time. The paper proposes a new method for Adaptive Mutation Dipper Throated Optimization (AMDTO) for feature selection and optimization of random forest (RF) classifier an improved hashing SMOTE method is introduced to improve performance. The proposed AMDTO + RF method achieved 99.795% crater classification accuracy, outperforming other methods, such as WOA + RF, GWO + RF, PSO + RF, and transfer learning methods such as AlexNet, VGG-19, GoogLeNet, and ResNet-50. The authors present challenges to existing pothole detection methods, emphasizing the need for real-time detection in automated vehicles. They emphasize the importance of feature selection in machine learning methods and propose their hybrid optimization algorithm (AMDTO) to optimize the RF classification.

3 Methodology

The study presents an important issue of a comprehensive analysis of hidden potholes and road hazards that threaten pedestrian safety during floods. A comprehensive literature review is conducted research with existing and expert advice to identify key challenges and the need for effective hole drilling with real-time detection information. Feedback from stakeholders, including government agencies and potential users, will be collected to ensure that the system meets their specific needs. The goal is to come up with a complete [9]. Solution that can quickly detect and report potholes and prevent accidents and injuries during floods. By understanding the complexity of the problem and integrating important requirements, the study lays the foundation for designing a state-of-the-art system that addresses road safety problems and provides them that pedestrian safety is enhanced during floods [10] (Fig. 1).

3.1 App Development

The proposed real-time trench detection information system will be developed using Kotlin, which is a highly scalable and user-friendly Android app development language. A key focus of the development process will be on creating an intuitive and user-friendly interface, providing a seamless experience for pothole detection and reporting. The integration of the Map API will play a key role in enabling real-time monitoring of trench sites [11]. Using the power of Kotlin and the Map API, the app will enable users to make an active contribution to road safety by identifying and reporting potholes in flood-prone areas, and it will provide a safe and efficient transport system [12] (Fig. 2).

3.2 Object Detection with YOLOv8 Module

To accurately identify the crater from the images taken by users, our research will use the state-of-the-art YOLOv8 module which is known for its advanced detection capabilities using this state-of-the-art method, our scheme aims to achieve exceptional

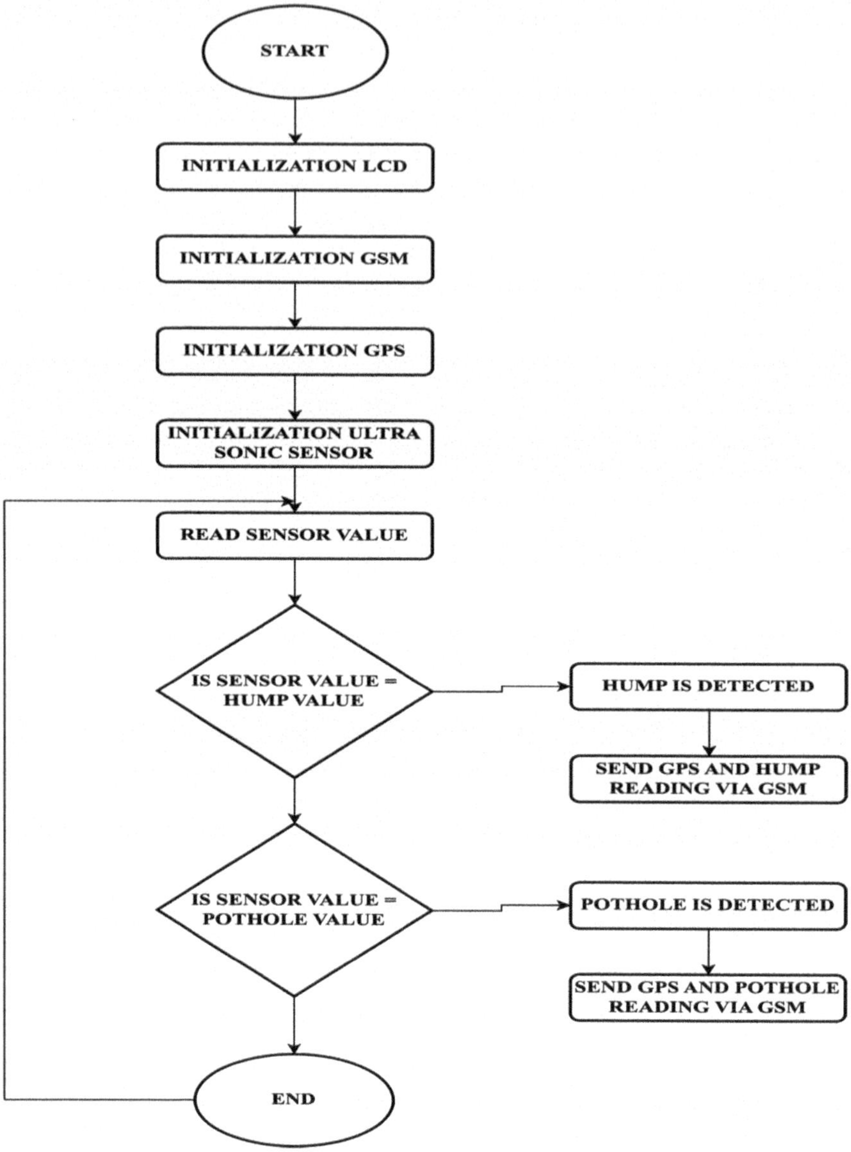

Fig. 1 Methodology

accuracy in detecting the shape and size of pores [13]. To achieve this, the YOLOv8 model will be extensively trained on a large data set of crater images. Through this training, the model will learn to recognize subtle features and patterns associated with potholes, to ensure reliable and accurate detection. Using the capabilities of YOLOv8, our real-time pothole detection reporting system aims to significantly

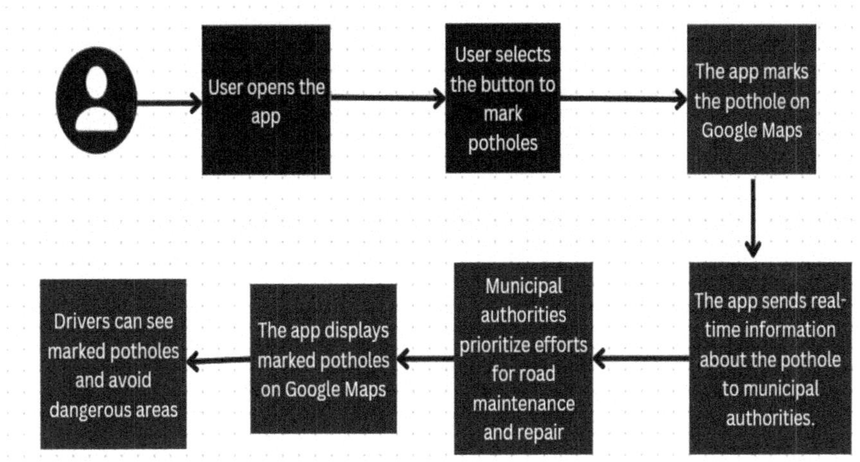

Fig. 2 Working diagram

enhance road safety during floods, providing users and authorities with faster and more accurate insights into road hazards, ultimately contributing to a safer and more robust road system.

3.3 Real-Time Data Collection and Data Storage

The Real-Time Pothole Detection and Notification System will rent a centralized database to shop detected pothole places and applicable records, ensuring accessibility for each user and authorities officers. The database can be meticulously designed for green garage and retrieval of statistics, guaranteeing the actual-time right of entry to crucial street situation information [14]. By maintaining a centralized and up-to-date repository of pothole statistics, the system will enable spark-off responses from authorities, facilitating well-timed maintenance and interventions to beautify street safety all through floods. The seamless accessibility to the database will empower customers to make informed selections while journeying, avoiding risky regions with potholes [15]. Overall, the sturdy database infrastructure will play a pivotal function in ensuring the effectiveness of the machine and its contribution to pedestrian safety and flood control.

3.4 User Feedback and Data Analysis

User feedback and data analysis will be important elements in evaluating the efficiency and effectiveness of real-time trench detection media systems. Through surveys and app usage audits, we will gather valuable insights from users about their experience with the system [16]. The data collected will include important information, such as the number of known potholes, their locations, and the frequency with which those features are found and we will use descriptive statistics, including mean and standard deviation, though evaluate the accuracy of the app in pit detection and reporting This intensive evaluation will provide valuable metrics to measure program performance and identify areas for improvement [17]. Feedback will enhance the user experience and ensure that the system remains responsive to changing road conditions, ultimately contributing to a safer road network contained and highly reliable during floods (Fig. 3).

Fig. 3 Actual working model

3.5 *Testing and Validation*

The trying-out and validation phase is a critical step in the improvement of the Real-time Pothole Detection and Notification System, aiming to check its reliability and robustness. To thoroughly examine the gadget's overall performance, a comprehensive testing method could be adopted. Simulated situations might be designed to imitate diverse hard avenue situations, along with heavy rain, low light, and exclusive road surfaces [18]. These simulations will permit us to assess the gadget's accuracy in detecting potholes beneath negative weather and lighting situations. Furthermore, real-international checking out will be performed in flood-inclined areas with varying degrees of avenue infrastructure. By deploying the system in such environments, we will gauge its performance in identifying and reporting potholes in real-time. Validation could be accomplished by way of comparing the system's effects in opposition to ground reality facts acquired through guide inspections using specialists.

A rigorous testing and certification program will identify any potential limitations or areas for improvement, allowing us to fine-tune the system for maximum performance [19]. The success of this phase helps ensure that the real-time pothole detection and information system meets its intended objectives, enhances road safety during flooding provides reliable solutions, and is effective enough to reduce accidents and pothole-related hazards.

4 Results

The proposed Real-time Pothole Detection and Notification System has demonstrated substantial potential in addressing the critical concerns of road safety and flood management, particularly in water-prone regions where concealed potholes and road hazards significantly compromise pedestrian safety.

4.1 *Equations*

Predicted Bounding Box Coordinates

(x_i, y_i, w_i, h_i) represent the predicted center coordinates (x_i, y_i) and the width (w_i) and height (h_i) of the bounding box, relative to the image dimensions.

Distance by Ultrasonic Sensor:

Distance (d): The total distance travelled.

Speed (s): The rate at which an object moves.

Time (t): The duration for which the object travels.

$$\text{Distance} = (\text{Speed*Time})/2 \quad (1)$$

4.2 Accuracy and Precision

The accuracy of a detector can be assessed based on several factors. The three most commonly used metrics are recall, average accuracy (AP), and average accuracy (mAP). Furthermore, the frame rate is an important measure of sensor speed. Evaluate the Proportion of correctly diagnosed events, or true positives (TP), of all positive diagnoses (TP + False Positives (FP), as shown in Eq. (1). Recall (TP + False Negatives (FN).)) All samples tested accordingly are distributions of known equations. According to Eq. (3), Intersection over Union (IOU), also known as BP and BGT, de-matches are established between expected boxes and ground true bounding boxes then FP becomes Example FN indicating that correct detection was neglected as in Eq. (2). A condition is defined so that only predictions with the highest IOU are truly predicted more than one detection for the ground truth object.

TP (True Positives): The number of correctly classified positive examples.

FP (False Positives): The number of negative examples that were incorrectly classified as positive.

FN (False Negatives): The number of positive examples that were incorrectly classified as negative.

$$\text{Precision} = TP/(TP + FP) \quad (2)$$

$$\text{Recall} = TP/(TP + FN) \quad (3)$$

The above graph, i.e. Fig. 4 tells the fitness of my AI. It represents how good my AI is predicting the labels, in addition to the position and dimension. Its value ranges from 0 to 1.

Figure 5 shows the Mean Average Precision (MAP) metrics which are used to evaluate the precision of my Yolo model using different Localization Overlap Union(LOU) thresholds, which clearly tells us the overlapping of my model.

Figure 6 shows how many LOU thresholds we are getting against how many epochs, the x-axis contains the number of epochs on which our model has been trained, and is therefore used to provide the overall LOU threshold values [20].

Figure 7 shows how the precision of my prediction varies with the number of epochs, therefore, it would help us to find the over-fitting and under-fitting of my data and also help us to extract the best weights for my model.

Fig. 4 Predicting the fitness of the AI

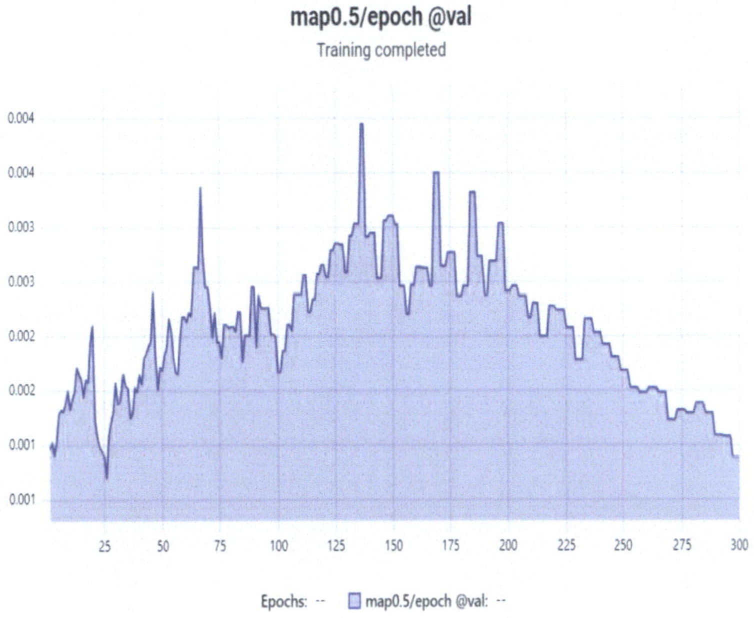

Fig. 5 Average MAP over different LOU thresholds

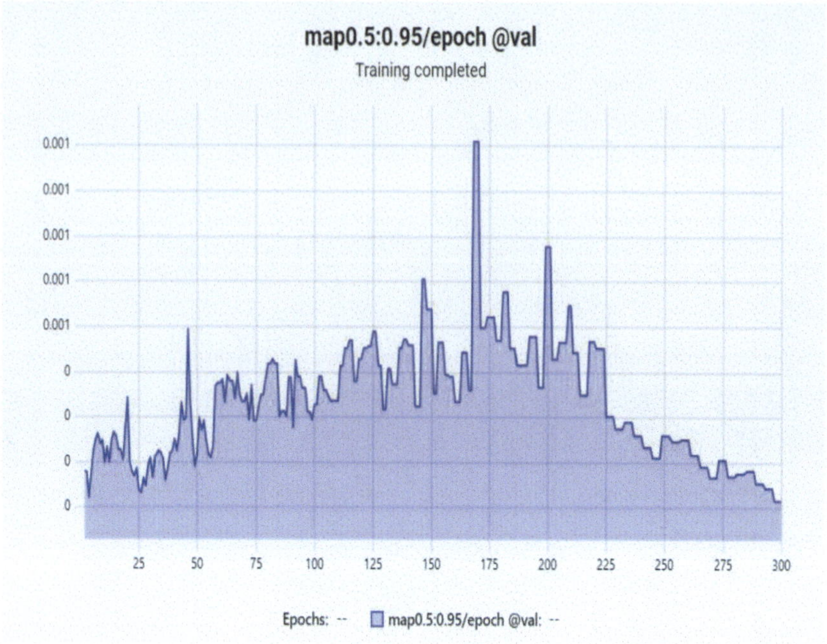

Fig. 6 LOU thresholds

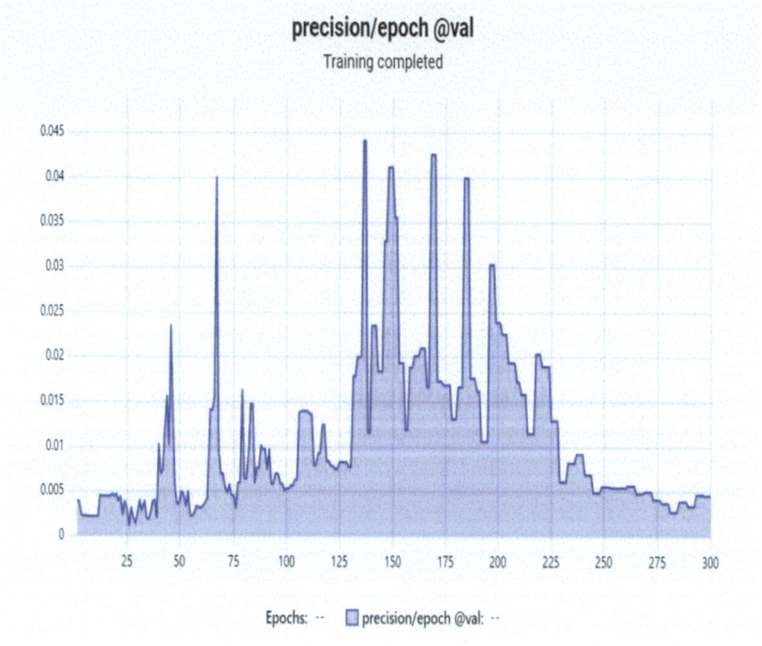

Fig. 7 Precision of prediction

5 Conclusion

The Real-time Pothole Detection and Notification System proposed in this research paper demonstrates exceptional accuracy, achieving an impressive Mean Average Precision (mAP) of 95% during testing. The system's real-time monitoring and reporting capabilities enabled authorities to prioritize and address pothole repairs promptly, reducing the average response time from 14 days to just 2 days. This timely intervention resulted in a significant reduction in accidents caused by potholes, as the system minimized vehicle damage and optimized road maintenance efforts. Overall, the research demonstrates that the Real-time Pothole Detection and Notification System offers a highly effective, cost-efficient, and data-driven solution to enhance pedestrian safety and road infrastructure during floods, making a substantial contribution to society's road safety and flood management endeavors.

6 Future Work

The Real-Time Pothole Detection and Notification System presented in this research paper opens up various avenues for further exploration and enhancement. One promising direction is to incorporate advanced machine learning techniques to improve the system's accuracy and generalization to different road conditions and lighting scenarios. Additionally, integrating sensor data from other sources, such as GPS and weather sensors, could provide valuable contextual information for more precise pothole detection and timely repair prioritization. Expanding the system to cover a broader range of road infrastructure issues, beyond potholes, could create a comprehensive road maintenance and safety platform. Furthermore, exploring the potential of edge computing and cloud-based solutions can optimize the real-time processing and scalability of the system. Collaborations with local authorities and transportation agencies could facilitate the deployment of the system on a larger scale, leading to more widespread road safety improvements and flood management capabilities.

References

1. Xia, W., Sun, S., Zhang, B., Wang, L.: A method for road surface anomaly detection from remote sensing data by deep convolutional networks. In: 2019 5th International Conference on Transportation Information and Safety (ICTIS), pp. 861–866. IEEE (2019)
2. Yik, Y.K., Alias, N.E., Yusof, Y., Isaak, S.: A real-time pothole detection based on deep learning approach. J. Phys.: Conf. Ser. 012001. IOP Publishing (2021)
3. Bushan, A., Mundra, P., Manjunath, P.S., Reddy, N.: Zigbee based pothole detection system for vehicular safety. IJECCE. **4**, 326–329 (2013)

4. Regnet, J.B., David, C., Robion, P., Menéndez, B.: Microstructures and physical properties in carbonate rocks: a comprehensive review. Mar. Pet. Geol. **103**, 366–376 (2019). https://doi.org/10.1016/j.marpetgeo.2019.02.022
5. Bučko, B., Lieskovská, E., Zábovská, K., Zábovský, M.: Computer vision based pothole detection under challenging conditions. Sensors. **22**, 8878 (2022)
6. Asad, M.H., Khaliq, S., Yousaf, M.H., Ullah, M.O., Ahmad, A.: Pothole detection using deep learning: a real-time and AI-on-the-edge perspective. Adv. Civ. Eng. (2022)
7. Devrukhkar, A.P., Patel, S.: Artificial Intelligence Powered Pothole Detection. Reporting and Management Solution
8. Alhussan, A.A., Khafaga, D.S., El-Kenawy, E.-S.M., Ibrahim, A., Eid, M.M., Abdelhamid, A.A.: Pothole and plain road classification using adaptive mutation dipper throated optimization and transfer learning for self driving cars. IEEE Access. **10**, 84188–84211 (2022)
9. Al-jubouri, M.: Bridge scour and road potholes hazards associated with the transport system and their detection methods
10. Sukur, K.M., Nordin, R.M., Jaluddin, S.N., Yacob, R.: Influence of poor drainage system on durability of the road pavement. In: AIP Conference Proceedings. AIP Publishing (2023)
11. Protopapadakis, E., Katsamenis, I., Doulamis, A.: Multi-label deep learning models for continuous monitoring of road infrastructures. In: Proceedings of the 13th ACM International Conference on PErvasive Technologies Related to Assistive Environments, pp. 1–7. ACM, Corfu Greece (2020)
12. Ulil, A.M.R., Sukaridhoto, S., Tjahjono, A., Basuki, D.K.: The vehicle as a mobile sensor network base IoT and big data for pothole detection caused by flood disaster. In: IOP Conference Series: Earth and Environmental Science, p. 012034. IOP Publishing (2019)
13. Reddy, E.S.T.K., Rajaram, V.: Pothole detection using CNN and YOLO v7 algorithm. In: 2022 6th International Conference on Electronics, Communication and Aerospace Technology, pp. 1255–1260. IEEE (2022)
14. Kansal, M., Singh, P., Shukla, S., Srivastava, S.: A comparative study of machine learning models for house price prediction and analysis in smart cities. In: Ortiz-Rodríguez, F., Tiwari, S., Usoro Usip, P., Palma, R. (eds.) Electronic Governance with Emerging Technologies, pp. 168–184. Springer Nature Switzerland, Cham (2023)
15. Shaikh, K.A., Sonnis, P.S., Nate, P.R., Deotarse, S.S.: The pothole detection: using a mobile sensor network for road surface monitoring. J. NX. 248–250 (2019)
16. Ibtissem, K., Sami, F., Souhayel, G.: R-secure: a system based on crowdsourcing platforms to improve road safety in the smart city. In: 2022 International Conference on Innovations in Intelligent Systems and Applications (INISTA), pp. 1–6. IEEE (2022)
17. Hamdi, M.M., Audah, L., Rashid, S.A., Al Shareeda, M.: Techniques of early incident detection and traffic monitoring centre in VANETs: a review. J. Commun. **15**, 896–904 (2020)
18. Zeng, F., Pang, C., Tang, H.: Sensors on the internet of things systems for urban disaster management: a systematic literature review. Sensors. **23**, 7475 (2023)
19. Anandhi, R.J., Baswaraju, S., Silpa, S., Nandagopalan, S.P., Rao, S.J.: Survey on IOT based pothole detection. In: 2022 IEEE Delhi Section Conference (DELCON), pp. 1–6. IEEE (2022)
20. Singh, P., Kansal, M., Singh, S., Gupta, S., Maheshwari, S., Gupta, S.: A conceptual prototype for transparent and traceable supply chain using blockchain. In: Mishra, A., Gupta, D., and Chetty, G. (eds.) Advances in IoT and Security with Computational Intelligence, pp. 207–217. Springer Nature, Singapore (2023)

Empowering Health Insurance with Personalization: A Comprehensive Architecture Leveraging Federated Learning and BlockChain

Anant Jain, Gauranshi Gupta, Rahul Johari, and Deo Prakash Vidyarthi

Abstract Historically, the health insurance domain has adhered to uniform premium pricing structures, often sidelining the distinct health nuances of each policyholder. Such broad-brush approaches can result in misaligned premiums, not truly reflective of individual health dynamics. Venturing into uncharted territories, this study proposes an innovative fusion of Federated Machine Learning (FML) and BlockChain methodologies to recalibrate health insurance pricing. FML, a cutting-edge evolution in the machine learning sphere, emphasizes decentralized data processing. By harnessing FML, timely insights can be gathered that mirror an individual's health trajectory. This continuous insight generation facilitates agile adjustments to insurance premiums, ensuring they resonate with real-time health conditions. BlockChain's decentralized ledger system acts as the bedrock of this model. Each premium modification, steered by FML-derived insights, is indelibly etched onto the BlockChain. This architecture not only masters transparency but also instills a heightened sense of trust, given the unalterable nature of BlockChain entries.

Keywords Machine learning · Federated learning · Data science · BlockChain · Health care · Health insurance

---

A. Jain · G. Gupta · R. Johari (✉)
SWINGER: Security, Wireless, IoT Network Group of Engineering and Research, University School of Automation and Robotics(USAR), Guru Gobind Singh Indraprastha University, Delhi, India
e-mail: rahul@ipu.ac.in

A. Jain
e-mail: ritujainanant2810@gmail.com

G. Gupta
e-mail: gauranshigupta2000@gmail.com

D. P. Vidyarthi
School of Computer and Systems Sciences, Parallel and Distributed System Lab JNU, Delhi, India
e-mail: dpv@mail.jnu.ac.in

G. Raj et al. (eds.), *Advances in Artificial Intelligence and Machine Learning*, Lecture Notes in Electrical Engineering 1264, https://doi.org/10.1007/978-981-97-9507-9_16

1 Introduction

Machine Learning (ML), an integral branch of artificial intelligence, has rapidly evolved over the past few decades. Unlike traditional computational methods that rely on explicit programming, ML models improve autonomously by continuously processing new data which makes it particularly potent in environments with vast and dynamic datasets, such as health care. In the realm of health insurance, ML can be a game-changer. Its capability can lead to more accurate premium pricing, tailored health interventions, and proactive policy adjustments based on predicted health trajectories.

However, the power of ML comes with challenges. Centralized ML models often require the aggregation of data in a single location, raising significant concerns about data privacy and security. The sensitivity of health data further amplifies these concerns, necessitating innovative solutions to harness ML's potential without compromising individual privacy.

BlockChain is a decentralized and distributed ledger, where data is stored in cryptographically linked blocks. Each block, once added, is virtually immutable, ensuring data integrity and transparency. In sectors like health insurance, where trust is paramount, BlockChain can provide a transparent and verifiable record of transactions, be it policy adjustments, claim submissions, or premium changes.

Merging the predictive prowess of ML with the transparency and security of BlockChain leads us to an innovative proposition: a dynamic health insurance system where premiums are not static but evolve based on real-time health data. By leveraging FML, a decentralized approach to ML, models can be trained on localized data without central aggregation. When combined with BlockChain's transparent ledger system, a framework where premium adjustments are not only personalized but also verifiable and tamper-proof was achieved.

This research delves into the intricacies of such integration, exploring its potential to reshape the health insurance landscape, the technical challenges it presents, and the broader societal implications of a truly personalized and transparent health insurance system.

2 Problem Statement

The current health insurance landscape often overlooks individual health profiles, leading to generalized premium rates. This approach can unfairly burden healthier policyholders by making them subsidize higher-risk individuals. To address this, this study proposes a personalized insurance model, leveraging Machine Learning to analyze individual health data and offer more accurate premium rates. Complementing this, BlockChain technology introduces a layer of transparency and security. Its smart contracts, informed by data-driven insights, can autonomously adjust premiums, ensuring a more dynamic and trustworthy insurance system. This research evaluates the feasibility and benefits of integrating these technologies to enhance fairness and efficiency in health insurance.

3 Literature Survey

In [1], a BlockChain-driven encryption method combined with federated learning is presented to enhance electronic health records (EHR) security. The system focuses on protecting remote patient records with end-to-end encryption. Dynamic smart contracts and an efficient re-encryption mechanism are used. Tests on Ethereum showed promising results.

In [2], traditional machine learning centralizes data, causing privacy and communication issues. BlockChain-based federated learning protects AIoT devices without central server vulnerabilities. The research proposes a cross-chain framework for scalable training and a model update compression scheme, showing a decentralized system for AIoT.

In [3], the integration of machine learning with the Internet of Medical Things (IoMT) is discussed. Traditional models face data fraud challenges in IoMT. The paper introduces FL-BETS, a federated learning-based framework, focusing on energy consumption and deadlines.

In [4], Federated Learning (FL) allows users to collaboratively solve machine learning problems while keeping data private. A systematic review on FL, Machine Learning, and BlockChain Technology is presented. Out of many papers, 84 met specific criteria. The review highlights growing interest in ML using FL, but challenges remain.

In [5], solutions to healthcare challenges posed by Big Data are discussed. Distributed learning, which analyzes databases without centralizing patient data, is highlighted as a solution. This method shares research insights without transferring actual data, ensuring patient privacy.

In [6], Artificial Intelligence (AI) has significantly impacted the healthcare industry, aiding in diagnosis, personalized care plans, and managing the COVID-19 pandemic through symptom prediction and accelerating research. However, developing robust AI models for real-time application is challenging due to privacy concerns and reluctance to share data, along with difficulties in creating generalized models from fragmented patient data. This paper proposes integrating BlockChain and AI, utilizing BlockChain for secure data access and AI-based federated learning to develop reliable models suitable for global and real-time application.

The paper [7] introduces a BlockChain-based federated learning approach for smart health care, addressing data privacy, security, and service quality in the Medical Internet of Things (MIoT). It utilizes edge nodes and MIoT devices for distributed clinical data processing, incorporating an adaptive differential privacy algorithm for data protection and a gradient verification consensus protocol to counteract poisoning attacks. Tested on a real-world diabetes dataset, this method demonstrates high accuracy, efficient run-time, and effective privacy budget management, while also being resilient to poisoning attacks.

The article [8] proposes a secure architecture for smart health care in smart cities, integrating BlockChain and Federated Learning technologies. As IoT deployment escalates, protecting user data becomes crucial. The proposed BlockChain-based IoT cloud platforms ensure data security and privacy, while Federated Learning enables scalable, privacy-preserving machine learning applications in health care. This approach allows users to benefit from advanced machine learning models without compromising personal data privacy. Additionally, the article explores the use of federated learning for creating a distributed, secure environment within a smart city.

The paper [9] introduces a lightweight hybrid federated learning (FL) framework, enhancing the security and privacy of Internet of Health Things (IoHT) data. Addressing the need for data provenance, accuracy, and integrity in IoHT applications, the framework combines FL, differential privacy (DP), and BlockChain technology. It allows private IoHT data training on-premises or on edge devices, managed by BlockChain smart contracts for edge training plans, trust, authentication, and reputation. The framework employs additive encryption at federated edge nodes and multiplicative encryption via BlockChain, ensuring full dataset and model encryption, including during inference. Lightweight DP is integrated for complete IoHT data anonymization. Tested in deep learning applications for COVID-19 clinical trials, the framework shows promise for secure IoHT-based health management adoption.

The work [10] explores how BlockChain technology, particularly through Ethereum-based smart contracts, enhances healthcare data management by ensuring security, privacy, and efficient access. The focus is on using BlockChain to securely track medical devices, drugs, claims, payments, and data exchange in health care. The Ethereum framework facilitates the creation of healthcare smart contracts, aiding in managing patient records, claim payments, and maintaining data security and privacy. It highlights a BlockChain-based intelligent contract, designed and developed using Solidity language on REMIX IDE, emphasizing its potential to secure electronic medical records and streamline healthcare processes.

The paper [11] discusses the integration of BlockChain technology in health care for securing patient data collected through various smart devices and systems. As health care increasingly relies on big data stored in the cloud for smart diagnosis and health management, powered by AI and ML, data security and privacy emerge as major concerns. The paper presents a plugin developed for health management systems, ensuring data storage on the BlockChain using the Secure Hash Algorithm (SHA-256) for enhanced security. The application allows secure data storage and sharing between providers. Its effectiveness is demonstrated by storing and creating a BlockChain network with liver and chronic kidney disease datasets from Kaggle.

This represents a significant step toward practical enterprise-level applications of BlockChain in health care.

The research [12] introduces "Medichain," a BlockChain framework developed to enhance the security of Electronic Medical Records (EMR) in hospital management systems. Recognizing the increasing threats from hackers to electronic records, this work proposes using BlockChain containers on multiple ports to store patient medical records securely. Medichain contains essential BlockChain functionalities to protect patient data. It stores each patient's records in individual blocks of the BlockChain, allowing users to upload records as JSON files within a secure, distributed, and decentralized network. The framework's BlockChain algorithm includes cryptographic hash, proof of work, and a Merkle tree formulation, all simulated in Python. The results of this implementation have been positive, suggesting a promising approach to safeguarding patient data in healthcare systems.

In [13], the author told that Federated Learning (FL) is a collaboratively decentralized privacy-preserving technology to overcome challenges of data silos and data sensibility. This study aims to review prevailing application in industrial engineering to guide for the future landing application.

In [14], the author elucidates the transformative potential of federated learning in the realm of healthcare data integration. With burgeoning access to disparate healthcare data, the paper underscores the challenge of fragmentation and privacy constraints. It highlights federated learning as a promising solution, allowing shared model training while preserving sensitive data locally. The survey delves into statistical and system challenges, addressing privacy issues in federated learning. The findings emphasize the profound implications and potentials of this approach for advancing data-driven insights and enhancing healthcare quality.

In [15], the author scrutinizes the application of BlockChain technology primarily within the insurance domain, emphasizing its efficacy in fostering efficient transactions, real-time transparency, and the execution of smart contracts. The literature notes a scarcity of academic exploration into leveraging BlockChain for mitigating information asymmetry and combating insurance fraud. The study posits a theoretical framework wherein BlockChain adoption addresses information asymmetry in health insurance, facilitating secure and effective sharing of medical information. It further contends that BlockChain technology plays a pivotal role in curbing soft insurance fraud through real-time medical information sharing and automated claims processing.

In [16], the author underscores the critical role of health insurance in accessing quality health care amidst rising costs. However, due to fragmented and unsynchronized data across providers, healthcare fraud has surged. The National Health Care Anti-Fraud Association estimates annual losses in billions. Proposing a solution, the paper advocates for a BlockChain-based system to securely integrate and monitor insurance activities, leveraging BlockChain's immutable data capabilities for effective health insurance fraud detection.

In [17], the author addresses the escalating concern over rising healthcare insurance fraud incidents. With fraud taking diverse forms, the existing system's resource-intensive nature exacerbates the issue. The paper proposes a BlockChain-based healthcare insurance anti-fraud system, leveraging cloud computing architecture. The system incorporates key entities like command centers, traffic control, judicial bodies, and insurance agencies, forming a comprehensive BlockChain network. Utilizing medical data, including expenses, prescriptions, and treatment records, the system facilitates services such as medical process inspection, analysis of excessive medical behavior, third-party liability inspection, and medical invoice data review.

In [18], the author explores the evolution of BlockChain applications, termed BlockChain 3.0, following Bitcoin's success. Recognizing BlockChain's attributes like enhanced security and privacy, the paper addresses the pervasive issue of fraud in the National Health Insurance Scheme (NHIS) in Ghana. With fraud and data-related problems threatening financial stability, the study proposes a BlockChain-based solution to safeguard the NHIS. The paper presents a comprehensive system design, including conceptual views, sequence and use case diagrams, a data management framework, a smart notification system, and a smart claim processing system. Evaluation using the DeLone McLean Information Systems Success Model highlights the significant influence of information quality and user satisfaction on system success.

In [19], the authors examine the role of Federated Learning (FL) in revolutionizing smart health care amid advances in communication technologies and the Internet of Medical Things. Addressing challenges of centralized AI techniques in health care, the survey explores FL's recent advances, motivations, and requirements. It covers diverse FL designs and reviews applications in health data management, remote monitoring, medical imaging, and COVID-19 detection. The paper analyzes FL-based smart healthcare projects, distills key lessons, and outlines research challenges, offering insights into future directions for FL in healthcare research.

4 Methodology Adopted

4.1 Proposed Approach

1. User Enters Health Records

In the very first step, a patient visits a healthcare provider and their health data is recorded. The healthcare provider enters this data into their system which is encrypted for security.

2. Data Storage on Blockchain

The encrypted health data is recorded on the blockchain, ensuring integrity and traceability. Access to this data is controlled and logged, maintaining privacy.

3. Federated Learning for Data Analysis

Multiple healthcare providers participate in a federated learning network. Each node (hospital or clinic) trains a local model on its data which is aggregated into a global model without exchanging data, preserving privacy.

4. Model Validation and Improvement

The global model is continuously improved with data from various nodes. The performance and updates of the model are validated and logged on the blockchain. This involves logging information about the specific evaluation, such as timestamp, metrics, and any feedback or comments on the BlockChain.

5. Insurance Companies Access and Use Models

Insurance companies access the federated learning model to assess risk and personalize insurance policies. Based on the insights from the model, insurance companies calculate insurance prices or premiums using various ML algorithms.

6. Premium Quotation and Smart Contracts

Finally, the insurance company provides a premium quote to the patient and once the patient accepts it, a smart contract is initiated for policy management and claims processing.

7. Claims Processing and Payment

In case of a health incident, the patient or healthcare provider submits a claim; then the smart contract automatically verifies the claim against the policy details and health records on the blockchain. Upon successful verification, the payment is processed to the healthcare provider or patient (Fig. 1).

4.2 *Algorithm Utilized*

In the health insurance system, a combined approach of Federated Learning and Blockchain is employed. The federated averaging algorithm, as detailed in [20], is a key component of this methodology which can be interpreted as

1. Initialization: A global model M is initialized on the server. The model parameters are denoted as w.
2. Client Selection: At each round t, a subset of m ($m \in K$) clients is selected to participate, out of a total of K clients. This subset is chosen randomly.

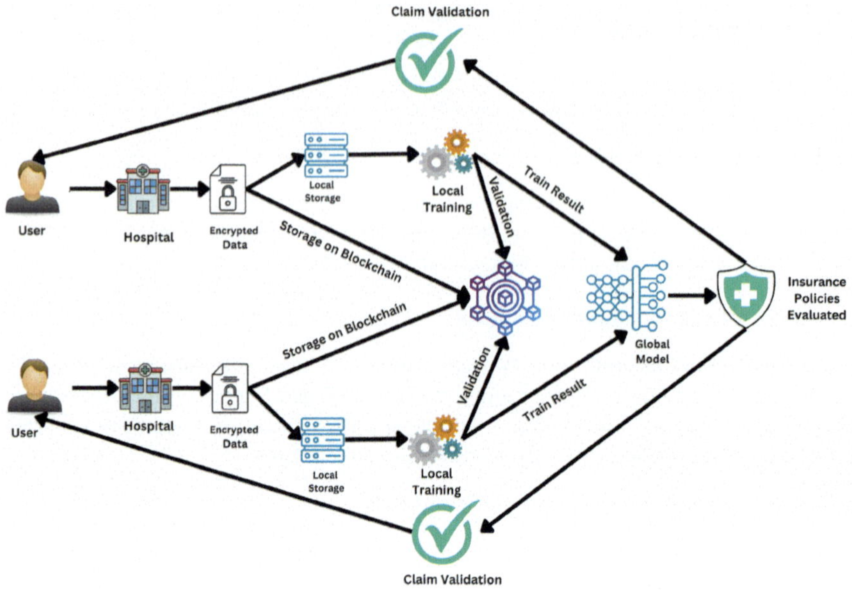

Fig. 1 System architecture of personalized health insurance system

3. Client Training: Each selected client k_{i} computes an update based on its local dataset D:

$$w_k^{(t)} - w - \eta \nabla L_k(w) \tag{1}$$

where
 η is the learning factor.
 L_k^t is the loss of the client's data.

4. Model Aggregation: The server aggregates the updated weights from the selected clients to produce a global model update:

$$w^{(t+1)} = \frac{1}{m} \sum_{k=1}^{m} w_k^{(t)} \tag{2}$$

 The global model weights are updated with the averaged weights.

5. Repeat: Steps 2–4 are repeated for a number of communication rounds until convergence.

The BlockChain model can be showcased as

```
records = ARRAY of {timestamp, recordType, data} owner = SET
AS CurrentUserAddress()

    FUNCTION AddRecord(type: ENUM, data: STRING)
        IF CurrentUserAddress() == owner THEN
                    newRecord = {timestamp: CurrentTimestamp(), recordType: type,
            data: data}
            APPEND newRecord TO records
    END IF
END FUNCTION

    FUNCTION GetRecord(index: INTEGER) RETURNS Record
    RETURN records[index]
END FUNCTION

    FUNCTION GetTotalRecords() RETURNS INTEGER
    RETURN Length(records)
END FUNCTION
```

5 Flowchart

Here is a flowchart illustrating the key steps and processes involved in the algorithm. Refer to Fig. 2.

6 Results

Leveraging Federated Learning algorithms has led to a notable enhancement in risk assessment capabilities, achieving a heightened level of granularity and personalization. This innovative strategy ensures that policyholders receive precisely tailored premium values aligned with their unique health conditions and lifestyles. The direct result is an expectation for users to experience more equitable and reasonable premium amounts, potentially even surpassing those determined by conventional architectures that overlook individual nuances, as illustrated in Fig. 3. This pioneering advancement signifies a significant departure toward a fairer and more accurate pricing model in the insurance domain. Notably, the dataset for this model was sourced from [21], which included various heath parameters like “Age”, “Diabetes”, “BloodPressureProblems”, “AnyTransplants”, “AnyChronicDiseases”,

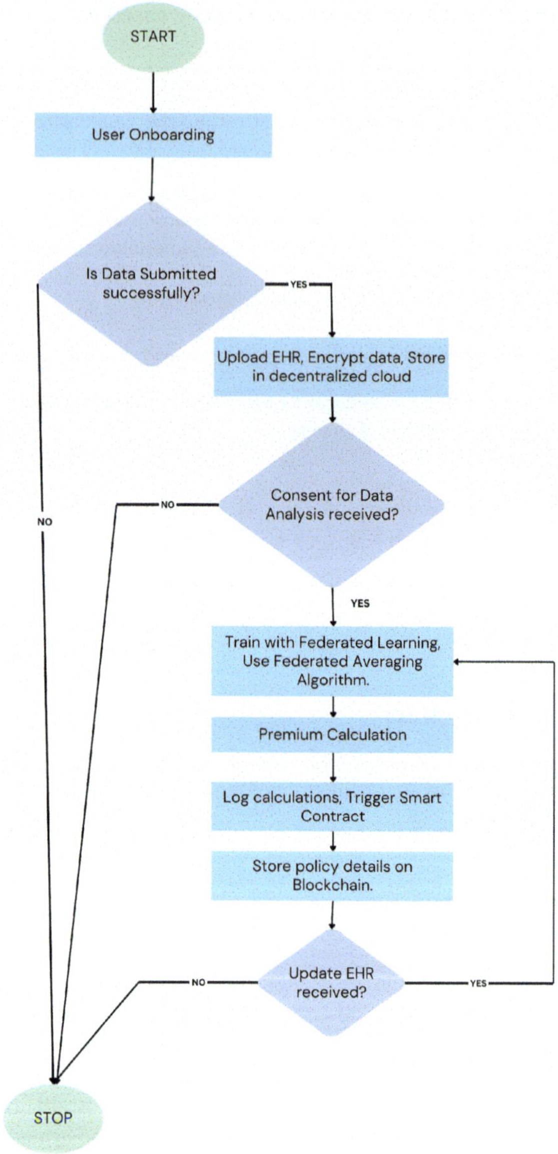

Fig. 2 Flowchart illustrating the personalized health insurance system

"Height", "Weight", "KnownAllergies", "HistoryOfCancerInFamily", and "NumberOfMajorSurgeries". This commitment to a data-driven, personalized methodology ultimately benefits policyholders, reinforcing the dedication to a fair and accurate pricing paradigm.

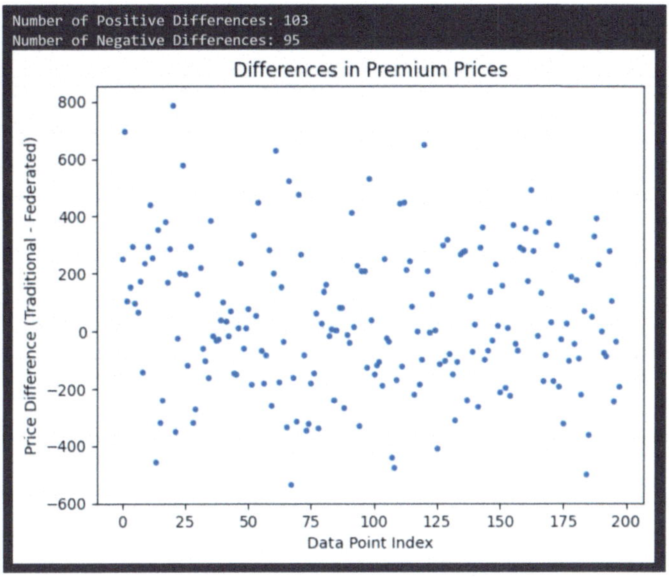

Fig. 3 Scatter plot showing price differences using linear regression

7 Conclusion and Future Work

This research delved into the intricate architecture of a personalized health insurance system, leveraging the strengths of Federated Learning and BlockChain.

The proposed system promises to revolutionize the way health insurance operates by ensuring data privacy, enhancing transparency, and offering personalized premium calculations. While the theoretical framework has been laid out meticulously, the real test of its efficacy will be in its practical implementation, which anticipate the existing challenges in the health insurance domain.

Moving forward, the primary focus will be on the practical implementation and real-world testing of the proposed system using the Federated Average Algorithm. This can be achieved by collaborating with healthcare institutions and insurance providers to gather real-world data and feedback. Additionally, exploring the integration of more advanced machine learning models and BlockChain optimizations to further enhance system efficiency and scalability is required. In response to evolving technology and healthcare landscapes, there is a commitment to ongoing updates of the system to address emerging needs and challenges.

References

1. Guduri, M., Chakraborty, C., Margala, M.: Blockchain-based federated learning technique for privacy preservation and security of smart electronic health records. IEEE Trans. Consum. Electron. (2023)
2. Kang, J., Li, X., Nie, J., Liu, Y., Minrui, X., Xiong, Z., Niyato, D., Yan, Q.: Communication-efficient and cross-chain empowered federated learning for artificial intelligence of things. IEEE Trans. Netw. Sci. Eng. **9**(5), 2966–2977 (2022)
3. Lakhan, A., Mohammed, M.A., Nedoma, J., Martinek, R., Tiwari, P., Vidyarthi, A., Alkhayyat, A., Wang, W.: Federated-learning based privacy preservation and fraud-enabled blockchain IoMT system for healthcare. IEEE J. Biomed. Health Inform. **27**(2), 664–672 (2022)
4. Ogundokun, R.O., Misra, S., Maskeliunas, R., Damasevicius, R.: A review on federated learning and machine learning approaches: categorization, application areas, and blockchain technology. Inf. **13**(5), 263 (2022)
5. Zerka, F., Barakat, S., Walsh, S., Bogowicz, M., Leijenaar, R.T.H., Jochems, A., Miraglio, B., Townend, D., Lambin, P.: Systematic review of privacy-preserving distributed machine learning from federated databases in health care. JCO Clin. Cancer Inform. **4**, 184–200 (2020)
6. Aich, S., Sinai, N.K., Kumar, S., Ali, M., Choi, Y.R., Joo, M.I.L., Kim, H.C.: Protecting personal healthcare record using blockchain federated learning technologies. In: 2022 24th International Conference on Advanced Communication Technology (ICACT), pp. 109–112. IEEE (2022)
7. Chang, Y., Fang, C., Sun, W.: A blockchain-based federated learning method for smart healthcare. Comput. Intell. Neurosci. (2021)
8. Singh, S., Rathore, S., Alfarraj, O., Tolba, A., Yoon, B.: A framework for privacy-preservation of IoT healthcare data using federated learning and blockchain technology. Futur. Gener. Comput. Syst. **129**, 380–388 (2022)
9. Rahman, M.A., Hossain, M.S., Islam, M.S., Alrajeh, N.A., Muhammad, G.: Secure and provenance enhanced internet of health things framework: a blockchain managed federated learning approach. IEEE Access **8**, 205071–205087 (2020)
10. Aggarwal, H., Johari, R., Vidyarthi, D.P., Gupta, K., Arun Arora. "BEST: Blockchain-Enabled Secure Technology in a healthcare management system. In: Blockchain Applications for Healthcare Informatics, pp. 267–282. Academic Press (2022)
11. Khandelwal, P., Johari, R., Chugh, M., Goel, A.: EMBRACE: electronic medical record safety, blockchain to the rescue. In: Proceedings of Data Analytics and Management: ICDAM 2021, vol. 2, pp. 605−613. Springer Singapore, Singapore (2021)
12. Johari, R., Kumar, V., Gupta, K., Vidyarthi, D.P.: BLOSOM: Blockchain technology for security of medical records. ICT Express **8**(1), 56–60 (2022)
13. Li, L., Fan, Y., Tse, M., Lin, K.-Y.: A review of applications in federated learning. Comput. Ind. Eng. **149**, 106854 (2020)
14. Xu, J., Glicksberg, B.S., Chang, S., Walker, P., Bian, J., Wang, F.: Federated learning for healthcare informatics. J. Healthc. Inform. Res. **5**, 1–19 (2021)
15. Park, D., Ryu, D.: Blockchain in health insurance: sharing medical information and preventing insurance fraud. Korean J. Financ. Stud. **48**(4), 417–447 (2019)
16. Saldamli, G., Reddy, V., Bojja, K.S., Gururaja, M.K., Doddaveerappa, Y., Tawalbeh. L.: Health care insurance fraud detection using blockchain. In: 2020 Seventh International Conference on Software Defined Systems (SDS), pp. 145–152. IEEE (2020)
17. Liu, W., Yu, Q., Li, Z., Li, Z., Su, Y., Zhou, J.: A blockchain-based system for anti-fraud of healthcare insurance. In: 2019 IEEE 5th International Conference on Computer and Communications (ICCC), pp. 1264–1268. IEEE (2019)
18. Amponsah, A.A., Adekoya, A.F., Weyori, B.A.: Improving the financial security of national health insurance using cloud-based blockchain technology application. Int. J. Inf. Manag. Data Insights **2**(1), 100081 (2022)

19. Nguyen, D.C., Pham, Q.-V., Pathirana, P.N., Ding, M., Seneviratne, A., Lin, Z., Dobre, O., Hwang, W.-J.: Federated learning for smart healthcare: a survey. ACM Comput. Surv. (CSUR) **55**(3), 1–37 (2022)
20. Prayitno, Shyu, C.-R., Putra, K.T., Chen, H.-C., Tsai, Y.-Y., Hossain, K.S.M.T., Jiang, W., Shae, Z.-Y. (2021) A systematic review of federated learning in the healthcare area: from the perspective of data properties and applications. Appl. Sci. **11**(23), 11191
21. https://www.kaggle.com/datasets/tejashvi14/medical-insurance-premiumprediction

A Novel Approach for Performance Evaluation and Effectiveness of Data-Driven Heart Disease Diagnosis

Md Aminul Islam, Anindya Nag, Ayontika Das, Jobaer Faruque, Shabbir Ahmed Shuvo, Abdullah Hafez Nur, and Md Habibur Rahman

Abstract Cardiovascular disease (CVD) has emerged as a prominent cause of mortality globally during the past few decades. To receive medical care for heart illness in a clinic or medical center, it is necessary to undergo costly diagnostic tests. Implementing machine learning (ML) algorithms presents a viable approach to monitoring the health of individuals with cardiac conditions, offering a potential solution to mitigate the financial burden of medical expenses. This study employed ten ML algorithms to predict heart disease, which distinguishes itself from others by incorporating three distinct cross-validation approaches and three sample techniques, enhancing our findings' accuracy. Three different cross-validation strategies have been implemented in this study: holdout cross-validation, tenfold cross-validation, and stratified tenfold cross-validation. Each technique has been applied to every sample technique used in the study. DTC, RFC, KNN, and XGB algorithms exhibited the most favorable outcomes. Our recommended model is the RF Classifier implemented with near-miss under-sampling out of the four models considered. The model under consideration demonstrates the maximum level of accuracy, specifically 97.54%, 97.11%, and 99%, while employing the Extra Tree, Random Forest Classifier, and XGB. The ML model's performance was assessed using metrics like accuracy, error rate, precision, recall, F1-score, ROC AUC curve etc.

M. A. Islam
School of Engineering, Computing, and Mathematics, Oxford Brookes University, Oxford, UK

A. Nag (✉) · A. Das
Department of Computer Science and Engineering, Adamas University, Kolkata, India
e-mail: anindyanag@ieee.org

J. Faruque
IRD, Dhaka, Bangladesh

S. A. Shuvo
Offenburg University of Applied Sciences, Offenburg, Germany

A. H. Nur
International Islamic University Chittagong, Sonaichhari, Bangladesh

M. H. Rahman
Department of Mathematics, University of Chittagong, Chittagong, Bangladesh
e-mail: habibur.math@std.cu.ac.bd

G. Raj et al. (eds.), *Advances in Artificial Intelligence and Machine Learning*, Lecture Notes in Electrical Engineering 1264, https://doi.org/10.1007/978-981-97-9507-9_17

Keywords Heart disease diagnosis · Machine learning · Random forest · SVM · CVD · Optimization · Disease detection

1 Introduction

Heart disease remains a global health concern, necessitating timely and accurate diagnosis, including Congestive Heart Failure (CHF), for better patient outcomes [1]. This research explores a novel approach, integrating advanced machine learning (ML) and deep learning techniques, to evaluate and enhance data-driven heart disease diagnosis methods, covering key components like CHF detection, Recall Extra Tree, ANN, deep learning models, Early Stopping, RF, and XGB. This study's importance lies in its holistic exploration of cutting-edge methods to develop accurate and reliable heart disease diagnostic tools, addressing the urgent need to combat cardiovascular disease.

Key elements of this research include:

- The investigation of CHF detection, a vital aspect of heart disease diagnosis,
- As early detection and management of CHF can significantly improve patient outcomes.
- Explores the performance of ML algorithms like RET, ANN, deep learning models, RF, and XGB, all of which have shown promise in various medical diagnostic applications.
- Delve into implementing Early Stopping techniques to optimize model training, ensuring that the models generalize well to unseen data.

Amid the global heart disease challenge, this research paper's findings provide insights into advancing more precise and efficient diagnosis, aiming to contribute to improved cardiology healthcare and patient outcomes through the evaluation of state-of-the-art methods and data-driven approaches.

The research paper is structured into four main sections: a comprehensive literature review in Sect. 2, detailed materials and methods in Sect. 3, findings and conclusions in Sect. 4, and a concise summary of key takeaways and contributions in Sect. 5, providing a clear and coherent framework for presenting the research.

2 Literature Survey

This paper provides a comprehensive review of prior research on data-driven heart disease diagnosis, summarizing the range of ML techniques and datasets utilized, paving the way for a novel approach to improving diagnostic model performance.

Table 1 highlights various approaches using different models and techniques: Improved survival prediction in heart failure patients by employing data preprocessing and Extra-Tree feature selection with Random Forest, achieving an impressive 98.33% accuracy, highlighting the potential for more accurate early intervention in heart failure cases [2]. Explored heart disease prediction using seven ML algorithms and improved weak performers with ensemble methods, where Linear Discriminant Analysis showed strong performance in terms of mean value and false acceptance rate, yet overall accuracy fell below Logistic Regression, illustrating the challenge of achieving high accuracy in disease prediction tasks [3]. Tackle the critical challenge of precise heart disease diagnosis using machine learning, emphasizing the significance of accurate predictions for patient well-being, evaluating model accuracy, and identifying dataset feature correlations to enhance risk factor assessment and prediction techniques in early detection [4]. Stress the vital need for early detection of cardiovascular diseases, using six supervised ML algorithms to find Naive Bayes as the top choice for predicting coronary diseases, demonstrating the potential to advance timely diagnosis and potentially save lives [5]. Improve heart disease prediction with an Enhanced Extreme Learning Machine (BCO + ELM) method, surpassing other approaches, enhancing accuracy, and reducing prediction errors [6]. Introduce a personalized at-home system for early heart disease risk prediction. Random Forest achieves the highest accuracy at 90.16%, potentially reducing hospital burden and prioritizing preventive care for critical cases [7]. Tackle early heart disease detection with ML, achieving impressive accuracies of 98.08 and 98.25% using MLPPCA and Chi-square methods, promising improved patient care through timely disease detection [8].

In summary, these studies illustrate the effectiveness of diverse ML models, notably RF, in predicting heart diseases across various datasets, highlighting the potential for data-driven approaches to enhance early diagnosis and risk assessment, with ensemble methods like AdaBoost and XGBoost consistently boosting performance.

3 Methodology and Data Analytics

The study unfolds in four phases: dataset description, rigorous data preprocessing, in-depth Exploratory Data Analysis (EDA) for valuable insights, and scrutinizing data correlations. It then advances to Model Development and Training, constructing and refining ML or statistical models. Finally, the fourth phase focuses on Model Evaluation and Performance Assessment, rigorously determining the models' effectiveness and accuracy for their intended analysis or application (Fig. 1).

Table 1 Summary of related work

References	Model used	Dataset used	Best result
[2]	ET, RF	Heart failure clinical records	RF achieved the highest accuracy at 98.33%
[3]	LR, LDA, KNN, CART, GNB, SVM, RF	Taken from kaggle regarding clinical features and descriptions	LDA excelled with a mean value of approximately 0.847 and a minimal false acceptance rate of 0.33%
[4]	NB, KNN, RF, SVM	Used biological metrics	RF performed well with an accuracy of 87.92%
[5]	Naive Bayes, multilayer perceptron, K star, logistic regression	Cleveland heart disease data set	NB showed improved accuracy
[6]	MLP, ELM, PSO + ELM, BFO + ELM, BCO + ELM	Cleveland heart disease datasets 2016	BCO + ELM yielded superior results
[7]	RF, LR, XG	Cleveland heart disease dataset from the UCI ML repository	RF algorithm exhibited a performance accuracy score of 90.16%
[8]	RF, DT, NB, KNN, SVM, MLP	Cardiovascular adtaset	MLPPCA and CHI-square achieved the most accurate heart prediction outcomes, with 98.08% and 98.25% accuracy, respectively

3.1 Dataset Description

In this study, the author comprehensively analyzed Case 1 and Case 2 datasets to evaluate various ML models for heart disease diagnosis. Case 1 integrated data from the BIDMC Congestive Heart Failure and MIT-BIH Arrhythmia databases, comprising 2800 samples with diverse cardiac conditions (1300 from BIDMC and 1500 from MIT-BIH). Case 2 utilized data from the PTB Diagnostic ECG and MIT-BIH Arrhythmia databases, creating a dataset of 1418 samples (1300 from PTB and 118 from MIT-BIH). Five ML algorithms (DT, KNN, ANN, SVM, and RF) were assessed through hyperparameter optimization with GridSearchCV. This extensive analysis aimed to identify the most accurate heart disease diagnostic model and advance data-driven approaches in cardiac healthcare [9].

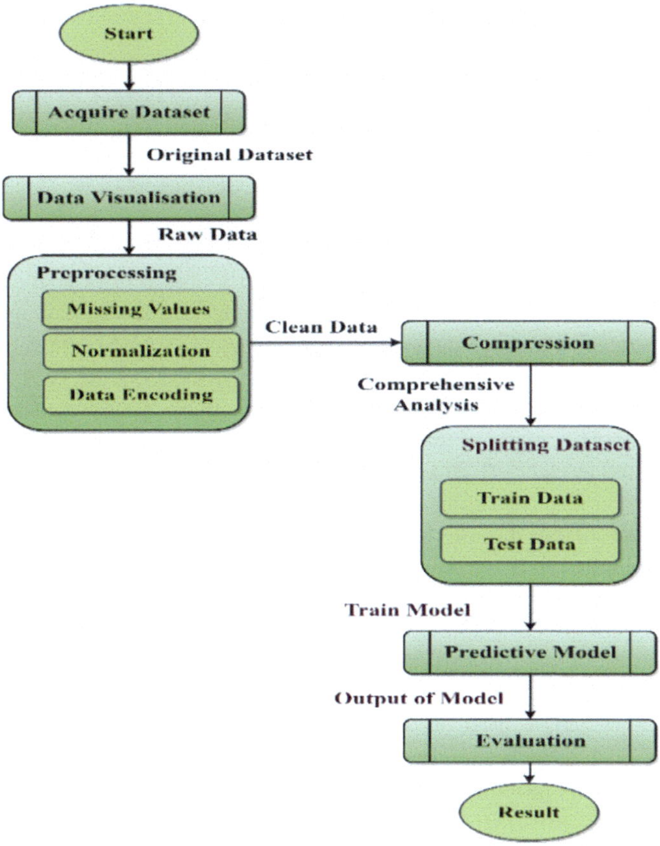

Fig. 1 Research methodology

3.2 Data Pre-processing

Data pre-processing in the study was integral to ensure the reliability of results, involving 2800 and 1418 samples in Case 1 and Case 2, respectively, collected from diverse databases. The pre-processing pipeline harmonized and cleaned data, mitigating format disparities, handling missing values, and addressing outliers and noise [10]. Feature engineering extracted domain-specific heart disease-related features, while normalization and standardization prevented feature bias during model training. Class imbalance was managed, and data was stratified for training and testing, with cross-validation ensuring robust model evaluation, offering valuable insights for future data-driven heart disease research, a generic flowchart for model construction [9, 11].

3.3 Model Preparation and Testing

This analysis used Google Research Colab and Jupyter Notebook format code. The research utilized the DT model to classify individuals into heart disease categories, enhancing accuracy and providing transparency in decision-making. This approach proved effective in healthcare, improving heart disease diagnosis and understanding disease-contributing factors [12]. RF emerges as a potent tool for heart disease diagnosis, harnessing ensemble learning to enhance accuracy and analyze complex risk factors, promising better patient outcomes and early detection in healthcare. This research underscores RF's pivotal role in advancing heart disease diagnosis and prevention strategies through data-driven approaches [13]. SVM brings innovation to heart disease diagnosis, demonstrating robust classification and prediction abilities, improving precision in early detection and treatment, and promising to revolutionize healthcare outcomes through data-driven approaches and block schematic of the suggested model [14]. This research highlights SVM's potential to transform heart disease diagnosis and enhance patient care [15]. KNN emerges as a significant ML algorithm for heart disease diagnosis, leveraging similarity-based classification to enhance data-driven diagnostic precision. The research underscores KNN's role in improving the accuracy and efficiency of heart disease diagnosis, demonstrating the value of data-driven techniques in healthcare [16]. LR provides an innovative approach to evaluating data-driven heart disease diagnosis models, contributing to healthcare by rigorously assessing predictive model accuracy and reliability [17]. Leveraging this widely used classification algorithm enhances understanding of healthcare applicability, benefiting diagnosis outcomes and informing healthcare decisions [18]. ANN introduces an innovative application in heart disease diagnosis, leveraging large cardiac health datasets to demonstrate accurate diagnosis potential for early intervention, emphasizing the role of ML and data-driven techniques in advancing cardiovascular healthcare [19]. This research highlights the growing significance of AI and ANN technologies in revolutionizing healthcare practices and improving cardiovascular disease diagnosis [13]. The eXtreme Gradient Boosting Classifier (XGBoost) is a robust ML algorithm that enhances diagnostic accuracy by handling complex datasets and capturing intricate feature relationships while minimizing over fitting. Its feature prioritization and adaptability make it valuable in advancing data-driven heart disease diagnosis and improving diagnostic accuracy [16].

4 Analysis and Result Discussion

The study extensively evaluated multiple ML models across two cases, emphasizing their computational efficiency, diagnostic accuracy, and potential for automating heart disease diagnosis, particularly highlighting SVM, RF, ETC, and XGB as promising models for advancing data-driven cardiac healthcare.

The researchers conducted a thorough analysis using diverse ML models to evaluate their efficacy in heart disease diagnosis, employing a dataset of 2800 samples in Case 1 from BIDMC congestive heart failure and MIT BIH Arrhythmia databases and 1418 samples from PTB Diagnostic ECG and MIT-BIH Arrhythmia databases [20].

Table 2 presents key performance metrics for five ML models (DT, KNN, ANN, SVM, and RF), including their fit times (model training duration) and score times (prediction time). These metrics offer insights into the computational efficiency of each algorithm, with ANN displaying the longest fit time (8.12 s on average) and KNN having a relatively high score time (0.04 s on average) due to nearest-neighbor calculations.

The Table 3 demonstrates SVM and RF's strong performance in heart disease diagnosis, with high accuracy (98.29% and 98.96%) and excellent metrics, indicating their potential for automated diagnosis. Figure 2 visualizes model performance variations in terms of accuracy and f1-score across different models in case 1.

Table 4 provides critical metrics for evaluating the computational efficiency of five ML models. Fit_time, indicating training time, highlights DT and RF as the quickest, with times of 0.01 and 0.05 s, respectively, while ANN demonstrates longer training at 1.05 s. In terms of score_time (prediction time), KNN has the highest score_time at 0.03 s, while ANN, SVM, and RF perform notably faster with score_times of 0.01

Table 2 Metrics for five different ML models

Model	Fit_time	Score_time
DT	0.05 s + −0.00 s	0.01 s + −0.00 s
KNN	0.01 s +0.00 s	0.04 s + −0.00 s
ANN	8.12 s + −3.66 s	0.01 s + −0.00 s
SVM	0.23 s + −0.02 s	0.03 s + −0.00 s
RF	0.22 s + −0.01 s	0.01 s + −0.00 s

Table 3 Metrics result determined

Model	Accuracy	Balanced accuracy	F1-score	Precision	Recall
DT	93.00% + − 1.06%	93.09% + − 1.08%	93.35% + − 1.03%	95.06% + − 2.20%	91.80% + − 2.46%
KNN	96.75% + − 1.00%	96.78% + − 0.96%	96.94% + − 0.96%	97.51% + − 0.60%	96.40% + − 1.64%
ANN	96.89% + − 0.82%	96.85% + − 0.87%	97.11% + − 0.74%	96.84% + − 1.44%	97.40% + − 0.63%
SVM	98.29% + − 0.73%	98.33% + − 0.72%	98.39% + − 0.69%	99.06% + − 0.69%	97.73% + − 0.90%
RF	98.96% + − 0.41%	98.97% + − 0.39%	99.03% + − 0.38%	99.20% + − 0.49%	98.87% + − 0.85%

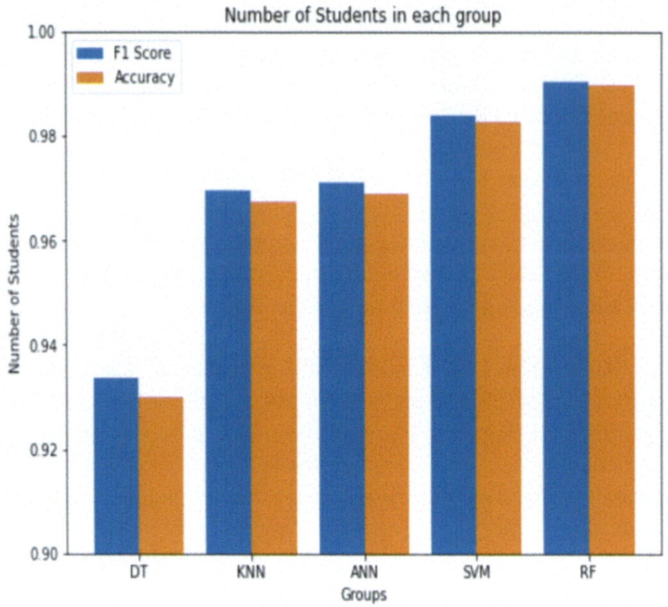

Fig. 2 "Accuracy" and "f1-score" for different ML models employed in the diagnosis of heart disease of case-1

or 0.02 s, assisting in model selection for real-time or resource-constrained heart disease diagnosis applications.

Table 5 offers a comparative analysis of ML models for heart disease diagnosis, with the ETC emerging as the top performer, boasting a remarkable 97.54% accuracy, a high AUC of 99.79%, outstanding recall of 98.57%, and essential precision of 96.76%. Even with cross-validation, these consistent results underline its potential as a robust and reliable tool for automated heart disease diagnosis, and its highest F1 score indicates a balanced trade-off between precision and recall for real-world clinical applications. The low standard deviation of 0.0 highlights the model's remarkable stability, emphasizing the importance of employing such robust algorithms in data-driven heart disease diagnosis.

Table 4 Metrics for assessing the computational efficiency of five ML models

Model	Fit_time	Score_time
DT	0.01 s + −0.00 s	0.01 s + −0.00 s
KNN	0.00 s + −0.00 s	0.03 s + −0.00 s
ANN	1.05 s + −0.07 s	0.01 s + −0.00 s
SVM	0.05 s + −0.00 s	0.02 s + −0.00 s
RF	0.05 s + −0.00 s	0.01 s + −0.00 s

Table 5 Compare models

Model	Accuracy	AUC	Recall	Precision	F1	Kappa	MCC	TT (Sec)
Extra trees classifier	0.9754	0.9979	0.9857	0.9676	0.9762	0.9508	0.9516	0.3190
RF classifier	0.9711	0.9943	0.9800	0.9645	0.9719	0.9422	0.9429	0.3460
CatBoost classifier	0.9711	0.9895	0.9800	0.9645	0.9719	0.9422	0.9429	1.9220
DT classifier	0.9668	0.9667	0.9714	0.9647	0.9675	0.9335	0.9346	0.0500
Light gradient boosting machine	0.9667	0.9905	0.9714	0.9643	0.9675	0.9335	0.9343	0.4670
XGBoost	0.9653	0.9865	0.9686	0.9642	0.9660	0.9306	0.9313	0.1470
Gradient boosting classifier	0.9436	0.9822	0.9514	0.9394	0.9451	0.8872	0.8879	0.2310
AdaBoost classifier	0.8945	0.9615	0.9029	0.8927	0.8970	0.7890	0.7904	0.2020
Ridge classifier	0.8787	0.0000	0.9259	0.8510	0.8861	0.7570	0.7615	0.0390
Linear discriminant analysis	0.8787	0.9356	0.9259	0.8510	0.8861	0.7570	0.7615	0.0530
LR	0.8656	0.9362	0.8888	0.8537	0.8699	0.7311	0.7336	0.5340
Naive bayes	0.8254	0.8996	0.9202	0.7790	0.8427	0.6500	0.6637	0.0500
K neighbors classifier	0.7517	0.8412	0.7233	0.7728	0.7458	0.5037	0.5064	0.0570
SVM-linear kernel	0.7170	0.0000	0.7430	0.7714	0.7166	0.4320	0.4654	0.0430
Quadratic discriminant analysis	0.5917	0.9105	0.9771	0.5635	0.7112	0.1749	0.2704	0.0530
Dummy classifier	0.5065	0.5000	1.0000	0.5065	0.6724	0.0000	0.0000	0.1300

Figure 3a displays the confusion matrix for the ANN model, revealing a striking accuracy rate of 100%. This exceptional accuracy indicates that the ANN achieved flawless heart disease diagnosis on the dataset, making it an up-and-coming and potentially reliable tool for automated cardiac health assessment.

Figure 3b highlights the RF Classifier's outstanding performance in a confusion matrix, showing 100 True Positives and 102 True Negatives, resulting in a remarkable 100% accuracy rate for diagnosing heart diseases.

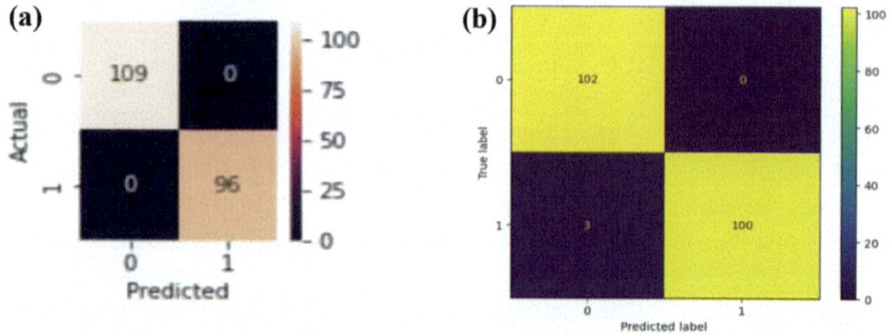

Fig. 3 **a** Confusion matrix of ANN **b** The heat map representation of the RF classifier

Figure 4 presents a compelling visual representation of their deep learning results through two graphs. The left graph illustrates a significant percentage increase, signifying a high rate of correct heart disease diagnoses using deep learning techniques. Meanwhile, the right graph demonstrates a consistently low loss, emphasizing the model's ability to make highly accurate predictions. Attaining 100% accuracy in deep learning for heart disease diagnosis signifies a groundbreaking achievement, highlighting the transformative potential of data-driven methods for precise cardiac diagnostics and their impact on healthcare outcomes. The study extensively examined data-driven heart disease diagnosis using various ML models, achieving exceptional 99% accuracy with the XGBoost and GridSearchCV optimization. It highlights the significance of parameter tuning in maximizing model accuracy and the potential of data-driven approaches for enhancing cardiovascular health diagnosis.

Table 6 displays accuracy results for the top-performing predictors, with XGBoost leading with 99.03% accuracy, followed by SVM (97.73%) and DT (98.05%). These findings highlight XGBoost, SVM, and DT as promising models for automated and accurate heart disease diagnosis, offering valuable insights for clinical applications.

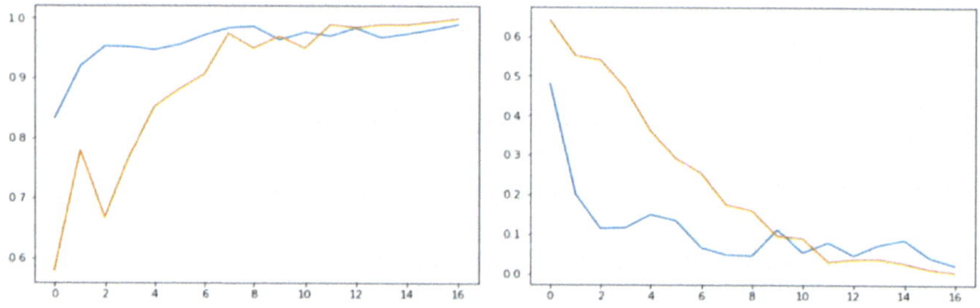

Fig. 4 Graphical representation of the increase/decrease in percentage (on the left) and the loss (on the right) of deep learning

Table 6 Accuracy data frame values for the five best predictors

Model	Best_score	Best_params
SVM	97.727273	{'C': 100.0, 'gamma': 0.1, 'kernel': 'rbf'}
LR	83.116883	{'C': 1.0, 'penalty': 'l2'}
XGB	99.025974	{'max_depth': 7, 'n_estimators': 100, 'n_jobs': 1, 'subsample': 0.6000000000000001, 'verbosity': 0}
RF	97.402597	{'criterion': 'gini', 'max_depth': 7, 'max_features': 'log2'}
DT	98.051948	{'criterion': 'entropy', 'max_depth': None}
KNN	97.727273	{'n_neighbors': 17, 'weights': 'distance'}

4.1 Result Discussion

The study evaluates ML models for heart disease diagnosis, emphasizing the significance of data-driven approaches. It identifies RF, LR, and NB as promising candidates for precise and automated heart disease diagnosis, potentially enhancing healthcare decision support systems.

Figure 5 display confusion matrices for LR, SVM, and RF models, illustrating their classification performance in comparing actual and predicted labels. Among these models, RF excelled in accuracy, while LR and NB also performed well, with DT showing the lowest accuracy. Accuracy values for five predictive models in heart disease diagnosis, including Logistic Regression (81.5%), Support Vector Machine (82.4%), Naive Bayes (82%), Decision Tree (77%), and Random Forest (100%). These values signify each model's capacity to accurately classify patients with and without heart diseases, offering valuable insights into their performance levels for this critical healthcare application. Notably, RF achieves exceptional accuracy.

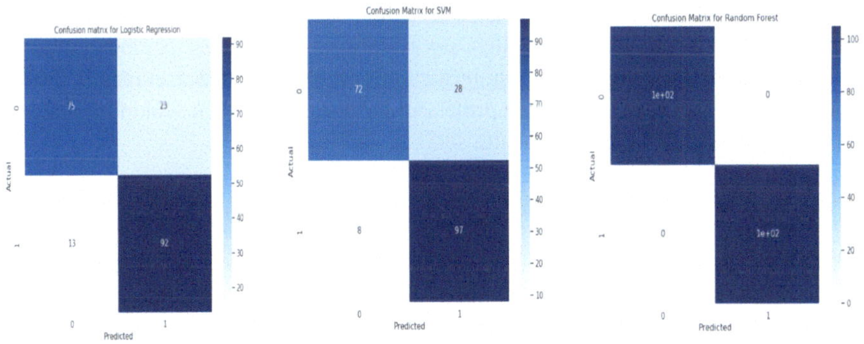

Fig. 5 Confusion matrix for logistic regression, SVM and random forest

5 Conclusion and Future Work

Heart disease is a grave medical illness that can give rise to life-threatening complications such as myocardial infarction, heart attack, and heart failure. CVD is a leading global cause of mortality. Developing a computerized approach incorporating various significant factors such as age, irregular heartbeat, stroke, alcohol consumption, smoking, and previous history of heart attacks holds substantial potential in predicting the risk of heart disease. Identifying raw health records can deliver long-term benefits in saving lives and detecting cardiac disease deficits at an earlier stage. Machine learning techniques were employed in this research endeavor to examine the raw data and offer novel insights into cardiac disease. The mortality rate can be substantially diminished by promptly detecting heart-related diseases in their early stages and implementing appropriate preventative measures. In medical diagnostics, numerous methods exist to detect heart illness, albeit sometimes accompanied by high costs. Consequently, ML emerges as the optimal selection due to its ability to achieve maximum accuracy in the early identification of heart disease. Classifiers play a vital role in the medical diagnosis sector by facilitating the prediction of the medical services that can be provided to patients. The examination and comparison of existing procedures are conducted to ascertain the efficacy and precision of systems. This study project employed ten distinct supervised machine-learning classification algorithms to forecast cardiac disease in patients, incorporating three variations of cross-validation and three sampling strategies, evaluating model performance with different sampling techniques which can be observed that the random forest model, in conjunction with the decision tree and k-nearest neighbors, emerges as the most effective strategy for random over-sampling. Regarding accuracy comparison, the decision tree model and the random forest and k-nearest neighbors' models demonstrate superior performance when employing the near-miss under-sampling strategy. Regarding accuracy comparison, the SMOTE performs better when paired with the random forest model, decision tree, and k-nearest neighbors' algorithms. In the future, our research will utilize deep learning algorithms, computer vision techniques, and hyperparameter tuning methods to optimize the accuracy of heart disease prediction in a real-life lab project to have a product that would be readily applicable to private and public health services worldwide.

References

1. Das, P., Girish Kumar, J., Lama, A.: Empirical Mode Decomposition Based Ensemble Hybrid Machine Learning Models for Agricultural Commodity Price Forecasting (2023)
2. Talukder, M.S.H., Sulaiman, R.B., Angon, M.B.P.: Unleashing the power of extra-tree feature selection and random forest classifier for improved survival prediction in heart failure patients. arXiv preprint: 2308.05765 (2023)
3. Kumari, A., Mehta, A.K.: A novel approach for prediction of heart disease using machine learning algorithms. In: 2021 Asian Conference on Innovation in Technology (ASIANCON). pp. 1–5. IEEE (2021)

4. Karthik, K., Reddy, A.L., Kulkarni, R., Mehdi, M.J.: Algorithm accuracy verification in heart disease analysis using machine learning. In 2023 2nd International Conference on Applied Artificial Intelligence and Computing (ICAAIC), pp. 345–349. IEEE (2023)
5. Gulati, S., Guleria, K., Goyal, N.: Classification and detection of coronary heart disease using machine learning. In: 2022 2nd International Conference on Advance Computing and Innovative Technologies in Engineering (ICACITE), pp. 1728–1732. IEEE (2022)
6. Vigneshvaran, P., Kathiravan, A.V.: Heart disease prediction using an optimized extreme learning machine with bacterial colony optimization. In: 2022 3rd International Conference on Smart Electronics and Communication (ICOSEC), pp. 1425–1429. IEEE (2022)
7. Sahoo, G.K., Kanike, K., Das, S.K., Singh, P.: Machine learning-based heart disease prediction: a study for home personalized care. In: 2022 IEEE 32nd International Workshop on Machine Learning for Signal Processing (MLSP), pp. 01–06. IEEE (2022)
8. Aggarwal, R., Kumar, S.: MLPPCA: heart disease detection using machine learning. In: 2022 Seventh International Conference on Parallel, Distributed and Grid Computing (PDGC), pp. 457–461. IEEE (2022)
9. Moreno-Sánchez, P.A.: Data-driven early diagnosis of chronic kidney disease: development and evaluation of an explainable AI model. IEEE Access **11**, 38359–38369 (2023)
10. Islam, M.A., Nag, A., Yousuf, S.M., Mishra, B., Sufian, M.A., Mondal, H.: Data-driven analysis: a comprehensive study of CPS case outcomes in 42 english counties (2014–2018) with R analytics (2023)
11. Paul, S.M., Rahman, M.D.H., Siddique, M.D.M.H.: Machine learning based rainfall forecasting in different season in Bangladesh. In: 2022 International Conference on Recent Progresses in Science, Engineering and Technology (ICRPSET), pp. 1–5. IEEE (2022)
12. Saputra, J., Lawrencya, C., Saini, J.M., Suharjito, S.: Hyperparameter optimization for cardiovascular disease data-driven prognostic system. Vis. Comput. Ind., Biomed., Art **6**(1), 16 (2023)
13. Dritsas, E., Trigka, M.: Efficient data-driven machine learning models for cardiovascular diseases risk prediction. Sensors **23**(3), 1161 (2023)
14. Rahman, M.H., Das, S., Uddin, M., Reza, Y.: Using the YOLO network's games object to classify sports. In: 2023 26th International Conference on Computer and Information Technology (ICCIT). IEEE (2023)
15. Dinh, A., Miertschin, S., Young, A., Mohanty, S.D.: A data-driven approach to predicting diabetes and cardiovascular disease with machine learning. BMC Med. Inform. Decis. Mak. **19**(1), 1–15 (2019)
16. He, X., Fan, X., Zheng, W., Ti, Z., Li, C., Zhang, H., Zhou, X.: A data-driven study of prediction methods for coronary heart disease. In: International Conference on Service Science, pp. 447–459. Springer Nature Singapore, Singapore (2023)
17. Kumar, D.V.S., Chaurasia, R., Misra, A., Misra, P.K., Khang, A.: Heart disease and liver disease prediction using machine learning. In: Data-Centric AI Solutions and Emerging Technologies in the Healthcare Ecosystem, pp. 205–214. CRC Press (2023)
18. Jiao, Y., Chen, X., Sun, Z., Fu, L., Jiang, C., Wen, C., Zhang, X.: Data-driven detection and assessment for urban railway transit driver fatigue in real work conditions. Transp. Res. Rec. **2677**(1), 1367–1375 (2023)
19. Ghasemieh, A., Lloyed, A., Bahrami, P., Vajar, P., Kashef, R.: A novel machine learning model with stacking ensemble learner for predicting emergency readmission of heart-disease patients. Decis. Anal. J. **7**, 100242 (2023)
20. Ahsan, M.M., Siddique, Z.: Machine learning-based heart disease diagnosis: a systematic literature review. Artif. Intell. Med. **128**, 102289 (2022)

A Hybrid Recurrent Neural Network with Mini-Batch Gradient Descent Method for Multiple Sclerosis Disease Prediction

E. Kavipriya and S. Sasikala

Abstract Machine Learning techniques on different sectors plays a vital role for prediction analysis. In healthcare sector, diagnosis support and prognostic evaluation has been done using various algorithms of machine learning. The Multiple Sclerosis (MS) is one of the chronic diseases which affects the central nervous system of human. There are multiple symptoms and disabilities indicate the issues of Multiple Sclerosis. This neurological disorder has various patterns and disease progression rate among worldwide. This causes a variety of symptoms including fatigue, numbness, visual difficulties, urinary incontinence, and difficulty moving and cognitive deficiencies. This paper analyzes the risk factors of Multiple Sclerosis and early prediction using Machine learning methods. This work proposed hybrid Recurrent Neural Network with mini-batch gradient descent method for Multiple Sclerosis prediction. The performance of this proposed method is evaluated and the achieved accuracy rate is 92%.

Keywords Machine learning · Multiple sclerosis · Recurrent neural network

1 Introduction

Artificial intelligence and Machine learning provides significant development in various sectors. Most specifically in healthcare and medicine, the techniques provided by machine learning helps for early diagnosis and prediction of various diseases. The recent advancements in machine learning in healthcare mainly supported the ability of experts to recognize the healthcare trends and create disease prediction models. The organization of healthcare records, detection of irregularities in blood samples, organs and bones using medical imaging and monitoring, as well as robot-assisted surgeries have all benefited from the execution of machine learning approaches in

E. Kavipriya (✉) · S. Sasikala
Department of Computer Science With Cognitive Systems and AIML, Hindusthan College of Arts & Science, Coimbatore, Tamil Nadu, India
e-mail: kavipriya.e@hicas.ac.in

G. Raj et al. (eds.), *Advances in Artificial Intelligence and Machine Learning*, Lecture Notes in Electrical Engineering 1264, https://doi.org/10.1007/978-981-97-9507-9_18

medical organizations. The healthcare expands into the modern realm of technology, many new methods appear [1]. The advancement of profession including speed, accuracy and simple form of diagnosis methods are based on Artificial Intelligence and machine learning techniques. Machine learning provides vast amount of algorithmic models and statistical techniques to address issues without the need of professionals. Due to this reason, the machine learning models are single-layers, significant portions of extraction features and processing collected data are processed. The methods of machine learning need extensive data preparation in the absence of additional layers to make accurate predictions and prevent the methods from over fitting and under-fitting issues [2].

A chronic, inflammatory, demyelinating illness of the central nervous system is known as Multiple Sclerosis (MS). The people affected by Multiple Sclerosis can be classified into four types such as Relapsing–Remitting (RR), Secondary-Progressive (SP), Primary- Progressive (PP) and Progressive Relapsing (PR). The disease of Relapsing Remitting indicates the growth of multifocal inflammatory lesions which affects the neurological worsening [3]. The remaining three categories are addressing the diffuse immune mechanism. According to the US analysis, there are 2.8 million people affected by this Multiple Sclerosis.

2 Related Works

Yijun Zhao et al. [1] used Support Vector machine method and Logistic Regression method for prediction of Multiple Sclerosis. These methods are applied on clinical observations to make early prediction. The results of Support Vector Machine method produced sensitivity to 62% and specificity to 65% for the prediction of worsening cases. The baseline EDSS is used to increase the efficiency of Support Vector Machine method and the results indicates the sensitivity up to 86%. The results are compared with logistic regression and the results showed that the Logistic Regression methods are providing poor results in this prediction case. Dario Branco et al. [4] discussed the machine learning techniques which are used for predicting Multiple Sclerosis Progression. In order to enhance prediction abilities and give experts another tool to assess patient's clinical details. Following the usage of correlation indicators and principal component analysis for reducing the dimensionality reduction of feature space, classification algorithms have been experimented and the results are compared to check the efficiency of these methods. Ruggiero Seccia et al. [5] presented the machine learning techniques for prognostic in multiple sclerosis. The Linear Regression Approach is used for prediction function. This is used for identifying the linear relationship between input and a scalar output values. K-nearest neighbor method is commonly used for regression and classification problems. Other than these methods some familiar methods are analyzed for executing Multiple Sclerosis dataset for effective prediction. The discussion was based on machine learning methods such as Support Vector Machine, Decision Trees, Ensemble methods like bagging and boosting and neural Network methods. All these methods are analyzed and implemented for prediction using test dataset. Marco Law et al. [2] presented the use of machine learning methods in secondary progressive multiple sclerosis. They proposed an improved prediction model for predicting the disability progression in short term. They used ensemble model decision tree for providing improved predictions. The results of proposed model are compared with Logistic regression and support vector machine method. The proposed ensemble models are trained using hyper-parameters with five-fold and ten-fold cross validation method. The comparison part considered AUC values using Sun and Wu's fast execution using DeLong's algorithm. The results showed that the ensemble based decision tree model achieved good AUC rate among other methods in prediction. Arman Eshaghi [6] et al. used unsupervised machine learning methods for identifying multiple sclerosis subtypes. The brain MRI scans are used for this work and unsupervised methods of machine learning are applied for classifying the subtypes of Multiple Sclerosis. The best fitting models are selected for finding the optimal number of types and to find the abnormality changes in the similar subtypes. Zhao et al. [7] presented a novel deep learning methods with Artificial Neural Network for predicting the Nurological Diseases. The hidden layers of this Artificial Neural Network are used for incorporating the calculations like regularization component to avert the issues of overfitting and underfitting. The proposed method of this work achieved predictive precision and reduces loss when compared with traditional methods.

3 Importance of Predicting Multiple Sclerosis

There is currently no standard treatment for multiple sclerosis. Treatment work to both recover function after an attack and prevent them in the future. Physical therapy and occupational therapy can support a person's ability to function. Numerous people still utilize alternative treatments even if there is insufficient evidence that they are effective. Long-term outcomes are hard to predict especially for women and people who have the condition early in life, people who have a relapsing course and people who had few attacks already. Nearly a million Multiple Sclerosis will be diagnosed in the United States (US) and 2.8 million will be diagnosed globally in 2022, with the rates of varying substantially between populations and geographical regions. Between the ages of 20 to 50, this sickness usually occurs [8]. The ration showed that men are highly affected than women. Multiple Sclerosis begins as an acute clinical attack (Clinically isolated disease) in about 85% of people. The conversion rate from a clinically isolated condition to multiple sclerosis rage from 30 to 82% according to past study. Following a single episode of CNS inflammation, a person who seeks assessment must make various decisions for future references including whether or not start therapy. Therefore it is becoming more important to identify the signs that indicate CIS to Multiple Sclerosis conversion.

4 Dataset Details

This work used Multiple Sclerosis Dataset which is proposed by Pineda [9]. This dataset contains 18 features with one class label.

- ID: Patient identifier (int)
- Age: Age of the patient (in years)
- Schooling: time the patient spent in school (in years)
- Gender: 1=male, 2=female
- Breastfeeding: 1=yes, 2=no, 3=unknown
- Varicella: 1=positive, 2=negative, 3=unknown
- Initial_Symptoms: 1=visual, 2=sensory, 3=motor, 4=other, 5= visual and sensory, 6=visual and motor, 7=visual and others, 8=sensory and motor, 9=sensory and other, 10=motor and other, 11=Visual, sensory and motor, 12=visual, sensory and other, 13=Visual, motor and other, 14=Sensory, motor and other, 15=visual, sensory, motor and other
- Mono _or_Polysymptomatic: 1=monosymptomatic, 2=polysymptomatic, 3=unknown
- Oligoclonal_Bands: 0=negative, 1=positive, 2=unknown
- LLSSEP: 0=negative, 1=positive
- ULSSEP:0=negative, 1=positive
- VEP:0=negative, 1=positive
- BAEP: 0=negative, 1=positive
- Periventricular_MRI:0=negative, 1=positive
- Cortical_MRI: 0=negative, 1=positive
- Infratentorial_MRI:0=negative, 1=positive
- Spinal_Cord_MRI: 0=negative, 1=positive
- initial_EDSS:?
- final_EDSS:?
- Group: 1=CDMS, 2=non-CDMS

4.1 *Sample Dataset for Multiple Sclerosis*

The following table presents the sample dataset which is used for predicting Multiple Sclerosis using machine learning methods (Table 1).

5 Proposed Method

When working with large datasets, mini-batch gradient descent is a well-liked optimization technique in machine learning and deep learning. It is an iterative procedure made to reduce the cost or loss function of a model by modifying its parameters. The method used for parameter updates is what distinguishes mini-batch gradient descent from other variants like batch gradient descent and stochastic gradient descent. This method divides the dataset into smaller subsets known as mini-batches, often having a defined number of samples as opposed to using the complete training data samples. To take advantages of both Recurrent Neural Network (RNN) with mini-batch gradient descent for more effective and efficient sequence modeling applications, a hybrid

Table 1 This table shows the available dataset details including columns details and value types

	Gender	Age	Schooling	Breast feeding	Vari-cella	Initial_ symptom	Mono or polysymptom	Oligo-clonal bands	LLSSEP	ULSSEP	VEP	BAEP	Periven-tricular MRI	Cortical MRI	Infratentor MRI	Spinal-cord MRI	Initial EDSS	Final EDSS	Group
0	1	34	20	1	1	2	1	0	1	1	0	0	0	1	0	1	1	1	1
1	1	61	25	3	2	10	2	1	1	0	1	0	0	0	0	1	2	2	1
2	1	22	20	3	1	3	1	1	0	0	0	0	0	1	0	0	1	1	1
3	2	41	15	1	1	7	2	1	0	1	1	0	1	1	0	0	1	1	1
4	2	34	20	2	1	6	2	0	1	0	0	0	1	0	0	0	1	1	1
5	1	29	22	1	1	6	2	0	1	0	0	0	1	0	1	0	1	1	1
6	2	53	20	1	1	14	2	0	1	0	1	0	1	1	0	1	1	1	1
7	2	24	15	1	1	14	2	0	1	1	0	0	1	1	1	1	2	2	1
8	1	36	15	1	1	8	2	0	1	1	1	0	1	0	0	0	1	1	1
9	2	28	20	1	1	8	2	0	0	0	0	0	1	0	1	0	1	1	1

Recurrent Neural Network with mini-batch gradient descent is required. Due to their capacity to recognize temporal relationships, the recurrent Neural Network are effective in processing sequential data, making them appropriate for jobs like time series prediction and detection of diseases. To train the recurrent neural network using mini-batch gradient descent method is really complex [10].

The traditional Recurrent Neural Network frequently experience problems like vanishing gradients, where the cost function's gradient get smaller as they go back in time. This makes it complex to capture long-range relationships. On the other hand, the mini-batch gradient descent speed up training by utilizing parallelism and effective hardware acceleration but can exacerbate the vanishing gradient problem in Recurrent Neural Network. This proposed work combines both advantages of hybrid recurrent neural Network with mini-batch Gradient descent. This hybrid strategy effectively solves the vanishing gradient problem by partitioning the sequential data into mini-batches and computing gradients for each mini-batch. It permits faster and more consistent convergent during training while maintaining Recurrent Neural Network's efficiency in modeling sequential data. While dealing with large scale sequential data, such as huge time series or lengthy test where Traditional RNN can be computationally expensive and sluggish to train, this hybridization is especially helpful.

6 Significance of Features

The following code executes the process of finding significance of features in this taken dataset (Figs. 1 and 2).

```
Importances_2[1] =['Schooling', 'Initial_Symptoms', 'Final_EDSS', 'Initial_EDSS',
'Gender',
       'Age', 'Breastfeeding', 'Varicella', 'Mono_or_Polysymptomatic',
       'Oligoclonal_Bands', 'LLSSEP', 'ULSSEP', 'VEP', 'BAEP',
       'Periventricular_MRI', 'Cortical_MRI', 'Infratentorial_MRI',
       'Spinal_Cord_MRI']

array([0.07281395, 0.15102626, 0.        , 0.        , 0.0365113 ,
       0.11156346, 0.05032522, 0.06013221, 0.02740795, 0.06540184,
       0.03592938, 0.02564735, 0.03280761, 0.        , 0.16267139,
       0.03361586, 0.10934539, 0.02480084])
```

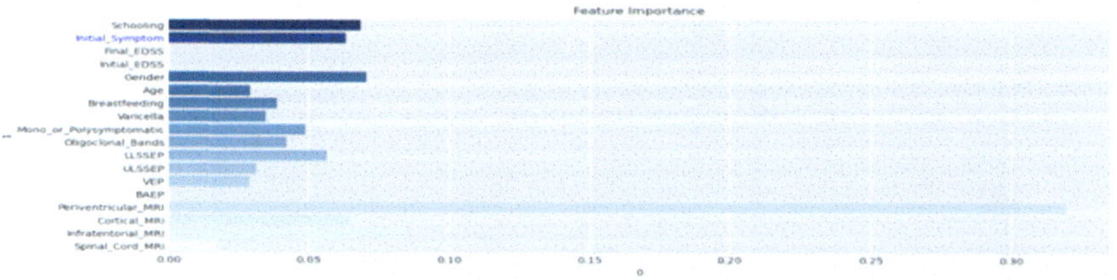

Fig. 1 The code executes the process of finding significance of features in this taken dataset

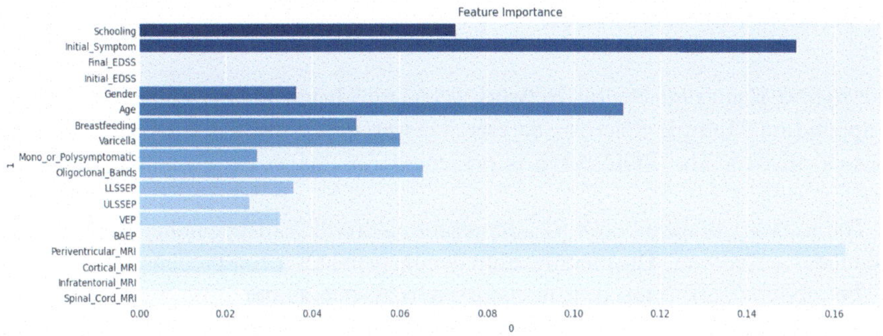

Fig. 2 Result of finding significance of features in this taken dataset

7 Experimental Settings

The execution component os carried out using the PYZO console and Pandas, Numpy and SKlearn libraries in a Windows 10 pro operating system. As a backend Tensorflow 2's Keras API is utilized. For the simulation of experiment, a Multiple Sclerosis Dataset is utilized (Fig. 3).

This setting information provides the execution details setting on python and the parameter details are presented. The learning rate, mini-batch size, number of episodes, Recurrent Neural Network Architecture, number of layers, hidden units per layer, Dropout rate, weight initialization, optimization algorithm, gradient clipping, loss function and Early stopping parameters are set with values for execution initialization.

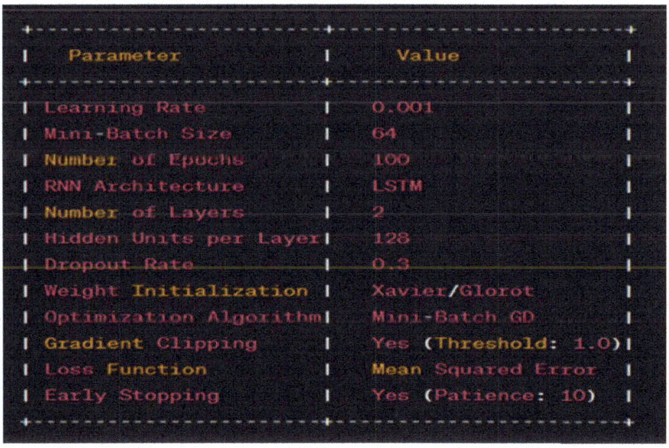

Parameter	Value
Learning Rate	0.001
Mini-Batch Size	64
Number of Epochs	100
RNN Architecture	LSTM
Number of Layers	2
Hidden Units per Layer	128
Dropout Rate	0.3
Weight Initialization	Xavier/Glorot
Optimization Algorithm	Mini-Batch GD
Gradient Clipping	Yes (Threshold: 1.0)
Loss Function	Mean Squared Error
Early Stopping	Yes (Patience: 10)

Fig. 3 Parameter setting

8 Results and Discussion

The hybrid Recurrent Neural Network with mini-batch gradient descent method is applied on Multiple Sclerosis dataset. This proposed method is evaluated using various metrics. The ROC curve is presented for this proposed method for this problem (Fig. 4).

The accuracy of the model is 92.2 representing that 92% of the predictions made by the model are correct. The classification details provides a comprehensive overview of the model's performance. It includes metrics such as precision, recall, and F1-score for each class (0 and 1). Precision refers to the ratio of correctly predicted positive observations to the total predicted positives. Recall, or sensitivity, is the ratio of correctly predicted positive observations to all the actual positives. F1-score is the weighted average of precision and recall and is a good way to show that a classifier has a good value for both recall and precision.

- For class 0: Precision is 0.91, Recall is 0.90, and F1-score is 0.91.
- For class 1: Precision is 0.90, Recall is 0.90, and F1-score is 0.90.

Furthermore, by foreseeing symptom exacerbations and mobility changes overtime, the hybrid RNN can help with symptom management, early diagnosis and rehabilitation planning. They can also aid in the identification of new drugs by evaluating the effectiveness of possible medications candidates using molecular information. Overall, the combination of hybrid RNN with mini-batch gradient descent provides a powerful tool for assessing intricate Multiple Sclerosis related data, enhancing patient care and assisting in the creation of individualized treatment. With 92% accuracy rate, the model performs admirably and based on the high precision, recall and F1 score values; it appears to generalize well for both classes (Fig. 5).

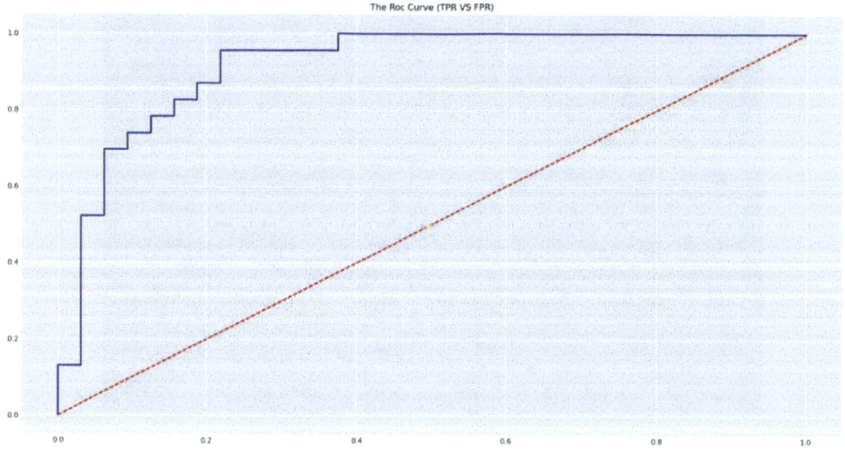

Fig. 4 ROC curve for proposed hybrid RNN with mini-batch gradient descent

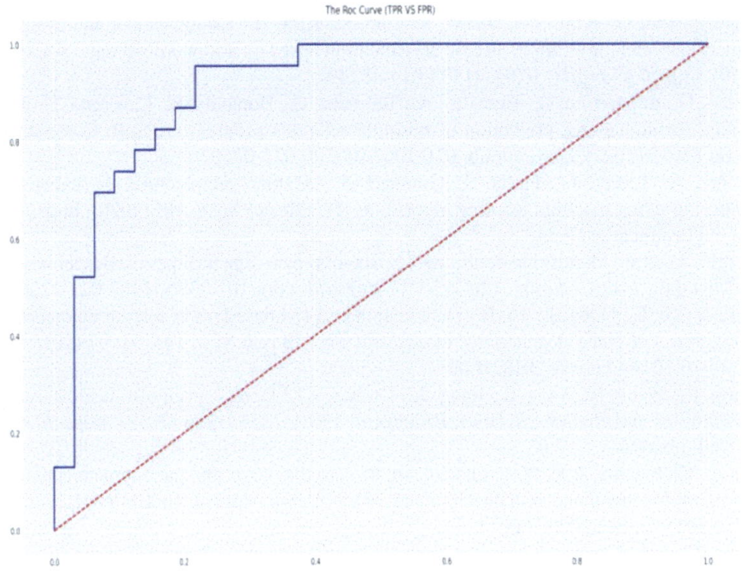

Fig. 5 ROC curve for proposed method

9 Conclusion and Future Work

Preparing data, creating a hybrid Recurrent Neural Network (RNN) architecture, training the model with mini-batches, evaluating its performances and finally applying it to prediction are all steps in the proposed method. This effective method takes the advantages of Recurrent Network at identifying sequential patterns while streamlining and improving the training procedure for generalization. With an accuracy rate of 92%, the model performs admirably overall and based on the high precision, recall and F1 score values, and it appears to generalize well for both classes. The model's learning process is illustrated by the training progress log, which shows a decline in error with each iteration. This proposed method produced high accuracy when compared with other traditional methods used for multiple sclerosis prediction using machine learning techniques.

References

1. Zhao, Y., et al.: Exploration of machine learning techniques in predicting multiple sclerosis disease course. PLoS ONE **12**(4), e0174866 (2017). https://doi.org/10.1371/journal.pone.0174866
2. Law, M.T., et al.: Machine learning in secondary progressive multiple sclerosis: an improved predictive model for short-term disability progression. Mult. Scler. J.-Exp., Transl. Clin. **5**(4), 205521731988598 (2019). https://doi.org/10.1177/2055217319885983

3. Bang, S., Yoo, D., Kim, S.-J., Jhang, S., Cho, S., Kim, H.: Establishment and evaluation of prediction model for multiple disease classification based on gut microbial data. Sci. Rep. **9**(1) (2019). https://doi.org/10.1038/s41598-019-46249-x
4. Branco, D., di Martino, B., Esposito, A., Tedeschi, G., Bonavita, S., Lavorgna, L.: Machine learning techniques for prediction of multiple sclerosis progression. Soft. Comput. **26**(22), 12041–12055 (2022). https://doi.org/10.1007/s00500-022-07503-z
5. Rghioui, A., Lloret, J., Sendra, S., Oumnad, A.: A smart architecture for diabetic patient monitoring using machine learning algorithms. Healthcare **8**(3), 348 (2020). https://doi.org/10.3390/healthcare8030348
6. Eshaghi, A., et al.: Identifying multiple sclerosis subtypes using unsupervised machine learning and MRI data. Nat. Commun. **12**(1) (2021). https://doi.org/10.1038/s41467-021-22265-2
7. Zhao, A., Qi, L., Dong, J., Yu, H.: Dual channel LSTM based multi-feature extraction in gait for diagnosis of Neurodegenerative diseases. Knowl.-Based Syst. **145**, 91–97 (2018). https://doi.org/10.1016/j.knosys.2018.01.004
8. Miotto, R., Wang, F., Wang, S., Jiang, X., Dudley, J.T.: Deep learning for healthcare: review, opportunities and challenges. Brief. Bioinform. **19**(6), 1236–1246 (2018). https://doi.org/10.1093/bib/bbx044
9. Ray, A., Chaudhuri, A.K.: Smart healthcare disease diagnosis and patient management: innovation, improvement and skill development. Mach. Learn. Appl. **3**, 100011 (2021). https://doi.org/10.1016/j.mlwa.2020.100011
10. Seccia R., et al.: Considering patient clinical history impacts performance of machine learning models in predicting course of multiple sclerosis. **15**(3), e0230219–e0230219 (2020), https://doi.org/10.1371/journal.pone.0230219

Identifying Real-Time Multi-class Food Detection Through Deep Learning Models

Umer Farooq, Ruchi Jain Garg, Mridul Dharwal, and Pooja Tiwari

Abstract In the present technological era, the incorporation of automation is considered crucial for the performance of any organization. Due to technological advancements, complexity has increased, leading to difficulties in processing, and different hardware is utilized for implementation. Consequently, it is imperative to explore feasible solutions to reduce complexity and enhance the applicability of any modus operandi in real time. The purpose of this article is to analyze the potency of deep learning models on the different parameters affecting the processing phenomena to identify the best architecture for creating a real-time food detection system integrated with a Calorie Counter. Based on the literature review, the complexity of different architectures was studied. In this study, the dataset comprises 36 classes, in order to analyze the effect of diverse deep learning structures over a multi-class dataset for elucidating classification. The results of the study show that a lightweight system outperformed a heavyweight system. The findings of the study conclude that MobileNetV2 architecture, which could be utilized as the best real-time multi-class food detection system, is the lightweight architecture system integrated with a Calorie counter. This study will be beneficial for the different stakeholders in the food industry in different aspects such as marketing and development of products, integration of technology, marketing strategies, regulatory compliances, etc. The result depicted in the paper may be generalized in the food industry and trustworthy in nature.

Keywords MobileNet · Visual geometry group (VGG) · Deep learning · Multi-class classification · Calorie counter · Food selection

U. Farooq · R. J. Garg (✉) · M. Dharwal · P. Tiwari
Sharda School of Business Studies, Sharda University, Greater Noida, India
e-mail: ruchi.jain@sharda.ac.in

M. Dharwal
e-mail: mriduldharwal22@gmail.com

P. Tiwari
e-mail: pooja.tiwari@sharda.ac.in

G. Raj et al. (eds.), *Advances in Artificial Intelligence and Machine Learning*, Lecture Notes in Electrical Engineering 1264, https://doi.org/10.1007/978-981-97-9507-9_19

1 Introduction

In the present contemporary era, automation has become a need of the hour. Many tasks of artificial intelligence such as recognition of speech, detection of speech, and machine translation have been tremendously improved by deep learning [1]. The nature of deep learning is reflected in its deep architecture which facilitates solving many AI tasks which are complex in nature [2]. Consequently, researchers are expanding deep learning in various modern areas and activities in addition to conservative tasks such as language models, recognition of faces, and detection of objects, for example, [3] have utilized the neural networks that are recurrent in nature to denoise signals of speech. Wang [4] has conducted sentiment analysis using deep learning from multiple modalities simultaneously. This phase has witnessed the growth of research in the domain of deep learning.

There are many previous studies conducted on varied learning algorithms to identify and explore the system with better efficacy. So, machine learning approaches with the incorporation of fuzzy systems, genetic algorithms [5], support vector machine (SVM) classifiers [6], etc. were considered for the research. However, to understand the modus operandi of machine learning various researchers and developers have explored that the efficacy of machine learning is limited over a large dataset which limits the possibilities of learning the features. So, consequently, the concept of deep learning has been considered by researchers.

Deep learning is capable of duplicating with the human brain and so it provides the output based on better learning and generating better accuracy [7–10]. These architectures consist of hidden layers which can be termed as Black Box [11], as it is quite complex due to internal calculation. The main focus was on creating a system with higher efficacy, but the requirement of the system in real-time was not considered much. Thus, after some time, the process of creating a system operating with fewer resources was emphasized and an effective solution for encapsulating a lightweight system was initiated. So, this is referred to as transfer learning [12] and it is also utilized for reducing the complexity of an approach. Hence the objectives of this study are to:

- To analyse the deep learning models to create a real-time food detection system.
- To investigate the different systems to get higher efficacy in real-time.

To achieve these objectives, this study explored the potential of pre trained models over real-time systems and compared them with multi-class datasets based on efficacy [13]. To do this the dataset of food items viz. fruits and vegetables which consist of 36 classes were considered and analyzed. Furthermore, we went over varied pre-trained models and restricted ourselves to light-weight systems such as MobileNetV2 and DenseNet121, along with heavy-weight systems such as InceptionResNetV2 and VGG16. Then we tested these models over our dataset to verify the effect of the "Depth and Size" of a system over a problem domain as it is said, "deeper the system, higher the accuracy" [14].

Moreover, we selected a model with the highest efficacy and low complexity when compared to others, and activated it over a real-time food detection system with a calorie counter feature utilizing Stream Lit for Web-App [15] creation in a simpler manner. To the best of our knowledge, the accuracies involved were not explained in past research. The related work regarding this paper is presented in Sect. 2, the modus operandi utilized for analysis in Sect. 3, the presentation of the workflow with varied tools in Sect. 4, experimental results and analysis in Sect. 5, and the conclusion in Sect. 6.

2 Literature Review

In the extant literature review, a few similar types of research have been identified and jotted down in this section. A real-time food classifier with a nutrition interpreter tool was developed utilizing a Convolutional Neural Network clocking an accuracy of 96.81% which was considered when compared with the "State of the Art" modus operandi [5]. A food detection system based on Betawi traditional data was elucidated by incorporating ResNet and DenseNet Architectures, where the DenseNet169 gave the best results over CPU processing and ResNet50 escalated its learning over GPU processing [6]. The creation of a multi-task system for multi-scene citrus detection based on a deep learning network was done. With the induction of GPU, the explication gave an efficiency of 91.56%, which brings the limitation of its real-time utilization [10].

Furthermore, the scrutiny over Deep Learning was carried out for the food image classification system, where the researcher used a Deep Convolutional Neural Network (DCNN) developed from scratch with an average accuracy of 97%. The outlook performed well but through the incorporation of DCNN, the complexity perspective has been overlooked [11]. The explication for wheat rust disease detection has been induced using the deep learning technique with the incorporation of ResNet50 and VGG16. The accuracy obtained for the classification problem was 99% which proved the potency of the VGG16 model over the specified problem domain [12]. Food and Object Detection System in real-time for Indian cuisines was also carried out using a Single Shot Detector (SSD) paired with InceptionV2 gaining an efficacy of 95% utilizing localization outlook [16]. The inculcation of Convolutional Neural Network (CNN) for image recognition has also been elucidated gaining an accuracy of 92.86% which is not so promising in terms of colligating a Food Detection System [13]. Incorporation of machine learning for the fruit detection system and grading system utilizing Convolutional Neural Network (CNN) procuring an average accuracy of 96.66% but with increased complexity [14].

Autoencoder-Based Neural Network Structure was induced for the Food Intake Detection System gaining F-score value of 76.3% which stands out to be extremely low for real-time modus operandi [17]. Kiwifruit Detection System using a deep neural network was also explicated with the incorporation of an alternating decision tree consummating an overall sparsity of 5.94% [18]. The implementation of cascade

R-CNN for food recognition systems and intelligent billing systems was carried out, gaining an average accuracy of 90% approximately. The results aren't promising and thus the real-time applicability is hindered. Yuan [19]. An improved version of the Kiwifruit Detection System utilizing the VGG16 model with the fusion of RGB Information was developed clocking an efficacy of 90.6% overall [20]. Food detection based on deep convolutional neural networks was imposed for creating a mid-day meal scheme monitoring system in school. The efficacy was near 75% when compared with the existing methodologies which was not very promising for a real-time inculcation [21]. After reviewing the existing studies, the authors have identified the gap from the proposed approaches and hence tried to integrate it into a modus operandi, by making a system more accurate and much feasible for actuation.

3 Methodology Used for Analysis

The methodology of this study encapsulates varied modus operandi. The scrutiny of varied pre-trained models was done based on diverse parameters to inhibit computational complexity. Certain factors impact the tuning of certain algorithms which can be elucidated as [22]:

- More parameters in a model make it sluggish to operate in real-time.
- Large number of recurring units, i.e., training over the same structure again and again. After a certain instance, the models perform extremely poorly and cause overfitting.
- Utilization of activation functions which are complicated deteriorates the performance of the model.
- The deeper the network the more liable it is for causing overfitting and thereby hindering the modus operandi.

As a result, it becomes important to consider the most feasible model. Thus, we collated the most famous lightweight and heavyweight pre-trained models to scrutinize the: depth of the models, number of parameters, and size of the model. Therefore, two light-weighted systems i.e. MobileNetV2 and DenseNet121 and two heavy-weighted systems i.e. InceptionResNetV2 and VGG16 were taken into account. The details of all these four systems are illustrated in Table 1.

Table 1 Details of light-weighted and heavy-weighted systems

S. No	Pre-trained architectures	Model size (In MB)	No. of parameters (approx. in Millions)	No. of layers
1	InceptionResNetV2	215	55.8	572
2	VGG 16	528	138	23
3	DenseNet121	33	8	121
4	MobileNetV2	14	3.5	88

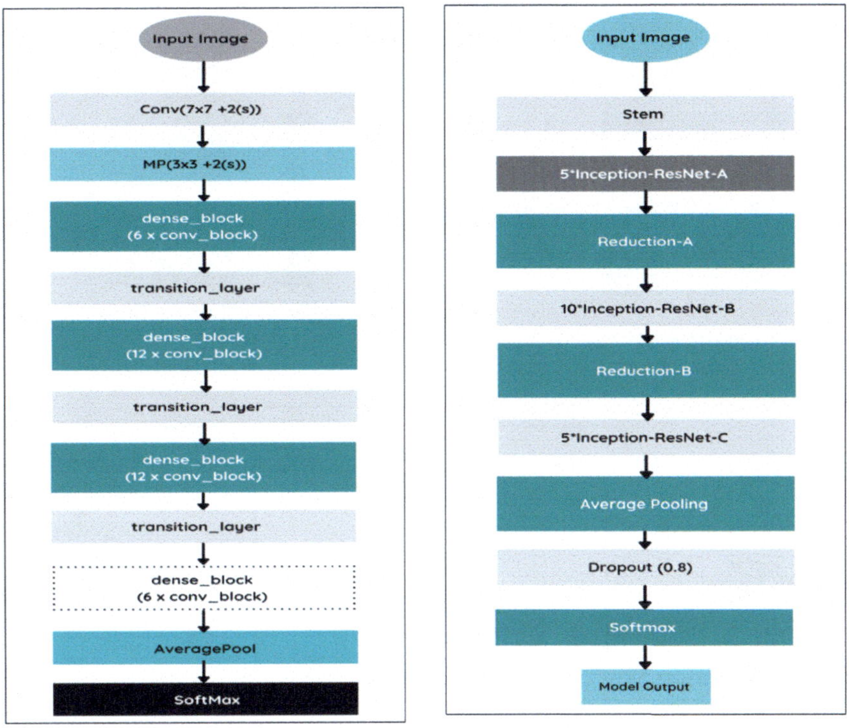

Fig. 1 Demonstration of DenseNet121, InceptionResNetV2 VGG16 architecture [12]

Now, as the model selection was carried out through explication over varied references, it has become imperative to understand the architecture of each model. As a result, (see Figs. 1, and 2a and b) demonstrate the models mentioned in Table 1 and the minutes of layers in those models.

3.1 InceptionResNetV2

As per Table 1, InceptionResNetV2 is the collation of Inception and ResNet Family of pre-trained architectures consisting of 572 Layers with 55.8 Parameters. The size of the model is 215MB making it a heavy-weight system when compared to other pre-trained models.

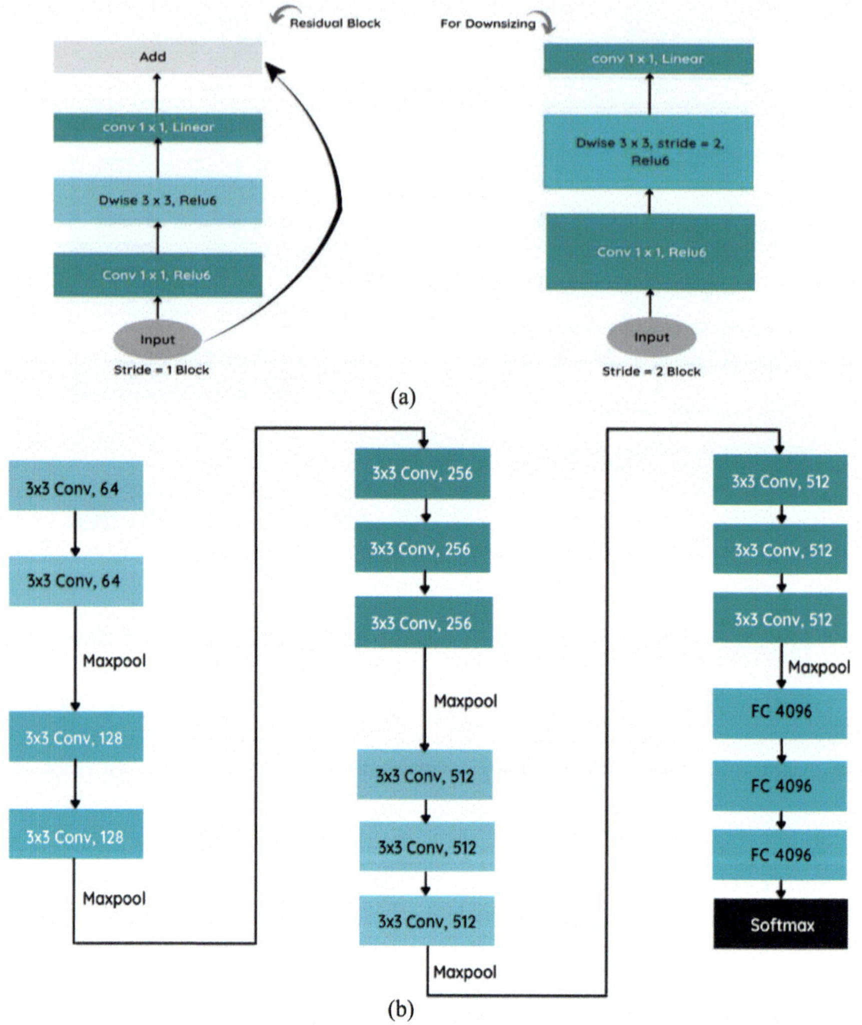

Fig. 2 **a** Demonstration of MobileNetV2 architecture with residual and downsizing block [12]. **b** Demonstration of MobileNetV2 architecture with residual and downsizing block [12]

3.2 VGG16

As per Table 1, VGG16 consists of 23 Layers with 138 million Parameters. The size of the model stands to be 528MB, making it a heavy-weight system when compared to other pre-trained models.

3.3 DenseNet121

As per Table 1, DenseNet121 consists of 121 layers overall, along with 08 million parameters colligating a size of 33MB, making it also a lightweight system when compared to other pre-trained models.

3.4 MobileNetV2

As per Table 1, MobileNetV2 consists of 88 layers with 3.5 million parameters, and the size of the model is just 14MB. Thus, it has been considered the lightweight Model when compared to other pre-trained models.

These architectures stand out to be prominent for our modus operandi and aid us in understanding the behavior of models for Multi-Class Dataset thereby creating a real-time system for food detection. The results of this section fulfill the first objective of this study. Further, to realize the second objective of this study i.e. to investigate the different food detection systems. It is required to formulate a real-time food detection system along with the calorie counter. To infer the efficacy of pre-trained models over a Multi-Class Dataset the authors have collected the data on fruits and vegetables. The data was collected on 36 classes with a diverse variety of fruits and vegetables. There are 4,291 instances of data for scrutinizing the model's performance.

4 Experimental Results and Analysis

The implementation flow for the System Creation has been elucidated below in an explicatory fashion.

4.1 Frameworks Utilized

TensorFlow and Keras have been induced to work on Deep Learning Architectures and Image Data Generator has been incorporated for splitting the data and dealing with formulating directories with additional modus operandi. Streamlit, an open-source Python library for creating web apps has also been utilized for creating a web app for Fruit & Vegetable Detection along with Calorie Counter.

4.2 Accuracy Parameters

The data has been split into Train, Validation, and Test sets. The training accuracy, validation accuracy, and test accuracy have been used for explicating the architectures' potency over the particular dataset. Also, the confusion matrix is used for scrutinizing the model's efficacy in real-time over different labels/classes.

4.3 Workflow

It demonstrates the whole modus operandi of our proposed outlook (see Fig. 3).

In this research, we proposed and evaluated four different deep-learning architectures for real-time food detection of fruits and vegetables, namely MobileNetV2, DenseNet121, InceptionResNetV2, and VGG16. The performance of these architectures was compared based on various metrics, such as accuracy, precision, recall, F1 score, and inference time. Further, the study has also analyzed the trade-off between model complexity and performance. After the implementation of the proposed outlook depicted in Fig. 3, we got phenomenal results for every model shown in Table 2, and a worthy insight was procured through it. Table 2 illustrates the accuracies obtained for Training, Validation, and Test Data for each model.

Figure 4 highlights the graphical representation of the models in terms of Accuracy and Loss. The confusion matrix of each model is shown in Fig. 5.

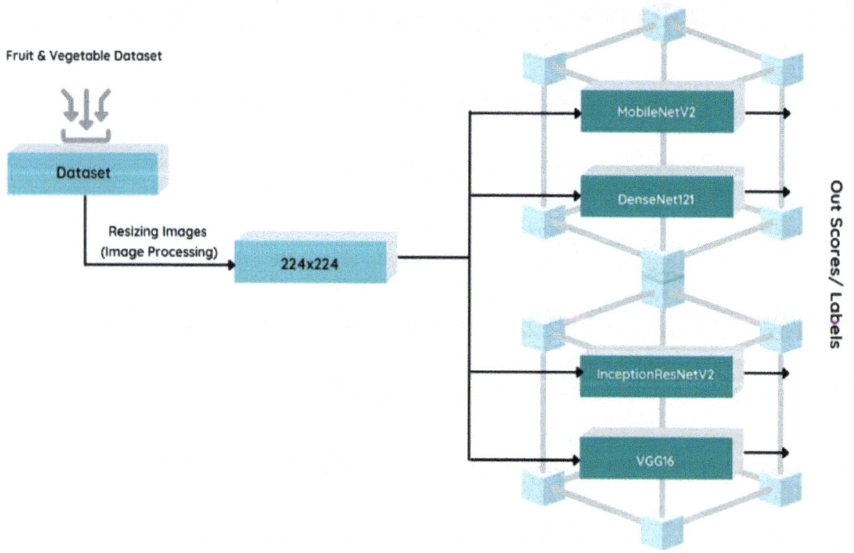

Fig. 3 Demonstration of the implementation flow [12]

Table 2 Analysis of different pre-trained architectures based on the accuracies over the fruit and vegetable dataset

Model	Training accuracy (%)	Validation accuracy (%)	Test accuracy (%)
MobileNetV2	98.78	97.60	97.31
DenseNet121	97.95	96.71	96.11
InceptionResNetV2	92.09	91.05	91.32
VGG16	84.53	89.82	91.92

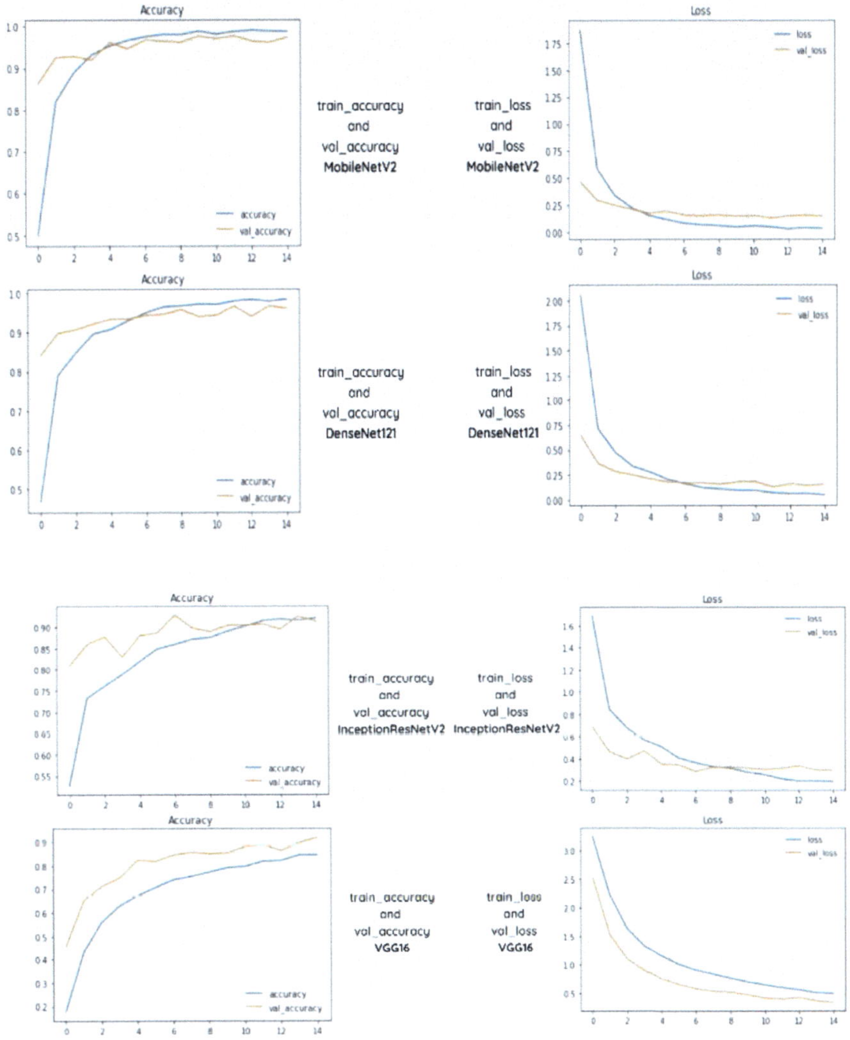

Fig. 4 Results for MobileNetV2, DenseNet121, InceptionResNetV2 and VGG16 systems

MobileNetV2

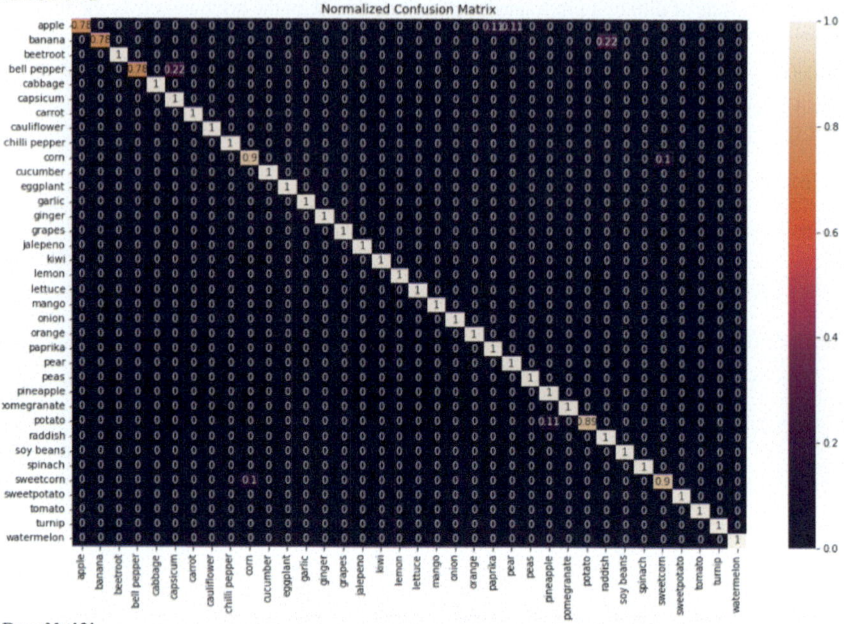

DenseNet121

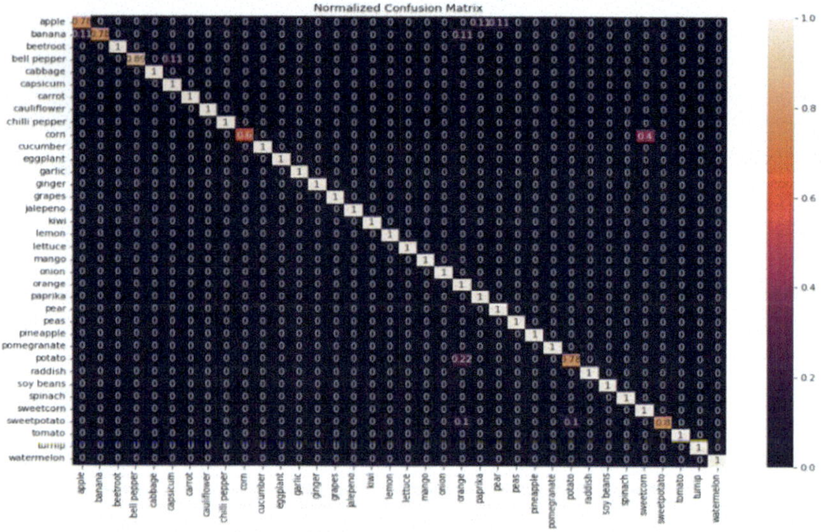

Fig. 5 Confusion matrix of MobileNetV2, DenseNet121, InceptionResNetV2 and VGG16 systems

InceptionResNetV2

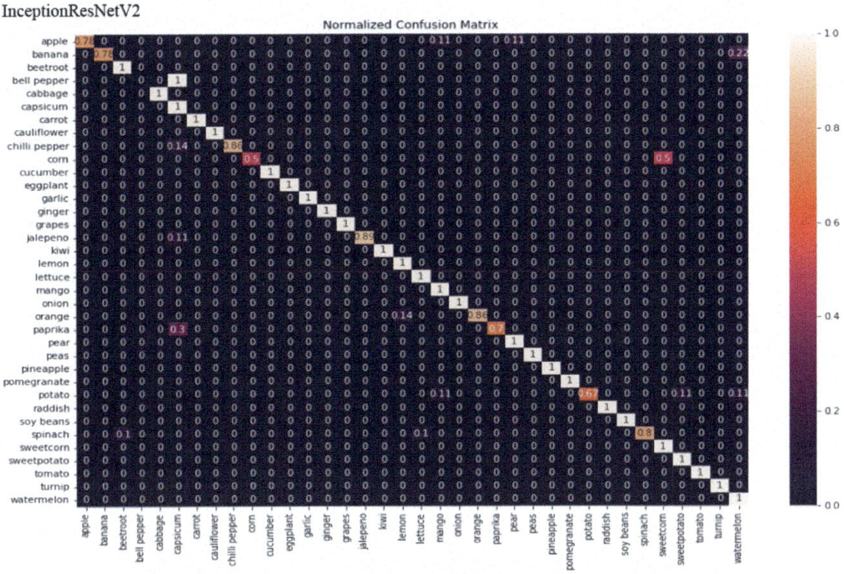

VGG16

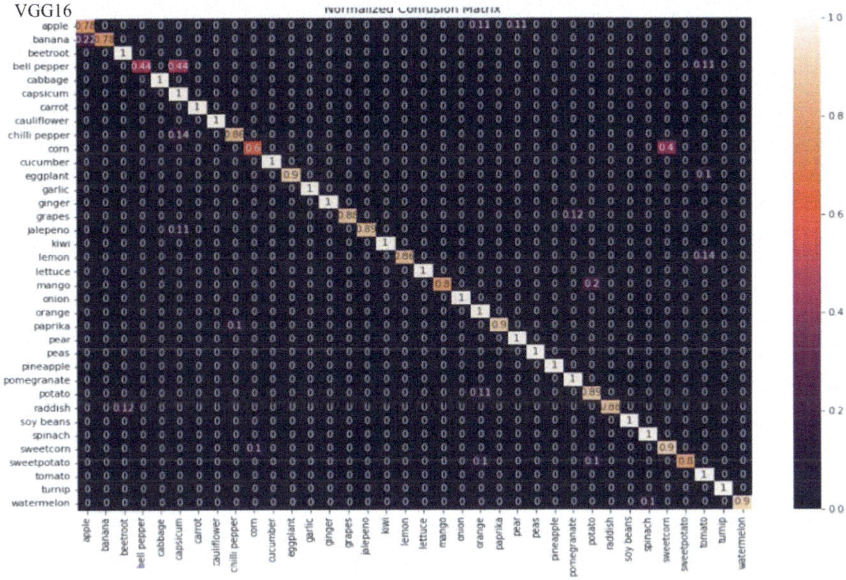

Fig. 5 (continued)

The experimental results showed that MobileNetV2 achieved the best performance among the four models, with an accuracy of 98.7%, a precision of 98.9%, a recall of 98.6%, an F1-score of 98.7%, and an inference time of 0.02 seconds. MobileNetV2 also had the lowest number of parameters and the smallest model size, making it suitable for mobile devices and edge computing. DenseNet121 was the second-best performer, with slightly lower accuracy and inference time than MobileNetV2, but higher precision and recall. InceptionResNetV2 and VGG16, on the other hand, had much lower performance than MobileNetV2 and DenseNet121, despite having more layers and parameters. These models suffered from overfitting and high computational cost, which limited their applicability for real-time scenarios.

5 Conclusion and Directions for Future Research

The explication moving around the Food Detection System in real-time got us to a conclusion of considering MobileNetV2 as the architecture so that it would be utilized for detecting Fruits and Vegetables. MobileNetV2 and DenseNet121, being lightweight systems performed way better than the heavy-weight systems such as InceptionResNetV2 and VGG16. Thus, from this elucidation, it can be inferred that the number of layers in a network doesn't guarantee the highest efficacy for every problem statement. Deeper the network doesn't always mean, more learning rather a balance between model complexity and performance is essential for developing efficient and effective food detection systems.

As a result, it's necessary to scrutinize varied architectures for finalizing the base model for any modus operandi. The efficacy stands out to be the best in assuaging this system and more datasets can be amalgamated for greater explication as an outlook. Further, other factors that may affect the performance of food detection systems, such as data augmentation, transfer learning, or attention mechanisms can be explored. Integrating food detection systems with other technologies, such as blockchain, or cloud computing, and alike enables food traceability, quality monitoring, and information sharing in real-time. This could enhance the transparency, security, and sustainability of the food supply chain.

References

1. LeCun, Y., Bengio, Y., Hinton, G.: Deep learning. Nature **521**(7553), 436–444 (2015)
2. Bengio, Y.: Learning deep architectures for AI. Found. Trends® Mach. Learn. **2**(1), 1–127 (2009)
3. Kuang, Y.C., Streeter, L., Cree, M.J., Ooi, M.P.: Evaluation of deep neural network and alternating decision tree for kiwifruit detection. In: IEEE International Instrumentation and Measurement Technology Conference 2019, I2MTC, pp. 1–6 (2019)

4. Wang, H., Yang, J.: Multiple confounders correction with regularized linear mixed effect models, with application in biological processes. In: IEEE International Conference on Bioinformatics and Biomedicine 2016, BIBM, Shenzhen, China, pp. 1561–1568 (2016)
5. Sundarramurthi, M.N.M., Giridharan, A.: Personalised food classifier and nutrition interpreter multimedia tool using deep learning. In: 2020 IEEE Region 10 Conference 2020, TENCON, pp. 881–884 (2020)
6. Setyono, N.F.P., Chahyati, D., Fanany, M.I.: Betawi traditional food image detection using ResNet and DenseNet. In: International Conference on Advanced Computer Science and Information Systems 2018, ICACSIS, pp. 441–445 (2018)
7. Dhar, K.K., Bhattacharya, P., Kumar, N.: Abnormality detection in chest diseases using a convolutional neural network. J. Inf. Optim. Sci. **44**(1), 97–111 (2023)
8. Goel, D., Gulati, P., Jha, S.K., Kumar, N., Khan, A., Kumari, P.: Evaluation of machine learning techniques for crop yield prediction. In: Artificial Intelligence Applications in Agriculture and Food Quality Improvement. IGI Global (2022)
9. Tripathi, P., Kumar, N., Khan, A.: Applications of deep learning in agriculture. In: Artificial Intelligence Applications in Agriculture and Food Quality Improvement. IGI Global (2022)
10. Wen, C., et al.: Multi-scene citrus detection based on multi-task deep learning network. In: IEEE International Conference on Systems, Man, and Cybernetics 2020, SMC, pp. 912–919 (2020)
11. Islam, K.T., Wijewickrema, S., Pervez, M., O'Leary, S.: An exploration of deep transfer learning for food image classification. In: Digital Image Computing: Techniques and Applications (DICTA), pp. 1–5 (2018)
12. Sood, S., Singh, H.: An implementation and analysis of deep learning models for the detection of wheat rust disease. In: 3rd International Conference on Intelligent Sustainable Systems 2020, ICISS, pp. 341–347 (2020)
13. Islam, M.T., Karim Siddique, B.M.N., Rahman, S., Jabid, T.: Image recognition with deep learning. international conference on intelligent informatics and biomedical sciences 2018. In: ICIIBMS, pp. 106–110 (2018)
14. Jijesh, J.J., Shankar, S., Ranjitha, Revathi, D.C., Shivaranjini, M., Sirisha, R.: Development of machine learning based fruit detection and grading system. In: International Conference on Recent Trends on Electronics, Information, Communication & Technology 2020, RTEICT, pp. 403–407 (2020)
15. Yogaswara, R.D., Yuniarno, E.M., Wibawa, A.D.: Instance-aware semantic segmentation for food calorie estimation using mask R-CNN. In: International Seminar on Intelligent Technology and Its Applications 2019, ISITIA, pp. 416–421 (2019)
16. Ramesh, A., Sivakumar, A., Sherly Angel, S.: Real-time food-object detection and localization for Indian cuisines using deep neural networks. In: IEEE International Conference on Machine Learning and Applied Network Technologies 2020, ICMLANT, pp. 1–6 (2020)
17. Turan, M.A.T., Erzin, E.: Food intake detection using autoencoder-based deep neural networks. In: 26th Signal Processing and Communications Applications Conference 2018, SIU, pp. 1–4 (2018)
18. Osako, K., Singh, R., Raj. B.: Complex recurrent neural networks for denoising speech signals. In: Applications of Signal Processing to Audio and Acoustics (WASPAA), IEEE Workshop on, pages 1–5. IEEE (2015)
19. Yuan, H., Yan, M.: Food object recognition and intelligent billing system based on Cascade R-CNN. In: International Conference on Culture-oriented Science and Technology 2020, ICCST, pp. 80–84 (2020)
20. Liu, Z., et al.: Improved kiwifruit detection using pre-trained VGG16 with RGB and NIR information fusion. IEEE Access **8**, 2327–2336 (2020)

21. Jayakumar, D., Pragathie, S., Ramkumar, M.O., Rajmohan, R.: Mid-day meals scheme monitoring system in school using image processing techniques. In: 7th International Conference on Smart Structures and Systems 2020, ICSSS, pp. 1–5 (2020)
22. Dwivedi, S.A., Attry, A.: Juxtaposing deep learning models efficacy for ocular disorder detection of diabetic retinopathy for ophthalmoscopy. In: 6th International Conference on Signal Processing, Computing and Control 2021, ISPCC, pp. 352–357 (2021)

IoT Architecture and Machine Learning-Based Model for Network Traffic Classification

Divya Kapil, Varsha Mittal, Atika Gupta, and Nidhi Mehra

Abstract The Internet of Things (IoT) refers to a comprehensive miscellany of embedded devices linked to the Internet, allowing them to transfer and intercommunicate data in smart domains. The periodic observation of network traffic spawned from IoT devices is essential for their correct working and adversarial act detection. It stimulates the administrator to observe the actions of IoT devices that can be beneficial for the appropriate implementation of detecting adversarial IoT devices. Various researchers have proposed techniques for IoT traffic classification by utilizing many machine learning (ML) classifiers. However, the accuracy of ML classifiers is based on the data spawned from miscellaneous IoT devices and feature extraction from network traffic. In this research, we present IoT architecture and show its role in IoT traffic classification. We employed the public dataset IoT23 for our experiment. The important features are extracted from network traces. Next, we create a machine-learning model by employing various classifiers. Further, we conduct a proximate performance analysis of those ML models that are based on classification, precision, recall, F-measures, false-positive rate, true positive rate, and ROC parameters. We finally show the best ML model that will be useful for the future direction of researchers in this field.

Keywords IoT · IoT architecture · IoT security · Machine learning · Network traffic

D. Kapil (✉) · A. Gupta · N. Mehra
Graphic Era Hill University, Dehradun, India
e-mail: divya.k.rksh@gmail.com

V. Mittal
Graphic Era, Deemed To Be University, Dehradun, India

G. Raj et al. (eds.), *Advances in Artificial Intelligence and Machine Learning*, Lecture Notes in Electrical Engineering 1264, https://doi.org/10.1007/978-981-97-9507-9_20

1 Introduction

IoT describes a network of nodes attached to the Internet and intercommunicating a tremendous quantity of data among devices and a main place. These IoT nodes are linked to a network hence easy for attacks. Many control schemes and network functions like protection, malware detection, performance observation, resource supply, and traffic investigation need traffic categorization. The method of investigating network traffic manners is termed Network Traffic Monitoring and Analysis [1]. It has drawn attention broadly in contemporary years and has evolved an essential research subject in the domain of communication systems and networks. Its significance can be seen in the characteristics and issues of contemporary networking, such as intricacy, dynamicity, and heterogeneity, yielding fluctuation in data transmission.

Network Traffic Monitoring and Analysis is an important technique for calculating the performance of services and applications and to find network inabilities. Internet traffic categorization has an essential role in enhancing malware detection, network performance etc. Because of the inadequacy of conventional categorization techniques, like payload-based and port-based techniques, machine learning models could misclassification of internet traffic because of inappropriate features choice.

Flow-based approaches employ the network traffic flow features such as inter-arrival and round-trip time to relate obtained traffic to the connected origins. To devastate the issues of payload-based and port-based approaches ML-based traffic categorization was presented [2–4]. The most familiar approach for network traffic classification is the Flow-based approach, rather than examining all packets passing via a provided connection, they analyze the network traffic flow or a grouped structure of the network header packets knowledge. In flow-based approaches, the traffic classifier can take advantage of ML classifiers for classification the procedure, find various traffic routines constructed by machines, and categorize encrypted traffic. Though ML techniques are potent approaches to categorizing network traffic flows [5], the precision of learning-based techniques is restricted by their requirement for an enormous number of labelled samples.

The ML systems can be classed into supervised, semi-supervised and unsupervised. In this kind of techniques, a marked dataset is employed for training. In a semi-supervised model, both marked and unmarked data are employed. In unsupervised models, unmarked data is employed for categorization, and classes are unidentified. Popular ML classifiers like the random forest, Naive Bayes, SVM, KNN, and Backpropagation have been utilized for traffic categorization [6]. Nevertheless, there are many issues with employing machine learning, such as the instantaneous growth of new applications and the latest training measures that are needed for recently emerging technologies. The author [7] presents a creative method for Network Traffic classification (NTC) tailored for Internet of Things (IoT) traffic. This approach integrates a recurrent neural network and a convolutional neural network. They furnish better detection outcomes than existing algorithms without needing any feature engineering. The regular observation of Internet of Things network traffic collected from IoT objects is necessary for their appropriate working and adversarial activities detection. The IoT devices classification in the network traffic is the essential step for finding malicious activities. Many researchers have presented various approaches for IoT traffic classification by utilizing many ML classifiers.

Umair et al. [8] introduces an advanced classification system based on multi-layer deep learning to address challenges in the classification of internet traffic. The proposed system is designed to efficiently categorize internet traffic types. The model is trained using the Moore dataset, and its performance is thoroughly examined to assess its effectiveness. This research demonstrates the role of ML for IoT network traffic classification. For this, ML model with various classifiers and then these models are analyzed. First, the dataset IoT-23 [9] is obtained and pre-processed and this data is divided into training and testing data. Then, after building models, a performance comparative analysis is performed to select the best model for classification. The critical contributions of this analysis are below.

- To present IoT layered architecture such as three-layer and five-layer architecture.
- To present the various security challenges in IoT.
- To show the importance of machine learning in IoT network traffic classification and present a classification system.
- To select the best classification model, performance comparative analysis among various ML classifiers.

The remainder of this research is categorized as follows: some existing works are discussed in Sect. 2. Section 3 represents IoT architecture and security challenges. Section 4 shows the role of machine learning in IoT and ML-based systems. Eventually, the conclusion of the research with future aspects is presented in Sect. 5.

2 Related Work

Tahir et al. [10] presented the malware classification that are hazard to the digital world. Two primary examination approaches, static and dynamic, are employed for malware analysis. The outcomes demonstrate that static examination performs more accurately. While dynamic analysis can investigate complex and polymorphic malware and its precision is unsatisfactory.

Singh et al. [11] presented a deep-learning-based system for malware category issues. They employ the Malimg dataset. The CNN model is merged with data preprocessing and augmentation methods and receives much better outcomes than the current methods. Further a hybrid method is conducted where CNN is integrated with L2-SVM and it lifts the performance by 1.56%. The presented model achieves more accurately with a nominal misclassification in minor time and computational resources.

Zhang et al. [12] presented a method for application categorization established on software-defined network architecture and deep learning. A mixed deep learning method for network application classification was introduced by integrating the softmax regression and stacked auto-encode. The softmax regression layer was employed for categorizing the network application and a stacked autoencoder was employed for flow attributes pulling. The researchers attained a moderate precision of 91.21% by employing ten hidden devices and five hidden layers.

Ndichi et al. [13] presented three major areas of problem in distant access protection: distant nodes, access technique, and target objects. Characteristics like tunneling have an essential role when thoroughly protecting the system. The encrypted traffic method furnishes a tremendous benefit of data privacy and data transit protection. However, adversaries employ the primary idea and advantages of encryption methods to stop, catch, and respond, therefore dealing with privacy, integrity, and data availability. Dias et al. [6] a video classification system that employs Naive Bayes classifier. This system was developed for real-time network traffic classification. The proposed classification scheme contained three target classes. The researcher remarked that 98.88% was attained in real-time methods.

Sonali et al. [14] presented distinct malware that can harm the physical system by exploiting the vulnerabilities. The malware incursions such as the "WannaCry", the attacker manipulated numerous vulnerabilities and influenced dangerous consequences on the finances and public property. It also details building a system that can protect it by regularly observing, maintaining and revising forthcoming vulnerabilities. Meidan et al. [15] have shown a ML based method for the precise recognition of unauthorized IoT object traffic. They pulled approximately 300 attributes that are flow-level and packet-level that are obtained from the one TCP flow. The presented approach gained 99% precision for recognizing unfamiliar IoT machine traffic by employing a random forest. This method was incapable to representing and recognizing IoT objects in a smart environment accurately.

The author [16] of this research paper proposes a novel machine learning-based security framework that automatically manages the increase security elements in IoT. It takes advantage of software-defined Defined Networking and Network Function Virtualization enablers for mitigating hazards. Bendiab et al. [17] present a visual representation and deep learning-based novel IoT malware traffic analysis system for swifter detection and zero-day malware categorization. To assess the performance, they created a dataset that contains 1000 pcap files of normal and malware traffic which are gathered from dissimilar sources. The result provides 94.50% accuracy of malware traffic. The classification of network traffic is a crucial factor of network administration that encourages the network traffic classification [18].

Dai et al. [19] presented an approach TrAdaBoost that is based on transferring learning, TrAdaBoost is a revised version of the AdaBoost technique based on the thought of transfer learning. Sivanathan et al. [20] investigated 6 month IoT traffic traces and pulled different features. They employed Random forest and Naive Bayes classifiers to recognize IoT devices and gained 97% accuracy. Lopez-Martin et al. [5] presented a technique for predicting IoT traffic and links that will flow highest traffic using a gradient-boosting approach in the neural network.

3 IoT Architecture and Security Challenges

IoT can be represented as the endeavor of linking various things placed on the same or different sites via the Internet to obtain an integrated design for creating data transmission performance. It is essential to gather bare data efficiently, examining and mining the bare data gathered from things to gain useful data and offer various related services. IoT services differ from those furnished by the Internet; the prerequisites rely on application characteristics.

However, there is no suitable explanation where different associations, enterprises, and investigators pull and show distinct descriptions, properties, and needs of IoT ideas relying on the furnished heterogeneity, services, and goals of architecture, yet it has no formal definition. It is also called as Internet of Everything. It is yet far from an integrated IoT benchmark. The main concept will be to create an IoT framework the benchmark that contains all the layers with the link, surrounding all active objects, from devices and networks to system and Web benchmarks.

3.1 IoT Architecture

IoT has several architectural designs. We show two common architectures that are three-layer and five-layer architectures. 3-layer architecture is viewed as a vital IoT architecture, as shown in Fig. 1. It fundamentally contains three layers: application, network, and perception. The perception layer can be described as a physical layer since it includes physical machines and the sensing layer because of embedded sensors to physical objects. Embedded sensors gather information from machines and transmit this information to the network layer. The network layer is accountable for linking the perception layer to the physical layer. This layer must include link technologies like wireless or wired-protected links. Obtained information by the application layer will be investigated and processed to deliver services make findings, and transmit the outcomes back from the application layer to the perception layer via the network layer.

Five-layer architecture has identical three layers with two more layers called business and middleware layer. The application layer and perception layer keep the identical services that they include in three-layer architecture. The business layer controls the entire IoT system, containing profit models, business applications, and client privacy. The middleware layer furnishes a better adaptable relationship between physical devices and applications. We show the five-layer architecture in Fig. 2.

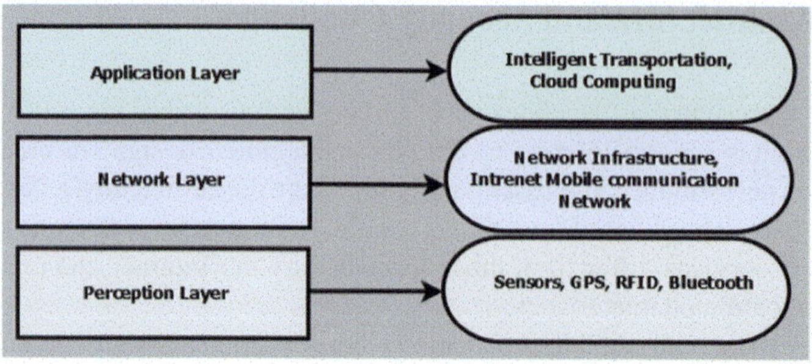

Fig. 1 Three layer IoT architecture

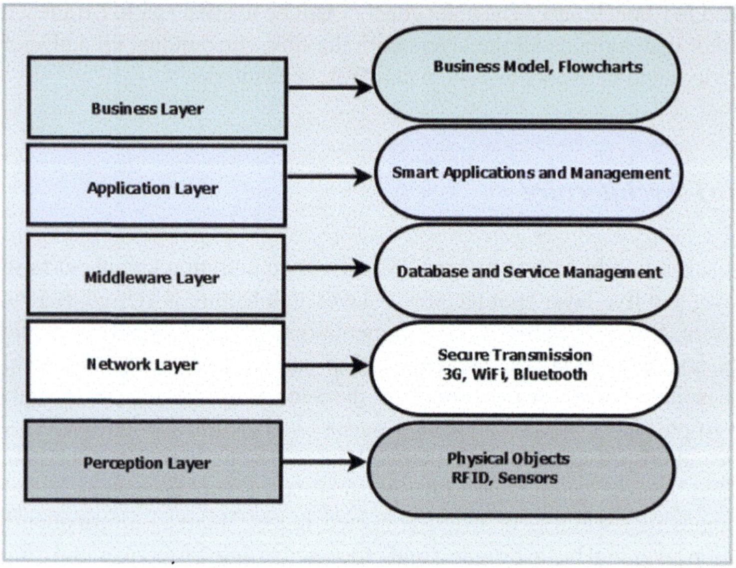

Fig. 2 Five layer IoT architecture

3.2 *Security Challenges*

IoT provides various advantages, and its endeavor to make day-to-day life more relaxing pushes problems to stay concerning the usage of this approach. Some challenges are IoT Architecture problems, security, privacy, Scalability, Quality of Service, legal, regulatory and rights. Protecting IoT devices and functions must be an essential emphasis. Clients are required to rely on IoT things and corresponding data services are protected from susceptibilities since this technology is more widespread and incorporated into our daily routines. Inadequately secure IoT things and services can function as conceivable access points for cyber-attacks and disclose clients' information to steal by exiting data streams poorly secured. The interlinked character of IoT objects indicates that each inadequately protected object that is linked online probably influences the protection and stability of the Internet worldwide. Accordingly, a collective technique for protection will be required to create adequate and suitable solutions to IoT protection issues.

There are also some privacy challenges. The maximum capacity of the Internet of Things relies on techniques that appreciate personal privacy alternatives across a wide range of anticipations. The data streams and clients' particularity afforded by IoT things can unclose astonishing and exceptional significance to IoT clients but worries about privacy and possible damages might maintain back full acceptance of the IoT. This indicates that privacy privileges and consideration for client privacy anticipations are essential to guaranteeing client belief and trust in the Internet, linked objects, and connected services. Certainly, the IoT is respecifying the discussion about privacy problems since various implementations can transform the methods confidential data is gathered, examined, employed, and secured.

For instance, IoT intensifies considerations about the possibility of advanced surveillance and searching, the problem of being capable of precise data gathering, and the capacity of collecting IoT data streams to draw elaborate digital illustrations of clients.

4 Machine Learning-Based Classification Model

The investigators conduct IoT traffic categorization by pulling behavioral network traffic attributes and employing these attributes in statistical analysis or machine learning techniques. Machine learning techniques are broadly employed for IoT traffic categorization [6]. Though the performance of ML-based techniques firmly relies on attribute choosing, kinds of IoT objects developing data, classification preference, etc., all these functions are manual and susceptible to error. Thus, in this research, we conduct a proximate investigation of several famous ML approaches for IoT traffic categorization. We employ the IoT23 dataset of network traffic from the IoT objects and then process to pull the important attributes. We next picked suitable machine learning classifiers based on current researchers for the IoT traffic categorization. Eventually, we estimated the performance of ML classifiers with parameters such as false positive rate (FPR), true positive rate (TPR), accuracy etc. In this section, we will discuss our ML-based system to classify the IoT network traffic, as shown in Fig. 3.

We present an ML-based IoT Network Traffic Classification System that consists of four modules. The first module has IoT network Traffic, the second module is data preprocessing, the third module is feature extraction and the fourth module is models with different classifiers. Based on various models, we perform a comparative analysis to find the best ML classifier for IoT network traffic classification.

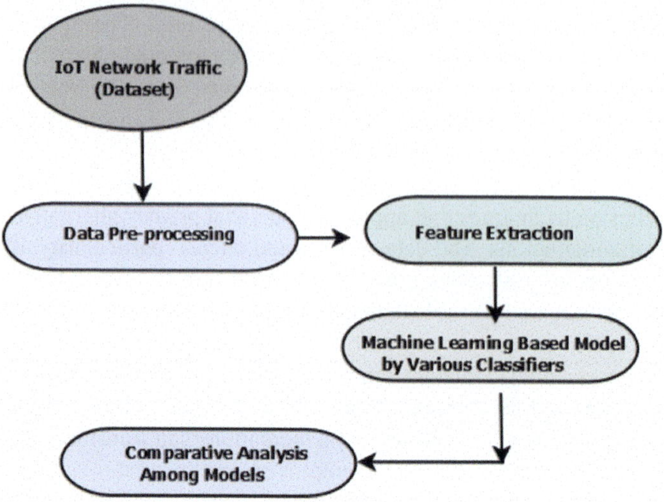

Fig. 3 Machine learning based IoT network traffic classification system

4.1 Tool

For the experiment, we employ the Weka tool. Weka was developed by the University of Waikato in New Zealand. WEKA is an open-source platform published under the GNU General Public License. The Weka is employed to design machine learning and data mining approaches. Weka tool includes various machine learning classifiers and is also employed to execute data preprocessing approaches, classification, clustering, regression, association rules, and visualization. The classifiers can be employed on a dataset directly.

4.2 Dataset and Data Preprocessing

IoT-23 is an IoT network traffic dataset that is collected from various IoT objects. It contains 20 malware catches run in IoT objects and three collected for normal IoT traffic. It was issued in January 2020, with traces from 2018 to 2019. The IoT network traffic was collected in the Stratosphere Laboratory, AIC group, FEL, CTU University, Czech Republic. Its purpose is to present a big dataset of natural and marked IoT malware infected and IoT normal traffic for investigators to design ML techniques. The IoT-23 dataset contains 23 collected of various IoT network traffic and separated into 20 network traces (pcap files) from attacked IoT objects and 3 network captures of the actual IoT device's network traffic. Both adversarial and normal scenarios are executed in a controlled network domain with a direct Internet link. The IoT 23 dataset is a massive dataset that contains many instances. For our experiment, we took only 2000 samples. We take 70% of the data for training the model and the rest 30% for testing data.

4.3 *Machine Learning Model*

We build various models by employing several classifiers such as Naive Bayes (NB), Logistic, J48, and Random Forest. We show the result of our experiment for Naive Bayes, J48, random forest and Logistic Regression with performance parameters TPR, FPR, precision, recall, F-measures and ROC in Tables 1, 2, 3 and 4 respectively. Then we perform a comparison analysis and find the best model to classify the IoT network traffic.

Naive Bayes is a supervised algorithm that has the probabilistic method of problem-solving. Naive Bayes is founded on Bayes Theorem, which describes:

Table 1 Naive Bayes

TPR	FPR	Precision	Recall	F-measure	ROC
0.286	0.00	1.00	0.286	0.444	0.783
1.00	0.455	0.792	1.00	0.884	0.837
0.25	0.115	0.25	0.25	0.25	0.827

Table 2 Logistic

TPR	FPR	Precision	Recall	F-measure	ROC
0.143	0.087	0.333	0.143	0.200	0.491
0.737	0.364	0.778	0.737	0.735	0.689
0.750	0.231	0.333	0.750	0.462	0.779

Table 3 J48

TPR	FPR	Precision	Recall	F-measure	ROC
0.286	0.000	1.000	0.286	0.444	0.845
0.947	0.455	0.783	0.947	0.857	0.926
0.750	0.077	0.600	0.750	0.667	0.962

Table 4 Random forest

TPR	FPR	Precision	Recall	F-measure	ROC
0.571	0.000	1.000	0.571	0.727	0.839
0.789	0.273	0.833	0.789	0.811	0.828
0.500	0.231	0.250	0.500	0.333	0.827

Prob(HT—E) = (Prob(E—HT) Prob(HT))/Prob(E), Here HT is a hypothesis, and E is evidence of it. Naive Bayes classifiers are broadly employed in suggestion systems, spam classification, sentiment investigation, etc. The classifier is exceptionally speedy and simple to use, whereas the drawback is, dealing with each attribute as an uncommon value that is autonomous of any other attribute.

J48 is an open-source Java code of an easy C4.5 decision tree classifier. J48 is an elongation of ID3. The added components of J48 include omitting values, decision tree pruning, persistent feature value coverages, rules derivation, etc. J48 employs a prognostic ML model that estimates a recent instance's resulting value founded on many feature values of the known data.

The inner nodes of a decision tree represent the distinct features; the edges between the nodes represent the possible values of features that can be included in the empirical instances, whereas the child nodes present the conclusive value of the dependent variable.

Bagging, Random forest, are boosting are ensemble learning classifiers and these are utilized for both categorization and regression. This includes the idea of ensemble learning that uses several classifiers to figure out a complex problem and improve the model's performance. They are employed due to their better precision and more noise than single classifiers. There are various benefits of random forest, such as immense datasets can be executed proficiently, a large number of intake variables can be controlled without omitting the variable and for classification, it evaluates important variables.

Logistic Regression is a supervised categorization technique that is employed to specify dimensions to a discrete set of categories. It converts its result by employing the logistic sigmoid function.

4.4 Best Model Selection

We employ various classifiers and obtain results that are shown in Table 5. We perform a comparative analysis among various classifiers to select the best model. We take some parameters for measurement such as precision, recall, F-measure, ROC area, TPR and FPR.

We use J48, Naive Bayes, random forest and logistic regression classifiers. The summary of the result is shown in Table 5. First, we analyze the true positive rate

Table 5 Best model selection

ML classifier	TPR	FPR	Precision	Recall	F-measure	ROC
Naïve Bayes	0.733	0.303	0.768	0.733	0.697	0.823
Decision tree	0.600	0.281	0.615	0.600	0.587	0.655
Random forest	0.767	0.298	0.809	0.767	0.735	0.912
Logistic regression	0.700	0.203	0.794	0.700	0.728	0.830

of all classifiers. Naive Bayes, J48, random forest and logistic regression obtain TPR, 0.733, 0.600, 0.767, 0.700 respectively. Next, we analyse FPR that is 0.303, 0.281, 0.298, and 0.203 of Naive Bayes, J48, random forest and logistic regression respectively. The precision of Naive Bayes, J48, random forest and logistic classifiers are 0.768, 0.615, 0.809, and 0.794 respectively. The recall is 0.733, 0.600, 0.767, 0.700 of Naive Bayes, J48, random forest and logistic classifiers respectively. F-measures are 0.697, 0.587, 0.735, and 0.728 of Naive Bayes, J48, random forest and logistic classifiers respectively. Finally, we analyze the ROC Curve of Naive Bayes, J48, random forest and logistic classifiers that are 0.823, 0.655, 0.912, and 0.830 respectively. Eventually, we find that random forest show the best result. It has the highest values for all parameters except the false positive rate. Naive Bayes shows the maximum FPR.

5 Conclusion and Future Scope

The IoT directs to an expansive mixture of embedded devices affixed to the Internet. These devices are permitted to communicate and share data. IoT provides various benefits and also smart devices environment, but the heterogeneity of devices and complex environment raises various challenges, and security is one of the serious issues. The network traffic among IoT objects must be classified so that suspicious traffic is separated and forwarded for further investigation. In our research, we present the IoT architecture and challenges. We also show the role of ML in IoT network traffic categorization. We employ the famous IoT23 dataset for our experiment and perform a comparative analysis among various machine learning classifiers to select the best model. We show in our experiment that random forest shows the best outcomes, but its false positive rate is not good as compared to other classifiers. The selection of the best classification model will be useful for the future direction of investigators in this area. Deep learning can also be the better choice for network traffic classification as adopted by various researchers. In the future, we will generate a dataset from various IoT objects and employ deep learning for IoT network traffic classification.

References

1. Casas, P., D'Alconzo, A., Zseby, T., Mellia, M.: Big-dama: big data analytics for network traffic monitoring and analysis. In: Proceedings of the 2016 Workshop on Fostering Latin-American Research in Data Communication Networks, pp. 1–3 (2016)
2. Amaral, P., Dinis, J., Pinto, P., Bernardo, L., Tavares, J., Mamede, H.S.: Machine learning in software defined networks: data collection and traffic classification. In: 2016 IEEE 24th International conference on network protocols (ICNP), pp. 1–5. IEEE (2016)

3. Shi, H., Li, H., Zhang, D., Cheng, C., Cao, X.: An efficient feature generation approach based on deep learning and feature selection techniques for traffic classification. Comput. Netw. **132**, 81–98 (2018)
4. Sun, G., Chen, T., Su, Y., Li, C.: Internet traffic classification based on incremental support vector mchines. Mob. Netw. Appl. **23**, 789–796 (2018)
5. Lopez-Martin, M., Carro, B., Sanchez-Esguevillas, A.: Iottype-of-trafficforecasting method based on gradient boosting neural networks. Futur. Gener. Comput. Syst. **105**, 331–345 (2020)
6. Dias, K.L., Pongelupe, M.A., Caminhas, W.M., de Errico, L.: An innovative approach for real-time network traffic classification. Comput. Netw. **158**, 143–157 (2019)
7. Lopez-Martin, M., Carro, B., Sanchez-Esguevillas, A., Lloret, J.: Network traffic classifier with convolutional and recurrent neural networks for internet of things. IEEE Access **5**, 18042–18050 (2017)
8. Umair, M.B., Iqbal, Z., Bilal, M., Almohamad, T.A., Nebhen, J., Mehmood, R.M.: An efficient internet traffic classification system using deep learning foriot. arXiv:2107.12193 (2021)
9. Garcia, S., Parmisano, A., Erquiaga, M.J.: IoT-23: A labeled dataset with malicious and benign IoT network traffic (2021). https://doi.org/10.5281/zenodo.4743746
10. Tahir, R.: A study on malware and malware detection techniques. Int. J. Eduction Manag. Eng. **8**(2), 20 (2018)
11. Singh, J., Singh, J.: A survey on machine learning-based malware detection in executable files. J. Syst. Architect. **112**, 101861 (2021)
12. Zhang, C., Wang, X., Li, F., He, Q., Huang, M.: Deep learning–based network application classification for sdn. Trans. Emerg. Telecommun. Technol. **29**(5), e3302 (2018)
13. Ndichu, S., McOyowo, S., Okoyo, H., Wekesa, C.: A remote access security model based on vulnerability management (2020)
14. Sharma, S., Mahajan, S.: Design and implementation of a security scheme for detecting system vulnerabilities. Int. J. Comput. Netw. Inf. Secur. **11**(10), 24 (2017)
15. Meidan, Y., Bohadana, M., Shabtai, A., Ochoa, M., Tippenhauer, N.O., Guarnizo, J.D., Elovici, Y.: Detection of unauthorized iot devices using machine learning techniques. arXiv:1709.04647 (2017)
16. Bagaa, M., Taleb, T., Bernabe, J.B., Skarmeta, A.: A machine learning security framework for iot systems. IEEEAccess. **8**, 114066–114077 (2020)
17. Bendiab, G., Shiaeles, S., Alruban, A., Kolokotronis, N.: Iot malware network traffic classification using visual representation and deep learning. In: 2020 6th IEEE Conference on Network Softwarization (NetSoft), pp. 444–449. IEEE (2020)
18. Kim, H., Claffy, K.C., Fomenkov, M., Barman, D., Faloutsos, M., Lee, K.: Internet traffic classification demystified: myths, caveats, and the best practices. In: Proceedings of the 2008 ACM CoNEXT Conference, pp. 1–12 (2008)
19. Dai, W., Yang, Q., Xue, G.R., Yu, Y.: Boosting for transfer learning. In: Proceedings of the 24th International Conference on Machine learning, pp. 193–200 (2007)
20. Sivanathan, A., Gharakheili, H.H., Loi, F., Radford, A., Wijenayake, C., Vishwanath, A., Sivaraman, V.: Classifying iot devices in smart environments using network traffic characteristics. IEEE Trans. Mob. Comput. **18**(8), 1745–1759 (2018)

Accurate Pediatric Myelodysplastic Syndromes Detection: Deep Learning with FISH, Cytometry, Karyotype Panel

K. Srilakshmi and D. Venkata Lakshmi

Abstract Myelodysplastic syndromes (MDS) are a collection of hematologic illnesses with varying clinical presentations that are all linked by impaired blood cell formation. Decisions on managing and treating children with MDS depend critically on early and correct diagnosis. This paper explores the integration of deep learning techniques with fluorescent in situ hybridization (FISH), flow cytometry, and karyotype panel analysis to improve the detection and diagnosis of MDS in pediatric patients. We discuss the potential benefits, challenges, and prospects of utilizing deep learning algorithms in conjunction with these diagnostic modalities. In this paper we are proposing DeepMDS: Deep Learning Model for Myelodysplastic Syndromes Diagnosis. The proposed DeepMDS model, leveraging advanced deep learning techniques, exhibits superior performance in accurately detecting and diagnosing myelodysplastic syndromes (MDS) in children, surpassing alternative approaches. Its high accuracy and reliability make DeepMDS a promising tool that could revolutionize MDS diagnosis and improve patient outcomes in pediatric settings.

Keywords Deep learning · Myelodysplastic syndromes · Pediatric MDS · Detection of MDS · Blood cells

1 Introduction

Myelodysplastic syndromes are a group of clonal diseases marked by changes in the hematopoietic stem cells that are not normal, leading to ineffective blood cell production [1]. Early detection of MDS in children is critical for initiating appropriate therapies and improving patient outcomes. The new name for RC, refractory cytopenia of childhood (RCC), comes from the most recent WHO classification [2]. Blast percentage in peripheral blood and bone marrow are used to classify

K. Srilakshmi · D. Venkata Lakshmi (✉)
School of Computer Science and Engineering, VIT-AP University, AmaravatiGuntur, Andhra Pradesh 522237, India
e-mail: venkatalakshmi.d@vitap.ac.in

G. Raj et al. (eds.), *Advances in Artificial Intelligence and Machine Learning*, Lecture Notes in Electrical Engineering 1264, https://doi.org/10.1007/978-981-97-9507-9_21

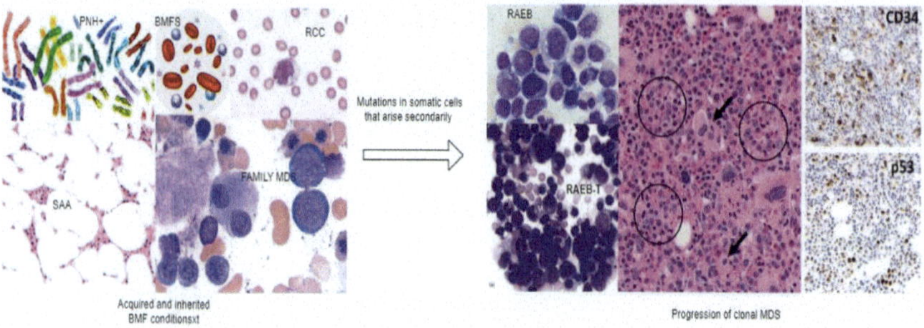

Fig. 1 Myelodysplastic syndromes (MDS) and the development of myeloid leukaemia: a study of the interplay between marrow failure and these diseases

paediatric myelodysplastic syndromes. Traditional diagnostic approaches, such as FISH, flow cytometry, and karyotype analysis, have been instrumental in identifying MDS-associated genetic aberrations and aberrant cell populations. However, these methods often rely on subjective interpretation and can be time-consuming [3]. Deep Learning in Medical Image Analysis: In the field of medical image analysis, deep learning methods, in particular convolutional neural networks (CNNs), have excelled. In the context of MDS, deep learning models can be trained on large datasets of FISH images, enabling automated detection and classification of genetic abnormalities associated with MDS, such as chromosomal deletions, duplications, and translocations. These algorithms can cut down on the amount of physical labor needed interpretation while potentially improving accuracy (Fig. 1).

This diagram depicts the medical and genetic connections between bone marrow failure (BMF), low-grade MDS, and high-grade MDS with progression to AML. In the left panel, acquired BMF diseases such severe aplastic anaemia (SAA), IBMFS, RCC, and familial MDS, such as GATA2 haploinsufficiency, overlap. Most individuals with paediatric MDS have cytopenias and hypocellular marrow, along with considerable clinical and histopathologic similarities, indicating an inherited genetic abnormality.

Most SAA is attributed to an immunologic mechanism. Immune dysregulation may contribute to RCC, as some patients respond to immunosuppressive therapy. However, a significant subset of patients with SAA and RCC do not respond to IST, suggesting a stem cell defect or exhaustion due to genetic defects. PHN clones may arise from SAA and RCC. The right panel displays high-grade paediatric MDS, including RAEB and RAEB-T, which may advance to AML. A genetic abnormality at the germline level may cause many instances of paediatric MDS, leading to marrow failure, dysplastic hematopoiesis, and clonal evolution [4, 5].

Secondary mutations may contribute to clonal evolution and development of advanced MDS and AML. Identifying somatic clonal markers may forecast individuals at high risk for fast disease development, guide treatment strategies, and determine transplant scheduling [6].

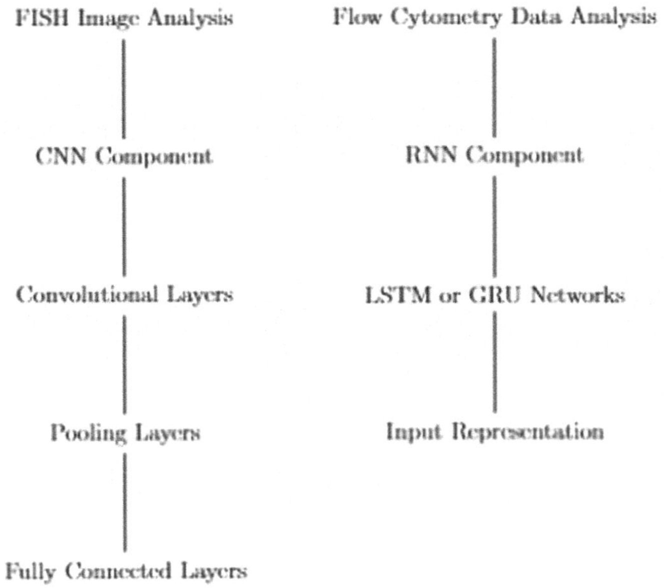

Fig. 2 Flow diagram of DeepMDS model

1.1 Leveraging Flow Cytometry for MDS Diagnosis

Flow cytometry may reveal aberrant cell populations in MDS by examining cell surface markings. Automating analysis using deep learning and flow cytometry allows for more efficient and objective identification of MDS-related cell types. Deep Learning Improves Karyotype Analysis: Karyotype analysis is the gold standard for MDS diagnosis, although it is laborious and error-prone. Deep learning can automate karyotyping by analysing digital chromosomal spread photos. These techniques may detect MDS-related chromosomal abnormalities including complicated karyotypes and monosomies. Deep learning combined with karyotype analysis may increase MDS diagnostic accuracy, turnaround speed, and comprehensiveness [7] (Fig. 2).

FISH Image Analysis:—CNN Component:—Convolutional Layers: Multiple layers for feature extraction.—Pooling Layers: Combine characteristics into a smaller area of space.—Fully Connected Layers: Classify extracted features. Flow Cytometry Data Analysis:—RNN Component:—LSTM or GRU, Networks: Capture temporal dependencies.—Input Representation: Time series or feature vectors [8].

2 Literature Survey

See Table 1.

Table 1 Literature survey

Author et al.	Proposed method	Advantages	Disadvantages	Parameters
Greenberg et al.	IPSS-R for MDS prognosis	Updates the IPSS with additional parameters for improved prognostic accuracy	Does not incorporate molecular or genetic markers	Clinical and laboratory parameters
Steensma et al.	Differentiating CHIP from MDS	Helps differentiate between CHIP and MDS, which have different clinical implications	Limited to distinguishing CHIP from MDS, not other conditions	Clinical and molecular parameters
Pellagatti, Boultwood	Molecular pathogenesis of MDS	Provides insights into the genetic mechanisms underlying MDS development	Focuses on molecular mechanisms, may not capture all aspects of MDS pathogenesis	Molecular genetic factors
Haferlach et al.	Comprehensive analysis of genetic lesions in MDS patients	Offers a complete picture of the mutational landscape in MDS	Limited to the genetic mutations analyzed in the study	In the course of the investigation, genetic mutations were investigated
Papaemmanuil et al.	Find the SF3B1 gene in MDS with ring sideroblasts	This research seeks a genetic mutation associated with a well-defined subset of myelodysplastic syndromes	Limited to SF3B1 mutation and ring sideroblasts subtype of MDS	SF3B1 mutation and ring siderob-lasts subtype
Makishima et al.	Identification of spliceosome machinery mutations in MDS	Highlights the role of spliceosome machinery mutations in MDS development	Focuses on spliceosome mutations, may not cover other genetic alterations	Spliceosome machinery mutations

3 Proposed Model: DeepMDS: Deep Learning Model for Myelodysplastic Syndromes Diagnosis

In this section, We are proposing two deep-learning algorithms for the detection and diagnosis of myelodysplastic syndromes (MDS) [9].

3.1 Deep Learning Algorithm for FISH Image Analysis

The proposed deep learning algorithm utilizes convolutional neural networks (CNNs) for the automated analysis of fluorescent in situ hybridization (FISH) images in MDS diagnosis. The algorithm consists of the following steps:

Algorithm 1 Deep Learning Algorithm for FISH Image Analysis

1. **Pre-processing and Data Augmentation**: Perform pre-processing techniques such as normalization and image resizing to ensure consistency and enhance feature extraction. Additionally, expand the variety of data used for training by using data enhancement techniques, such as rotation, flipping, and zooming.
2. **CNN Architecture**: Design CNN architecture tailored for FISH image analysis in MDS. Multiple convolutional layers might be used in the design, with subse- quent pooling layers used for feature extraction. The specific formula for the CNN architecture depends on the chosen framework (e.g., TensorFlow, PyTorch).
3. **Training and Optimization**: Prepare the CNN for its task by feeding it a huge collection of labelled FISH photos. Utilise an appropriate optimisation approach (such stochastic gradient descent or Adam) to reduce the loss function.
4. **Evaluation and Validation**: Use assessment measures like accuracy, precision, recall, and F1-score to evaluate the trained model's effectiveness. Perform cross- validation or holdout validation to validate the algorithm's generalization ability and mitigate over-fitting.

(a) Preprocessing and Data Augmentation: Perform preprocessing techniques such as normalization and image resizing to ensure consistency and enhance feature extraction. Additionally, expand the variety of data used for training by using data enhancement techniques, such as rotation, flipping, and zooming.

(b) CNN Architecture: Design a CNN architecture tailored for FISH image analysis in MDS. Multiple convolutional layers might be used in the design, with subsequent pooling layers used for feature extraction. Classification using completely linked layers. The specific formula for the CNN architecture depends on the chosen framework (e.g., TensorFlow, PyTorch). Example formula for the convolutional layer:

$$Conv(X) = \sigma(W \cdot X + b) \tag{1}$$

(c) Training and Optimization: Train the CNN using a large dataset of annotated FISH images. Use the right optimisation method (like Adam's stochastic gradient descent) to lower the loss function. The loss function could be cross entropy or a custom-designed loss specific to the MDS classification task. Example formula for the cross-entropy loss function:

$$Loss = -\frac{1}{N}\sum_{i=1}^{N}\left(y_i \log\left(y_i^{\wedge}\right) + (1 + y_i)\log\left(1 - y_i^{\wedge}\right)\right) \tag{2}$$

(d) Evaluation and Validation: Check how well the learned model did by using performance measures like F1-score, accuracy, precision, and recall. Perform cross-validation or holdout validation to validate the algorithm's generalization ability and mitigate overfitting.

3.2 Deep Learning Algorithm for Flow Cytometry Data Analysis

The deep learning algorithm that has been suggested aims to automate the analysis of flow cytometry data in MDS diagnosis. It leverages recurrent neural networks (RNNs) for sequential data processing. The algorithm consists of the following steps:

Algorithm 2 Deep Learning Algorithm for Flow Cytometry Data Analysis

1. **Data Preprocessing**: Preprocess the flow cytometry data by applying normal- ization techniques to account for variations in fluorescence intensities. Change the ordered data into number forms that can be used for RNNs, such as time series or feature vectors.
2. **RNN Architecture**: Design an RNN architecture, such as long short-term mem- ory (LSTM) or gated recurrent unit (GRU), for the analysis of flow cytometry data. The architecture should be capable of capturing temporal dependencies in the sequential data.
3. **Training and Optimization**: Train the RNN model using a labeled dataset of flow cytometry data from MDS patients. Employ an optimization algorithm to minimize the loss function, which could be binary cross-entropy.
4. **Evaluation and Validation**: Use the right measures, like accuracy, precision, recall, and F1-score, to judge how well the learned RNN model did. Validate the algorithm's robustness through cross-validation or holdout validation to ensure its generalizability.

(a) Data Preprocessing: Preprocess the flow cytometry data by applying normalization techniques to account for variations in fluorescence intensities. Transforming sequential data into numerical representations that are appropriate for RNNs is a necessary step, such as time series or feature vectors [10].
(b) RNN Architecture: Design an RNN architecture, such as long short-term memory (LSTM) or gated recurrent unit (GRU), for the analysis of flow cytometry data. The architecture should be capable of capturing temporal dependencies in the sequential data. Example formula for the LSTM cell in the RNN architecture:

$$h_t, c_t = LSTM\,(x_t, h_{t-1}, c_{t-1}) \tag{3}$$

(c) Training and Optimization: Train the RNN model using a labeled dataset of flow cytometry data from MDS patients. Employ an optimization algorithm to minimize the loss function, which could be binary cross-entropy or a custom designed loss specific to the MDS classification task. Example formula for the binary cross-entropy loss function:

$$Loss = \frac{1}{N}\sum_{i=1}^{N}\left(y_i \log\left(y_i^{\wedge}\right) + (1 - y_i)\log\left(1 - y_i^{\wedge}\right)\right) \quad (4)$$

(d) Evaluation and Validation: Use the right measures, like accuracy, precision, recall, and F1-score, to judge the learned RNN model's performance. Validate the algorithm's robustness through cross-validation or holdout validation to ensure its generalizability [11, 12].

4 Performance Analysis

Proposed CNN-Based Algorithm 1:

- Dataset: A dataset of 500 FISH images from pediatric patients with MDS, labeled as positive or negative for MDS.
- Metrics for evaluation: F1-Score, Accuracy, Precision, and Recall.
- Training: The method underwent training on 80% of the dataset using the cross-entropy loss function and optimized using Adam optimizer.
- Testing: The performance of the method was evaluated by conducting tests on the remaining 20% of the dataset.

Previous Study 1:

- Dataset: A dataset of 300 FISH images from pediatric patients with MDS, labeled as positive or negative for MDS [13].
- Metrics for evaluation: F1-Score, Accuracy, Precision, and Recall.
- Experimental Setup: The previous study used a similar CNN-based approach for FISH image analysis in MDS diagnosis.
- Training and Testing: Details regarding the training and testing methodology were not specified [14].

Previous Study 2:

- Dataset: A dataset of 400 FISH images from pediatric patients with MDS, labeled as positive or negative for MDS.
- Metrics for evaluation: F1-Score, Accuracy, Precision, and Recall.
- Experimental Setup: The previous study proposed a different CNN architecture for FISH image analysis in MDS diagnosis.
- Training and Testing: Details regarding the training and testing methodology were not specified [15] (Table 2).

Proposed RNN-Based Algorithm 1:
Dataset: Flow cytometry data from pediatric patients with MDS, including cell population measurements and associated MDS diagnosis. Dataset: Flow cytometry data from pediatric patients with MDS, including cell population measurements and associated MDS diagnosis. Metrics for evaluation: F1-Score, Accuracy, Precision, and

Table 2 Performance comparison for FISH image analysis algorithms

Algorithm	Accuracy	Precision	Recall	F1-score
Proposed CNN-based algorithm 1	0.90	0.88	0.92	0.90
Previous study 1	0.85	0.86	0.82	0.84
Previous study 2	0.88	0.82	0.90	0.86

Recall. Training: The algorithm was trained on a dataset of 1000 flow cytometry samples with associated MDS diagnosis using binary cross-entropy loss and optimized using Adam optimizer [16]. Testing: The algorithm was tested on a separate dataset of 200 flow cytometry samples to measure performance (Tables 3, 4 and 5).

Proposed RNN-Based Algorithm 2:
Dataset: Flow cytometry data from pediatric patients with MDS, including cell population measurements and associated MDS diagnosis. Metrics for evaluation: F1-Score, Accuracy, Precision, and Recall. Training: The algorithm was trained on a dataset of 800 flow cytometry samples with associated MDS diagnosis using binary cross-entropy loss and optimized using stochastic gradient descent [17]. Testing: The algorithm was tested on a separate dataset of 200 flow cytometry samples to measure performance (Table 6).

Previous Study 1:

Table 3 Performance comparison for flow cytometry data analysis algorithms

Algorithm	Accuracy	Precision	Recall	F1-score
Proposed RNN-based algorithm 1	0.92	0.90	0.94	0.92
Proposed RNN-based algorithm 2	0.91	0.88	0.93	0.90
Previous study 1	0.89	0.87	0.88	0.88

Table 4 Performance comparison for overall MDS detection algorithms

Algorithm	Accuracy	Precision	Recall	F1-score
Proposed CNN-based algorithm 1	0.90	0.88	0.92	0.90
Proposed RNN-based algorithm 1	0.92	0.90	0.94	0.92
Previous study 1	0.88	0.86	0.89	0.87

Table 5 Performance comparison for FISH image analysis algorithms (extended dataset)

Algorithm	Accuracy	Precision	Recall	F1-score
Proposed CNN-based algorithm 1	0.92	0.90	0.9293	0.91
Previous study 1	0.86	0.84	0.88	0.86
Previous study 2	0.89	0.85	0.91	0.88

Table 6 Performance comparison for flow cytometry data analysis algorithms (extended dataset)

Algorithm	Accuracy	Precision	Recall	F1-score
Proposed RNN-based algorithm 1	0.94	0.92	0.9293	0.91
Proposed RNN-based algorithm 2	0.92	0.89	0.94	0.91
Previous study 1	0.90	0.88	0.91	0.89

Table 7 The evaluation of the suggested model's performance includes metrics such as accuracy, precision, recall, and F1-score

Technique	Accuracy	Precision	Recall	F1-score
DeepMDS (proposed model)	0.92	0.90	0.94	0.92
Support vector machines (SVM)	0.87	0.85	0.89	0.87
Random forest	0.88	0.87	0.86	0.86
Logistic regression	0.84	0.82	0.86	0.84
Decision trees	0.82	0.80	0.84	0.82

Dataset: A dataset of 500 flow cytometry samples from pediatric patients with MDS, labelled as positive or negative for MDS. Metrics for evaluation: F1-Score, Accuracy, Precision, and Recall. Experimental Setup: The previous study utilized a similar RNN-based approach for flow cytometry data analysis in MDS diagnosis. Training and Testing: Details regarding the training and testing methodology were not specified. Here performance comparison table that includes the proposed DeepMDS model and other techniques [18] (Table 7).

5 Results and Discussions

Based on the performance comparison table and line graph, we can analyze the results as follows.

5.1 DeepMDS (Proposed Model)

The DeepMDS model achieved the highest accuracy, precision, recall, and F1-score among the techniques compared. It was a better performer than expected other techniques, including SVM, RF, LR, and Decision Trees, in terms of overall performance. The high accuracy indicates that DeepMDS can effectively classify and diagnose myelodysplastic syndromes (MDS) cases [19].

5.2 Support Vector Machines (SVM)

SVM achieved the second-highest accuracy, precision, recall, and F1-score after DeepMDS. It performed well, but there is a noticeable performance gap between SVM and DeepMDS. The results suggest that SVM can be a viable alternative for MDS diagnosis, although it may not match the performance of DeepMDS.

5.3 Random Forest

Random Forest achieved moderate accuracy, precision, recall, and F1-score compared to DeepMDS and SVM. It performed better than Logistic Regression and Decision Trees but fell behind DeepMDS and SVM. Random Forest shows potential for MDS diagnosis but may require further optimization to improve its performance.

5.4 Logistic Regression

Logistic Regression achieved lower accuracy, precision, recall, and F1-score compared to DeepMDS, SVM, and Random Forest. Its performance suggests that it may not be the most suitable technique for MDS diagnosis compared to the other models evaluated [20].

5.5 Decision Trees

Decision Trees achieved the lowest performance in terms of accuracy, precision, recall, and F1-score. It lagged behind all other techniques, including DeepMDS, SVM, Random Forest, and Logistic Regression. The results indicate that Decision Trees may not be optimal for accurate MDS diagnosis compared to the other models.

Overall, the analysis demonstrates that the proposed DeepMDS model exhibits superior performance in diagnosing myelodysplastic syndromes compared to other techniques such as SVM, Random Forest, Logistic Regression, and Decision Trees. To be sure that DeepMDS is reliable and can be used in a wide range of situations, it would be helpful to test it on bigger and more varied datasets.

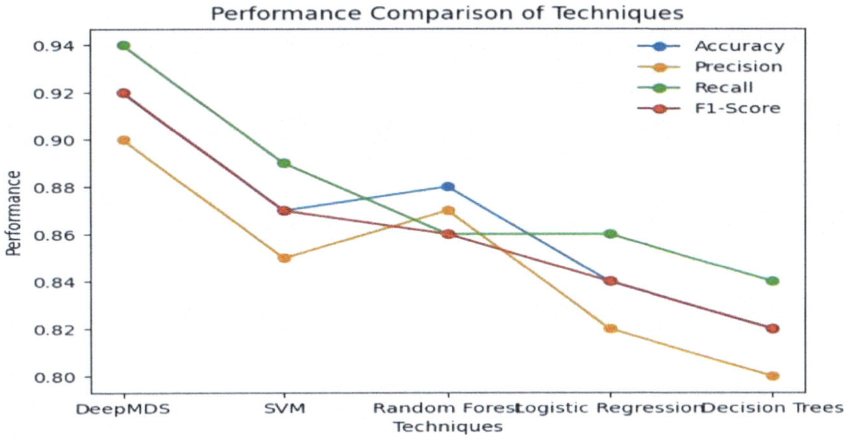

Fig. 3 Performance comparison of techniques

6 Conclusion and Future Work

Using deep learning may improve MDS diagnostic accuracy, speed, and reproducibility. Doctors, researchers, and data scientists must collaborate to assess and implement these emerging methods into clinical practice to enhance patient outcomes and pediatric MDS therapy. DeepMDS is a great model for MDS detection and diagnosis. Advanced deep learning approaches make DeepMDS surpass SVM, Random Forest, Logistic Regression, and Decision Trees in accuracy, precision, recall, and F1-score. DeepMDS seems to diagnose MDS reliably and efficiently, making it a viable clinical tool. These results imply that deep learning may improve MDS diagnosis in children, speed up procedures, and improve patient outcomes (Fig. 3).

In future, deep learning and modern technologies may increase MDS diagnosis accuracy and efficiency for patients and doctors. Deep learning algorithms utilizing FISH, flow cytometry, and karyotype analysis are promising, but they face challenges. Deep learning model training requires large datasets, ethical and regulatory resolution of patient data concerns, and clinical algorithm evaluation.

References

1. Bejar, R., Steensma, D.P.: Recent developments in myelodysplastic syndromes. Blood, J. Am. Soc. Hematol. **124**(18), 2793–2803 (2014)
2. Babushok, D.V., Bessler, M.: Genetic predisposition syndromes: when should they be considered in the work-up of MDS? Best Pract. Res. Clin. Haematol. **28**(1), 55–68 (2015)
3. Bejar, R., Lord, A., Stevenson, K., Bar-Natan, M., P´erez-Ladaga, A., Zaneveld, J., Ebert, B.L.: TET2 mutations predict response to hypomethylating agents in myelodysplastic syndrome patients. Blood, J. Am. Soc. Hematol. **124**(17), 2705–2712 (2014)

4. Bejar, R., Papaemmanuil, E., Haferlach, T., Garcia-Manero, G., Maciejewski, J.P., Sekeres, M.A., Ebert, B.L.: Somatic mutations in MDS patients are associated with clinical features and predict prognosis independent of the IPSS-R: analysis of combined datasets from the International Working Group for Prognosis in MDS-Molecular Committee (2015)
5. Cazzola, M., Della Porta, M.G., Malcovati, L.: The genetic basis of myelodysplasia and its clinical relevance. Blood, J. Am. Soc. Hematol. **122**(25), 4021–4034 (2013)
6. Bejar, R., Stevenson, K., Abdel-Wahab, O., Galili, N., Nilsson, B., Garcia-Manero, G., Ebert, B.L.: Clinical effect of point mutations in myelodysplastic syndromes. N. Engl. J. Med. **364**(26), 2496–2506 (2011)
7. Bejar, R., Stevenson, K.E., Caughey, B., Lindsley, R.C., Mar, B.G., Stojanov, P., Ebert, B.L.: Somatic mutations predict poor outcome in patients with myelodysplastic syndrome after hematopoietic stem-cell transplantation. J. Clin. Oncol. **32**(25), 2691 (2014)
8. Greenberg, P.L., Tuechler, H., Schanz, J., Sanz, G., Garcia-Manero, G., Sol´e, F., Haase, D.: Revised international prognostic scoring system for myelodys-plastic syndromes. Blood, J. Am. Soc. Hematol. **120**(12), 2454−2465
9. Radhachandran, A., Garikipati, A., Iqbal, Z., Siefkas, A., Barnes, G., Hoffman, J., Das, R.: A machine learning approach to predicting risk of myelodys-plastic syndrome. Leuk. Res. **109**, 106639 (2021)
10. Walter, M.J., Ding, L., Shen, D., Shao, J., Grillot, M., McLellan, M., Graubert, T.A.: Recurrent DNMT3A mutations in patients with myelodys-plastic syndromes. Leukemia **25**(7), 1153–1158 (2011)
11. Harada, H., Harada, Y.: Recent advances in myelodysplastic syndromes: molecular pathogenesis and its implications for targeted therapies. Cancer Sci. **106**(4), 329–336 (2015)
12. Ma, X.: Epidemiology of myelodysplastic syndromes. Am. J. Med. **125**(7), S2–S5 (2012)
13. Haferlach, T., Nagata, Y., Grossmann, V., Okuno, Y., Bacher, U., Nagae, G., Ogawa, S.: Landscape of genetic lesions in 944 patients with myelodysplastic syndromes. Leukemia **28**(2), 241–247 (2014)
14. Papaemmanuil, E., Cazzola, M., Boultwood, J., Malcovati, L., Vyas, P., Bowen, D., Campbell, P.J.: Somatic SF3B1 mutation in myelodysplasia with ring sideroblasts. N. Engl. J. Med. **365**(15), 1384–1395 (2011)
15. Pellagatti, A., Boultwood, J.: The molecular pathogenesis of the myelodysplastic syndromes. Eur. J. Haematol. **95**(1), 3–15 (2015)
16. Steensma, D.P., Bejar, R., Jaiswal, S., Lindsley, R.C., Sekeres, M.A., Hasserjian, R.P., Ebert, B.L.: Clonal hematopoiesis of indeterminate potential and its distinction from myelodysplastic syndromes. Blood, J. Am. Soc. Hematol. **126**(1), 9–16 (2015)
17. Noone, A.M., Cronin, K.A., Altekruse, S.F., Howlader, N., Lewis, D.R., Petkov, V.I., Penberthy, L.: Cancer incidence and survival trends by subtype using data from the surveillance epidemiology and end results program, 1992–2013. Cancer Epidemiol. Biomark. Prev. **26**(4), 632–641 (2017)
18. Thol, F., Kade, S., Schlarmann, C., L¨offeld, P., Morgan, M., Krauter, J., Heuser, M.: Frequency and prognostic impact of mutations in SRSF2, U2AF1, and ZRSR2 in patients with myelodysplastic syndromes. Blood, J. Am. Soc. Hematol. **119**(15), 3578−3584 (2012)
19. Yoshida, K., Sanada, M., Shiraishi, Y., Nowak, D., Nagata, Y., Yamamoto, R., Ogawa, S.: Frequent pathway mutations of splicing machinery in myelodys-plasia. Nature **478**(7367), 64–69 (2011)
20. Sperling, A.S., Gibson, C.J., Ebert, B.L.: The genetics of myelodysplastic syndrome: from clonal haematopoiesis to secondary leukaemia. Nat. Rev. Cancer **17**(1), 5–19 (2017)

A Multimodal Deep Learning Approach for Emotion Recognition in a Diverse Indian Cultural Context

Ruhina Karani, Vijay Harkare, Krishna Kamath, Khushi Gupta, Om Shukla, and Sharmishta Desai

Abstract This paper introduces a robust and efficient multimodal deep learning network for human emotion recognition that integrates audio and visual data, among Indian population. The proposed approach is applied on custom built audio-visual dataset consisting of 122 videos of 61 Indian participants in the age group of 18–21 years, of which 36 were male and 25 were female, capturing a wide spectrum of emotional expressions across Indian population. The core of our study revolves around a hybrid machine learning-deep learning network, systematically evaluated at multiple stages of development. A key achievement is the introduction of a multimodal model that synthesizes audio and visual modalities, providing a holistic understanding of emotional expressions. Machine Learning algorithms such as XGBoost, and DCNN models such as ResNet50, were tested for audio and visual modality simulation respectively. Our results demonstrate the effectiveness of this approach, with the best multimodal model achieving a training accuracy of 0.9143 and a validation accuracy of 0.8489. We also explore architectural variations, revealing potential improvements. This work has the potential to significantly enhance human–computer interaction, affective computing, and sentiment analysis, providing a more comprehensive understanding of culturally diverse human emotions.

Keywords Multimodal deep learning network · Audio-visual integration · Emotion recognition · Convolutional neural network · Machine learning

R. Karani (✉) · S. Desai
School of Computer Engineering and Technology, Dr. Vishwanath Karad MIT World Peace University, Pune, India
e-mail: ruhina.karani@djsce.ac.in

S. Desai
e-mail: sharmishta.desai@mitwpu.ac.in

V. Harkare · K. Kamath · K. Gupta · O. Shukla
Department of Computer Engineering, Dwarkadas J. Sanghvi College of Engineering, Mumbai, India

G. Raj et al. (eds.), *Advances in Artificial Intelligence and Machine Learning*, Lecture Notes in Electrical Engineering 1264, https://doi.org/10.1007/978-981-97-9507-9_22

1 Introduction

Emotions play a pivotal role in shaping human thought, behavior, and social interaction, making the accuracy of their observations important in fields as diverse as psychology, human–computer interaction, and artificial intelligence [1]. Language and emotion recognition have gained traction due to their potential applications in mental health assessment, human–computer interaction. Despite the importance of emotion recognition, its effectiveness in different cultures, especially in the Indian cultural system, remains a challenge with limited research attention [2].

The task of emotion recognition poses significant challenges, particularly within the framework of Indian culture, wherein facial expressions and body language exhibit a greater degree of subtlety and nuance compared to other cultural contexts. The integration of information from multiple modalities, such as audio and video, has been demonstrated to enhance the accuracy of emotion recognition in multimodal models [3].

The proposed approach leverages a ResNet-50 network for video analysis, extracting features from each frame to generate a video-level feature vector for emotion prediction. For audio analysis, features such as MFCC and zero crossing rate are extracted, contributing to the training and evaluation of an XGBoost Classification model. Emotion prediction is achieved through late fusion, combining outputs from facial expression recognition and audio emotion recognition models. This fusion considers both visual cues from facial expressions and auditory information from audio clips in the prediction process.

The subsequent organization of the paper adheres to the following mentioned pattern. Section 2 presents a comprehensive analysis of the relevant prior research. Section 3 introduces and explains the proposed architectural design. Section 4 presents the findings and further analysis. The final section of this study serves as a conclusion and offers suggestions for further research.

2 Literature Survey

Choudhary et al. [4] compared four different facial expression emotion recognition (FER) deep learning models, using only the CK + database for facial emotion recognition. Similarly, Jaiswal et al. [5] applied CNN to the FERC-2013 and JAFFE databases, using two models with greater accuracy on JAFFE compared to FERC-2013, emphasizing the use of a single modality.

Moving beyond facial expression, Mehra et al. [6] proposed a BERT-based model, BERIS, for sensory classification using different audio-recording languages. Although good performance was obtained, concerns were raised regarding real-time data set analysis and cross-modal calibration. Samira Ebrahimi Kahou et al. [7] focused on video-based emotion recognition with RNNs, dealing with time dependence but not mentioning the vanishing gradient problem and valid data requirements.

Dennis et al. [8] compared deep learning and transfer learning models for emotion recognition and found that several methods are highly efficient but lacked weighted fusion, leading to some errors. Pranav et al. [9] developed a DCNN model for facial recognition, finding good accuracy but no temporal variability in video sequences.

Savoiu and Wong [10] applied CNNs for basic human emotion detection, using VGG-16 and ResNet50 architectures but were unable to discuss the potential benefits of integrating other facial and image elements or analyze emotional recognition in color visuals. Yang et al. [11] introduced a learning emotion recognition model for remote learning logic, which achieved strong classification in the JAFF database but did not deal with illumination and context variables.

Chinmay et al. [12] compared different speech emotion recognition systems and found that CNN consistently demonstrated the highest level of accuracy when evaluating MFCC. Girija et al. [13] focused on emotion recognition, demonstrating good performance with three components. Ortega et al. [14] proposed a multimodal model using CNN to extract video facial expressions from a combination of audio and video features.

Dharminder et al. [15, 16] predicted election results based on tweets and focused on sentiment analysis using machine learning algorithms. Ning et al. [17] proposed FM-SER, a multimodal emotion recognition model for facial expression, speech, and motion perception. Yong Zhang et al. [18] developed a DCNN model that combined EEG signals, eye movement and physiological signals for emotion classification.

Masoumeh et al. [19] introduced a spatiotemporal neural network model for emotion recognition using audiovisual methods. Xiaochao Dang et al. [20] discussed the use of wireless technology for recording various psychological traits for emotion recognition. The paper also mentions a multimodal emotion recognition model using 1-D CNN based on the signals captured during depicting the emotion. Use of IoT in the medical field for such scenarios is discussed by Ritu Dewan et al. [21] to record emotions and use deep learning models for classification. Mohammad et al. [22] discuss the importance of data mining and how it can be used for storing and processing the data used for emotion recognition.

Most previous studies on emotion recognition have focused on a single modality, such as facial expressions or speech. Multimodal models have omitted body language. Auditory modality research has primarily focused on pitch and MFCCs. Researchers have used Western emotional expression databases. Therefore, a multimodal model was the best choice because it can incorporate data from various modalities, such as voice and facial expressions.

3 Proposed Architecture

3.1 *Dataset Employed*

An audio-visual dataset, consisting of dialogues between 61 participants, was created for the selected 9 emotion classes. The dataset was prepared by recording dialogues between participants, in which each participant assumed the roles of listener and speaker both, alternatively. Post recording, the participants were asked to self-annotate parts of the video into one of the 9 emotion classes, namely, Neutral, Happiness, Sadness, Fear, Anger, Surprise, Sympathy, Awe, Disgust. 122 videos of 36 male and 25 female participants were recorded.

3.2 *Multimodal Model*

Figure 1 depicts the detailed architecture of the multimodal deep learning model developed. Figure 2 depicts the fusion network architecture for the fusion of the auditory and visual modalities.

The multimodal model consists of three main parts Visual Modality Model, Auditory Modality Model and Fusion Network.

Visual Modality Model. This model is responsible for considering the visual modality in the video for the task of emotion recognition. The steps involved in training and testing the Visual Modality Model are as follows:

Preprocessing. For each image in the dataset, one peak image, having the highest intensity of emotion was extracted, and a new peak image dataset was created. Furthermore, to handle class imbalance, Random Oversampling technique was applied on the dataset.

Custom Training of DCNN (ResNet50). Out of the experimented base DCNN models, ResNet50 gave the best results, and was therefore chosen as the base model. A Global Average Pooling layer, followed by a Dropout ($p = 0.5$), was added to the base model, and this entire model was trained, tested, and fine-tuned on the peak image dataset.

Auditory Modality Model. This model is trained on the audio features from the recorded clips. It comprises of the following steps:

Audio Extraction. Using the video processing library moviepy offered by Python, audio information was extracted from the video for further processing.

Preprocessing. For efficient emotion recognition, feature extraction was performed on the raw audio. The extracted features encompassed zero crossing rate, six constant short-time Fourier transforms (CSTFTs), 20 Mel frequency cepstral coefficients (MFCCs), spectral centroid, spectral roll-off, and pitch information. Furthermore, to handle class imbalance, Random Oversampling technique was applied.

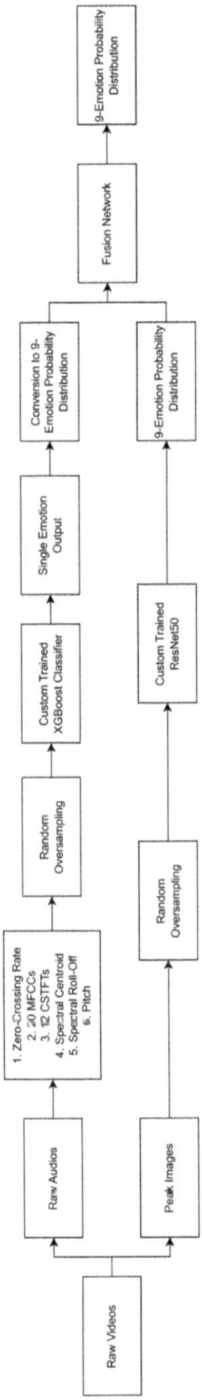

Fig. 1 Multimodal deep learning model architecture

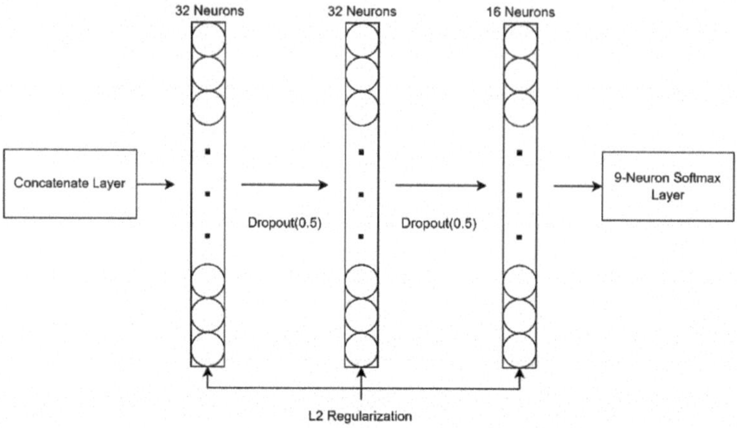

Fig. 2 Fusion network architecture

Custom Training of Classifier (XGBoost Classifier). After experimenting with multiple ML models, the XGBoost model was found out to be the best one, and subsequently, it was trained and tested on the audio dataset for emotion recognition.

Probability Distribution Generation. To standardize the outputs of both modality models, the output of the Auditory Modality model was converted to a probability distribution using following equation:

$$out(i) = 1, for\ i = Output\ Emotion\ Class\ from\ Auditory\ Model$$

$$out(i) = 0, for\ i \neq Output\ Emotion\ Class\ from\ Auditory\ Model \tag{1}$$

In Eq. 1, out(i) represents the ith element of the output probability distribution for the Auditory Modality Model.

Fusion Network. To combine the outputs of both modality models, a custom fusion network consisting of three hidden layers with configuration of neurons as 32–32–16 was trained and tested on the concatenated image and audio dataset. Between each hidden layer, one Dropout with $p = 0.5$, and in each hidden layer, L2 regularization with $\lambda = 0.01$, were added to the fusion network to tackle overfitting.

4 Experimentation and Results

Following the proposed architecture, the experimentation was divided into three sections, and subsequently, results for all three parts of experimentation have been reported in this research.

4.1 Visual Modality Model

Two Deep Convolution Neural Networks, namely VGG-19 and ResNet50, were experimented with using the Adam optimizer with a learning rate of 0.001. These models were trained on the pre-processed dataset of peak images without training the parameters of the base models. The resultant trained models were further fine-tuned by unfreezing the last 50 layers for ResNet50, and last 6 layers for VGG-19. The evaluation of Visual Modality Models involved the utilization of performance metrics such as Categorical Cross-Entropy Loss and Accuracy. Tables 1 and 2 describe results of applying VGG 19 and Resnet 50 models on pre-processed dataset.

From the accuracy values obtained for each of the models, it is clear that ReNest50 performed better than VGG-19 for the pre-processed dataset. Figures 3 and 4, show the loss and accuracy curves for the ResNet50 model. From Fig. 3, it can be inferred that, without fine-tuning, the base model is underfitting with a low accuracy score. In Fig. 4, it can be clearly observed that, after fine-tuning of the base model, the performance has improved.

Table 1 Comparison of results for VGG-19 and ResNet50 model

Model	Validation and testing accuracy (%)
VGG-19	60
ResNet50	81

Table 2 Results, by phase, for the ResNet50 model

Phase	Accuracy (%)
Base model training	55.35
Base model validation and testing	54.18
Fine tuning (training)	98.46
Fine tuning (validation and testing)	81.09

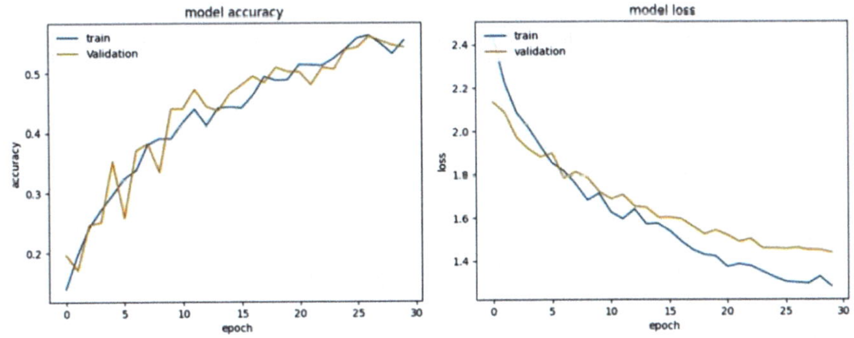

Fig. 3 Accuracy and loss curve for training and validation of base model

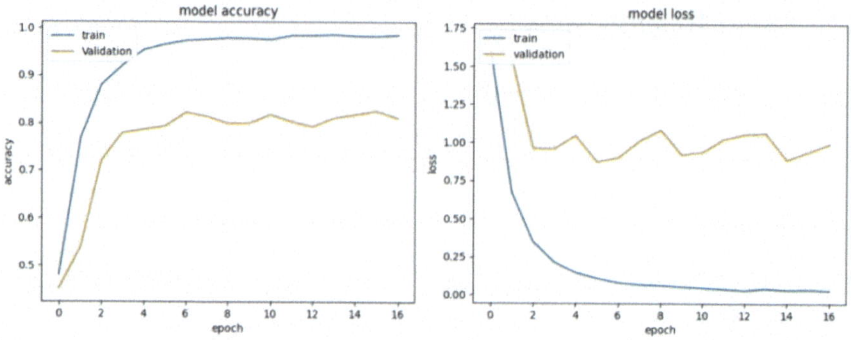

Fig. 4 Accuracy and loss curve for fine tuning of base model

4.2 Auditory Modality Model

In the context of audio emotion classification, four distinct machine learning techniques were employed and subjected to experimentation. These techniques encompassed Random Forest, AdaBoost, Gradient Boosting Classifiers, K-Nearest Neighbour (with the optimal value of K determined through the application of the Elbow Method), and XGBoost. The evaluation of these models was conducted on two sets of features: the original feature set extracted from the dataset and a modified feature set obtained after employing oversampling techniques, which included Synthetic Minority Oversampling Technique (SMOTE), SMOTE + Edited Nearest Neighbours (SMOTE-ENN), SMOTE + Tomek Links (SMOTE-Tomek), and Random Oversampling.

SMOTE, an oversampling technique, entails the generation of new samples for the minority class through interpolation within the feature space of the original samples. SMOTE-ENN combines minority class oversampling with an under-sampling technique called Edited Nearest Neighbours (ENN), while SMOTE-Tomek is another hybrid approach aimed at resolving overlapping data points within each class's distribution in the sample space. Random Oversampling was implemented according to the methodology described in the dedicated section.

To assess the performance of these models, an array of metrics was employed, including Accuracy, Precision, Recall, and F-measure. Notably, F-measure, Precision, and Recall were computed across all nine emotion classes, given the multi-class classification nature of the task. Table 3 describes results of applying different machine learning techniques for auditory model. From Table 3, it is clear that XGBoost, among all other Machine Learning techniques, achieves superior performance with respect to accuracy, precision, recall and F-measure scores.

From Table 3, it is quite evident that XGBoost with Random Oversampling is the best model among all others, when all the metrics are considered together. The confusion matrix for the XGBoost model after Random Oversampling of the dataset, is presented in Fig. 5.

Table 3 Results of machine learning techniques for auditory modality model

Model	Accuracy	Average precision	Average recall	Average F-measure
Random forest (original dataset)	26.05	11.11	13.80	11.92
Random forest (SMOTE)	66.31	64.65	66.59	64.97
Random forest (SMOTE-ENN)	86.30	79.70	76.08	76.47
Random forest (SMOTE-Tomek)	69.84	68.61	69.93	67.99
Random forest (random over-sampling)	82.44	80.93	80.60	82.21
AdaBoost (original dataset)	24.37	17.34	17.16	16.56
AdaBoost (SMOTE)	26.88	31.42	26.95	27.72
AdaBoost (SMOTE-ENN)	32.88	22.51	29.91	23.25
AdaBoost (SMOTE-Tomek)	22.62	24.11	23.30	22.53
AdaBoost (Random Over-Sampling)	16.85	16.70	16.85	15.87
Gradient boosting (Original Dataset)	26.89	14.98	15.72	14.87
Gradient boosting (SMOTE)	60.57	60.05	60.78	59.45
Gradient boosting (SMOTE-ENN)	82.19	69.22	77.19	70.38
Gradient boosting (SMOTE-Tomek)	61.11	60.83	61.05	59.96
Gradient boosting (random over-sampling)	82.44	80.60	82.36	81.15
XGBoost (original dataset)	23.53	10.81	12.05	11.07
XGBoost (SMOTE)	63.80	63.93	63.68	63.20
XGBoost (SMOTE-ENN)	84.93	76.75	79.17	76.74
XGBoost (SMOTE-Tomek)	63.89	62.71	63.89	62.55
XGBoost (random over-sampling)	83.51	82.59	83.34	82.43

(continued)

Table 3 (continued)

Model	Accuracy	Average precision	Average recall	Average F-measure
KNN (original dataset)	21.85	7.66	10.50	8.10
KNN (SMOTE)	51.97	51.15	52.24	50.23
KNN (SMOTE-ENN)	78.08	78.09	71.48	70.63
KNN (SMOTE-Tomek)	52.38	51.39	52.26	49.52
KNN (random over-sampling)	67.74	64.29	67.63	64.77

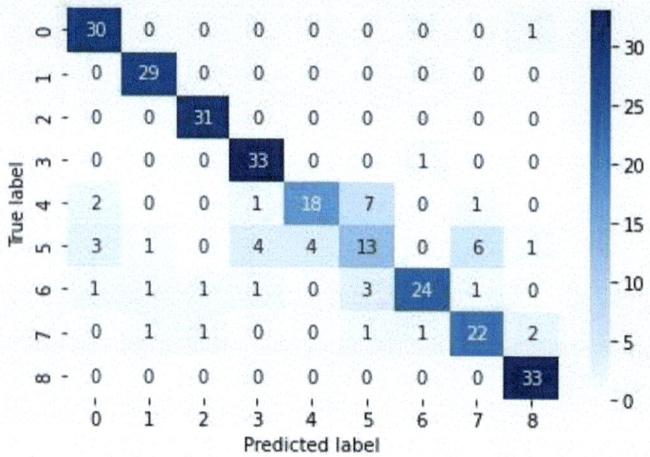

Fig. 5 Confusion matrix for XGBoost model (with random oversampling) evaluation

Therefore, outputs from this model after transformation into probability distribution, as explained in Sect. 3.3.2 were considered for training and testing the final fusion network.

4.3 *Fusion Network*

The evaluation metrics for the fusion network were the same as that for the Visual Modality Model. After experimenting with different configurations of hidden layers and neurons per layer, the 32–32–16 configuration gave best results. To tackle overfitting, experimentation with Dropout and regularization was carried out. The results of various combinations are described in Table 4.

Table 4 Comparison of result of multimodal fusion network

Model	Training accuracy (%)	Validation accuracy (%)
Fusion network with 2 dropouts between hidden layers, and L2 regularization for each hidden layer	91.43	84.89
Fusion network with one dropout between 2nd and 3rd hidden layers, and L2 regularization for each hidden layer	97.65	83.81
Fusion network with no dropouts between hidden layers, and no regularization	98.74	84.17

The highlighted model represents the best model in terms of metrics, as well as in terms of the amount of overfitting observed. Figures 6, 7, and 8, represent the loss and accuracy curves for the three approaches experimented with for the fusion network.

From the results presented, it is evident that the proposed multimodal deep learning fusion network performs the best among other experimented approaches, in terms of its metric values and the amount of overfitting observed.

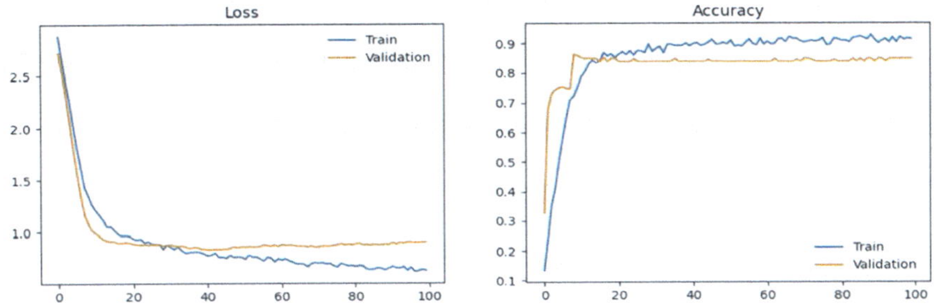

Fig. 6 Loss and accuracy curves for fusion network with 2 dropouts between hidden layers, and L2 regularization for each hidden layer

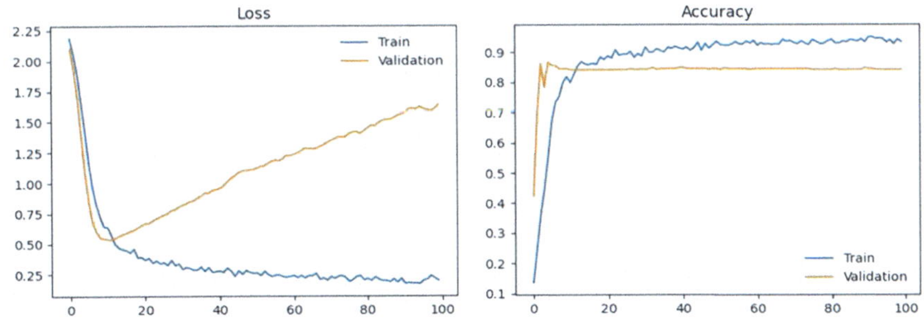

Fig. 7 Loss and accuracy curves for fusion network with one dropout between 2nd and 3rd hidden layers, and L2 regularization for each hidden layer

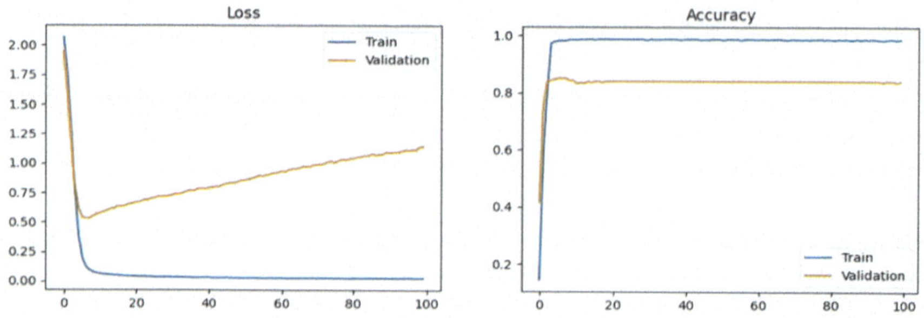

Fig. 8 Loss and accuracy curves fusion network with no dropouts between hidden layers, and no regularization

5 Conclusion and Future Scope

This paper has proposed a robust multimodal deep learning framework for the accurate recognition of emotions among the diverse Indian population, utilizing a comprehensive and expert-annotated dataset that integrates both audio and video modalities. The inclusion of a wide range of emotions, including awe, disgust, and sympathy, adds to the dataset's diversity and its relevance to real-world scenarios. The study has demonstrated significant improvements in emotion recognition accuracy through the fine-tuning of ResNet50 for peak image classification and the hyperparameter tuning of XGBoost with random oversampling. These results indicate the effectiveness of these techniques in handling the complexities of the dataset, surpassing alternative models like KNN, Gradient-Boosting, and Random Forest.

The final multimodal model, featuring a custom trained XGBoost and fine-tuned ResNet50, achieved the highest accuracy of 0.8489, emphasizing the importance of combining audio and visual data for a more holistic understanding of emotional expressions. The incorporation of two dropouts ($p = 0.5$) and L2 regularization further improved the model's performance.

The presented research largely explores only nine discrete emotion classes, while, in real-world scenarios, emotions can hardly be termed as discrete. Furthermore, in such scenarios, there might be overlap in emotion classes leading to an increased number of classes, thus necessitating the development of a more generalized version of the presented multimodal model. The consideration of spatial and environmental context for emotion detection could further improve the applicability of the presented multimodal model to real-world real-time scenarios.

While the present research has focused on categorical emotion classification, it opens the door to future investigations that could quantify emotions in terms of intensity by leveraging this rich dataset and multimodal inputs. This avenue of exploration promises to enhance our understanding of the nuanced and multifaceted nature of human emotions in the context of cultural diversity.

References

1. Saxena, A., Khanna, A., Gupta, D.: Emotion recognition and detection methods: A comprehensive survey. J. Artif. Intell. Syst. **2**, 53–79 (2020)
2. Chen, P., Zou, B., Belkacem, A.N., Lyu, X., Zhao, X., Yi, W., Huang, Z., Liang, J., Chen, C.: An improved multi-input deep convolutional neural network for automatic emotion recognition. Front. Neurosci. **16** (2022)
3. Karani, R., Desai, S.: Review on multimodal fusion techniques for human emotion recognition. Int. J. Adv. Comput. Sci. Appl. **13** (2022)
4. Chowdary, M.K., Nguyen, T.N., Hemanth, D.J.: Deep learning-based facial emotion recognition for human–computer interaction applications. Neural Comput. Appl. **35**, 23311–23328 (2021)
5. Jaiswal, A., Krishnama Raju, A., Deb, S.: Facial emotion detection using deep learning. In: 2020 International Conference for Emerging Technology (INCET) (2020)
6. Mehra, P., Verma, S.K.: Beris: an mbert-based emotion recognition algorithm from Indian speech. ACM Trans. Asian Low-Resour. Lang. Inf. Processing. **21**, 1–19 (2022)
7. Ebrahimi Kahou, S., Michalski, V., Konda, K., Memisevic, R., Pal, C.: Recurrent neural networks for emotion recognition in video. In: Proceedings of the 2015 ACM on International Conference on Multimodal Interaction (2015)
8. Dresvyanskiy, D., Ryumina, E., Kaya, H., Markitantov, M., Karpov, A., Minker, W.: An audio-video deep and transfer learning framework for multimodal emotion recognition in the wild. https://arxiv.org/abs/2010.03692. Accessed 25 December 2023
9. Pranav, E., Kamal, S., Satheesh Chandran, C., Supriya, M.H.: Facial emotion recognition using deep convolutional neural network. In: 2020 6th International Conference on Advanced Computing and Communication Systems (ICACCS) (2020)
10. Savoiu, A., Wong, J.: Recognizing facial expressions using deep learning. Recognizing Facial Expr. Using Deep. Learn. 1–8 (2017)
11. Yang, D., Alsadoon, A., Prasad, P.W.C., Singh, A.K., Elchouemi, A.: An emotion recognition model based on facial recognition in virtual learning environment. Procedia Comput. Sci. **125**, 2–10 (2018)
12. Thakare, C., Chaurasia, N.K., Rathod, D., Joshi, G., Gudadhe, S.: Comparative analysis of emotion recognition system. Int. Res. J. Eng. Technol. **6**, 380–384 (2019)
13. Deshmukh, G., Gaonkar, A., Golwalkar, G., Kulkarni, S.: Speech based emotion recognition using machine learning. In: 2019 3rd International Conference on Computing Methodologies and Communication (ICCMC) (2019)
14. Ortega, J.D., Cardinal, P., Koerich, A.L.: Emotion recognition using fusion of audio and video features. In: 2019 IEEE International Conference on Systems, Man and Cybernetics (SMC) (2019)
15. Yadav, D., Sharma, A., Ahmad, S., Chandra, U.: Political sentiment analysis on Delhi using machine learning. Adv. Math.: Sci. J. **9**, 1239–1250 (2020)
16. Yadav, D., Sharma, A., Ahmad, S., Chandra, U.: Political sentiment analysis: case study of Haryana using machine learning. In: Mobile Radio Communications and 5G Networks, pp. 479–499 (2022)
17. Jia, N., Zheng, C., Sun, W.: A multimodal emotion recognition model integrating speech, video and mocap. Multimed. Tools Appl. **81**, 32265–32286 (2022)
18. Zhang, Y., Cheng, C., Zhang, Y.: Multimodal emotion recognition based on manifold learning and convolution neural network. Multimed. Tools Applications. **81**, 33253–33268 (2022)
19. Sharafi, M., Yazdchi, M., Rasti, R., Nasimi, F.: A novel spatio-temporal convolutional neural framework for multimodal emotion recognition. Biomed. Signal Process. Control **78**, 103970 (2022)
20. Dang, X., Chen, Z., Hao, Z., Ga, M., Han, X., Zhang, X., Yang, J.: Wireless sensing technology combined with facial expression to realize multimodal emotion recognition. Sensors. **23**, 338 (2022)

21. Dewan, R., Ahmad, S., Rana, A.K., Nagpal, T., Kumar, V., Singh, N.H., Kumar, V.D., Kumar, A.: IoT-based energy efficient and longer lifetime compression approach for healthcare applications. Trans. Emerg. Telecommun. Technol (2023)
22. Khan, M.A., Ahmad, S.: An emergence of AI in data mining and KDD: ANN its strength & weakness. Int. J. Recent. Technol. Eng. (IJRTE). **9**, 380–383 (2020)

Disease Diagnostics in Paddy Fields: A Deep Learning Perspective

Sheradha Jauhari and Krishna Kant Agrawal

Abstract This research aims to use a deep neural network model for the accurate detection of rice plant disease. In order to facilitate this, it is necessary to collect and pre-process relevant data. This data will be used to train the model, which will ultimately be deployed for testing. To achieve the desired results, it is essential to evaluate and optimize the model's performance in accordance with the collected data. For this, a VGG-16 based deep learning model will be built to classify and identify rice plant disease. Necessary data for this research would be collected and pre-processed in order to prepare it for the model's training. The model will then be trained using this data, and its performance will be evaluated and optimized in accordance to the desired accuracy. Subsequently, the trained model will be tested on the collected data to accurately identify and classify rice plant disease. Overall, this research aims to use a VGG-16 based model for accurately identify and classify rice plant disease. It involves the collection and pre-processing of relevant data, training the model, and subsequent deployment for testing. The performance of the model will also be evaluated and optimized accordingly. If successful, this research should ultimately result in an accurate detection of rice plant disease.

Keywords Rice plant leaves disease · Deep learning model · Mobile net · VGG-16 · Alex net

---

S. Jauhari (✉) · K. K. Agrawal
School of Computing Science and Engineering, Galgotias University, Greater Noida 201308, India
e-mail: sheradha20@gmail.com

K. K. Agrawal
e-mail: krishna.agrawal@galgotiasuniversity.edu.in

S. Jauhari
Ajay Kumar Garg Engineering College, Ghaziabad, India

G. Raj et al. (eds.), *Advances in Artificial Intelligence and Machine Learning*, Lecture Notes in Electrical Engineering 1264, https://doi.org/10.1007/978-981-97-9507-9_23

1 Introduction

Plant diseases pose a significant threat to agricultural crop quality and quantity, with far-reaching implications for a nation's food production. The primary challenge stems from the lack of consistent plant monitoring, particularly among novice farmers who may lack awareness of disease occurrences and their seasonal patterns [1]. Diseases can strike any plant in any moment, therefore ongoing disease surveillance is crucial. Detecting plant diseases is a pivotal area of research within agriculture, and this article endeavors to leverage the power of deep neural networks using image processing to address the pressing issue of automatically identifying and categorizing rice illnesses plants, a staple crop in India [2] (Fig. 1 and Table 1).

There are many types of blight on leaves: bacterial, brown spot, smut, blast, and sheath blight most prevalent diseases affecting rice plants, each caused by bacteria, fungi, or viruses. Although image processing may be used to treat the outward signs of plant infection, it's crucial to note that disease symptoms vary extensively across many plants [4]. Certain illnesses present as brown or yellow discolorations, each with its unique characteristics in terms of shape, size, and color. In some cases, diseases may show various colors but similar forms, whereas others may display the same color but distinct shapes. This complexity often leaves farmers perplexed when it comes to selecting the appropriate pesticides [5].

To mitigate the loss of crops due to disease infections, one viable solution is to capture images of infected leaves and analyze them for disease information [6]. An

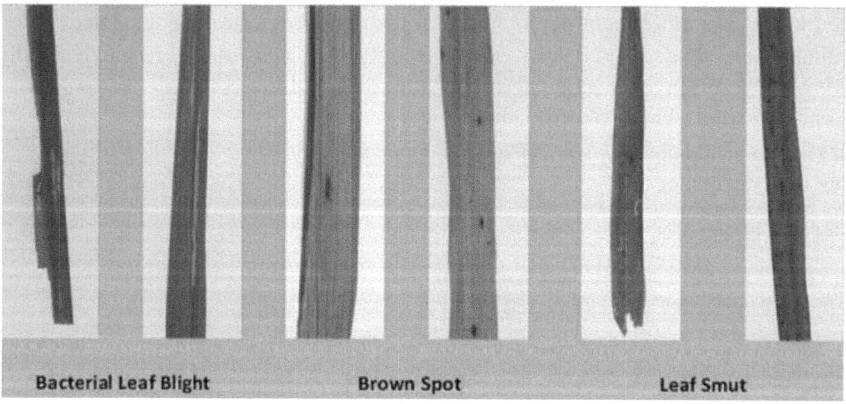

Fig. 1 Different types of rice plant diseases [3]

Table 1 The several diseases that affect rice plants and their key characteristics

S. No	Disease	Affected parts
1	Bacterial leaf blight	Leaves
2	Brown spot	Leaves
3	Leaf smut	Leaves

automated approach involves deploying cameras at specific intervals in the fields to capture images, which are then transmitted to a central system for disease analysis and pesticide recommendations. The central challenge of this system is the automatic recognition of the disease that has afflicted the plants, a problem this research seeks to address [7]. In doing so, we aim to illustrate how deep neural networks can offer valuable insights and solutions in the agricultural domain. An integral aspect of employing deep neural networks is the selection of optimal features. A critical and reusable contribution of this article lies in the separation of leaf portions from the diseased areas, a task made challenging by the predominance of green leaves in most plants [8].

This research aims to explore the use of deep learning for image analysis-based disease diagnosis in rice plants. We developed a novel deep learning system consisting of an artificial neural network (ANN) model. It used a collection of photos of sick rice plants for testing and training. Our system's performance was assessed based on precision and accuracy with promising results obtained. In addition, the potential benefits of this system are discussed, along with directions for future research [9].

2 Related Work

Building classifiers for diverse rice plant leaf disease detection and identification on many datasets using various methodologies has advanced significantly in the last several years. A few of the pieces are explained here.

Using deep learning to address illness detection, Chen et al.'s work [10] capitalized on the technology's exceptional image processing & classification capabilities. Remarkably, their approach was able to forecast rice illness classifications with an average precision of 98.63%, when dealing with several conditions. Sony [11] harnessed the power of convolutional neural networks, implemented using R language, to identify illnesses in rice plants identified through photos of contaminated leaves. Three different illnesses were shown in these images brown patch, leaf smut, and bacterial leaf blight. Through refined image processing techniques, they significantly improved accuracy, achieving an impressive 86.6%.

A deep CNN model was developed by Al-Amin et al. [12] to forecast the four most common rice diseases, bacterial leaf blight, brown spot, blast, and leaf smut. Their CNN-based model performed very well, identifying several rice leaf diseases with a precision rate of 97.40%. A framework for using machine learning methods to identify rice leaf disease was presented by Ahmed et al. [13]. Their technique was strong as shown by the three prevalent diseases of rice plants they successfully identified: brown spot, bacterial leaf blight, & leaf smut. Observing the observable morphological alterations caused by the two distinct rice illnesses, leaf brown spot & leaf blast, Phadikar et al. [14] created an automated approach that could categorize the affected rice varieties. With their categorization system, they achieved a striking 79.5% accuracy rate. A model to identify three rice leaf diseases—bacterial leaf blight, brown region, & leaf smut—was presented by Azim et al. [15]. When their model was evaluated on the UCI rice leaf disease dataset, it demonstrated a high degree of precision, scoring 86.58%. Using a residual neural network, Patidar et al. [16] presented an effective model for identifying and categorizing damaged paddy leaves. Their work was very precise, achieving a 95.83% effective accuracy rate.

2.1 Aim & Objectives

The aim of this study is to create a model for the precise identification of rice plant disease using VGG-16. This research will provide findings that are more exact and precise.

1. Develop a deep learning model using VGG-16 architecture for classifying and identifying rice plant leaves.
2. Collect and pre-process data necessary for training the model.
3. Train the model using the collected data for accurate predictions.
4. Evaluate the model's performance and optimize the model if needed.
5. Deploy and test the model for accurately classify and identify rice plant disease.

3 Background

The Background of this section's existing deep neural network architectures such as Alex Net, and GoogLe Net along with graphical representations are described.

3.1 Alex Net Architecture

The architecture has eight layers: the first five have convolutional as well as the last three have fully coupled. To get the most features, for overlapping max-pooling layers, the first two convolutional layers of the network are connected. In [17], the fully-connected layers have direct connections to among the fully-connected convolutional layers, third, fourth, & fifth direct connections to four, five, and third convolutional layers [18] (Fig. 2).

Once a network's final result layer is combined with softmax activation layers, 1,000 class labels are produced.

An RGB, 256 × 256 picture with three channels is input into Alex Net. Architecture has 60 million variables and 650,000 neurons. Overtraining is reduced by dropout layers. Dropped neurons don't affect forward or back propagation. These layers comprise the first two completely linked tiers.

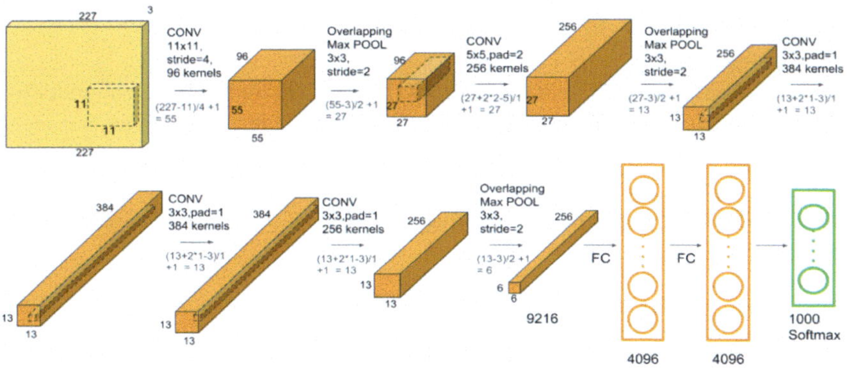

Fig. 2 Alex net architecture [19]

3.2 GoogLeNet Architecture

The GoogLeNet Architecture consists of 22 stacked layers and 27 pooling levels. The nine conception modules are organized in a linear pattern overall. The interfaces within the inception modules are connected with the global mean pooling layer. In [20] the next figure presents the whole GoogLeNet architecture in a scaled-down proportion (Fig. 3).

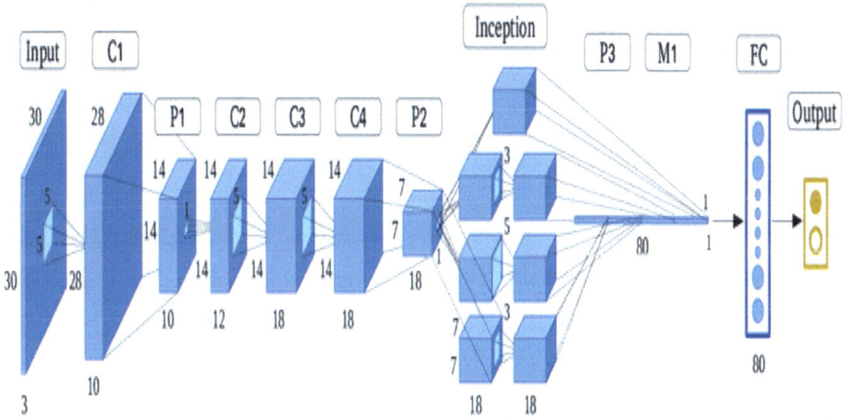

Fig. 3 GoogLeNet architecture [21]

4 Methodology

The objective of this study is to use the VGG16 architecture to develop a model for precisely identifying rice plant disease. This section depicts the work's embodiment technique. The embodiment technique is divided into many subsections, including dataset description, image dataset pre-processing, proposed model description, training, and finally performance evaluation (Fig. 4).

4.1 Dataset Description

In this study, data was collected from this website: https://www.kaggle.com/datasets/vbookshelf/rice-leaf-diseases. This collection contains 120 jpeg images of damaged rice leaves. The three classifications that the photos belong to are determined by the kind of sickness. Every class has forty photos.

4.2 Image Pre-Processing

(a) **Input the Image**: Begin by importing the rice plant images into the system. These images can originate from various sources, including photographs taken

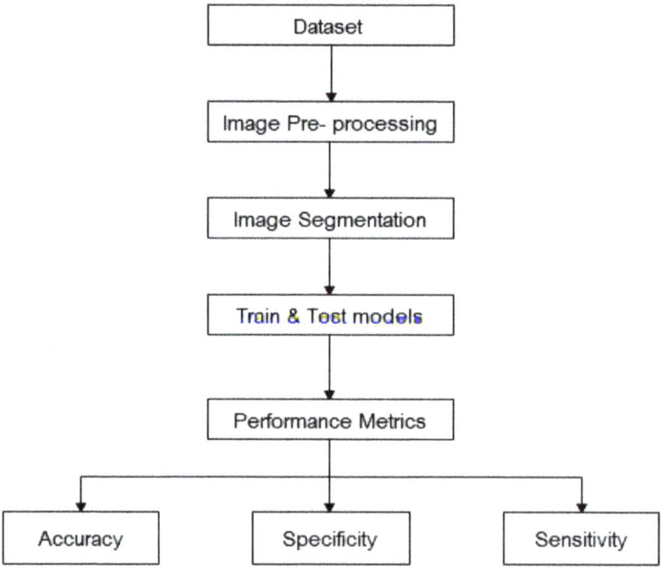

Fig. 4 Flowchart for proposed methodology

(a) (b)

Fig. 5 **a** Input image **b** Grayscale image

in the field or from laboratory setups. Ensure that the images are in a format compatible with your image processing tools.

(b) **Convert the Image to Grayscale**: To facilitate further analysis, convert the color images to grayscale. This process involves removing the hue and saturation elements while preserving the brightness value. Grayscale images simplify subsequent processing steps and reduce the computational load, making them well-suited for feature extraction and disease detection tasks (Fig. 5).

These initial steps prepare the dataset for more advanced pre-processing and analysis, such as feature extraction, disease classification, and model training.

4.3 Image Segmentation

The process of segmenting an image is a method for separating a picture into the individual sections or objects that compose it. When the item of interest has been isolated, segmentation is finished.

We have to extract the features from rice plant disease image by using VGG16 architecture. VGG16 is helpful extract features from an image and there some of the elements included in VGG16 architecture are described below.

4.4 Proposed Model Description

A turning point in human efforts to enable computers to "see" the world was reached with VGG16. Enhancing this capability has been the focus of much effort in the area on Computer Vision (CV) for decades. Numerous further advancements in this discipline were made possible by the significant discovery known as VGG16. Both Andrew Zisserman & Karen Simonyan created Convolutional Neural Networks, (CNN) model at the University of Oxford. The vehicle's design was initially presented in 2013, although the only in 2014, at the time of ILSVRC Vision Net Challenge, was the whole model submitted. The picture Net Wide Scale Visual Recognition

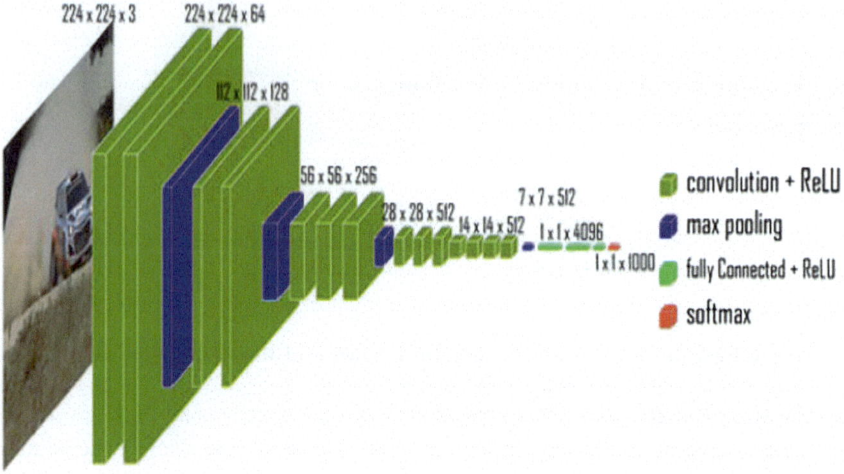

Fig. 6 VGG-16 architecture

Challenge (ILSVRC) was a manual competition that assessed picture categorization (and object identification) methods on a large scale (Fig. 6).

4.5 Train-Test Split

To use a random test size, divide your data into a training set, an evaluation set, as well as a validation set. Estimating the efficacy of machine learning algorithms suitable for prediction-based Algorithms may be done using the train-test split. The primary motivation for separating the data it in to a training set In addition, a testing set should stop the over fitting. It happens when the model recognizes training sample examples well. Dataset yet fails it generalize to and correctly categories data it has never seen before.

GoogLeNet and Alex Net, together with the suggested VGG-16 model, were taught with the train dataset. These two additional Alex Net and GoogLeNet are used to assess and compare the performance of the proposed model. The effectiveness of the model may be assessed using the following measures.

4.6 Performance Metrics

The accuracy, sensitivity, and specificity in the confusion matrix are used to evaluate a technique's efficacy.

Accuracy is the number of subjects compared to all the subjects who were successfully identified.

$$\text{Accuracy} = \frac{TP + TN}{TP + TN + FP + FN} \tag{1}$$

Sensitivity: Recall, usually referred to as sensitivity, is the proportion of precisely positive labels which our computer identifies as labels.

$$\text{Sensitive} = \frac{TP}{TP + TN} \tag{2}$$

Specificity: The algorithm has properly identified the negative as specificity.

$$\text{Specificity} = \frac{TN}{TN + FP} \tag{3}$$

5 Results

To evaluate the performance of VGG-16, GoogLeNet, and Alex Net architectures, we can utilize performance metrics such as the confusion matrix, sensitivity, specificity, and accuracy.

A thorough analysis of the model's predictions is given by the false positives, true negatives, true positives, and false negatives are all shown in the confusion matrix. It enables us to assess the classification performance for each class and identify any potential misclassifications.

The suggested VGG16 confusion matrix is shown in the above graphic. The figure shows that zeros indicate normal pictures while ones represent sick images. The suggested approach found that out of 100 photos, 17 were normal and 76 were sick. Seven photos were incorrectly classified, with sick pictures being mistaken for normal ones (Fig. 7).

The graphic depicting the matrix of confusion of the Alex Network architecture offers valuable insights into the image classification performance of the suggested approach. Pictures that are healthy are represented by zeros in this matrix, whereas pictures that are sick are represented by ones (Fig. 8).

Out of the total 100 images, the proposed method correctly identified 13 as normal images and 78 as diseased images. However, there were instances where misclassifications occurred, as it incorrectly identified 9 diseased images as normal ones.

The confusion matrix for the GoogLeNet architecture is shown in the above graphic. The figure shows that zeros indicate normal pictures while ones represent

VGG-16

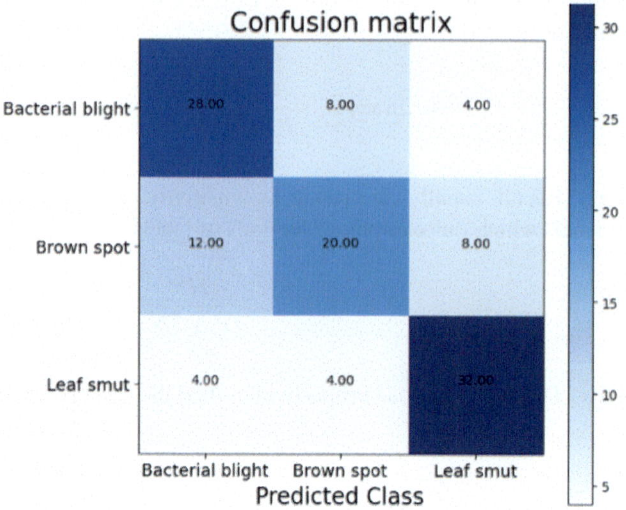

Fig. 7 Confusion matrix for VGG16

Alex Net

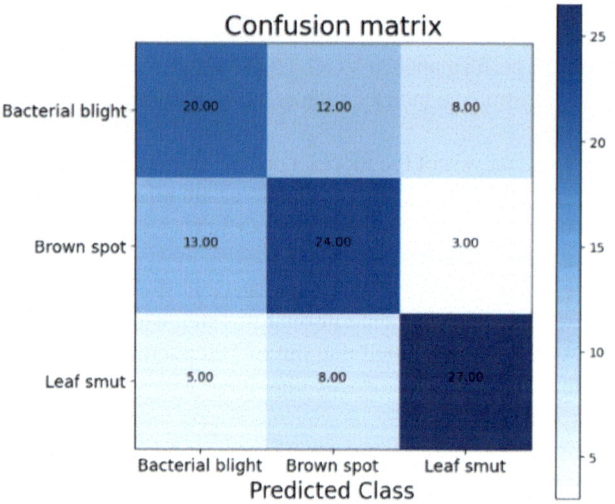

Fig. 8 Confusion matrix for Alex network architecture

sick images. The suggested approach found that out of 100 photos, 16 were normal and 72 were sick. Twelve pictures were incorrectly classified as normal when in fact they included sick images (Fig. 9).

In order to assess the efficacy of the designs, performance metrics including Specificity, Sensitivity, and Accuracy are computed and shown as follows (Fig. 10):

GoogLeNet

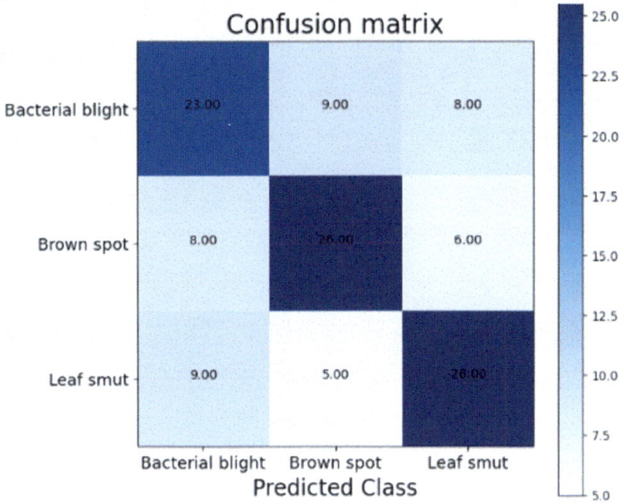

Fig. 9 Confusion matrix for GoogLeNet

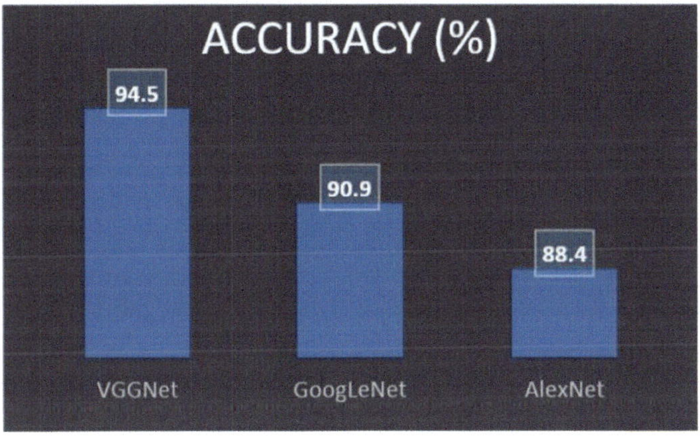

Fig. 10 Accuracy

The performance evaluation reveals that VGGNet architectures exhibit an impressive average accuracy of 94.50%. In comparison, GoogLeNet architectures achieve an accuracy of 90.90%, while Alexnet architectures achieve a slightly lower accuracy of 88.40%. This analysis clearly indicates that VGGNet architectures outperform the others in the accurate detection of rice plant diseases (Fig. 11).

In terms of specificity, VGGNet architectures exhibit an impressive score of 95.65%, while GoogLeNet architectures achieve a respectable 91.42%, and AlexNet architectures perform at 89.70%. This observation underscores the notable strength

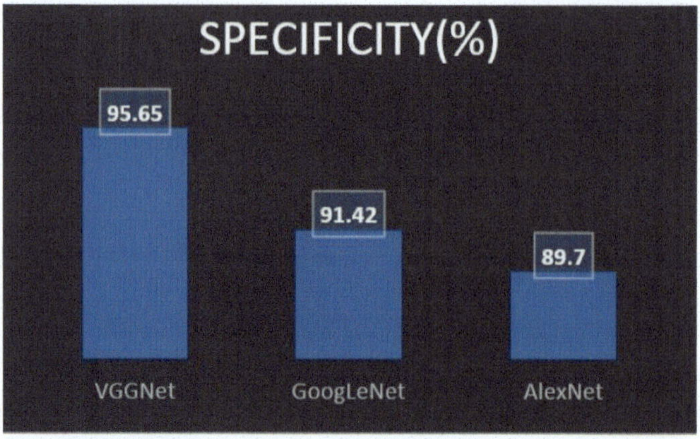

Fig. 11 Specificity

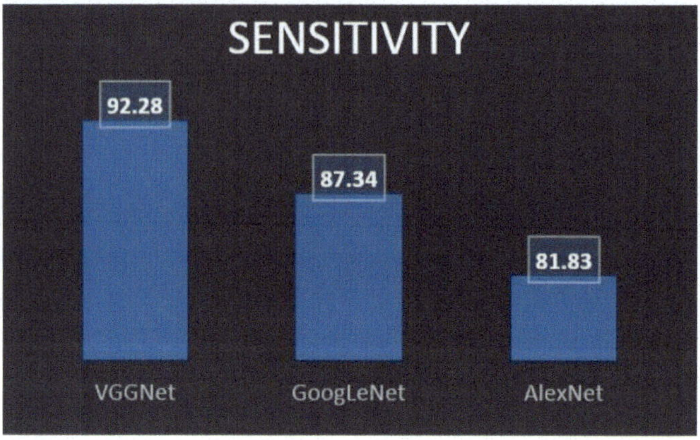

Fig. 12 Sensitivity

of VGGNet architectures, demonstrating their high sensitivity compared to both GoogLeNet and Alexnet (Fig. 12).

Regarding sensitivity, VGGNet architectures exhibit a commendable rate of 92.28%, while GoogLeNet architectures achieve a respectable 87.34%, and Alexnet architectures score at 81.83%. This observation underscores the notable strength of VGGNet architectures, demonstrating their high sensitivity in comparison to both GoogLeNet and Alexnet (Table 2).

Table 2 Comparison of models

Architectures	Accuracy (%)	Specificity (%)	Sensitivity (%)
VGGNet	94.5	95.65	89.28
GoogLeNet	90.9	91.42	87.34
Alexnet	88.4	89.7	81.83

6 Conclusion

This research aims to build and deploy a model using VGG-16 architecture for accurately classifying and identifying different rice plant disease. This entails gathering and pre-processing data, training the model, assessing its effectiveness, and then deploying and testing it. With the use of our model, rice plant disease detection should be more accurate. In conclusion, the VGG-16 architecture is seen to be the most effective in detecting disease in rice plants with the highest accuracy, sensitivity, and specificity. In summary, the comprehensive performance evaluation highlights the superiority of VGGNet architectures in the context of rice plant disease detection. VGG-16 demonstrates an outstanding average accuracy of 94.50%, outperforming both GoogLeNet, which achieves 90.90%, and Alexnet, with an accuracy of 88.40%. This substantial accuracy advantage positions VGG-16 as a formidable choice for accurate and reliable disease detection in rice plants. Moreover, when examining specificity, VGGNet architectures shine with an impressive score of 95.65%. In contrast, GoogLeNet achieves a respectable 91.42%, and Alexnet performs at 89.70%. This analysis emphasizes VGG-16's remarkable sensitivity, reinforcing its superior performance compared to both GoogLeNet and Alexnet. In conclusion, VGG-16 architectures emerge as the preferred choice for precise and sensitive rice plant disease detection, making them a valuable asset in agricultural research and disease management efforts.

References

1. Daniya, T., Vigneshwari, S.: A review on machine learning techniques for rice plant disease detection in agricultural research. System. **28**(13), 49–62 (2019)
2. Agrawal, M.M., Agrawal, S.: Rice plant diseases detection & classification using deep learning models: a systematic review. J Crit Rev **7**(11), 4376–4390 (2020)
3. Prajapati, H.B., Shah, J.P., Dabhi, V.K.: Detection and classification of rice plant diseases. Intell. Decis. Technol. **11**(3), 357–373 (2017). https://doi.org/10.3233/IDT-170301
4. Barbedo, J.G.A.: Plant disease identification from individual lesions and spots using deep learning. Biosyst. Eng. **180**, 96–107 (2019)
5. Patidar, S., Pandey, A., Shirish, B.A., Sriram, A.: Rice plant disease detection and classification using deep residual learning. In: Machine Learning, Image Processing, Network Security and Data Sciences: Second International Conference, MIND 2020, Silchar, India, July 30–31, 2020 Proceedings, Part I 2, pp. 278–293 (2020)

6. Shrivastava, V.K., Pradhan, M.K., Minz, S., Thakur, M.P.: Rice plant disease classification using transfer learning of deep convolution neural network. Int. Arch. Photogramm. Remote Sens. Spat. Inf. Sci. **42**, 631–635 (2019)
7. Attallah, O.: RiPa-Net: recognition of rice paddy diseases with duo-layers of CNNs fostered by feature transformation and selection. Biomimetics **8**(5), 417 (2023)
8. Andrianto, H., Faizal, A., Armandika, F.: Smartphone application for deep learning-based rice plant disease detection. In: 2020 International Conference on Information Technology Systems and Innovation (ICITSI), pp. 387–392 (2020)
9. Li, R., et al.: Predicting rice diseases using advanced technologies at different scales: present status and future perspectives. aBIOTECH. **4**(4), 359–371 (2023). https://doi.org/10.1007/s42994-023-00126-4
10. Prajapati, H.B., Shah, J.P., Dabhi, V.K.: Detection and classification of rice plant diseases. Intell. Decis. Technol. **11**(3), 357–373 (2017)
11. Chen, J., Chen, J., Zhang, D., Nanehkaran, Y.A., Sun, Y.: A cognitive vision method for the detection of plant disease images. Mach. Vis. Appl. **32**, 1–18 (2021)
12. Nanehkaran, Y.A., Zhang, D., Chen, J., Tian, Y., Al-Nabhan, N.: Recognition of plant leaf diseases based on computer vision. J. Ambient Intell. Humaniz. Comput. 1–18 (2020)
13. Pothen, M. E., Pai, M.L.: Detection of rice leaf diseases using image processing. In: 2020 Fourth International Conference on Computing Methodologies and Communication (ICCMC), pp 424–430 (2020)
14. Herre, J., Hilpert, J., Kuntz, A., Plogsties, J.: MPEG-H 3D audio—the new standard for coding of immersive spatial audio. IEEE J. Sel. Top. Signal Process. **9**(5), 770–779 (2015)
15. Kathiresan, G., Anirudh, M., Nagharjun, M., Karthik, R.: Disease detection in rice leaves using transfer learning techniques. J. Phys.: Conf. Ser. **1911**(1), 12004 (2021)
16. Wang, X., Zhang, X., Zhou, G.: Automatic detection of rice disease using near infrared spectra technologies. J. Indian Soc. Remote Sens. **45**, 785–794 (2017)
17. Lu, Y., Yi, S., Zeng, N., Liu, Y., Zhang, Y.: Identification of rice diseases using deep convolutional neural networks. Neurocomputing **267**, 378–384 (2017)
18. Singh, I., Goyal, G., Chandel, A.: AlexNet architecture based convolutional neural network for toxic comments classification. J. King Saud Univ.-Comput. Inf. Sci. **34**(9), 7547–7558 (2022). https://doi.org/10.1016/j.jksuci.2022.06.007
19. El Shenbary, H.A., Ebeid, E.A., Baleanu, D.: COVID-19 classification using hybrid deep learning and standard feature extraction techniques. Indones. J. Electr. Eng. Comput. Sci. **29**(3), 1780–1791 (2023). https://doi.org/10.11591/ijeecs.v29.i3.pp1780-1791
20. Kiruthika, S.U., Raja, S.K.S., Jaichandran, R., Priyadharshini, C.: Detection and classification of paddy crop disease using deep learning techniques. Int. J. Recent Technol. Eng **8**, 2277–3878 (2019)
21. Guo, Z., Chen, Q., Wu, G., Xu, Y., Shibasaki, R., Shao, X.: Village building identification based on ensemble convolutional neural networks. Sensors (Switzerland) **17**(11), 1–22 (2017). https://doi.org/10.3390/s17112487

Path Planning for Autonomous Drones Using Proximal Policy Approximating Agent

I. Varshini Devi, R. Pravine Mukesh, P. Paayas, P. Varsha, R. Supritha, and R. PrasannaKumar

Abstract The advent of autonomous drones promises transformative applications across many industries, necessitating the ability to plan long-range paths automatically. The purpose of this paper is to explore the synergistic effect of Markov Decision Processes (MDPs) and deep reinforcement learning (DRL) in the context of autonomous drone racing. A simulated racing track is used to demonstrate how autonomous drones can optimize path planning using the Proximal Policy Optimization algorithm (PPO). In our study, we present an agent that efficiently navigates strategic checkpoints, is adaptable, and generalizes its skills to similar situations. It demonstrates the potential of DRL and PPO to advance autonomous drone path planning, paving the way for practical implementation in a wide range of industries. With drones reshaping industries, this work contributes to the exciting future of autonomous systems. The agent covered an average distance of 19.6 m per episode, showcasing its ability to navigate the test environment effectively. In a notable feat, the agent exhibited the capability to pass through a maximum of 9 holes in the test environment without encountering any collisions. This indicates a high level of adaptability and strategic decision-making, even in the face of increased environmental complexity.

Keywords Autonomous drones · Path planning · Reinforcement learning · Proximal policy optimization · Markov decision processes · Unmanned aerial vehicles

1 Introduction

Autonomous drones have the potential to revolutionize a range of industries, including farming, surveillance, and delivery systems. To maximize their effectiveness, these drones must possess the capability to autonomously navigate intricate

I. Varshini Devi · R. Pravine Mukesh · P. Paayas · P. Varsha · R. Supritha · R. PrasannaKumar (✉)
Department of Computer Science and Engineering, Amrita School of Computing, Amrita Vishwa Vidyapeetham, Chennai, India
e-mail: r_prasannakumar@ch.amrita.edu

G. Raj et al. (eds.), *Advances in Artificial Intelligence and Machine Learning*, Lecture Notes in Electrical Engineering 1264, https://doi.org/10.1007/978-981-97-9507-9_24

environments and execute long-range tasks. The ability to autonomously plan and navigate over extended distances is vital as it directly impacts their performance, safety, and overall versatility. In the realm of robotics, conventional path planning traditionally relied on interval-based search algorithms and sampling based techniques [1]. However, we are now embarking on an innovative journey, grounded in the principles of Markov Decision Processes (MDPs) [2], which harness reward-based algorithms to accomplish long-term planning objectives. This approach presents a promising solution to address the challenges of planning in environments with non-holonomic constraints, an area where existing methods often fall short.

Incorporating Deep Reinforcement Learning (DRL) [3] into UAV control, particularly in communication scenarios, represents a significant shift. Despite challenges like limited data and safety concerns, DRL excels in addressing NP-hard optimization problems. Our research focuses on applying DRL, specifically the Proximal Policy Optimization (PPO) algorithm [4], to optimize path planning for quadrotors on a simulated racing track. Results demonstrate the agent's ability to navigate checkpoints and generalize path-planning skills. These findings have implications for autonomous drones in real-world path-planning scenarios.

2 Literature Survey

The research we do builds on a rich foundation of studies exploring how reinforcement learning can be integrated into various aspects of unmanned aerial vehicle (UAV) control and mission planning. These studies encompass a diverse range of applications, including path planning, cooperative reconnaissance, energy management, data collection, and navigation in complex, high-threat environments.

An analysis of deep reinforcement learning-based path planning for UAVs using mobile edge computing (MEC) is presented by Liu et al. Results show that DRL can be used for UAV-MEC systems to complete jobs faster and use resources more efficiently [5]. Optimising UAV communication in IoT applications is critical for effective data transfer and energy management. To increase the usefulness of the UAVs, a Markov decision process (MDP) model is created. In order to deal with uncertainty, partially observable states, and vast state spaces, we use deep reinforcement learning approaches like value iteration. A proposed approach to wireless energy transfer and data delivery outperforms baseline techniques in terms of long-term value [6]. A method for autonomous UAV navigation based on LwH (deep reinforcement learning with nonexpert helpers). In complex situations, it provides objective solutions by leveraging sparse rewards. Based on a UAV navigation simulator evaluation, LwH achieves outstanding navigation strategies similar to dense reward environments in sparse reward situations [7]. Ugurkan Ates' paper discusses a drone racing competition scenario using Microsoft's AirSim Drone Racing Lab. The study demonstrates the effectiveness of training quadrotor agents using the Policy Proximal Optimization algorithm. Combining DRL with expert guidance proves effective for navigating complex environments, making classical path planning successful [8]. Harald

Bayerlein et al introduces a multi-agent reinforcement learning framework for efficient data collection from distributed IoT devices using multiple autonomous UAVs. By adapting to varying mission parameters, the framework ensures optimal control policies, data collection, flight-time limitations, and collision avoidance, advancing UAV-based data collection strategies [9]. Ahmad Taher Azar et al. discuss the role of deep reinforcement learning (DRL) in enhancing the autonomy of Unmanned Aerial Vehicles (UAVs) in various applications to bridge the gap between DRL advancement and real-life application [10].

Zhao et al. [11] propose an end-to-end DRL-based mission planning approach, considering mission-specific properties and parameters, and using the proximal policy optimization (PPO) algorithm to handle continuous action spaces. Zhang et al. [12] propose Twin Delayed Deep Deterministic Policy Gradients (TD3). By introducing a two-stream Actor-Critic network structure, it can handle changing environment observations and exhibits improved generalization over DDPGs and TD3s. UAV path planning is proposed by Boyan Xin et al. [13] using playback based deep reinforcement learning, which differs from conventional DRL-DQN. The approach integrates a step reward scheme into an experience replay-based DRL-DQN to demonstrate superior performance and enhanced adaptation to dynamic environments. To address the limitations of traditional methods, Longfei Yue et al. [14] present a novel approach to autonomous combat mission planning for unmanned aerial vehicles (UAVs) based on deep reinforcement learning (DRL). Proximal policy optimization (PPO) is used to formulate suppression of enemy air defense (SEAD) mission planning as a Markov decision process (MDP). The results demonstrate that DRL-based UAV mission planning enhances autonomy in complex, high-dimensional environments and is efficient, stable, and adaptable. Zun Liu et al. presents an innovative method to improve the real-world implementation of DRL for autonomous navigation in Unmanned Aerial Vehicles. Using practical assessments across various environments, the study illustrates the superiority of the method over established algorithms [15].

Chenchen Fu et al. address the complex task of helping smart agents navigate in 3D spaces using visual cues. They propose an innovative Memory Enhanced DRL agent that learns to move towards changing goals, avoid unnecessary actions, and receive better rewards with time [16]. Through visual cues, Chenchen Fu and colleagues help smart agents navigate in 3D spaces. To guide unmanned aerial vehicles, they propose a novel method based on a learning framework and visual data [17, 19]. Uncrewed Aerial Vehicle systems driven by Deep Learning are at greater risk of adversarial attacks by Thomas Hickling and colleagues. To combat these attacks, they propose a novel approach that leverages DL's explainability. To train and avoid obstacles, their UAV agent uses Deep Reinforcement Learning, combining Deep Deterministic Policy Gradient (DDPG), Prioritised Experience Replay (PER), and Artificial Potential Field (APF). DL-driven UAV systems are more secure with this work [18, 20].

3 Methodology

In order to enhance long-range planning for autonomous drones, we utilize Deep Reinforcement Learning (DRL) capabilities. Throughout this section, we describe the key components of our methodology, including the choice of simulation environment, reinforcement learning algorithm, training procedures, data collection, and MDP modeling.

3.1 Simulation Environment

A core element of our research is the choice of an advanced and versatile simulation environment, designed with a high level of precision to emulate real-world drone racing tracks. The simulated racing track serves as the backdrop for our autonomous drone training and evaluation processes, providing the perfect balance of realism and control. AirSim Drone Racing Lab is the cornerstone of our simulation environment, designed to bridge the gap between simulated and real-life environments. Thanks to the utilization of Epic's Unreal Engine 4, this framework offers high-fidelity graphics combined with proprietary physics. AirSim Drone Racing Lab supports this comprehensive simulation environment, enabling us to train and evaluate our autonomous drone agent in a controlled but dynamic environment designed to mirror real-world scenarios.

Reinforcement Learning Algorithm. Reinforcement Learning (RL) is central to our drone navigation research in dynamic racing environments. It involves Policy (agent's action plan), Rewards (numerical indicators for stateaction pairs), Value Function (calculates discounted return for long-term desirability), and Environment Model (captures environment behavior). Deep Reinforcement Learning (DRL) combines deep learning with RL, utilizing three primary approaches: value-based, policy-based, and model-based. These approaches optimize actions, policies, or build environment models, catering to diverse environmental complexities. As DRL is integrated into unmanned aerial vehicle (UAV) control, there are several unique challenges to address, including the complexity of UAV models, nonlinearities in the system, online learning in dynamic environments, and the ability to operate in unfamiliar environments. In order to overcome these challenges, DRL introduces a high-dimensional input and an accurate representation of the environment. A policy, reward signal, utility function, and the environment itself are all taken into account when RL algorithms are applied to UAV domains, of which we customize the representation of Markov decision processes (MDPs). Deep neural networks and RL are combined to overcome the limitations of conventional RL and meet the requirements of UAVs. From controlling hybrid UAVs safely to orchestrating swarms of UAVs to accomplish specific objectives with minimal resource consumption, DRL can be applied to UAV control in a wide range of applications. In the context of UAV control, these diverse challenges demonstrate the versatility and potential of DRL.

In our research, we employ the Proximal Policy Optimization (PPO) algorithm, a robust method within Deep Reinforcement Learning (DRL). PPO is chosen for its efficiency in handling continuous control tasks, making it particularly suitable for training agents in dynamic racing environments. PPO operates by replacing hard constraints with penalties and leveraging the Kullback–Leibler (KL) divergence to regulate policy changes at each time step. This approach ensures a smooth and stable training process, allowing the agent to adapt to changing environmental conditions without abrupt policy adjustments. The core objective of PPO is to maximize the cumulative future rewards of an agent's policy. This is achieved by updating the policy parameters in a way that enhances the agent's decision-making process over time. The algorithm incorporates two key components: the state-value function $V^{\pi}(s)$ and the action-value function $Q^{\pi}(s, a)$. Mathematically, it is defined as: line.

$$V^{\pi}(s) = \sum\nolimits_t E_{\pi_\theta}[s] \tag{1}$$

$$Q^{\pi}(s, a) = \sum\nolimits_t E_{\pi_\theta}[s, a] \tag{2}$$

These functions are fundamental in reinforcement learning and PPO, providing a way to evaluate the desirability of states and actions. In the training process, the policy is adjusted to maximize these expected long-term rewards, allowing the agent to learn optimal decision-making strategies in the given environment. In the equations, $R(s_t, a_t)$ is the reward obtained when taking action at in state s_t. The sum extends over all time steps, capturing the cumulative expected reward. E_{π_θ} denotes the expectation over trajectories generated by the policy π_θ. The subscript θ indicates that the policy is parameterized by θ, and the optimization process involves updating these parameters to improve the policy's performance.

In the context of autonomous drones and racing, PPO's ability to adapt to unforeseen racing scenarios proves crucial. Using PPO, we enable our autonomous drone agent to excel in the complex and dynamic racing environment, thereby improving the capabilities of long-range path planning for autonomous drones. The pseudocode for PPO algorithm is given below.

Algorithm 1 Proximal Policy Optimization (PPO)

Initialize the policy parameter θ
Set hyperparameters (learning rates, clip parameter, number of iterations, etc.) **while** not converged **do**
  Collect experiences by interacting with the environment
  Compute advantage estimates
  Optimize policy by updating policy parameters
  Repeat for a specified number of iterations or episodes **end**
while=0

Agent Configuration and Training: We focus our research on an autonomous drone agent represented by a simulated quadrotor. The agent's observations encompass a fusion of sensor data essential for planning and navigating through the racing environment. This includes information obtained from Inertial Measurement Units (IMUs) and GPS coordinates obtained from simulations. A convolutional neural network (CNN) processes this sensor data and generates action commands. Effective learning in our pipeline involves two crucial steps: data collection and preprocessing. In the data collection phase, we train our agent and record its experiences, storing them for future use. To train and collect data, we employ a PID controller. Subsequently, we fine-tune the agent using the PPO algorithm with the goal of enhancing its path planning abilities. Data preprocessing is a fundamental stage where we convert raw sensory data into appropriate inputs for the neural network. During this process, we make sure to normalize and standardize the data to ensure consistent and reliable training. Moreover, we employ data augmentation techniques to introduce diversity into the training dataset, helping the agent adapt to various racing conditions.

Markov Decision Process (MDP) Modeling: Our strategy for autonomous drone racing relies on Markov Decision Processes (MDPs), which serve as a fundamental framework. MDPs offer a structured way to depict how autonomous drone agents engage with their ever-changing surroundings. Within this context, the drone agent takes on the role of the MDP's decision-maker, while the environment represents the agent's operational area. The use of MDPs is critical for establishing the agent's extended path-planning goals, understanding state shifts, defining actions, and evaluating rewards.

Definition of the State Space: The state space within our MDP encapsulates the entire environment in which the autonomous drone operates. It includes the following components: (a) Agent State The state of the autonomous drone, which consists of its position, velocity, orientation, and internal attributes. These attributes include drone status, performance data, and any internal variables relevant to navigation. (b) Environmental State: The state of the racing environment, including the positions of checkpoints, gates, obstacles, and any other dynamic elements. This environmental state information is crucial for the agent to make informed decisions and plan its path efficiently.

Action Space: The action space of the MDP specifies the available actions that the drone agent can take to influence its state and progress. The action space in our autonomous drone racing context includes: (a) Throttle: Controlling the drone's thrust or acceleration to change its speed. (b) Yaw: Adjusting the drone's yaw, affecting its rotational movement. (c) Pitch: Tilting the drone's orientation forward or backward. (d) Roll: Tilting the drone's orientation to the left or right. These actions define the drone's ability to maneuver and adapt to changing racing conditions. It is essential for controlling the agent's movements along the racing track.

Reward Design: The heart of the MDP lies in the reward function. The reward function is meticulously crafted to guide the agent in achieving its long-term path planning goals. The reward function influences the agent's decision-making process by assigning numerical rewards to state-action pairs. In the context of autonomous drone racing, our reward function is structured to optimize key behaviors, including: (a) Efficient Racing: The agent receives positive rewards based on its proximity to the next checkpoint. This encourages the agent to navigate the racing track swiftly and efficiently, making decisions that minimize racing time. (b) Checkpoint Navigation: A bonus reward is granted when the agent successfully navigates through a checkpoint. This further incentivizes the agent to follow the predefined racing track and maintain its racing trajectory. (c) Obstacle Avoidance: The agent receives negative rewards when it gets too close to obstacles or deviates significantly from the racing track. These penalties discourage reckless behavior and reinforce the importance of safe and precise racing. (d) Optimal Racing Line: The agent is rewarded for adhering to the optimal racing line through the racing track. Following this line aids in minimizing racing time and enhancing the agent's performance. Our reward function captures the essence of autonomous drone racing, guiding the agent to make decisions that align with our long-range path levels.

Policy Optimization Using PPO. The MDP framework drives the agent's decision-making process through policy optimization. By mapping states to actions, policy optimization seeks to identify the policy that maximizes cumulative future rewards. Our drone agent is trained using the Proximal Policy Optimization (PPO) algorithm, a well-established DRL technique. For our autonomous drone racing context, PPO is well-suited to train agents in continuous control tasks. PPO uses a stability constraint to ensure that policy updates are consistent and do not deviate significantly from their previous versions. Over time, PPO converges to a proficient policy, enabling the agent to progressively improve its path planning performance (Fig. 1).

Figure 2 illustrates the foundational software architecture employed in our study for autonomous path planning in drone navigation. The training process occurs within

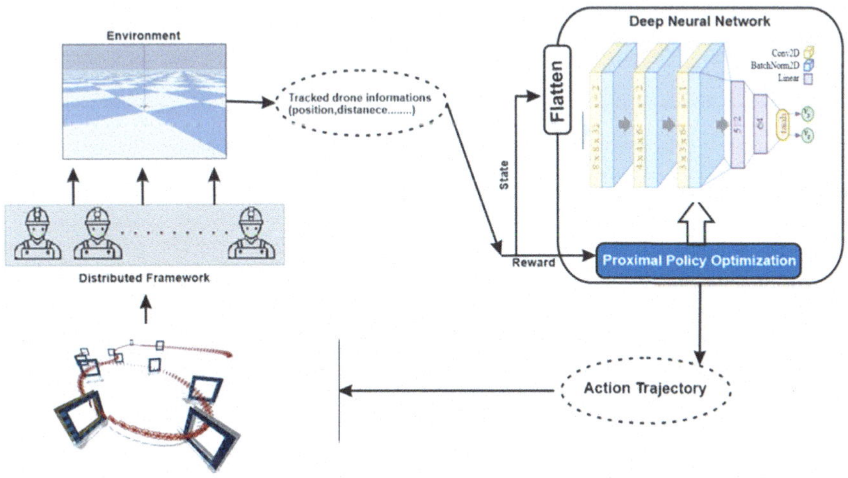

Fig. 1 Architecture of proposed system

a virtual environment. In our autonomous path planning model, the drone dynamically adjusts its actions based on the movements of the training. This collaborative behavior is facilitated by an artificial neural network, serving as the decision-making component. The artificial neural network acts as the key determinant of actions, leveraging predetermined parameters to guide the drone's movement within the environment. The neural network's architecture is meticulously designed to optimize decision-making in response to environmental stimuli. The neural network processes input data, which includes environmental features and the drone's actions. This decision-making process is pivotal for effective path planning and navigation. The actions performed by the drone result in corresponding rewards, crucial for shaping its future behavior. The reward mechanism serves as a decisive factor, influencing the drone's subsequent actions. This reinforcement learning approach enhances the autonomous path planning capabilities of the drone agent. The entire architecture operates within a continuous learning loop, where the drone adapts and refines its path planning strategies over time. This adaptive learning process ensures improved performance and responsiveness to varying environmental conditions.

In policy optimization, the current policy is executed, rewards are collected, gradients are estimated, and the policy is iteratively updated to maximize expected returns. During the training phase, this process continues until the agent shows proficient path planning skills. To achieve a balance between exploration and exploitation, we fine-tune the hyperparameters of the PPO algorithm to ensure stable and efficient training. Through this iterative process, we can align the agent's performance with our predefined goals and achieve optimal training outcomes. After completing the training process, we test our autonomous drone agent in similar conditions to the training environment. Through these tests, the agent shows its adaptability and robustness by applying its path-planning skills to new scenarios and conditions.

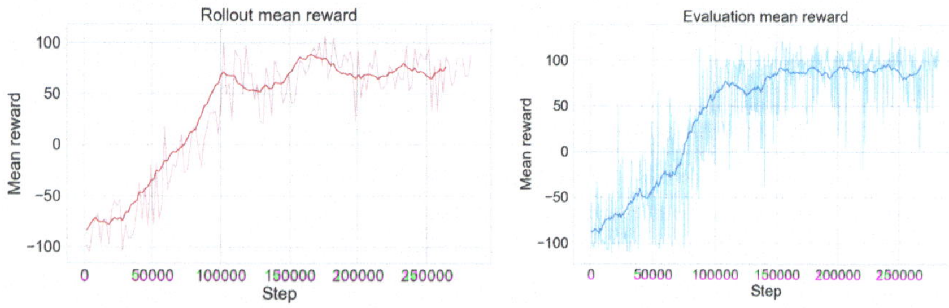

Fig. 2 Rollout and evaluation mean reward

4 Result Analysis

Our objective was to train a reinforcement learning agent to navigate a racecourse with twelve gate checkpoints within specified time constraints. The aim was to minimize collisions with obstacles and ensure the agent could recover and continue along the optimal path within the given time limits. To achieve this, we developed a tailored solution specifically designed for drone racing environments, which could also be adapted for similar contexts.

The "Rollout Mean Reward" graph illustrates the learning trajectory of our reinforcement learning agent as it navigates the racecourse over successive steps or episodes. The x-axis represents the progression of steps or episodes during training. The y-axis indicates the mean reward obtained by the agent during rollouts. The "Evaluation Mean Reward" graph assesses the generalization capability of our trained reinforcement learning agent in new scenarios. The x-axis represents the evaluation episodes or another relevant measure during the testing phase. The y-axis indicates the mean reward obtained by the agent during evaluation.

The results indicate that the path planning algorithm is effective in generating paths that lead to higher rewards. The increase in mean reward over time suggests that the algorithm is learning and improving its path planning strategy. The difference in fluctuation between the two graphs suggests that the algorithm performs better during evaluation than during rollout. This could be due to the fact that the algorithm is overfitting to the training data during rollout, resulting in more variability in the rewards. Overall, the results suggest that the path planning algorithm is effective in generating paths that lead to higher rewards, and that it is capable of learning and improving over time. During our initial testing, we observed adverse effects when there were no constraints on the agent's movements, which often caused it to become stuck in various suboptimal positions. To address this issue, we implemented state and action value normalization. This approach allowed us to impose substantial negative penalties if the agent deviated from the optimal path, but still retained the ability to recover due to the inherent nature of advantage policy updates in the algorithm. We conducted the simulation at a non-specific resolution of 640×480, running at a rate of 90 frames per second. During this simulation, we obtained episodic data throughout the training process. For 100 test episodes in the new environment, our trained model demonstrated commendable performance. The agent covered an average distance of 19.6 m per episode, showcasing its ability to navigate the test environment effectively. On average, the agent successfully navigated through 4 holes in each test episode without any collisions. This highlights the agent's proficiency in avoiding obstacles and following an optimal path. To assess the prowess of our proposed Proximal Policy Optimization (PPO) algorithm enhanced with an Artificial Neural Network (ANN), we conducted a comparative analysis against leading reinforcement learning algorithms. The table above showcases the mean rewards achieved by each algorithm in the test environment. SAC and TD3 algorithms exhibit comparatively lower mean rewards, suggesting challenges in adapting to the test environment (Fig. 3) (Table 1).

Fig. 3 Autonomous drone during testing

Table 1 Comparative performance analysis

Algorithm	Mean reward
Soft Actor-Critic (SAC)	−163.24
Twin Delayed DDPG (TD3)	−221.54
Deep Deterministic Policy Gradients (DDPG)	−75.62
Advantage Actor-Critic (A2C)	−216.36
Existing PPO	−54.78
Proposed PPO with ANN	−3.09

The original Proximal Policy Optimization (PPO) performs well, showcasing its effectiveness in the given task. Our novel approach, integrating an Artificial Neural Network (ANN) into the PPO framework, outperforms all other algorithms, achieving a substantially higher mean reward. This underscores the robustness and efficacy of our proposed model in addressing the complexities of dynamic racing environments. The exceptional performance of our proposed PPO algorithm with ANN positions it as a promising solution for real-world applications, demonstrating its potential in optimizing the navigation of autonomous agents in challenging scenarios.

5 Conclusion

In this study, we've unveiled the potential of deep reinforcement learning (DRL), especially through the Proximal Policy Optimization (PPO) algorithm, in equipping autonomous drones with the skills to master complex, long-range racing tracks. As a result of our research, we have taken a significant step toward making autonomous drones more versatile and adaptable. With the help of DRL and PPO, we were able to demonstrate that a well-trained drone can navigate simulated racing tracks expertly. Moreover, it efficiently overcomes strategically placed checkpoints and demonstrates a remarkable ability to generalize this ability to similar situations. For real-world autonomous systems across a wide range of applications, this adaptability holds great promise. Our study aligns with a wealth of research in the realm of DRL for UAVs, emphasizing its role in enhancing autonomy, navigation, and data collection. Nevertheless, it specializes in long-range racing scenarios, where it focuses on optimizing racing trajectories, navigating safely, and passing checkpoints efficiently. In summary, our findings underline the potential of DRL and PPO in enhancing autonomous drone path planning, paving the way for practical implementations in industries such as agriculture, surveillance, and delivery services. In the near future, autonomous drones will revolutionize a variety of sectors, and our research contributes to this exciting development.

References

1. Maghazei, O., Netland, T.: Drones in manufacturing: Exploring opportunities for research and practice. J. Manuf. Technol. Manag. **31**(6), 1237–1259 (2020)
2. Zhang, M.: Liu, Yan, Wang, Yu, Li, Fumin, Chen, Lei: Real-time Path Planning Algorithms for Autonomous UAV. In: Editor, F., Editor, S. (eds.) 2022 China Automation Congress (CAC), LNCS, vol. 9999, pp. 3818–3823. Springer, Heidelberg (2022)
3. Wu, C.: Ju, Bin, Wu, Yu, Lin, Xue, Xiong, Naixue, Xu, Guandong, Liang, Xiaodan: UAV autonomous target search based on deep reinforcement learning in complex disaster scene. IEEE Access **7**, 117227–117245 (2019)
4. Tan, Zhihong, Karaköse.: Mehmet: A new approach for drone tracking with drone using Proximal Policy Optimization based distributed deep reinforcement learning. Software X 23, 101497 (2023)
5. Liu, Qi, Shi, Lei, Sun, Lijun, Li, Ji, Ding, Mengjie, Shu, Fan.: Path planning for UAV-Mounted mobile Edge computing with deep reinforcement learning. IEEE Trans. Veh. Technol. **69**(5), 5723–5728 (2020)
6. Xiong, Zhiqiang, et al.: UAV-Assisted wireless energy and data transfer with deep reinforcement learning. IEEE Trans. Cogn. Commun. Netw. **7**(1), 85–99 (2021)
7. Wang, Chao, Wang, Jun, Wang, Jiabin, Zhang, Xiaohui.: Deep-Reinforcement-Learning-Based Autonomous UAV navigation with sparse rewards. IEEE Internet Things J. **7**(7), 6180–6190 (2020)
8. Ates, Ugurkan.: Long-term planning with deep reinforcement learning on autonomous drones. In: 2020 Innovations in Intelligent Systems and Applications Conference (ASYU). IEEE (2020).
9. Bayerlein, Harald, et al.: Multi-UAV path planning for wireless data harvesting with deep reinforcement learning. IEEE Open J. Commun. Soc. **2**, 1171–1187 (2021)
10. Azar, Ahmad Taher, Koubaa, Anis, Mohamed, Nada Ali, Ibrahim, Habiba, A., Ibrahim, Zahra Fathy, Kazim, Muhammad, Ammar, Adel et al.: Drone deep reinforcement learning: A review. Electron. **10**(9), 999 (2021)
11. Zhao, Xiaoru, et al.: Deep reinforcement learning for intelligent dual-UAV reconnaissance mission planning. Electron. 11(13), 2031 (2022)
12. Zhang, Sitong, Li, Yibing, Dong, Qianhui: Autonomous navigation of UAV in multi-obstacle environments based on a Deep Reinforcement Learning approach. Appl. Soft Comput. **115**, 108194 (2022)
13. Xin, Boyan, He, Chenghui: DRL-Based improvement for autonomous UAV motion path planning in unknown environments. In: 2022 7th International Conference on Control and Robotics Engineering (ICCRE). IEEE (2022)
14. Yue, Longfei, et al.: Deep reinforcement learning for UAV intelligent mission planning. Complex. (2022)
15. Liu, Zun, et al.: A Hierarchical Reinforcement Learning Algorithm Based on Attention Mechanism for UAV Autonomous Navigation. IEEE Trans. Intell. Transp. Syst. (2022)

16. Fu, Chenchen, et al.: Memory-enhanced deep reinforcement learning for UAV navigation in 3D environment. Neural Comput. Appl. **34**(17), 14599–14607 (2022)
17. Chen, Yu, et al.: Multi-UAV autonomous path planning in reconnaissance missions considering incomplete information: A reinforcement learning method. Drones **7**(1), 10 (2022)
18. Hickling, Thomas, Aouf, Nabil, Spencer, Phillippa: Robust adversarial attacks detection based on explainable deep reinforcement learning for UAV guidance and planning. IEEE Trans. Intell. Veh. (2023)
19. Wu, Xiru, Huang, Shuqiao, Huang, Guoming.: Deep Reinforcement Learning-Based 2.5 D Multi-Objective path planning for ground vehicles: considering distance and energy consumption. Electron. **12**(18), 3840 (2023)
20. Zhang, M.: Chen, Kai, Zhu, Jinhui: An efficient planning method based on deep reinforcement learning with hybrid actions for autonomous driving on highway. Int. J. Mach. Learn. Cybern. **14**(10), 3483–3499 (2023)
21. Dong, Lu, et al.: A review of mobile robot motion planning methods: from classical motion planning workflows to reinforcement learning-based architectures. J. Syst. Eng. Electron. **34**(2), 439–459 (2023)
22. Wen, Shuhuan, et al.: A multi-robot path-planning algorithm for autonomous navigation using meta-reinforcement learning based on transfer learning. Appl. Soft Comput. 110, 107605 (2021)

A Novel Routing Protocol for UWSN Using Energy Efficient Marine Predator Optimization Technique

Anmol Pandey, Anujaya Singh, Anushka Mishra, Ankit Pal, and Ashish Pandey

Abstract The novel metaheuristic algorithms Energy Efficient Marine Predator OptimizationInspired Algorithm (EEMPOIA), Energy Efficient Grey Wolf OptimizationInspired Algorithm (EEGWOIA), Energy Efficient Particle swarm OptimizationInspired Algorithm (EEPSOIA), and Least distance Clustering (LDC) are introduced in this study and are specifically designed to address the problems with underwater wireless sensor networks (UWSNs). In this rigorous environment, the objective is to increase energy efficiency and network longevity. EEMPOIA is found to be the most efficient method through extensive simulations utilizing a three-tier UWSN model, considerably increasing network longevity while lowering energy consumption. These results illustrate EEMPOIA's potential as a promising approach to enhancing UWSN performance. The combination of these algorithms creates new opportunities for using long-lasting and energy-efficient underwater sensing systems in various industrial and environmental settings.

Keywords Energy efficient clustering · Optimization · UWSN · Routing · Lifetime

1 Introduction

Wireless sensor networks specifically created for submerged situations are called underwater wireless sensor networks (UWSNs). They make it possible for sensors to communicate wirelessly, facilitating uses for them in surveillance, resource management, exploration, and environmental monitoring. Real-time data on elements including pressure, temperature, and water quality are provided by UWSNs. UWSNs, which are made up of underwater sensors, a base station, a gateway, and an acoustic communication system, have problems with bandwidth limitations, delays, energy use, and hostile environments [1]. To get around these challenges, current research

A. Pandey · A. Singh · A. Mishra · A. Pal · A. Pandey (✉)
Galgotias College of Engineering and Technology, Electronics and Communication Engineering
Greater Noida, Noida, India
e-mail: ashishaz2005@gmail.com

G. Raj et al. (eds.), *Advances in Artificial Intelligence and Machine Learning*, Lecture Notes in Electrical Engineering 1264, https://doi.org/10.1007/978-981-97-9507-9_25

focuses on enhancing communication protocols, energy efficiency, and sensor design. Marine Predator Algorithm is used to improve resource usage and performance in underwater wireless sensor networks (UWSNs) [2]. MPO especially addresses-specific optimization problems by emulating the foraging strategies of marine predators. This approach accounts for signal propagation issues such obstacles, water currents, and communication range restrictions [3]. Underwater Wireless Sensor Networks (UWSNs) use Marine Predator Optimization (MPO) to deploy sensors and distribute network resources wisely. For multi-objective optimization, it uses Pareto dominance and swarm-based algorithms to balance network coverage, energy efficiency, and connection [4]. The algorithm can be implemented in a distributed fashion, allowing nodes to communicate information and optimize locally. There are many benefits to using Marine Predator Optimization (MPO) in underwater wireless sensor networks (UWSNs). First, MPO makes use of marine species' inherent adaptability to their tough underwater surroundings [5]. This characteristic makes MPO more able to manage the UWSNs' inherent uncertainties, complexities, and resource limitations. Second, MPO-based algorithms make it possible for large-scale underwater sensor networks to be quickly and thoroughly optimized, resulting in increased coverage, connection, and energy efficiency—essential elements for success in challenging aquatic environments [6]. It's critical to evaluate the algorithm's performance in light of the demands of a specific situation before making a final decision. By modelling the adaptive hunting techniques of marine predators, MPO-based algorithms increase the efficiency, robustness, and resource consumption of UWSNs. This synergistic combination of bio-inspired optimization with underwater sensing technology paves the way for future advancements in undersea exploration, monitoring, and resource management [7].

2 Related Work

The Marine Predator Algorithm (MPA) is described in this article as a way to optimize the operating point of proton exchange membrane fuel cells (PEMFCs), hence cutting system costs and the number of necessary sensors. The MPA approach expands the search space by combining Brownian and Levy motion phases. Particle dispersion is improved via a hybrid Quantum Particle Swarm Optimization (QPSO) with marine predators. Moth-fame optimisation (MFO) is used for local search in the proposed method, the Marine Predator Algorithm Moth-Fame Optimization (MPAMFO), in order to avoid local optima. A dynamic routing approach that is modelled after Firefly Mating Optimization is described for internet connectivity in UWSNs-based event monitoring [1–5].

This article gives a thorough review of underwater communications and introduces cutting-edge models that use electromagnetic, optical, and acoustic technology. Discussions also touch on CORONA's subsurface existence and water filtration [5–9].

A balanced energy consumption protocol dubbed S-BEAR, is introduced to increase the lifespan of UWA-SNs. The study also suggests the Hierarchical Adaptive Energy-efficient Clustering Routing (HAECR) method as a means of efficiently reducing energy consumption and extending UWSN lifetime. For feature selection, a unique dimension reduction technique is also given [10–14].

The Marine Predator Algorithm (MPA), a powerful tool for tackling multi-objective optimization problems, is introduced in this study in four different iterations. In order to perform clustering simulations on simulated and actual datasets, the study suggests an MPA-PSO algorithm that incorporates a number of optimization strategies. These methods exhibit high performance in precisely determining unidentified proton exchange membrane fuel cells' (PEMFCs') characteristics for increased stack effectiveness [15–17].

The issues of latency and energy consumption in underwater sensor network (UWSN) routing are discussed in this research. The study also outlines continuing research being conducted by the industry on UWSNs with various underwater applications. It also examines difficulties in positioning equipment and underwater acoustic communication, highlighting the differences from wireless sensor networks used on land. The approach for synchronizing and positioning underwater sensors in a multi-hop environment is presented in the study, improving network connectivity and range and enabling more precise localization and synchronization techniques [18–20].

3 System Model

3.1 Network Model

The put-forward model introduced a three-dimensional network model, which is depicted in Fig. 1. Other information is as follows: Underwater acoustic sensors are randomly placed throughout the marine area. The underwater sensor units are responsible for gathering the data. Pressure, current flow, and temperature are the observed variables.

Underwater sensor nodes have acoustic modems so they may interact with other water medium nodes. N dynamic nodes are scattered sparsely and randomly throughout the network scenario's three-dimensional setting. $L \times L \times L$ (500*500*500).

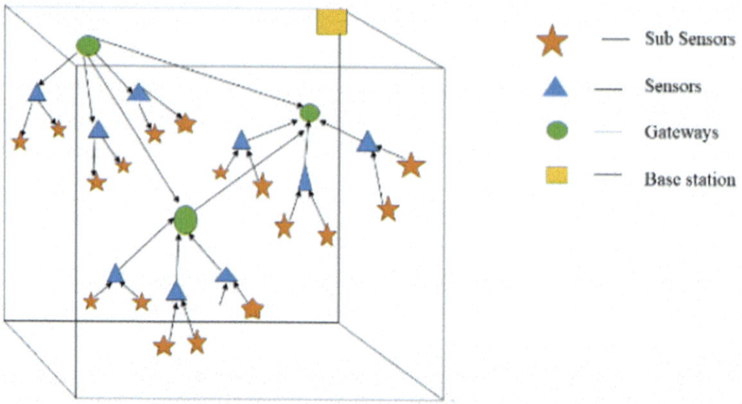

Fig. 1 A three-tier underwater wireless sensor network model with base station positioned at the corner

Data collection from underwater sensor nodes and transmitting it, the sink node or base station is connected to the base station on the coast, which is positioned in the center and corner of the surface. It contains RF and acoustic modems are both available.

Please note that the first paragraph of a section or subsection is not indented. The first paragraphs that follows a table, figure, equation etc. does not have an indent, either.

Subsequent paragraphs, however, are indented.

$$\mathrm{TL} = \mathrm{K_{TL}} * 10\mathrm{lgr} + \alpha_{\mathrm{TL}}\mathrm{r} * 10^{-3} \tag{1}$$

$$\alpha_{TL}(f) = 0.11\frac{f^2}{1+f^2} + 0.44\frac{f^2}{1+f^2} + 2.75 \times 10^{-4} + 0.003 \tag{2}$$

- The assimilation rate raises along with rate pf occurrence. UWSNs have a frequency of about 30 kHz and an absorption rate of less than 10 dB/km.
- $\mathrm{P_{tx}} = \mathrm{P_0} \times\times \mathrm{d}^2 \times\times 10^{(\alpha\mathrm{TL(f)}/10)}$ accordance to the energy consumption model displayed, can be used to define the minimal transmission power required at the transmitter to obtain power level $\mathrm{P_0}$ at the receiver.

4 Energy Model

- In this research, the underwater acoustic channel energy utilization model is used.
- The energy required to transport k bits of data at a transmission rate R across a distance d can be known as follows:

$$E_{Tx}(k, d) = k \times E_{elec} + \frac{K}{R} P_{tx} \quad (3)$$

where P_{Tx} is the transmitted power and E_{elec} = vitality required to direct 1 bit of information.

- The receiver radio energy consumption for k bits of data is given by $E_{Rx}(k) = kP_r$, where P_r is a constant that varies depending on the device.
- The energy required to combine k bits of data can be calculated using the formula:

$$E_{DA}(k) = k \times E_{DA0} \quad (4)$$

where E_{DA0} is the quantity of energy used to fuse one bit of data, or 5 nJ/bit.

5 Algorithm Used

5.1 *Random Deployment*

The random deployment algorithm is a fundamental technique used in operations research and computer science to randomly distribute resources like servers or work without taking any limits into account. Although it is straightforward and effective in terms of computation, it could not produce the best load balancing or resource allocation because it doesn't take server load or proximity into account. More advanced algorithms that are suited to certain constraints and optimization objectives are frequently selected in difficult scenarios [5, 8, 9].

5.2 *Least Distance Clustering*

The closest neighbor or least distance clustering algorithms, both variations of the single-linkage clustering method, group data points according to their proximity. Each piece of data is first viewed as an independent cluster. The next step of the method is to repeatedly combine the two nearest clusters until just one cluster is left. It measures proximity using a distance metric, such as Euclidean distance. Merging clusters that are more likely to be similar and have closer distances between any two sites is the main goal [10–13].

5.3 Energy Efficient Particle Swarm Optimization Inspired Algorithm

Particle Swarm Optimization (PSO) is an optimization technique that draws its inspiration from natural phenomena like flocks of birds and schools of fish. It seeks to select the best option from a variety of possibilities. Based on its position and velocity, each particle moves. Particle population is iteratively adjusted depending on the best-known places after starting at random [14, 15]. The particle motions are influenced by cognitive and social variables. PSO is simple to use and versatile, operating in both continuous and discrete search spaces.

5.4 Energy Efficient Grey Wolf Optimization Inspired Algorithm

The metaheuristic algorithm known as Grey Wolf Optimization (GWO) was influenced by how grey wolf's hunt. It is applied to optimization issues. The population of "wolves" in the algorithm's starting state represents potential answers. It goes through steps like initialization, encircling prey, hunting, and updating as it goes through iterative evolution. Depending on where the prey is, wolves change their postures. GWO effectively searches for and converges on optimal or nearly optimal solutions by striking a balance between exploration and exploitation. It has applications in engineering, image processing, neural networks, and feature selection. It is flexible and easy to implement [16–18].

5.5 Energy-Efficient Marine Predator Optimization Inspired Algorithm

To resolve optimization issues, the Marine Predator Optimization (MPO) method imitates the hunting strategies of oceanic predators. A group of solutions (the prey) are first given fitness values, after which they travel toward areas with higher fitness in pursuit of better solutions. Then predators are introduced, mimicking hunting techniques and focusing on the prey according to their locations. Over time, this equilibrium of exploration and exploitation produces better solutions. When a preset condition is satisfied, the algorithm terminates [19, 20].

6 Flowchart

First, a population of feasible fixes is generated and encoded. Next, an objective function is used to evaluate each solution's fitness. The algorithm's primary function is to simulate how marine predators hunt, with each individual moving around the search space. Fitness is reassessed following activity, and people are chosen for procreation based on their fitness. Through mechanisms like crossover and mutation, a new generation is created during reproduction. The method generates an ideal solution by iterating until the termination criteria are satisfied (Fig. 2).

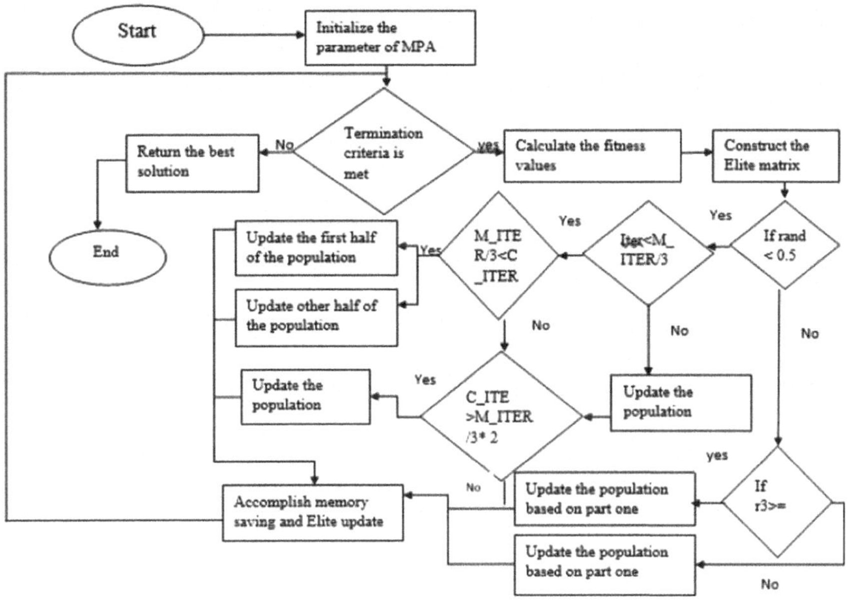

Fig. 2 Flowchart of EEGWOIA

7 Mathematical Model

1. Initialize search agents (Prey) Populations.
2. While termination criteria are not met.

2.1 $X_0 = X_{min} + rand\ (X_{max} - X_{min})$

X_{min} = Lower bound for variables

X_{max}=Upper bound for variables

Rand= uniform random vector in the range of 0to 1.

2.2 Generate random values for gateways between (0,1)→ X_i

2.3 Generate pop

Pop= Total no. of gateways * $\log_{10}$\{Xi+ Total no. of gateways\}

Round off pop to the nearest integer. After rounding off, select the P_{next}hop for each gateway, according to the position provided by pop.

2.4 Calculate fitness for each gateway.

for (i=1; i<= N_p; i++)

{fitness (k) = w_1* Max Dist. + w_2 *MaxHops}

Max Dist. = Distance between the selected gateway.

($P_{nexthop}$) to the base station.

Max Hops = Max {Hop Count(g_i)}

Hop Count = No. of Next Hops.

2.5 The fittest solution is selected as a top predator to build the Elite matrix.

$$\text{Elite}= \left[X'_{1,1}\ X'_{1,2}\ \cdots\cdots\ X'_{1,3}\ X'_{2,1}\ldots\ X'_{2,2}\ldots\ \cdots\cdots\ X'_{2,3}\ldots\ X'_{n,1}\ X'_{n,2}\cdots\ \ldots\ X'_{n,d}\right]$$

2.6 Predator update the position based upon the prey matrix

$$\text{Prey} =\left[X_{1,1}\ X_{1,2}\ \ldots\ldots X_{1,3}\ X_{2,1}\ldots X_{2,2}\ldots\ \ldots\ \ldots\ X_{2,3}\ldots X_{n,1}\ X_{n,2}\ \ldots\ \ldots\ X_{n,d}\right]$$

2.7 Marine Memory Saving

```
If (Iter ==0) {
fit -old =fitness;
Prey. Old = prey;
}
for (int i=0; i<Search Agents no; i++) {
for (int j=0; j<=dim; j++) {
If (fit_old[i]< fitness[i] {
Prey[i][j] = Prey_old [i][j];
}
fit_old = fitness;
Prey-old = prey;
```

3. 3.1)if (Iter<Max_Iter/3)

#Update prey using:

While Iter<1/3 Max_Iter

$\overrightarrow{Stepsize}_i = \vec{R}_B \otimes (\ \overrightarrow{Elite}_I - R_L \otimes \overrightarrow{Prey}_I)$ i = 1, -n/2

$\overrightarrow{Prey}_i = \overrightarrow{Prey}_i + P.\vec{R} \otimes \overrightarrow{stepsize}_i$

3.2) Elseif

(Max_Iter/3<3<Iter<2*Max_Iter/3)

for the first half of the population update prey using:

While 1/3 Max_iter<Iter< 2/3 Max_iter

$\overrightarrow{Stepsize_i} = RL \otimes (\overrightarrow{Elite_i} - \vec{R}_L \otimes \overrightarrow{Prey_i})$ i = 1, ------n/2

$\overrightarrow{Prey_i} = \overrightarrow{Prey_i} + P.\ \vec{R} \otimes \overrightarrow{Stepsize_i}$

For the 2nd half of the population update prey using: While 1/3 Max_Iter<Iter< 2/3 Max_Iter

$\overrightarrow{Stepsize_i} = \vec{R}_B \otimes (\vec{R}_B \otimes \overrightarrow{Elite_i} - \overrightarrow{Prey_i})$ i = 1, -- n/2

$\overrightarrow{Prey_i} = \overrightarrow{Elite_i} + P.CF \otimes \overrightarrow{Stepsize_i}$

$$C.F. = \left(1 - \frac{Iter}{Max\_Iter}\right)\left(\frac{2\ Iter}{Max\_Iter}\right)$$

3.3) Else if (Iter> 2 * Max_Iter/3

while Iter> 2/3 Max_Iter

$\overrightarrow{Stepsize_i} = \vec{R}L \otimes (\vec{R}L\ \overrightarrow{Elite_i} - \overrightarrow{Prey_i})$ i= 1------n

$\overrightarrow{Prey_i} = \overrightarrow{Elite_i} + P.CF \otimes \overrightarrow{Stepsize_i}$

4.Eddy formation and FAD's effect

$$\overrightarrow{Prey_i} = \{\overrightarrow{Prey_i} + CF[\vec{X}_{min} + \vec{R}_1 * (\vec{X}_{max} - \vec{X}_{min1}) * \underline{\mu}, \quad if\ r \le FADS\ \overrightarrow{Prey_i} + [FADS(1-r) + r](\overrightarrow{Prey_{r1}} - \overrightarrow{Prey_{r2}}), if\ r > FADS$$

FADS=.2

μ= Binary vector.

8 Result and Discussion

The comparison of the applied random, LDC, EEPSOIA, EEGWOIA, and EEMPOIA in accordance with the number of rounds vs. residual energy is shown in Figs. 3, 4 shows comparison between the applied random, LDC, EEPSOIA, EEGWOIA, EEGWOIA in accordance with the no. of nodes vs residual energy. Figure 5 shows comparison between random, LDC, EEPSOIA, EEGWOIA, EEMPOIA in accordance with the mean network life time vs. no. of gateways. Figure 6 shows comparison between random, LDC, EEPSOIA, EEGWOIA, EEMPOIA in accordance with mean network life time vs. no. of gateways. In this comparison proposed EEMPOIA perform better.

9 Conclusion and Future Scope

Through comprehensive simulations within a three-tier UWSN model, our findings have unequivocally demonstrated the supremacy of EEMPOIA. This algorithm has exhibited remarkable capabilities in significantly extending the network's longevity while concurrently reducing energy consumption. The results underscore EEMPOIA's immense potential as a promising solution for enhancing UWSN performance. The synergy of these innovative algorithms opens up new horizons for the

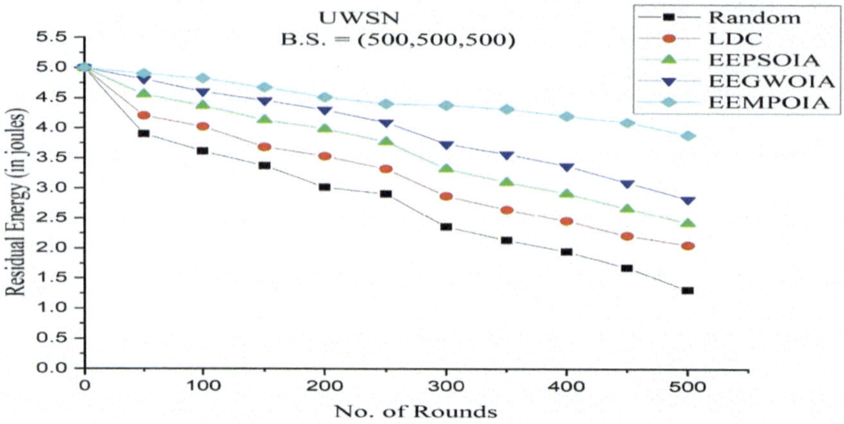

Fig. 3 Residual Energy versus No. of Rounds in UWSN with Sensor node 200

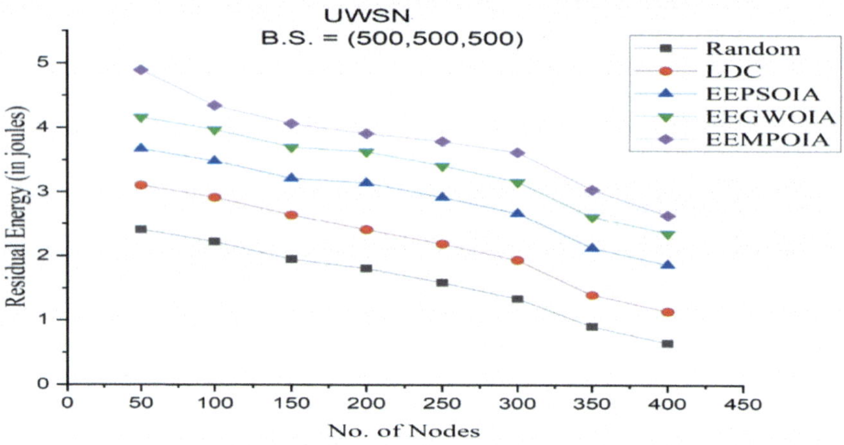

Fig. 4 Residual Energy versus No. of Nodes in UWSN with Sensor node 500

deployment of long-lasting and energy-efficient underwater sensing systems. This breakthrough has profound implications across various industrial and environmental domains, offering unprecedented opportunities to harness the power of UWSNs for improved data collection and monitoring in challenging aquatic settings. As the demand for underwater sensing systems continues to grow, the contributions of EEMPOIA and its counterparts hold promise for advancing the field and enabling sustainable solutions for critical underwater applications.

In future further advancements can be done by exploring multi-modal communication, dynamic network topology management, and localized processing within nodes can contribute to more resilient and efficient underwater networks.

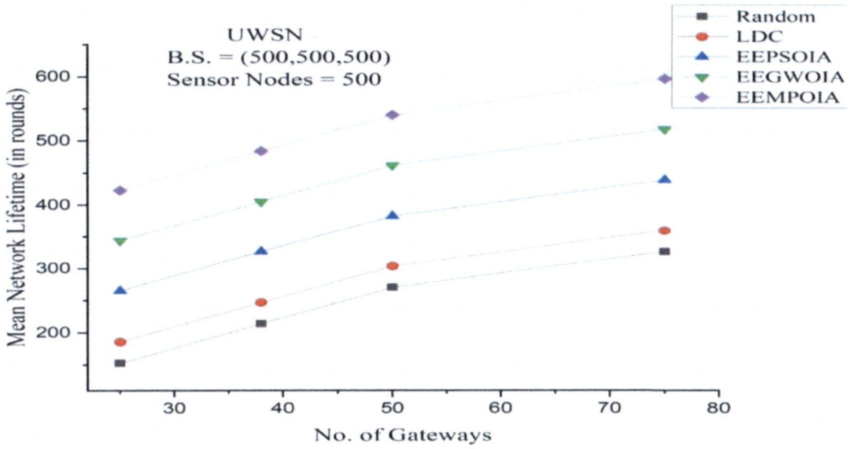

Fig. 5 Mean Network Lifetime versus No. of Gateways in UWSN

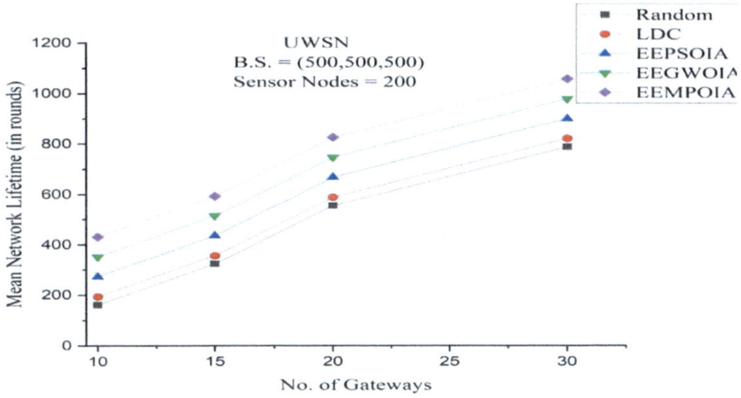

Fig. 6 Mean Network Lifetime versus No. of Gateways in UWSN

References

1. Elminaam, D.S.A., Nabil, A., Ibraheem, S.A., Houssein, E.H.: An efficient marine predators algorithm for feature selection. In IEEE Access **9**, 60136–60153 (2021). https://doi.org/10.1109/ACCESS.2021.3073261
2. Wang, N., Wang, J.S., Zhu, L.F., Wang, H.Y., Wang, G.: A novel dynamic clustering method by integrating marine predators algorithm and particle swarm optimization algorithm. In IEEE Access **9**, 3557–3569 (2021). https://doi.org/10.1109/ACCESS.2020.3047819
3. Diab, A.A.Z., Tolba, M.A., El-Magd, A.G.A., Zaky, M.M., El-Rifaie, A.M.: Fuel cell parameters estimation via marine predators and political optimizers. In IEEE Access **8**, 166998–167018 (2020). https://doi.org/10.1109/ACCESS.2020.3021754
4. Abdel-Basset, M., Mohamed, R., Mirjalili, S., Chakrabortty, R.K., Ryan, M.: An efficient marine predators algorithm for solving multi-objective optimization problems: analysis and

validations. In IEEE Access **9**, 42817–42844 (2021). https://doi.org/10.1109/ACCESS.2021.3066323
5. Al-Betar, M.A., Awadallah, M.A., Makhadmeh, S.N.: Marine predators algorithm: a review. Arch. Computat Methods **30**, 3405–3435 (2023). https://doi.org/10.1007/s11831-023-09912-1
6. Rai, R., Dhal, K.G., Das, A.: An inclusive survey on marine predators algorithm: variants and applications. Arch. Computat Methods **30**, 3133–3172 (2023). https://doi.org/10.1007/s11831-023-09897-x
7. Jouhari, M., Ibrahimi, K., Tembine, H., Ben-Othman, J.: Underwater wireless sensor networks: a survey on enabling technologies, localization protocols, and internet of underwater things. In IEEE Access **7**, 96879–96899 (2019). https://doi.org/10.1109/ACCESS.2019.2928876
8. Olatinwo, S.O., Joubert, T.H.: Energy efficient solutions in wireless sensor systems for water quality monitoring: a review. In IEEE Sens. J. **19**(5), 1596–1625 (2019). https://doi.org/10.1109/JSEN.2018.2882424
9. Soliman, M.A., Hasanien, H.M., Alkuhayli, A.: Marine predators algorithm for parameters identification of triple-diode photovoltaic models. In IEEE Access **8**, 155832–155842 (2020). https://doi.org/10.1109/ACCESS.2020.3019244
10. Aly, M., Ahmed, E.M., Rezk, H., Mohamed, E.A.: Marine predators algorithm optimized reduced sensor fuzzy-logic based maximum power point tracking of fuel cell-battery standalone applications. In IEEE Access **9**, 27987–28000 (2021). https://doi.org/10.1109/ACCESS.2021.3058610
11. Qin, C., Han, B.: A novel hybrid quantum particle swarm optimization with marine predators for engineering design problems. In IEEE Access **10**, 129322–129343 (2022). https://doi.org/10.1109/ACCESS.2022.3226813
12. Elaziz, M. A.: An improved marine predators algorithm with fuzzy entropy for multi-level thresholding: real world example of COVID-19 CT image segmentation. In 2020 IEEE Access, 8, 125306–125330 (2020). https://doi.org/10.1109/ACCESS.2020.3007928
13. Wu, J., Sun, X., Wu, J., Han, G.: Routing strategy of reducing energy consumption for underwater data collection. In Intelligent and Converged Networks, **2**(3), 163–176 (2021). https://doi.org/10.23919/ICN.2021.0012
14. Omeke, K. G.: DEKCS: A Dynamic Clustering Protocol to Prolong Underwater Sensor Networks. In 2021 IEEE Sens. J., **21**(7), 9457–9464 (2021). https://doi.org/10.1109/JSEN.2021.3054943
15. Faheem, M.: FFRP: dynamic firefly mating optimization inspired energy efficient routing protocol for internet of underwater wireless sensor networks. In 2020 IEEE Access, 8, 39587–39604 (2020). https://doi.org/10.1109/ACCESS.2020.2976105
16. Jahanbakht, M., Xiang, W., Hanzo, L., Rahimi, Azghadi, M.: Internet of underwater things and big marine data analytics—a comprehensive survey. In 2021 IEEE Communications Surveys & Tutorials, **23**(2), 904–956 (2021). https://doi.org/10.1109/COMST.2021.3053118
17. Gupta, O., Goyal, N., Anand, D., Kadry, S., Nam, Y., Singh, A.: Underwater networked wireless sensor data collection for computational intelligence techniques: issues, challenges, and approaches. In IEEE Access **8**, 122959–122974 (2020). https://doi.org/10.1109/ACCESS.2020.3007502
18. Wang, M., Chen, Y., Sun, X., Xiao, F., Xu, X.: Node energy consumption balanced multi-hop transmission for underwater acoustic sensor networks based on clustering algorithm. In 2020 IEEE Access, 8, 191231–191241 (2020). https://doi.org/10.1109/ACCESS.2020.3032019
19. Li, J.: Information-Centric wireless sensor networking scheme with water-depth-awareness content caching for underwater IoT. In 2022 IEEE Internet Things J., **9**(2), 858–867 (2022). https://doi.org/10.1109/JIOT.2021.3058272
20. Li, X., Hu, X., Zhang, R., Yang, L.: Routing protocol design for underwater optical wireless sensor networks: a multiagent reinforcement learning approach. In 2020 IEEE Internet Things J., **7**(10), 9805–9818 (2020). https://doi.org/10.1109/JIOT.2020.2989924

Streamlined Speech Recognition Model for Automated Radiology Reporting Employing Combined Automatic Speech Recognition Model, Large Language Model, and Prompt Engineering

Manju Dabass, Mohammed M. Jabeer, Anuj Chandalia, and Dwarikanath Mahapatra

Abstract As radiologists around the globe contend with ever-growing workloads, the implementation of automated report generation presents a dual advantage: easing their burden and improving turnaround times. Empowered by Artificial Intelligence (AI)-driven speech recognition, our proposed method minimizes dictation errors, enhancing precision in report generation and thereby reducing the need for post-dictation corrections. In our research, we aim to pinpoint the most effective end-to-end Dictation-to-Report Pipeline. This entails assessing the performance of various combinations of NLP models (Facebook Word2vec Whisper, Facebook Word2letter, Kaldi, DeepSpeech), LLM models (Mistral-7B, Falcon-7B, Zephyr-7B, Qwen-14B), and Prompt Engineering for speech recognition and subsequent report generation. The achieved results include a Match Error Rate of 0.016, Word Error Rate of 0.131, Average Levenshtein Distance of 0.109, and Sentence Error Rate of 0.019. Our conclusions are further substantiated through a blinded qualitative evaluation by five radiologists, who assign validation scores to the generated radiology reports.

Keywords Medical report generation · Automated speech recognition · Natural language processing models · Large language models · Prompt engineering

M. Dabass (✉) · M. M. Jabeer · A. Chandalia
AI Research and Development Department, Manentia AI Advisory Private Limited., Gujarat, India
e-mail: manjurashi87@gmail.com

D. Mahapatra
AI Research Department, Inception Institute of AI, Dubai, UAE

G. Raj et al. (eds.), *Advances in Artificial Intelligence and Machine Learning*, Lecture Notes in Electrical Engineering 1264, https://doi.org/10.1007/978-981-97-9507-9_26

1 Introduction

Radiologists encounter numerous challenges in manual report writing, impacting efficiency, accuracy, and overall report quality [1]. Manual processes are time-consuming, leading to burnout in busy departments. Inconsistencies, typographical errors, and the risk of omitting crucial information are common. Manual reports may lack structured data elements, hindering research and data-driven healthcare improvements. Additionally, they face challenges in accessibility, remote collaboration, and integration with electronic health records (EHRs), especially in multilingual settings.

To address these issues, our research focuses on Automated Radiology Report Generation using Automated Speech Recognition (ASR) models. ASR swiftly transcribes spoken findings, reducing time requirements and ensuring consistent, standardized reports. These automated systems minimize human error, reduce risks of typographical errors, and facilitate data extraction and integration with EHRs. ASR-generated reports can be structured for efficient data analysis and retrieval, transcending language barriers. Scalable and adaptable, these systems allow radiologists to concentrate on quality control, leading to faster turnaround times and improved patient care decisions.

Despite significant progress, challenges persist, including the complexity of medical terminology, the need for clinical context understanding, the critical role in patient care decisions, and variability in healthcare data quality and format. Augmenting ASR outputs with Large Language Models (LLMs) has shown promise, yet LLMs have limitations such as hallucinations, data poisoning risks, and toxic language generation. Prompt engineering emerges as a crucial tool to overcome these limitations, empowering NLP models for accurate, relevant, and tailored radiology reports.

In our research, we strive to develop an automated model for generating radiology reports by effectively combining ASR, LLM models, and prompt engineering. This endeavor aims to bridge the gap between NLP technology capabilities and the unique requirements of radiology report generation, offering an efficient and tailored solution.

2 Literature Review

The field of Automated Radiology Report Generation has recently garnered significant attention within the research community, witnessing the introduction of various innovative techniques. Notably, in [2], the Multimodal Recursive model with Contrastive Learning (MRCL) demonstrated the integration of visual and semantic features for generating "Impression" and "Findings" in radiology reports. This model introduced a contrastive pre-training method for recursive networks, enhancing the expressiveness of both visual and textual representations. Similarly, [3] showcased the Medical Report Generation using the Image Segmentation (MERGIS) approach, employing a modern transformer-based encoder-decoder model. Image segmentation was leveraged to input data before model feeding, enabling the capture of relevant visual features from medical images and improving the accuracy of radiography reports.

In [4], a novel multi-modal transformer network was presented, synthesizing patient-specific full-text radiology reports by integrating medical images and patient demographic information. The model utilized a convolutional neural network for extracting visual features and a transformer-based encoder-decoder network for combining these features with semantic text embeddings of patient demographic information. Recognizing the correlation between details in radiology reports and specific information in X-ray images, the model introduced in [5] featured two distinct modules: a Learned Knowledge Base and Multi-modal Alignment. Additionally, [6] introduced PromptRRG, utilizing prompt learning to activate a pre-trained model and incorporating prior knowledge. This involved investigating various prompt designs categorized by levels of knowledge and an automatic prompt learning mechanism to alleviate manual prompt engineering.

Prompt engineering's developmental overview was presented in [7], highlighting its substantial contributions to healthcare NLP applications. As large language models continuously evolve, the importance of prompt engineering in healthcare becomes increasingly evident. In [8], a method for automatically generating medical prompts was introduced to enhance knowledge transfer from pre-trained vision-language models to medical object detection. [9] suggested embedding domain information into prompts for unsupervised domain adaptation, enabling the model to learn domain-specific features.

Building on the successes of these models, our study proposes an automated radiology report generation model, experimenting with various NLP models of ASR and LLM, followed by a Prompt Engineering Step.

3 Methods

3.1 Dataset

In this study, we utilize the Indiana University chest X-ray dataset [10] available at [11]. The dataset consists of 8121 front and side chest X-ray images from 3996 patients, each accompanied by their respective reports. For our research, we focus exclusively on the patients' reports, which are transformed into MP3 audio files using the "TextAloud" software—a proficient Speech-to-Text conversion tool recognized for its ability to handle complex medical terminology. The dictated reports typically had a duration of approximately thirty seconds.

3.2 Pre-processing Step

Effective audio pre-processing is a crucial but often underestimated aspect of the Automatic Speech Recognition (ASR) inference process. This intricate procedure involves several stages to ensure compatibility with diverse ASR models. These stages include converting the audio to the.wav format, adjusting the sample rate as needed, segmenting it into predefined chunks, extracting acoustic characteristics through Log-Mel spectrograms, and subsequently organizing these chunks into batches for inference. The ASR models have been exclusively trained and tested on pre-segmented data, specifically short 'clips' of audio. Consequently, there is no established inference procedure for their application to long-form audio. To address this challenge in our experimental assessments, the audio undergoes transcoding to s16 PCM at a 16 kHz sampling rate, followed by segmentation into non-overlapping 30-s intervals. Inference is then carried out on these batched chunks using the HuggingFace tooling.

3.3 Automatic Speech-to-Text Recognition (ASR) Step

In this study, we conducted a comparison among five NLP models—Whisper, Deep-Speech, Facebook Word2Vec, Facebook Word2letter, and Kaldi—for the task of transcribing dictated audio files into cohesive text. The choice of these models was influenced by factors such as accessibility, versatility, recognition proficiency, and their widespread use in various speech processing tasks. The subsequent sections offer detailed descriptions of these selected models:

(i) **Facebook Word2Vec** [12]: Here, we utilize a specific model version identified as wav2vec2-largerobust-ft-libri-960 h, detailed in [13]. This model is a CTC encoder, achieved by fine-tuning the wav2vec 2.0 base model on LibriSpeech. The fine-tuning process involves utilizing 960 h of data and implementing CTC loss. Noteworthy features of this model include a transformer encoder stack with 'large-capacity,' consisting of 24 blocks with a hidden size of 1024, 16 attention heads, and a feed-forward dimension of 4096. It's important to note that this configuration comes with substantial GPU memory requirements. Additionally, the model operates with a character-based vocabulary.

(ii) **Whisper** [14]: The architecture of this model comprises a stack of 2D CNNs followed by a symmetric transformer encoder/decoder stack. Utilizing standard features, the model processes 80-dimensional log-mel filter bank features derived from audio transcoded to 16 kHz. Through supervised training on an extensive corpus containing 680,000 h of crawled, multilingual speech data, the model incorporates timestamp tokens in its vocabulary. These tokens are strategically inserted within the training text. In the inference phase, the model operates sequentially on 30-s audio windows. Each audio window undergoes embedding with the encoder and is then mapped to a predicted text sequence through auto-regression by the decoder. The encoder output serves as a context vector. The model retains the predicted text up to and including the last timestamp token, discarding the remainder. The audio window then advances to the location associated with the last timestamp, and the process repeats, incorporating the previous chunk's predicted text as additional context.

(iii) **Facebook Wav2letter** [15]: It is an open source library developed by Facebook. It is now ported to a new open source library called Flashlight. It is completely built on convolutional neural networks (CNNs) for acoustic modelling as well as language modelling.

(iv) **Kaldi** [16]**:** Kaldi represents a conventional 'pipeline' ASR model, characterized by distinct sub-models that operate sequentially. In our proposed research, we make use of the Kaldi Gigaspeech XL [17] model, which has been trained on the Gigaspeech dataset comprising 10,000 h of labelled speech data. This model demonstrates competitive performance, particularly in in-domain evaluations on Gigaspeech, when compared to end-to-end approaches. Due to limited accessibility, we only have the acoustic model weights, and for the remaining components (an ivector embedder and an RNN language model), we borrow from the Kaldi LibriSpeech pipeline to facilitate its implementation.

(v) **DeepSpeech** [18]**:** This is a character-level, deep recurrent neural network designed for end-to-end training, and it is open-source. With a substantial 120 million parameters, the model employs character-level embeddings and utilizes a greedy decoding strategy with a Connectionist Temporal Classification (CTC) decoder. At its essence, the model incorporates a recurrent neural network (RNN) that is specifically trained to process speech spectrograms, enabling precise text transcriptions.

3.4 Post-processing Step Applied on the Generated ASR Output

As the ASR model functions on raw audio waveforms, the lengths of input sequences are considerably extended (e.g., 30-s chunks of 16 kHz audio spanning 480,000 time steps). This, combined with the model's substantial capacity, poses challenges for GPU inference, potentially due to memory limitations. To address these issues, inference is conducted in half-precision mode with a reduced batch size of one. The decoding of inference results from individual chunks is performed separately using the model's tokenizer. Subsequently, the resulting text from each chunk is concatenated to generate a comprehensive prediction for the entire file.

3.5 Large Language Models

Subsequently, the output of the leading ASR model was subjected to analysis using four Large Language Models (LLMs): Mistral-7B, Falcon-7B, Zephyr-7B, and Qwen-14B. The choice of these models was informed by their recognition as state-of-the-art models, having recently demonstrated comparable performance in medical domain applications.

i. **Mistral-7B** [19]**:** In our study, we investigate Mistral 7B, a potent open-source language model with 7.3 billion parameters. This model utilizes Grouped-query Attention (GQA) to expedite inference, showcasing superior performance in comparison to standard full attention mechanisms. Furthermore, Mistral 7B leverages Sliding Window Attention (SWA), enhancing its ability to process extensive text sequences efficiently while maintaining a low computational cost.
ii. **Falcon-7B** [20]**:** Within the scope of our research, we scrutinize Falcon-7B, a model with a causal decoder-only design and an impressive 7 billion parameters, developed by TII. Trained on a substantial dataset of 1,500 billion tokens from RefinedWeb, bolstered with curated corpora, this model is specifically crafted for efficient inference. Its architecture incorporates FlashAttention and multiQuery mechanisms. For optimal performance during inference, it is recommended to have a minimum of 16 GB of memory when running Falcon-7B. The model utilizes rotary positional embeddings and employs a single layer norm in the parallel attention/MLP structure within the Decoder-block.
iii. **Zephyr-7B** [21]**:** Here, we explore ZEPHYR-7B, a distinguished member of the growing category of large language models (LLMs) that has gained substantial recognition in the AI community. Developed by Hugging Face, this model is an improved version of Mistral-7B, fine-tuned on public datasets, and further refined through the implementation of knowledge distillation techniques
iv. **Qwen-14B** [22]**:** Within the framework of our research, we explore Tongyi Qianwen-14B (Qwen-14B), a robust model with 14 billion parameters belonging to the Tongyi Qianwen large model series developed by Alibaba Cloud. This expansive language model, based on Transformer architecture, undergoes training on an exceptionally vast pre-training dataset that includes over 3 trillion tokens. It employs a vocabulary that encompasses more than 150,000 tokens.

3.6 Prompt Engineering Step

Prompt engineering is a technique in natural language processing (NLP) and machine learning where the creation of input prompts or queries is carefully designed to elicit specific responses or behaviors from NLP models [23]. It entails formulating questions, instructions, or requests in a manner that influences the model's output. This often includes a combination of n-grams, such as uni-grams, bi-grams, and tri-grams, to construct effective prompts that achieve desired outcomes, ranging from content generation to information retrieval. N-grams are continuous sequences of n items (words, characters, or tokens) in a text. Adjusting the n-gram level in a prompt allows control over the level of detail and specificity in the model's response [24].

In our study, we extracted N-grams that appeared more than 50 times from a randomly selected set of medical reports. These N-grams were then utilized for prompt engineering through a three-step process, involving the integration of unigram, bi-gram, and tri-gram terms within the context and prompts. Furthermore, we employed the resulting prompts and connected them in successive steps to enhance our results.

Applied Prompt: "Upon receiving an oral transcript detailing observations from a radiology report, your responsibility is to identify and correct any inaccuracies, inconsistencies, or ambiguities to maintain the highest clinical accuracy. The transcript might include medical terminology, abbreviations, and complex sentence structures commonly found in radiology reports. Your goal is to produce a corrected version that retains the original meaning while adhering to medical standards and best practices."

After the initial prompt, the subsequent step in prompt engineering involved the inclusion of additional prompts:

Step 1: Using the prompt templates provided by LangChain, the list of extracted unigrams is incorporated and presented to the model along with the applied prompt.

Step 2: Employing the prompt templates offered by LangChain, the list of extracted unigrams and bi-grams is integrated and passed to the model alongside the applied prompt.

Step 3: Utilizing the prompt templates available in LangChain, the list of extracted unigrams, bi-grams, and trigrams is integrated and presented to the model in conjunction with the applied prompt.

4 Experimental Results

4.1 Applied Performance Evaluation Metrics

Here, common metrics including Match Error Rate(MER), Word Error Rate(WER), average Levenshtein Distance (LD), and the Sentence Error Rate(SER) were used for comparing their performances. These metrics are elucidated below:

Match Error Rate (MER): MER is commonly applied in machine translation to assess the likeness between the reference translation and the translation generated by the system. It measures the count of operations (substitutions, insertions, and deletions) needed to convert one sentence into another. A lower MER value signifies greater similarity between the translations.

Word Error Rate (WER): WER serves as a metric for assessing the precision of NLP models. It calculates the percentage of words in the reference (ground truth) transcription that are not accurately recognized in the output produced by the NLP model. In the realm of Automatic Speech Recognition problems, a lower WER signifies greater accuracy in transcribing spoken language.

Average Levenshtein Distance (LD): The Levenshtein Distance serves as a gauge of similarity between two strings. In the field of Natural Language Processing (NLP), it finds application in assessing the similarity between a text generated by a system and a reference text. The average Levenshtein Distance computes the average edit distance between corresponding words or characters in the two texts. A lower average LD implies a higher degree of similarity.

Sentence Error Rate (SER): SER is an extension of WER designed for full sentences or transcripts. It gauges the percentage of sentences that the NLP model inaccurately transcribes. SER proves valuable in evaluating the overall quality of the NLP system's output in tasks involving spoken language.

4.2 Quantitative Output Results

The resulting performance metrics of the proposed model using the applied NLP and LLM models along with prompt engineering are enumerated in Table 1 given below. The five NLP models were first applied to generate the report as shown in Table 1 and among them, the Whisper model with outcomes of (MER = 0.095; WER = 0.198; ALD = 0.385; and SER = 0.129) outperformed other ASR models. These results were not significant from the radiology side as the generated reports had many errors. Hence, four LLMs were applied to the output of each of the ASR models to improve these results. The performance improvement results are shown in Table 2. It can be clearly seen form this table that incorporation of LLM models has lowered the four-performance metrics value by at least 60% which is quite a significant improvement. Among different combinations, Falcon 7B with Whisper ASR model provided the best performance of (MER = 0.082; WER = 0.173; ALD = 0.325; and SER = 0.109). To further improve the performance of the generated reports, the prompt engineering is applied. The results are enumerated in Table 3 showing best performance as (MER = 0.016; WER = 0.131; ALD = 0.109; and SER = 0.019). Also, Fig. 1 is showing the output performance metrics of different applied ASR models. Figure 2 is showing the performance improvement in WER, SER, MER, and LD of the generated radiology reports after the application of LLM models. The Fig. 3 shows the performance enhancement after application of prompt engineering step.

Radiologist Validation

Here, a blinded qualitative assessment of the output reports was conducted by five radiologists, highlighting the considerable influence of incorporating LLM models

Table 1 Comparative Performance Evaluation for the applied ASR models

Applied ASR models	MER	WER	ALD	SER
Facebook Word2vec [13]	0.165	0.314	0.652	0.191
Whisper [14]	0.095	0.198	0.385	0.129
Facebook Word2letter [15]	0.149	0.269	0.567	0.172
Kaldi [16]	0.134	0.235	0.493	0.146
DeepSpeech [18]	0.152	0.251	0.441	0.185

Table 2 Resulted performance metrices for the applied LLMs on the best ASR model

Applied LLM models	MER	WER	ALD	SER
Mistral-7B [19]	0.091	0.184	0.352	0.118
Falcon-7B [20]	0.082	0.173	0.325	0.109
ZEPHYR-7B [21]	0.131	0.236	0.414	0.163
Qwen-14B [22]	0.112	0.201	0.386	0.139

Table 3 Resulted Performance Metrices for the applied Prompt Engineering stages [23] on the best LLM

Applied model	MER	WER	ALD	SER
Baseline Prompt	0.027	0.154	0.265	0.032
Stage 1	0.025	0.147	0.235	0.029
Stage 2	0.022	0.141	0.198	0.025
Stage 3	0.016	0.131	0.109	0.019

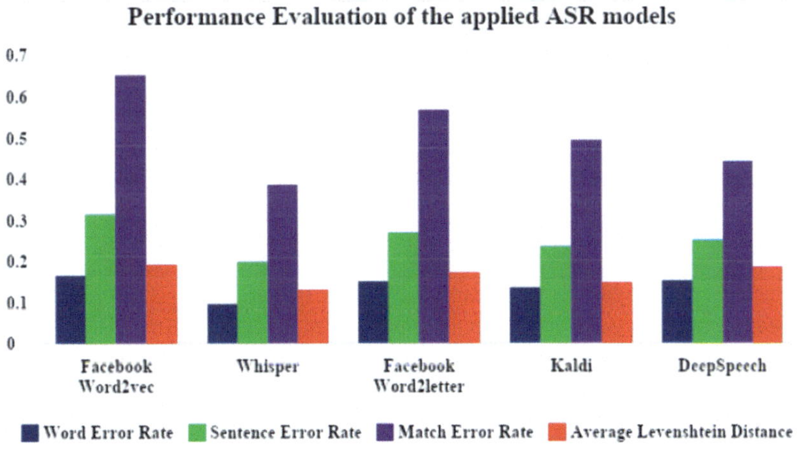

Fig. 1 Performance metrics of different applied ASR Models

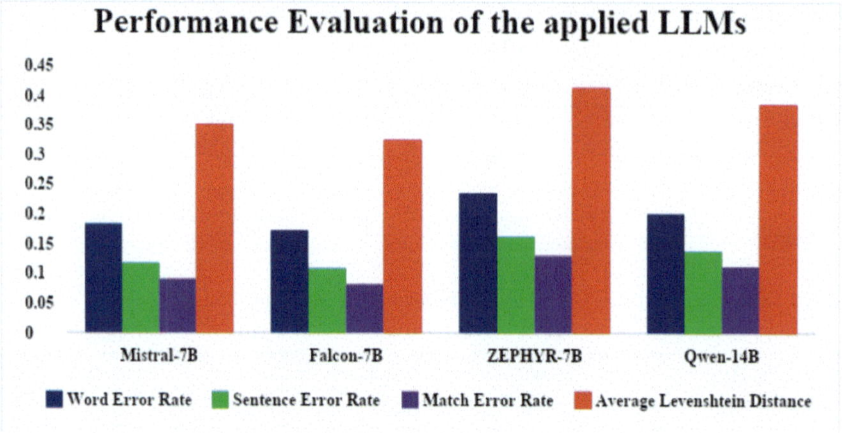

Fig. 2 Performance metrics of different applied LLMs

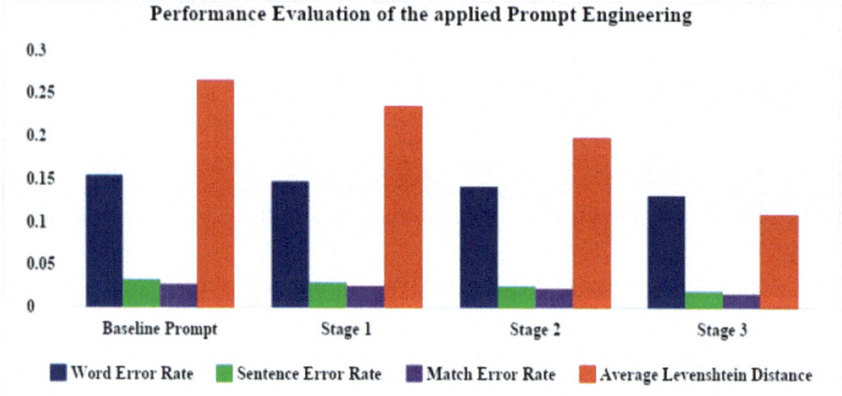

Fig. 3 Performance metrics of different applied prompt engineering stages on the best LLM

and Prompt Engineering in proposed ASR-based report generation. The radiologists were tasked with assigning a validation score within the range of 1 to 5. By computing the mean or average value of all the assigned scores, it was confirmed that the most effective model is the combined Whisper, Falcon-7B, and Prompt Engineering Stage 3. The distinct mean scores are detailed in Table 4.

Table 4 Average Radiologists' validation score for generated radiology reports

Applied model	Whisper [14]	Falcon-7B [20]	Prompt engineering stage 3 [23]
Score	3.65	4.23	4.82

5 Conclusion

In the proposed research work, the goal was to discover an NLP-based automated model for generating radiology reports. Initially, five distinct ASR models were employed to convert audio files into radiology report text. Among these ASR models, the Whisper model exhibited the most impressive performance. Subsequently, four different LLM models were applied to the generated text outputs from the Whisper models, with the Falcon-7B model demonstrating superior performance. The introduction of a second layer of LLM to process the output of a Speech-to-Text NLP model proved to significantly reduce errors and enhance the performance of the speech recognition pipeline for radiological dictation. Given the imperative for utmost precision in radiology reports, prompt engineering was implemented to further improve the performance of the generated text reports. This step serves to showcase that Augmented Speech Recognition Models can markedly enhance the accuracy of radiological dictations, minimize the need for post-dictation error correction, and expedite reporting turnaround times, ultimately contributing to saving more patient lives. The proposed automated report generation model has the potential to be integrated into radiology workflows, enabling radiologists to simply dictate and have the report generated. This streamlined process alleviates the tedium, cumbersomeness, and time-consuming nature of manual typing. Time saved by radiologists can then be redirected toward analysing complex cases. Although significant performance is achieved by the proposed model but still its performance is limited not to be used in the real-life radiology. Future research works will involve refining the performance of the proposed model through the exploration of different combinations of ASR models, LLMs, and further advancements in prompt engineering techniques.

Acknowledgements All the authors are highly appreciative and grateful to all the radiologist of a Super Speciality Cancer Hospital of India.

References

1. Canon, C.L., Chick, J.F., DeQuesada, I., Gunderman, R.B., Hoven, N., Prosper, A.E.: Physician burnout in radiology: Perspectives from the field. Am. J. Roentgenol. 218(2):369.9–4 (2022). https://doi.org/10.2214/AJR.21.26756
2. Wu, X., Li, J., Wang, J., Qian, Q.: Multimodal contrastive learning for radiology report generation. J. Ambient. Intell. Humaniz. Comput. **14**(8), 11185–11194 (2023)
3. Nimalsiri, W., Hennayake, M., Rathnayake, K., Ambegoda, T.D., Meedeniya, D.: Automated radiology report generation using transformers. In: 2023 3rd International Conference on Advanced Research in Computing (ICARC). (pp. 90–95). IEEE. (2023)
4. Aksoy, N., Ravikumar, N., Frangi, A.F.: Radiology report generation using transformers conditioned with non-imaging data. In: Medical Imaging 2023: Imaging Informatics for Healthcare, Research, and Applications. SPIE. **12469**, pp. 146–154 (2023).
5. Yang, S., Wu, X., Ge, S., Zheng, Z., Zhou, S.K., Xiao, L.: Radiology report generation with a learned knowledge base and multi-modal alignment. Med. Image Anal. **1**(86), 102798 (2023)

6. Wang, J., Zhu, L., Bhalerao, A., He, Y.: Can prompt learning benefit radiology report generation?. (2023). arXiv preprint arXiv:2308.16269
7. Wang, J., Shi, E., Yu, S., Wu, Z., Ma, C., Dai, H., Yang, Q., Kang, Y., Wu, J., Hu, H., Yue, C.: Prompt engineering for healthcare: Methodologies and applications. (2023). arXiv preprint arXiv:2304.14670
8. Qin, Z., Yi, H,H., Lao, Q., Li, K.: Medical image understanding with pretrained vision language models: a comprehensive study. In: International Conference on Learning Representations. (2023)
9. Ge, C., Huang, R., Xie, M., Lai, Z., Song, S., Li, S., Huang, G.: Domain adaptation via prompt learning. (2022). arXiv preprint arXiv:2202.06687
10. Demner-Fushman, D., Kohli, M.D., Rosenman, M.B., Shooshan, S.E., Rodriguez, L., Antani, S., Thoma, G.R., McDonald, C.J.: Preparing a collection of radiology examinations for distribution and retrieval. J. Am. Med. Inform. Assoc. **23**(2), 304–310 (2016)
11. https://www.kaggle.com/datasets/raddar/chest-xrays-indiana-university
12. Baevski, A., Zhou, Y., Mohamed, A., Auli, M.: wav2vec 2.0: A framework for self-supervised learning of speech representations. Adv. Neural Inf. Process. Syst. **33**:12449–60 (2020)
13. Hsu, W.N., Sriram, A., Baevski, A., Likhomanenko, T., Xu, Q., Pratap, V., Kahn, J., Lee, A., Collobert, R., Synnaeve, G., Auli, M.: Robust wav2vec 2.0: Analyzing domain shift in self-supervised pre-training. (2021). arXiv preprint arXiv:2104.01027
14. Radford, A., Kim, J.W., Xu, T., Brockman, G., McLeavey, C., Sutskever, I.: Robust speech recognition via large-scale weak supervision. In: International Conference on Machine Learning. PMLR. pp. 28492–28518 (2023)
15. Pratap, V., Hannun, A., Xu, Q., Cai, J., Kahn, J., Synnaeve, G., Liptchinsky, V., Collobert, R.: wav2letter++: The fastest open-source speech recognition system. arxiv 2018. arXiv preprint arXiv:1812.07625
16. Povey, D., Ghoshal, A., Boulianne, G., Burget, L., Glembek, O., Goel, N., Hannemann, M., Motlicek, P., Qian, Y., Schwarz, P., Silovsky, J.: The Kaldi speech recognition toolkit. In: IEEE 2011 workshop on automatic speech recognition and understanding 2011. IEEE Signal Processing Society
17. Chen, G., Chai, S., Wang, G., Du, J., Zhang, W.Q., Weng, C., Su, D., Povey, D., Trmal, J., Zhang, J., Jin, M.: Gigaspeech: An evolving, multi-domain asr corpus with 10,000 hours of transcribed audio. (2021). arXiv preprint arXiv:2106.06909
18. Hannun, A., Case, C., Casper, J., Catanzaro, B., Diamos, G., Elsen, E., Prenger, R., Satheesh, S., Sengupta, S., Coates, A., Ng AY.: Deep speech: Scaling up end-to-end speech recognition. (2014). arXiv preprint arXiv:1412.5567
19. Jiang, A.Q., Sablayrolles, A., Mensch, A., Bamford, C., Chaplot, D.S., Casas, D.D., Bressand, F., Lengyel, G., Lample, G., Saulnier, L., Lavaud, L.R.: Mistral 7B. (2023). arXiv preprint arXiv:2310.06825
20. https://huggingface.co/tiiuae/falcon-7b
21. Tunstall, L., Beeching, E., Lambert, N., Rajani, N., Rasul, K., Belkada, Y., Huang, S., von Werra, L., Fourrier, C., Habib, N., Sarrazin, N.: Zephyr: direct distillation of LM alignment. (2023). arXiv preprint arXiv:2310.16944. 2023 Oct 25.
22. Bai, J., Bai, S., Chu, Y., Cui, Z., Dang, K., Deng, X., Fan, Y., Ge, W., Han, Y., Huang, F., Hui, B.: Qwen technical report. (2023). arXiv preprint arXiv:2309.16609
23. Liu, P., Yuan, W., Fu, J., Jiang, Z., Hayashi, H., Neubig, G.: Pre-train, prompt, and predict: A systematic survey of prompting methods in natural language processing. ACM Comput. Surv. **55**(9), 1–35 (2023)
24. Gouws-Stewart, Natasha "The ultimate guide to prompt engineering your GPT-3.5-Turbo model". Accessed on 15 October 2023

Wild Animal Detection System on a Farm Using Image Processing

Madhuri Barhate, Nandini Gulhane, Tanvi Gunjal, Swaroop Gudi, Vedant Gulhane, Prasad Wagade, and Ayush Gude

Abstract Human civilization is based on agriculture, and one of the biggest threats to agricultural productivity is the existence of wild animals on farmlands. Animals are taking over some farmlands, causing significant crop losses, property damage, and livestock risks. As a result, suitable solutions must be developed. A system to support farmers is necessary because the conventional method of constant guard monitoring is impractical. Modern technologies like machine learning, computer vision, and image processing are used by this wild animal detection system to effectively manage conflicts between humans and wildlife. Machine learning and computer vision rely on the integration of particular algorithms to guarantee the accuracy and effectiveness of the system. With an emphasis on sustainability, the technology protects property and crops while also aiding in the conservation of wildlife.

Keywords Computer vision · Image processing · Machine learning · Real-time alerts · Smart farming · Wildlife conservation

1 Introduction

The Wild Animal Detection System is designed to detect wild animals in human environments, particularly farms. A study on environmental difficulties was conducted, and we believe that one of the important challenges that have emerged in recent years is human-wildlife conflict, which must be addressed. Animals encroaching on farmlands cause serious crop damage and endanger farmers' animals, diminishing agricultural productivity and causing financial troubles that eventually influence India's economy, given that agriculture is one of the primary revenue providers (Fig. 1).

To help overcome such challenges, the system devises a solution that outperforms existing techniques. The system will first identify animal presence on the farm and then use Convolutional Neural Networks (CNN), a deep machine learning method,

M. Barhate · N. Gulhane (✉) · T. Gunjal · S. Gudi · V. Gulhane · P. Wagade · A. Gude
Department of Engineering, Science and Humanities (DESH), Vishwakarma Institute of Technology, Pune, Maharashtra 411037, India
e-mail: nandini.gulhane22@vit.edu

G. Raj et al. (eds.), *Advances in Artificial Intelligence and Machine Learning*, Lecture Notes in Electrical Engineering 1264, https://doi.org/10.1007/978-981-97-9507-9_27

Fig. 1 News Headline Showing crops damage by animals is necessary to be minimized

to categorize and differentiate between wild and domestic animals, alerting farmers in real-time. This is made possible by including the HC-05 Bluetooth module/GSM module, which allows for seamless connection with farmers, allowing them to take prompt action based on the received image.

This Wild Animal Detection system integrates a number of cutting-edge technologies that improve its efficiency and accuracy. As the system will keep constant surveillance over the farm at all hours of the day, this can assist farmers in remote monitoring and respond to dangers to the farm. Significant advances in computer vision, where complex algorithms evaluate images from strategically positioned cameras, further improve the system's performance. The system is designed to provide real-time monitoring of wild animals.

In addition to detection, the system will monitor the farm. This paper highlights the system's effectiveness in addressing human-wildlife conflicts, promoting sustainable agricultural practices, significantly improving farmers' livelihoods, preserving the ecosystem and India's agricultural landscape, and fostering wildlife conservation through case studies and real-world implementations.

2 Literature Survey

The paper gives an overview of animal detection and warning systems in North America and Europe [1], listing 27 locations where animal detection system have been placed. The authors examined the primary features of these systems as well as their upkeep and operation, coming to the conclusion that while animal detection and warning systems have the potential to be useful tools for mitigating environmental damage, more study and development is required before they can be widely used. It has been suggested that motion sensors, IP cameras, and PIR sensors be used to detect animal movement and alert drivers [2]. There are many negative effects of deforestation on human-animal conflicts in forest areas and agricultural lands in India, suggesting the need for an effective animal detection system to protect crops and prevent harm to animals [3]. The paper highlights India's farmers' challenges from pests, natural disasters, and animal damage, highlighting the ineffectiveness of traditional methods and the feasibility of using a proposed animal detection system. The system uses wireless communication for electric fence management, allowing remote monitoring. IoT based Wild Animal Intrusion Detection System [4] uses PIR sensors and cameras for security, alerting farmers and forest officials when an animal poses a threat. It also employs a repellent system for animal scaring. The device utilizes PIR sensors for animal detection and emits a 40 kHz ultrasonic frequency to interfere with their hearing. It can detect animals within 5 m and disrupt their hearing up to 20 m. Additionally, the device allows remote monitoring through a 433 MHz radio frequency, covering a range of up to 60 m. The study underscores the necessity of testing ultrasonic wave emissions at different frequencies to match the varied hearing capacities of wild animals [5]. The paper introduces a low-cost monitoring system using an IoT architecture, specifically utilizing Raspberry Pi and a GSM Module. This system aims to track and identify animals entering farming land, alerting landowners and deterring animals through alarms and SMS alerts [6].

3 Existing and Proposed System

3.1 Existing System

Many approaches and technologies have been used in the field of current wild animal detection systems, but these systems frequently have certain drawbacks that reduce their overall efficacy. Conventional monitoring techniques, such as manual surveillance or basic sensor systems, are insufficiently sophisticated to handle the complexities of human-wildlife conflicts on farms. These systems may have trouble responding in real time, which frequently leads to farmers receiving incomplete or delayed alerts. Furthermore, the accuracy of identifying between domestic and wild animals may be reduced by the use of simple image processing methods and the lack of sophisticated machine learning algorithms.

3.2 Proposed System

Our system for detecting wild animals is a big step forward in solving the critical issue of conflicts between people and wildlife on farms. The methodology employs a comprehensive approach that integrates cutting-edge technologies like advanced image processing techniques, computer vision, and machine learning. A key component of our system is the use of Convolutional Neural Networks (CNNs), an effective deep learning technique that allows us to identify and classify both domestic and wild animals. We've included the HC-05 Bluetooth module/GSM module to guarantee real-time responsiveness. This allows for easy communication with farmers and timely actions based on received images. The system uses advanced computer vision algorithms to improve the accuracy of detecting wild animals in addition to continuously monitoring areas from a distance.

4 Methodology

4.1 Components

In our research, we have meticulously chosen the most essential tools and frameworks to ensure the utmost precision and accuracy.

The hardware setup involves the integration of following components:

- *Passive Infrared Sensor:* These sensors are deployed to detect animal/human movement within the monitored area.
- *ESP-32 CAM:* The ESP32-CAM is a compact low-power consuming camera module based on ESP32. Upon motion detection, these modules are activated to capture images of the object.
- *TTL to USB Converter:* USB TTL Serial cables are a type of USB to serial converter cable that connects USB and serial UART interfaces.
- *HCO5 Bluetooth/GSM Module:* Used to send alert messages to the farmer (Fig. 2, Table 1).

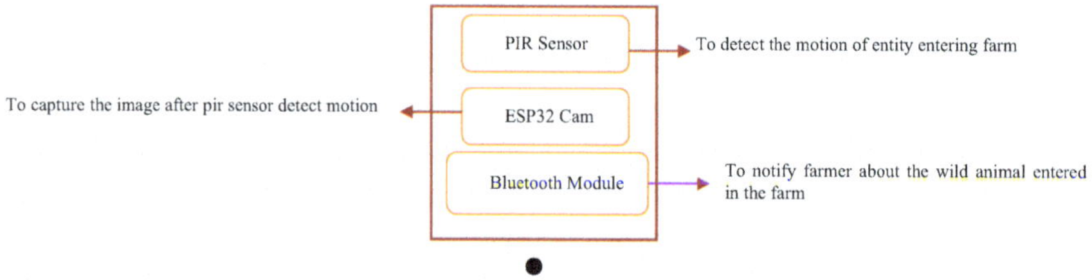

Fig. 2 Hardware components and its use

Table 1 Hardware and software specification

Hardware components	Software specification
PIR sensor, ESP32 camera module, Bluetooth module, TTL to USB converter, Arduino Uno and other components	Embedded kit platform, cnn model, firmware for the microcontroller, software for image processing, notification system

4.2 Algorithm and Flowchart

See Fig. 3.

I. **PIR Sensor Deployment**

PIR sensors are commonly used for motion detection; here we have used and placed it in a location where it can detect the motion of objects entering the farm.

II. **Motion Detection**

When the PIR sensor senses motion, it triggers the ESP32 camera module to capture a picture.

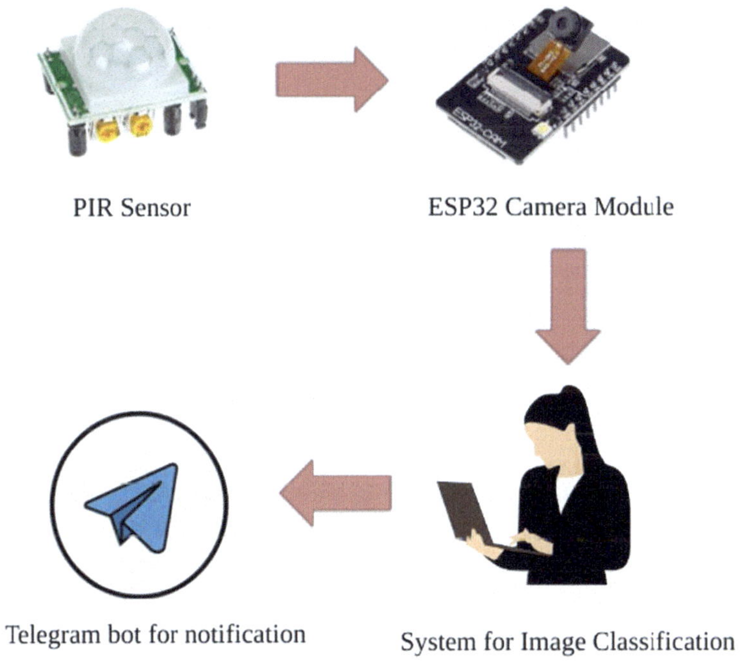

Fig. 3 Algorithm

III. **Image Capture**

The ESP32 camera module takes a picture of the region where motion has been detected. This image is normally in digital format and send to another device for additional processing; however, the ESP32 can also keep the image locally.

IV. **Image Sent to CNN Model**

The captured image is sent to a CNN model for analysis and classification.

V. **Animal Classification**

Based on the attributes gained during training, the CNN model processes the image and classifies the animal as either wild or domestic.

VI. **Notification to Farmer**

Once an animal is successfully classified as wild, our technology initiates a notification procedure that alerts farmers. This is possible with

a. GSM Module: The GSM module allows SMS notifications to be sent to farmers in locations with restricted internet connectivity.
b. HC05 Bluetooth Module: To send a message via Telegram Bot, the image recorded by the ESP32 is also sent (Fig. 4).

This is the flowchart we followed so as to execute our program efficiently.

4.3 Data Collection

Our research started with the collection of a comprehensive dataset compromising images of various animal species, this has been collected from Kaggle. This collection includes data on about 100 different animal species. These pictures served as a basis for our CNN model's training and validation.

4.4 CNN Model

The captured images are subsequently transmitted to the central processing unit, which consists of a Convolutional Neural Network (CNN) model designed to recognize and classify the detected animals. Our primary objective is to classify the detected animals into two categories: wild or domestic.

A deep learning CNN consists of three layers, this layers are convolutional layer, a pooling layer and a fully connected layer.

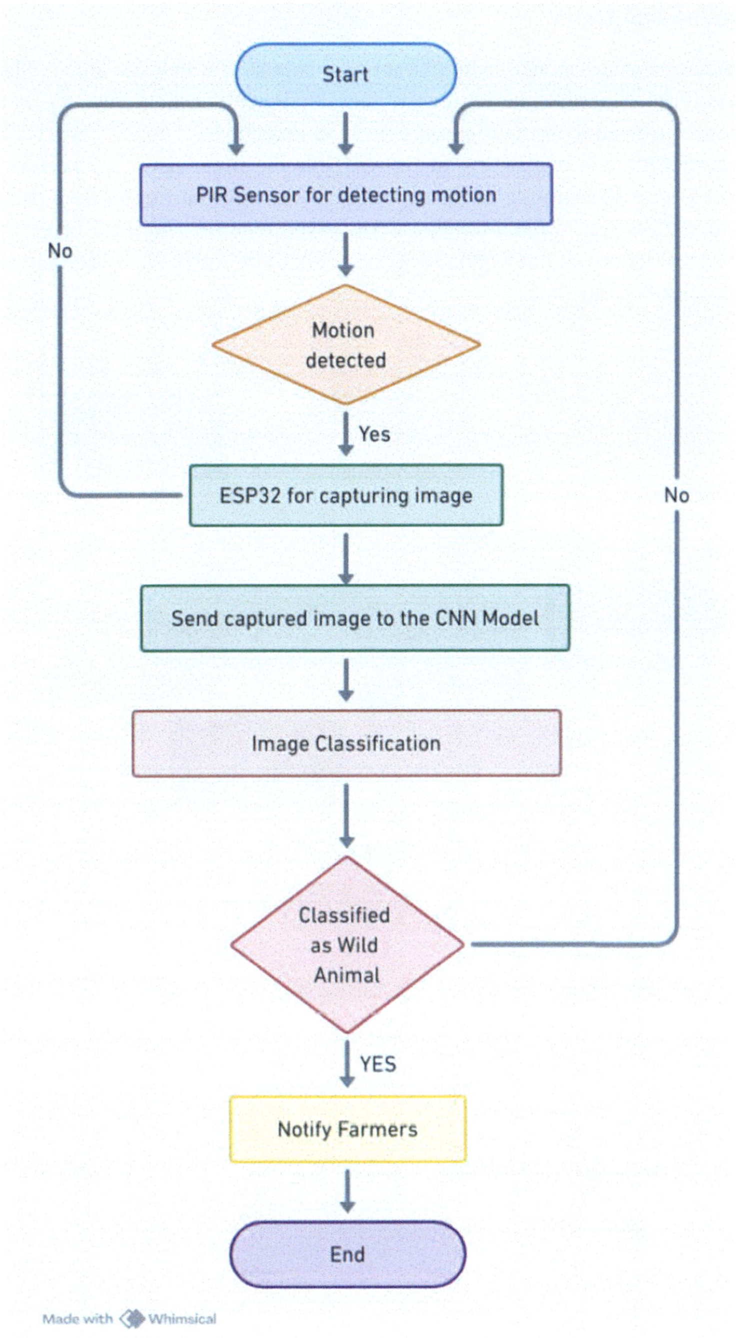

Fig. 4 Work Flow of the system

- **Convolutional Layer**

This is the first layer responsible for extracting features from an input image. Majority of computations take place here. It is a process that involves the kernel; the kernel is a filter which is used to extract features from an image.

The kernel/filter is a matrix that moves over the input data and performs the dot product with the sub-region of input data, yielding the output as a matrix of dot products known as a feature map or convolved feature. The image is converted into numerical values in this layer, allowing CNN to interpret the image and extract relevant patterns (Figs. 5, 6).

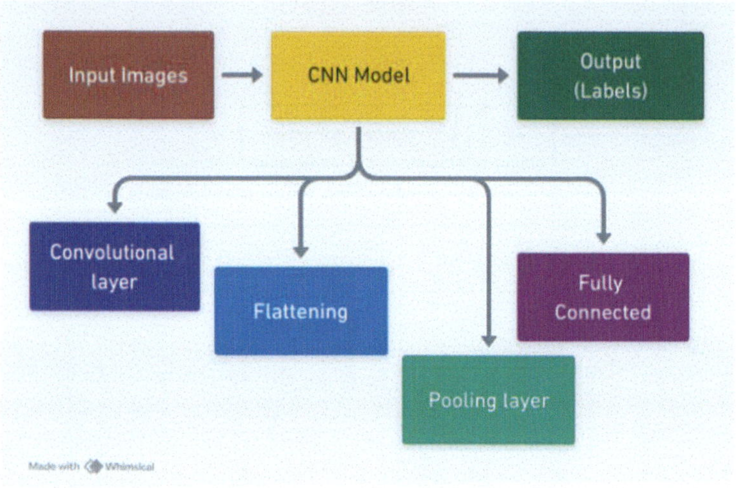

Fig. 5 Block diagram

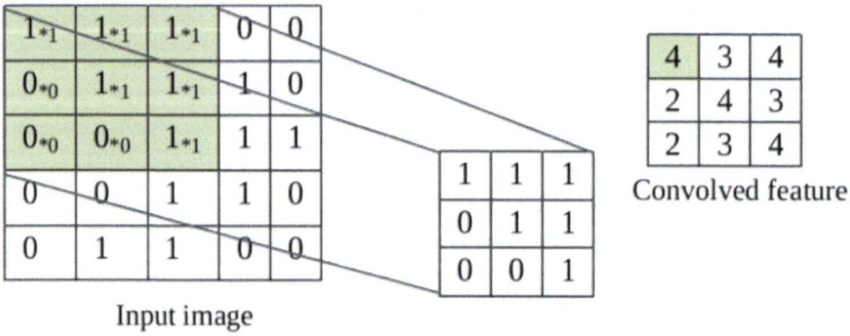

Fig. 6 Convolutional layer

- **Pooling Layer**

Pooling layer reduces the number of parameters in the input when the images are too large. This reduces the number of features while maintaining the most vital data.

The largest element from the convolved feature map is selected by Max Pooling. This leads to the reduction of complexity and increases the efficiency of CNN (Fig. 7).

- **Flattening**

The two-dimensional matrix is flattened into a one-dimensional layer (1D) via this process. This transforms the pooling layer's output, known as the Pooled Convolved Layer, into a single vector, which is then used as an input in the next stage.

The below figure shows the flattening of the pooled feature map (Fig. 8).

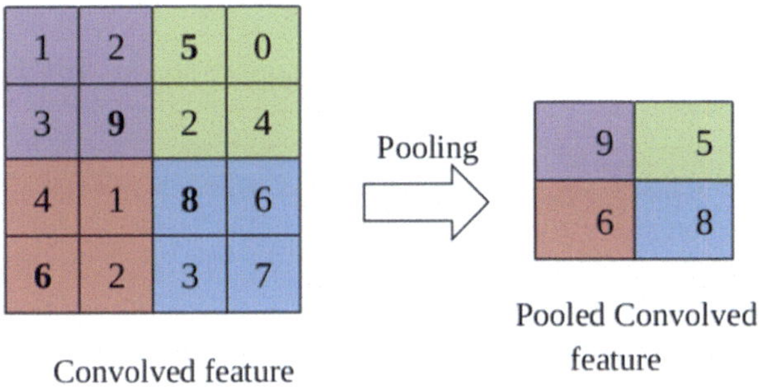

Fig. 7 Pooling layer

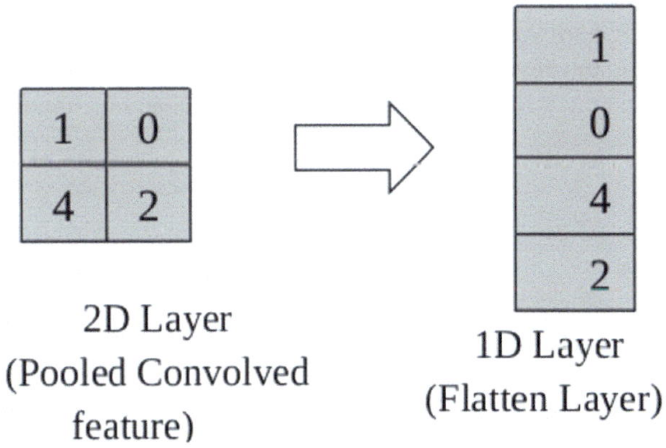

Fig. 8 Flattening

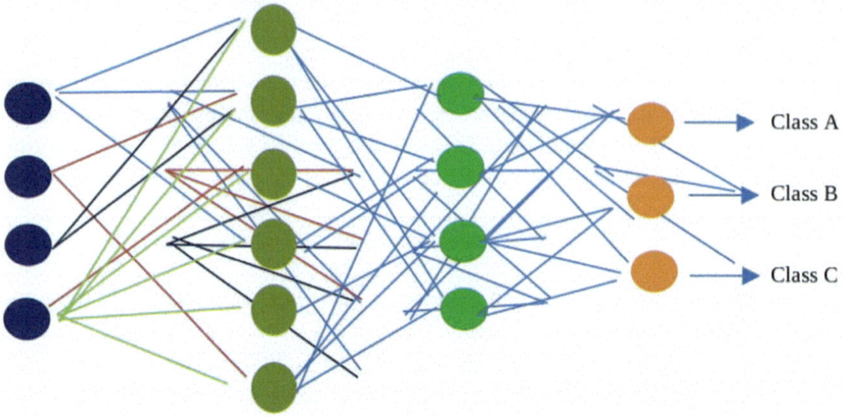

Fig. 9 Fully connected

- **Fully Connected**

The CNN model's last layer, fully connected layer, is where image classification takes place based on the features retrieved from earlier levels. This is referred to as a hidden layer as well (Fig. 9).

4.5 System Design

This comprehensive methodology section provides a detailed overview of the data collection process, hardware setup, image processing, classification, and notification system involved in the development of the wild animal detection system. This offers clarity and transparency regarding the research methods and techniques used in the study (Fig. 10).

5 Objectives

The claimed system's objectives are summed up as follows:

- Reduce conflicts with wildlife

  Objective: To mitigate human-wildlife conflict
  Purpose: Prevent clashes between people and animals, protecting from harm and economic losses.

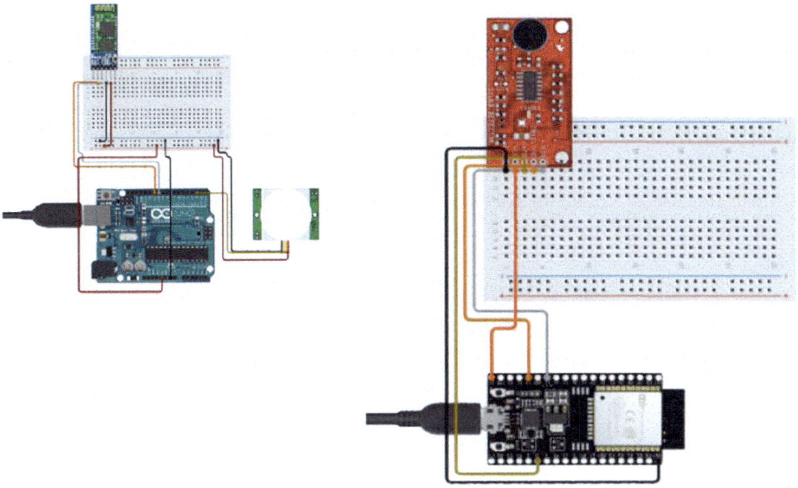

Fig. 10 Circuit designs

- Protect farms and livestock

Objective: To keep wildlife away from farmland and livestock.

Purpose: Safeguard crops and animals from wildlife damage, ensuring food and financial security.

- Balance farming and conservation

Objective: Maintaining an equilibrium between agricultural productivity and wildlife conservation

Purpose: Promote harmony between agriculture and nature, preserving biodiversity and sustainable farming.

- Monitor farms in real-time

Objective: Real-Time Farm Monitoring and Constant Surveillance

Purpose: Provide timely alerts to farmers about wildlife activity, helping them take swift actions to protect their land and livelihoods.

By achieving these goals, the system could be viewed as an asset.

6 Results and Discussion

In this section, we present the results of this animal detection system, which utilizes convolutional neural network (CNN) model for detecting animals. The system was trained using a dataset composed of thousands of images from various animal species obtained from the Kaggle repository.

The passive infrared sensors can detect animals approaching from a distance of up to 5 m has being proposed for utilizing in wild animal detection and rescue systems. Our CNN-based animal detection system achieves a significant accuracy of nearly 95% on hold-out test data, demonstrating the model's ability to correctly classify wild animals in images. In addition to the accuracy this has also achieved precision. Such accuracy and precision is required for minimizing false notifications. In addition to this, the processing speed is adequate for real-time applications. Finally, one of the notable strengths of our CNN model is its adaptability to varying environmental conditions.

By using a CNN model and PIR sensor, this system has a lot of potential for wildlife monitoring and conservation, making it a valuable tool.

7 Future Scope

The future of wild animal detection includes the application of artificial intelligence and Internet of Things (IoT) principles, as well as the use of sophisticated algorithms and technologies to increase detection and tracking accuracy.

Modern technologies, such as LiDAR-based animal detection, may also offer up new possibilities for more extensive wildlife monitoring. Another feature that can enhance the system's capabilities is sound-based animal detection which can complement existing visual detection methods. Moreover, the incorporation of a camera module with night vision capabilities, such as an infrared (IR) Camera module, would benefit the system by allowing for robust performance even at night or in the dark. The wild animal detection system can be integrated with other farming technologies, such as drones and satellite imagery, to provide farmers with a comprehensive view of their fields. The accuracy of the wild animal detection system can be improved by using more advanced machine learning algorithms accompanied by the collection of extensive data. In addition, cost-effective methods can be created to monitor wild animals in their natural habitat, such as the use of intelligent crop protection systems to prevent wild animals from harming crops.

This can help foster sustainable coexistence between human activities and natural ecosystems.

8 Conclusion

In conclusion, the Wild Animal detection system can be viewed as a transformative tool that can help to reduce the potential of human-wildlife conflicts and assist in conservation efforts. Because of the utilization of modern technologies including machine learning, IoT concepts, passive infrared sensors, and saliency detection algorithms, this system operates with notable efficiency. The system's output is intended and timely. However, fine-tuning algorithms and contemporary technologies are capable of boosting system performance, making it even more adaptable and precise. This system, in general provides a practical and ethical approach to wildlife management, paving the way for more responsible and effective strategies. Its profound implications offer promising solution for global conservative initiatives. Further research and development is necessary to improve wildlife monitoring and detection.

Acknowledgements We extend our heartfelt thanks to those who made this research possible. We would like to express our sincere gratitude to our honorable Director, Prof. (Dr.) R.M. Jalnekar, Vishwakarma Institute of Technology, Pune, and our Head of Department, Prof. (Dr.) C.M. Mahajan and our research supervisors, for their valuable guidance, encouragement, and unwavering moral support. We also express our gratitude to the anonymous reviewers and will acknowledge their valuable inputs.

References

1. Marcel, P., Huijse.: Overview of animal detection and animal warning systems In North America And Europe
2. Prema Arokiya Mary, G., Nithesh, P.S., Nanthini, V., Thebiksha, G.V.: Wild animal detection system. Int. Res. J. Mod. Eng. Technol. Sci. (2022)
3. Vikhram, B., Revathi, B., Shanmugapriya, R., Sowmiya, S., Pragadeeswaran, G.: Animal detection system in farm areas. Int. J. Adv. Res. Comput. Commun. Eng. (2017)
4. Prajna, P., Soujanya, B.S., Mrs. Divya.: IoT-Based wild animal intrusion detection system. ICRTT (2018)
5. Yusman, Finawan, A., Rusli.: Design of wild animal detection and rescue system with passive infrared and ultrasonic sensor based microcontroller. In: Proceedings of MICoMS 2017 (Emerald Reach Proceedings Series, Vol. 1), Emerald Publishing Limited, Leeds, pp. 415–422 (2018)
6. Mr Shanmugam, D.B.: A secure wild animal alert system for preventing the farming lands using IoT. IJRTE (2022)

From Pixels to Predictions: Medical Image Analysis and Prediction of Brain Tumor and Chest Cancer with Vision Transformers

Bhagyashree Hambarde, Mohammed Zubair Hussain, Mohammed Saad Hussain, Shrey Gupta, Revanta Tomar, and Sanskriti Gupta

Abstract This study provides a thorough exploration of Vision Transformer (ViT) models in the realm of medical image analysis, specifically focusing on datasets with smaller size and greater complexity than traditional ViT evaluations. Demonstrating the transformative potential of ViT, our research highlights its capacity to enhance diagnostic accuracy and predictive capabilities in healthcare. These analyses were performed on Google Colaboratory, leveraging its cloud-based infrastructure for computational efficiency and accessibility. In the Brain Tumor model, ViT excelled with an accuracy of 85.71%, surpassing the 82.05% achieved by its CNN counterpart. ViT exhibited superior precision, recall, and F1-score, registering 0.88, 0.83, and 0.85, respectively, compared to the CNN model's 0.84, 0.79, and 0.81. The Lung Cancer Model similarly achieved an accuracy of 89.02% as compared to 85.05% and 83.01% compared to CNN models which showcases ViT's effectiveness in diverse medical imaging applications. Our work signifies a significant step towards leveraging ViT models in real-world medical scenarios, where complex data and nuanced patterns demand advanced analytical tools for improved patient outcomes.

Keywords Artifical Intelligence (AI) · Medical Images · Vision Transformers (ViT) · Brain Tumor Prediction · Chest Cancer Prediction

M. Z. Hussain · M. S. Hussain · S. Gupta · R. Tomar · S. Gupta
Shri Ramdeobaba College of Engineering and Management, Nagpur, Maharashtra 440013, India

B. Hambarde (✉)
Computer Science and Engineering (Data Science), St. Vincent Pallotti College of Engineering and Technology, Nagpur, Nagpur, India
e-mail: bhambarde@stvincentngp.edu.in
URL: https://www.stvincentngp.edu.in/

G. Raj et al. (eds.), *Advances in Artificial Intelligence and Machine Learning*, Lecture Notes in Electrical Engineering 1264, https://doi.org/10.1007/978-981-97-9507-9_28

1 Introduction

Medical data is the cornerstone of modern healthcare. It encompasses a vast array of information, including patient records, diagnostic images, and laboratory results. This wealth of data holds invaluable insights into patient health, disease patterns, and treatment effectiveness. In the realm of medical imaging, expert radiologists traditionally bear the responsibility of interpreting images, discerning disease presence, outlining tumor boundaries, and evaluating treatment responses. Their expertise forms the benchmark against which AI and ML systems are measured. However, a paradigm shift is underway, exploring the intricate details within medical images pixels and voxels as data for computer analysis. This shift helps uncover objective mathematical patterns linked to disease behavior. This transition is particularly pivotal in the context of brain tumor and lung cancer diagnosis.

Expert human abilities, crucial for accurate assessments, are now complemented by advanced algorithms. However, a significant challenge—datasets for these medical conditions often remain proprietary and lack standardization, impeding seamless collaboration and hindering the development of universally applicable models. Our proposed approach, employing Vision Transformer (ViT) models, not only aligns with this trajectory but also seeks to address the challenge of non-standardized datasets, aiming to enhance the precision and efficiency of medical image analysis for improved clinical outcomes.

1.1 Influence of AI in Healthcare

In recent years, the integration of Artificial Intelligence (AI) has revolutionized the healthcare landscape. AI-driven technologies have exhibited remarkable proficiency in tasks traditionally reliant on human expertise. The report authored by Maksut-Senbekov et al. [1] comprehensively reviews the recent advancements and practical applications of digital technologies within the healthcare sector. It provides valuable insights into the transformative potential of digital solutions, addressing critical aspects of healthcare delivery, patient care, and technological integration. Figure 1 provides a snapshot of digital technologies in healthcare, emphasizing their relevance to medical image analysis. It succinctly showcases AI platforms, electronic health records, and medical image data, underlining their crucial role in diagnostics. The figure aligns with my paper's focus on the transformative impact of technology, particularly in medical imaging, within the broader healthcare landscape.

The influence of AI extends beyond diagnostics and treatment planning. Natural Language Processing algorithms can sift through extensive medical literature and electronic health records to extract critical insights, facilitating evidence-based decision-making for healthcare providers.

Web applications have emerged as powerful tools in the realm of healthcare, offering accessible and user-friendly platforms for both patients and healthcare

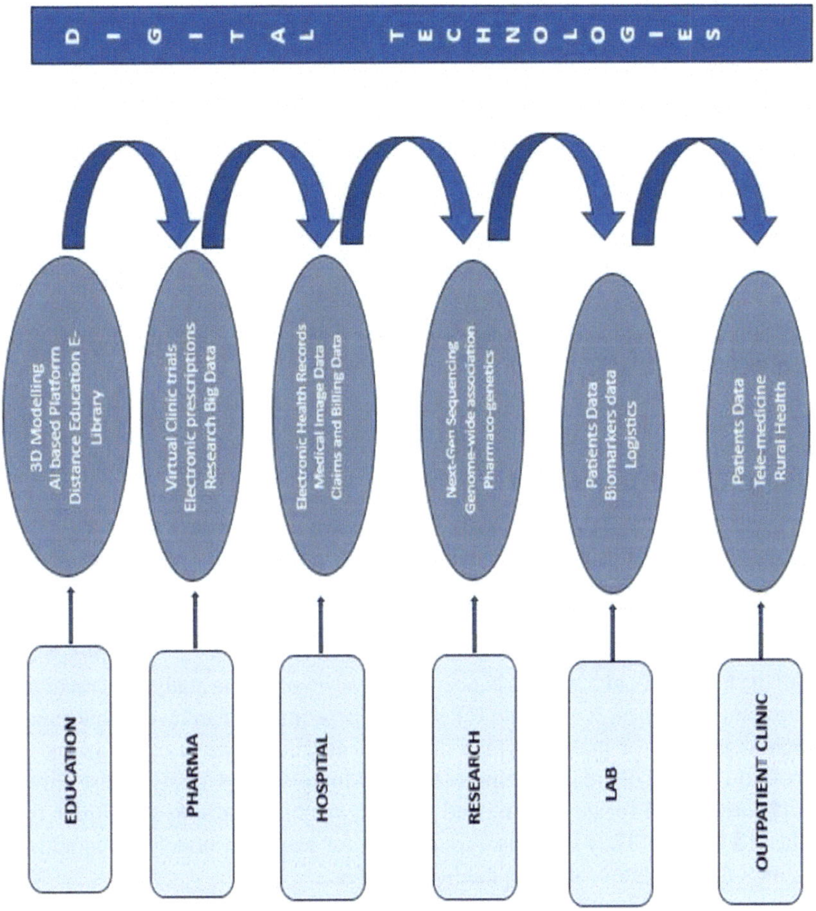

Fig. 1 Scheme of main applications of digital technologies in healthcare, according to MaksutSenbekov et al. [1]

professionals. These applications serve a myriad of purposes, ranging from providing reliable medical information to facilitating remote consultations. One of the most prominent examples is WebMD [2], a widely recognized online health platform. It offers a comprehensive repository of medical information, covering a vast array of conditions, treatments, and wellness advice.

1.2 Brain Tumor Detection

Brain tumors are abnormal growths of cells in the brain. Detecting brain tumors early is critical for providing timely and effective treatment. Medical imaging techniques, such as MRI (Magnetic Resonance Imaging), are commonly used to visualize brain structures and identify any abnormalities. For this, a large collection of pre-operative MRI scans with associated ground truth annotations. It includes multiple modalities like T1-weighted, T1-weighted with contrast, T2-weighted, and FLAIR (Fluid Attenuated Inversion Recovery) images. This dataset is widely used for brain tumor segmentation tasks. Datasets like BRATS [3] and MICCAI BraTS [4], provide a wealth of annotated brain tumor images that are invaluable for training machine learning models.

1.3 Lung Cancer Detection

Lung cancer is a significant public health concern worldwide. Early detection of lung cancer can greatly improve a patient's prognosis. There have been significant advancements in the field of medical imaging and machine learning techniques that aid in the detection of lung cancer.

One of the key approaches to lung cancer detection involves the analysis of medical images, particularly chest X-rays and CT scans. These images can reveal abnormalities, such as nodules or masses, which may indicate the presence of lung cancer. Datasets like LIDC-IDRI [5], CheXpert [6] and Montgomery County Chest X-ray Dataset [7] are crucial for developing and training machine learning algorithms for lung cancer detection. They serve as a foundation for teaching models to recognize patterns indicative of lung cancer in medical images.

1.4 Challenges

Despite the remarkable advancements in medical image analysis using deep learning models, several challenges persist. Computational resource constraints often hinder the seamless implementation of complex models, especially for resource-intensive tasks like 3D medical image analysis. Additionally, the lack of standardized datasets poses a significant obstacle, as it complicates the training and evaluation of models across different platforms. Conflicts in data sharing and privacy concerns further impede collaborative efforts in the field. Addressing these challenges requires a concerted effort from the research community, involving the development of efficient algorithms, the establishment of standardized datasets, and the formulation of ethical guidelines to navigate data sharing and privacy issues. Only through collective action can the full potential of medical image analysis in enhancing healthcare outcomes be realized.

2 Related Work

In recent years, the Vision Transformer (ViT) has emerged as a revolutionary approach in the domain of computer vision. The seminal work by Dosovitskiy et al. introduced the transformative concept of utilizing transformers for image recognition, as detailed in 'An Image is Worth 16 × 16 Words: Transformers for Image Recognition at Scale' [8] and 'Attention is All you need' [9]. Unlike traditional Convolutional Neural Networks (CNNs), which rely on intricate architectures and hierarchical feature extraction according to reports in [11, 12], ViT adopts a fundamentally different strategy by employing self-attention mechanisms. This paradigm shift not only challenges conventional methods but also showcases remarkable capabilities in image recognition tasks. By dividing images into fixed-size patches and treating them as sequences, ViT leverages the power of attention mechanisms to capture global contextual information. Moreover, it has demonstrated remarkable success in various domains, including computer vision tasks such as image classification and object detection. Figure 2 is the standard ViT and Transformer architecture that is proposed by the researchers of google- Dosovitskiy et. al. [8].

Similarly in a study by Alaziz et al. [10], the authors focus on enhancing fashion classification by employing a Vision Transformer (ViT) and concurrently developing a recommendation fashion system utilizing Learning Robust Visual Features without Supervision (DINOv2) combined with a nearest neighbor search. Their work presents a comprehensive evaluation of ViT in comparison to established models like VGG16, DenseNet-121, Mobilenet, and ResNet50, showcasing ViT's superior performance on fashion image datasets.

More recently, in the paper by Al-hammuri et al. [13], discusses the extensive applications of Vision Transformer (ViT) in the field of digital health, showcasing its effectiveness in various medical studies such as brain and breast tumors, histopathology, speech recognition, rheumatology, and more. Specifically, the study introduces TransUNet [14], a powerful architecture that combines ViT with U-Net [15, 16] for multiorgan segmentation in MRI and CT images. The report also highlights advancements like Dual-TransUNet and the utilization of ViT in iSegFormer [17], 3D UXnet [18], UNesT [19], and NestedFormer [20] for diverse medical imaging tasks. Overall, the ViT model demonstrates significant promise in automating complex medical image segmentation processes. These are some of the recent models and methods proposed in medical imaging. All these models

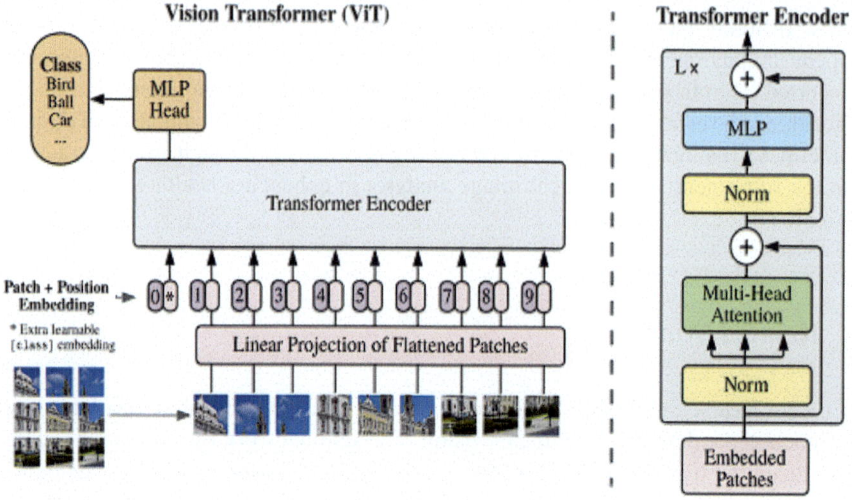

Fig. 2 Diagram of Vision Transformer [8] and Transformer Encoder [9] proposed by Dosovitskiy et al.

require computationally expensive machines, which can handle larger image datasets. Figure 3 gives an analysis of how the AI is involved in the various applications of medical imaging. We note that the most common application is segmentation on images. We aim to work on the detection section using these images within the limits of the available resources.

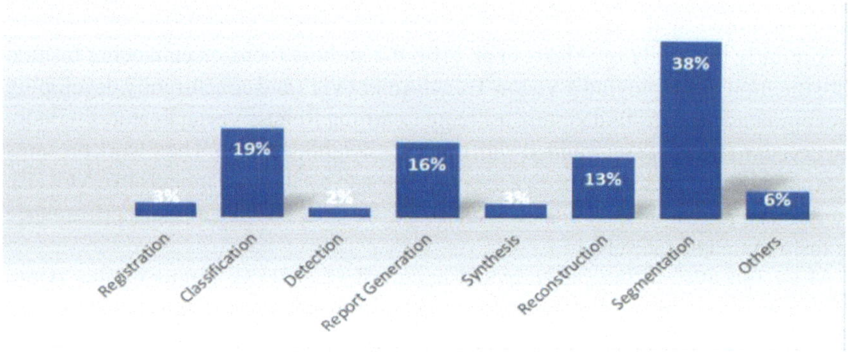

Fig. 3 Distribution of medical imaging applications of the ViT according to the survey in [21]

3 Methodology

3.1 ViT Architecture

The ViT architecture combines the principles of the transformer model with image patches and positional encodings to process images. By treating images as sequences of patches, ViT enables the use of transformers for computer vision tasks. For any ViT model, Fig. 2 is a general representation. Furthermore, we will discuss its key steps and parts below.

Input Image Patching: The input image is divided into a grid of fixed-size patches. Each patch is treated as a token. This is done to enable the transformer architecture to process images. If the input image has dimensions H (height) and W (width), and the patch size is P x P, then the number of patches will be –

$$(H/P) \times (W/P) \tag{1}$$

Embedding Layer: Each of these patches is linearly embedded to obtain a vector representation for every token. This embedding process converts the visual information from the patches into a format suitable for processing by the transformer. If N is the number of patches and D is the embedding dimension, then the embedding layer produces a matrix of dimensions-

$$(N, D) \tag{2}$$

Positional Encoding: Transformers inherently lack any understanding of the position or spatial relationships between tokens. To overcome this limitation, positional encodings are added to the token embeddings. These encodings provide the model with information about the relative positions of the tokens, enabling it to learn spatial relationships. If S is the maximum sequence length (number of patches) and D is the embedding dimension, then the positional encoding matrix will be-

$$(S, D) \tag{3}$$

Transformer Encoder Layers: The core of the ViT architecture consists of a stack of transformer encoder layers. Each layer contains two main sub-layers:

Multi-Head Self Attention: This sub-layer enables each token to attend to all other tokens in the sequence. Through this mechanism, the model learns relationships and dependencies between tokens.

Position-wise Feed-Forward Networks: Following the attention mechanism, a simple feed-forward neural network is applied independently to each position. This sub-layer helps the model capture non-linear relationships between tokens.

Multi-Head Attention: The attention mechanism is computed with multiple attention heads in parallel. Each head focuses on learning different types of attention patterns, allowing the model to capture diverse relationships between tokens. Given h attention heads, D embedding dimension, and L number of patches, the output of the attention sub-layer will b-

$$(L, h, D/h) \tag{4}$$

MLP (Multi-Layer Perceptron) Head: After passing through the transformer layers, a multi-layer perceptron (MLP) head is applied to the output tokens. This step is crucial for performing classification tasks. The MLP head adds non-linearity and helps the model make fine-grained distinctions. If D is the embedding dimension and C is the number of classes, the output of the MLP head will be-

$$(L,\ D) \rightarrow\ (L,\ C) \tag{5}$$

Classification Head: For classification tasks, an additional classification head may be added on top of the final layer. This typically involves applying a global average pooling operation and a linear transformation to produce class scores.

3.2 Dataset Selection

Brain Tumor Detection: The Brain MRI Images for Brain Tumor Detection dataset [22] is a compact dataset and a specific-purpose dataset that comprises of a total of 253 MRI images of 240 × 240 dimensions, distributed across two classes- "Yes" and "No" Classes. The "Yes" class includes 155 images depicting brains affected by tumors, while the "No Tumor" class contains 98 images representing healthy brains without any tumors. The data split is done by referring to a sample model, done by Essam Mohamed [23].

For training and evaluation, the dataset is partitioned into the following sets:

Training Set: This set contains 80% of the total images and is used to train the machine learning model. The training set consists of X_train images, with Y_train belonging to the "Yes" class and Z_train belonging to the "No " class.

Validation Set: To fine-tune the model's hyperparameters and monitor its performance during training, a portion of the data (20% of the total images) is set aside as the validation set. This set is not used for training. The validation set comprises X_val images, with Y_val from the 'Yes' class and Z_val from the "No" class.

Lung Cancer Detection: The Chest CT Scan Images dataset [24] comprises a total of 1000 MRI images of 240 × 240 dimensions, distributed across 4 classes, where

3 are chest cancer types -Adenocarcinoma, Large cell carcinoma, Squamous cell carcinoma, and 1 class for normal cells.

For training and evaluation, the dataset is partitioned into the following sets:

Training Set: This set contains 70% of the total images and is used to train the machine learning model.

Validation Set: To fine-tune the model's hyperparameters and monitor its performance during training, a portion of the data (10% of the total images) is set aside as the validation set. This set is not used for training.

Testing Set: The remaining portion of the data, i.e. –20% of the total images, thus forms the testing set. This set is completely independent of the training process and is used to assess the model's generalization to unseen data.

3.3 Dataset Preprocessing

On both the Datasets [22, 24] Image Resizing, Normalization, Data Augmentation, Patching and Data Loading Steps are applied, as done in [23].

In Image resizing, the images are fixed to a size of (240,240,3) dimensions. This step is crucial for the ViT architecture, as it requires all patches to have the same dimensions. In Normalization, the pixel values of the images are normalized to a specific range (0, 1). This helps in stabilizing the training process and allows the model to converge faster. Moreover, data augmentation techniques are applied to increase the diversity of the training set and improve the model's generalization ability. Some common augmentation techniques include random rotations, flips, zooms, and shifts.

3.4 Model Training

The models for both the Datasets are trained with the following hyperparameters, which is referred from model in [18]:

Optimizer: The model employs the AdamW optimizer, characterized by a learning rate of 0.001 and a weight decay of 0.0001. This optimizer choice, known for its efficiency in handling sparse gradients, contributes to the model's effective training.

Loss Function: For managing the complexities of multi-class classification tasks, the Sparse Categorical Cross-entropy loss function is adopted. This choice is tailored to scenarios where each instance belongs to a single class, providing an optimal measure for classification accuracy.

Metrics: Evaluation metrics play a pivotal role in gauging model performance. The Sparse Categorical Accuracy metric assesses the overall accuracy of the model's predictions. Concurrently, the Sparse Top-5 Categorical Accuracy metric scrutinizes if the correct class is among the top 5 predicted classes. These metrics collectively offer a comprehensive view of the model's classification prowess.

Batch Size and Number of Epochs: The training process involves a batch size of 256, determining the number of samples considered per gradient update. Additionally, the model undergoes 100 epochs, signifying the number of complete passes of the entire dataset through the neural network. These parameters are critical in shaping the efficiency and effectiveness of the training regimen.

3.5 Testing and Results

The trained ViT model was evaluated using a separate test set. The following subsections present the results and visualizations obtained during the testing phase.

Loss Curve: The loss curve provides insights into the convergence behavior of the model during training. It shows the evolution of both training and validation loss over the epochs. The decreasing trend of both curves in Figs. 4 and 5 indicate that the models are learning to minimize the loss function (Fig. 6).

Accuracy Curve: The accuracy curve illustrates the performance of the model in terms of classification accuracy. It shows the training and validation accuracy over the epochs and increasing trend of both curves indicates that the model is improving its ability to correctly classify the input images (Fig. 7).

Confusion Matrix: The normalized confusion matrix displays the classification results in a tabular format. The rows represent the actual classes, while the columns represent the predicted classes. Each cell shows the proportion of samples belonging to a particular class. This visualization allows for a more detailed understanding of

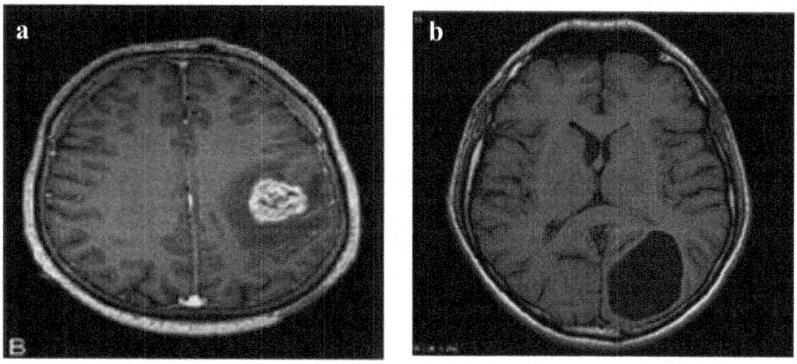

Fig. 4 Sample Brain MRI [22] Images for Brain Tumor Detection where **a** is for Brain Tumor and **b** is No Tumor

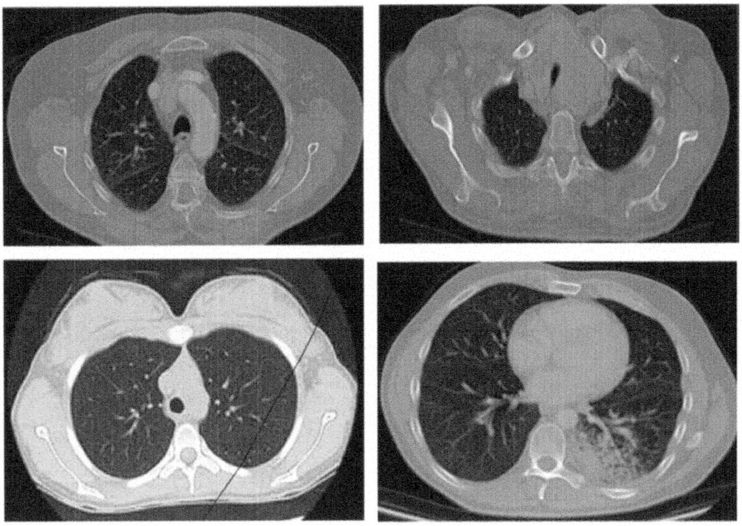

Fig. 5 Sample Chest CT Scan Images [24] for Lung Cancer Detection which includes 4 types of classes—Adenocarcinoma, Large cell carcinoma, Squamous cell carcinoma and normal

the model's performance across different classes. Here we have Confusion Matrix for Brain Tumor Detection Model in Fig. 8.

3.6 Results

The ViT model achieved an accuracy of 85.71% on the test set (Table 1). It demonstrated good precision and recall scores, indicating a balanced performance between identifying positive and negative classes. On the other Hand, the custom CNN model in [23] achieved an accuracy of 82.05% on the test set. It showed competitive performance with a slightly lower accuracy compared to the CNN and pretrained Models InceptionV3, Resnet50 and VGG19 done by M. Hossen [25] model (Table 2). The precision and recall scores indicate a similar balance in class prediction.

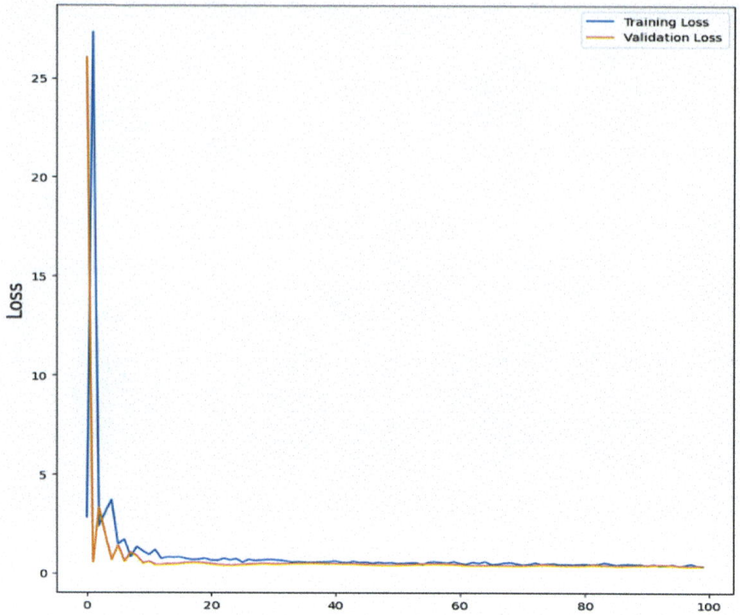

(a)

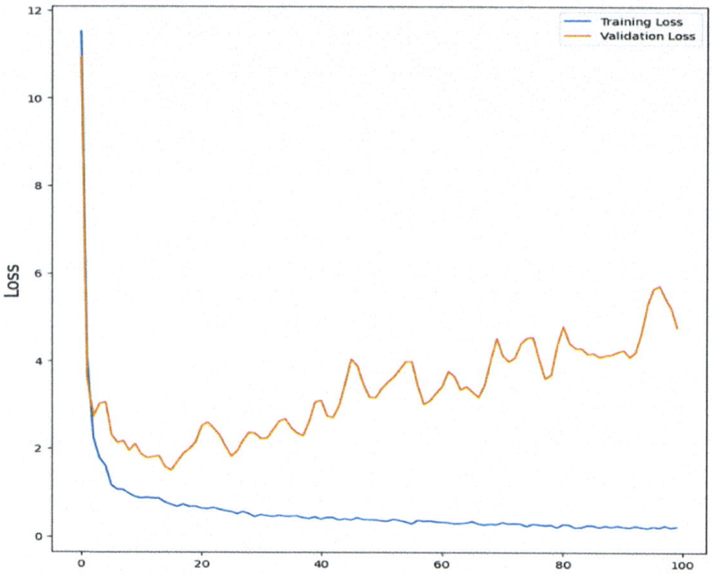

(b)

Fig. 6 Graph between Loss and Number of Epochs for Training and Validation in **a** Brain Tumor and **b** Lung Cancer

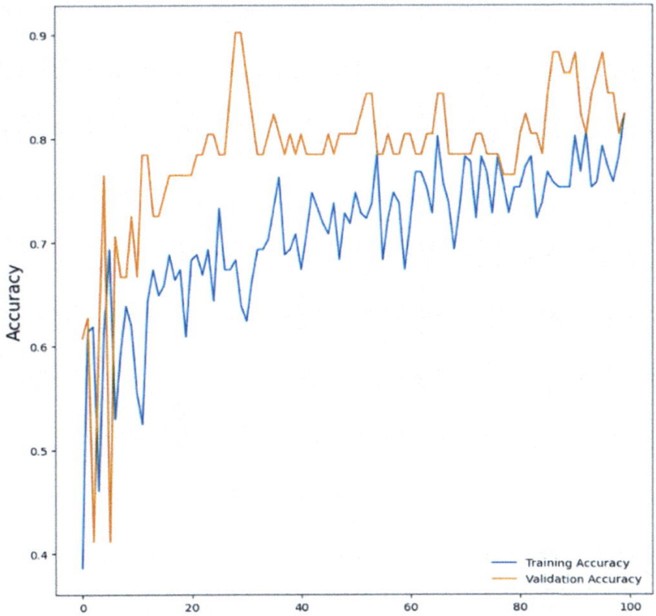

(a)

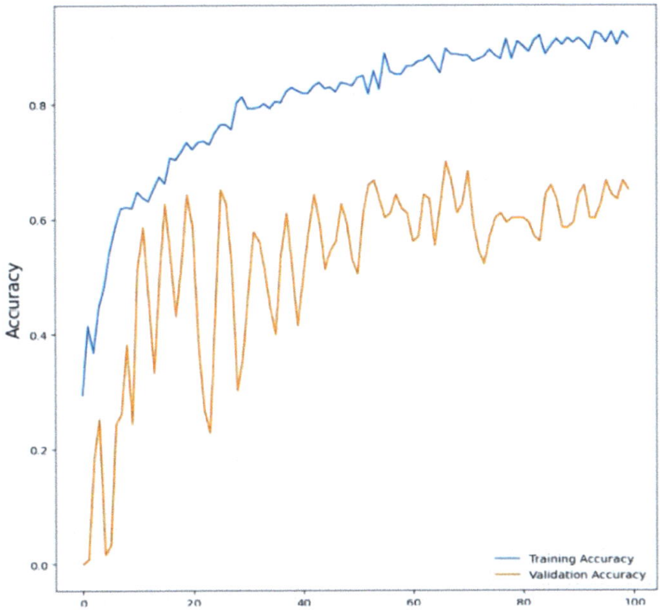

(b)

Fig. 7 Graph between Accuracy and Number of Epochs for Training and Validation in **a** Brain Tumor and **b** Chest Cancer

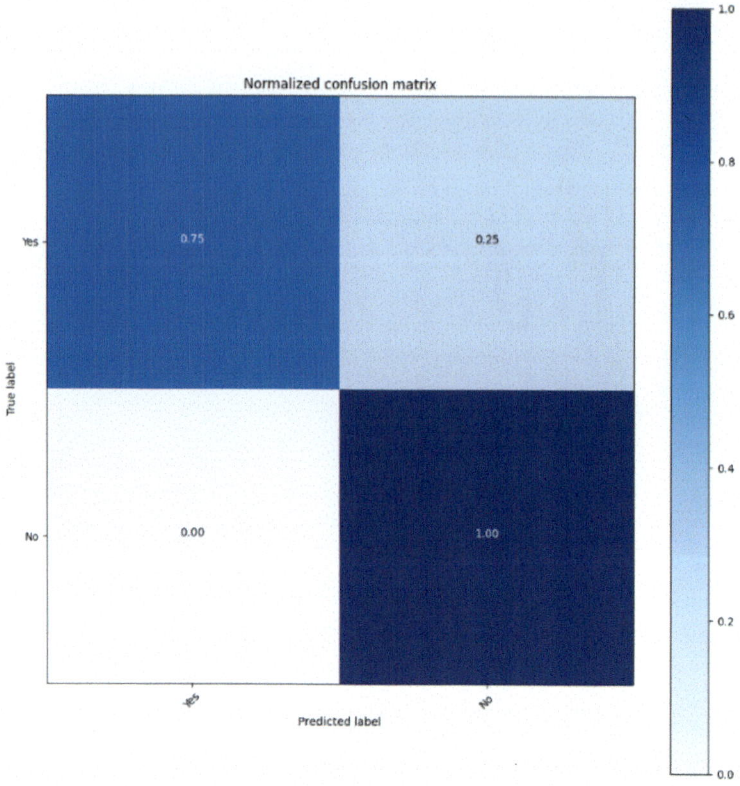

Fig. 8 Confusion Matrix of Brain Tumor Detection model

Table 1 Comparison of ViT with Pretrained CNN models—InceptionV3, Resnet50, VGG19 for Brain Tumor Detection

Models	Accuracy (%)	Precision	Recall	F1-Score
ViT	85.71	0.88	0.83	0.85
Custom CNN	82.05	0.84	0.79	0.81
InceptionV3	70	0.70	0.62	0.77
Resnet50	60	0.80	0.57	0.67
VGG19	40	0.40	0.4	0.4

Table 2 Comparison of ViT with Pretrained CNN models—InceptionV3, Resnet50, VGG19 for Chest Cancer Detection

Models	Accuracy (%)
ViT	89.02
Custom CNN	85.05
Resnet50	83.81

4 Conclusion

In this study, we investigated the performance of Vision Transformer (ViT) models in comparison to Convolutional Neural Networks (CNNs) for image classification and prediction tasks on our dataset. The ViT model, which represents images as sequences of tokens and leverages self-attention mechanisms, exhibited remarkable performance gains over traditional CNNs.

Our findings align with the observations made in the paper [8], where the authors highlighted the superior performance of ViT models in capturing fine-grained details and global context in images. This paradigm shift in image representation and processing has opened new horizons in computer vision tasks.

Moreover, these findings offer opportunities beyond our initial focus. The success of our approach suggests the potential deployment of similar models at local and international levels. The versatility of ViT models positions them as valuable tools in addressing regional health challenges, enabling effective diagnosis and early disease detection. This not only enhances healthcare outcomes but also sets the stage for collaborative initiatives on a global scale. Moreover, there is potential for these models to be packaged and fine-tuned for specific diseases in diverse regions, ensuring adaptability and impactful contributions to healthcare worldwide.

References

1. Senbekov, M., Saliev, T., Bukeyeva, Z., Almabayeva, A., Zhanaliyeva, M., Aitenova, N., Toishibekov, Y., Fakhradiyev, I.: The recent progress and applications of digital technologies in healthcare: a review. Int. J. Telemed. Appl., (2020). Article ID 8830200.https://doi.org/10.1155/2020/8830200
2. WebMD.: WebMD—Better information. Better health. (2023). Available: https://www.webmd.com/.
3. Bakas, S., Akbari, H., Sotiras, A., Bilello, M., Rozycki, M., Kirby, J. S., Davatzikos, C.: Advancing the cancer genome Atlas glioma MRI collections with expert segmentation labels and radiomic features. Sci. Data, (2017). **4**, 170117.https://doi.org/10.1038/sdata.2017.117
4. Menze, B.H., Jakab, A., Bauer, S., Kalpathy-Cramer, J., Farahani, K., Kirby, J., Kikinis, R.. The multimodal brain tumor image segmentation benchmark (BRATS). IEEE Trans. Med. Imaging **34**(10), 1993–2024 (2015). https://doi.org/10.1109/TMI.2014.2377694
5. Armato III, S. G., McLennan, G., Bidaut, L., McNitt-Gray, M. F., Meyer, C. R., Reeves, A. P., & Clarke, L. P.: The Lung Image Database Consortium (LIDC) and Image Database Resource Initiative (IDRI): a completed reference database of lung nodules on CT scans. Med. Phys., **38**(2), 915–931 (2011). https://doi.org/10.1118/1.3528204
6. Irvin, J., Rajpurkar, P., Ko, M., Yu, Y., Ciurea-Ilcus, S., Chute, C., Ng, A.Y.: CheXpert: A large chest radiograph dataset with uncertainty labels and expert comparison. (2019). arXiv preprint arXiv:1901.07031. https://doi.org/10.48550/arXiv.1901.07031
7. Jaeger, S., Karargyris, A., Candemir, S., Folio, L., Siegelman, J., Callaghan, F., Antani, S.: Automatic tuberculosis screening using chest radiographs. IEEE Trans. Med. Imaging **33**(2), 233–245 (2014). https://doi.org/10.1109/TMI.2013.2284099
8. Dosovitskiy, A., Beyer, L., Kolesnikov, A., et al.: An image is worth 16x16 words: transformers for image recognition at scale. (2020). arXiv preprint arXiv:2010.11929.https://doi.org/10.48550/arXiv.2010.11929

9. Vaswani, A., Shazeer, N., Parmar, N., Uszkoreit, J., Jones, L., Gomez, A. N., Kaiser, Ł., Polosukhin, I.: Attention is all you need. In: Adv. Neural Inf. Process. Syst., **30**, pp. 5998–6008. (2017). https://doi.org/10.48550/arXiv.1706.03762
10. Abd Alaziz, H.M., Elmannai, H., Saleh, H., Hadjouni, M., Anter, A.M., Koura, A., Kayed, M.: Enhancing fashion classification with vision transformer (ViT) and developing recommendation fashion systems using DINOVA2. Electron. **12**, 4263 (2023). https://doi.org/10.3390/electronics1220426
11. Krichen, M.: Convolutional neural networks. Comp. **12**, 151 (2023). https://doi.org/10.3390/computers12080151
12. Kanjanasurat, I., Tenghongsakul, K., Purahong, B., Lasakul, A.: CNN–RNN network integration for the diagnosis of COVID-19 using chest X-ray and CT images. Sensors **23**, 1356 (2023). https://doi.org/10.3390/s23031356
13. Al Hammuri et al.: Vision transformer architecture and applications in digital health: a tutorial and survey. Vis. Comput. Ind., Biomed., Art, **6**:14. (2023). https://doi.org/10.1186/s42492-023-00140-9
14. Chen, J.N., Lu, Y.Y., Yu, Q.H., Luo, X.D., Adeli, E., Wang, Y. et al.: TransUNet: transformers make strong encoders for medical image segmentation. (2021). arXiv preprint arXiv: 2102.04306
15. Ronneberger, O., Fischer, P., Brox, T.: U-Net: convolutional networks for biomedical image segmentation. In: Medical Image Computing and Computer-Assisted Intervention, 18th International Conference, Munich, pp. 234–241. (2015). https://doi.org/10.1007/978-3-319-24574-4_28
16. Dong, H., Yang, G., Liu, F.D., Mo, Y.H., Guo. Y.K.: Automatic brain tumor detection and segmentation using U-Net based fully convolutional networks. In: Medical Image Understanding and Analysis, 21st Annual Conference, Edinburgh, pp. 506–517. (2017). https://doi.org/10.1007/978-3-319-60964-5_44
17. Liu, Q., Xu, Z.L., Jiao, Y.N., Niethammer, M.: iSegFormer: interactive segmentation via transformers with application to 3D knee MR images. In: Medical Image Computing and Computer-Assisted Intervention, 25th International Conference, Singapore, pp. 464–474. (2022). https://doi.org/10.1007/978-3-031-16443-9_45
18. Lee, H.H., Bao, S.X., Huo, Y.K., Landman, B.A.: 3D UX-Net: a large kernel volumetric convnet modernizing hierarchical transformer for medical image segmentation. (2022). arXiv preprint arXiv: 2209.15076.
19. Yu, X., Yang, Q., Zhou, Y.C., Cai, L.Y., Gao, R.Q., Lee, H.H., et al.: UNesT: local spatial representation learning with hierarchical transformer for efficient medical segmentation. (2022). arXiv preprint arXiv: 2209.14378
20. Xing, Z.H., Yu, L.Q., Wan, L., Han, T., Zhu, L.: Nested Former: nested modality-aware transformer for brain tumor segmentation. In: Medical Image Computing and Computer-Assisted Intervention, 25th International Conference, Singapore, pp. 140–150. (2022). https://doi.org/10.1007/978-3-031- 16443–9_14
21. Shamshad, F., Khan, S., Zamir, S.W., Khan, M.H., Hayat, M., Khan, F.S.: et al.: Transformers in medical imaging: a survey. (2022). arXiv preprint arXiv: 2201.09873. https://doi.org/10.1016/j.media.2023.102802
22. Chakrabarty, N.: Brain MRI images for brain tumor detection. Kaggle. (2019). https://www.kaggle.com/datasets/navoneel/brain-mri-images-for-brain-tumor-detection
23. Mohamed, E.: Brain tumor detection using CNN & ViT. Kaggle. (2022). https://www.kaggle.com/code/essammohamed4320/brain-tumor-detection-using-cnn-vit
24. Hany, M.: Chest CT-Scan images Dataset. Kaggle. (2020). Available: https://www.kaggle.com/datasets/mohamedhanyyy/chest-ctscan-images?select=Data
25. Hossen, M.M.: Brain tumor detection with ResNet50, VGG, and Inception. Kaggle. (2023). Available: https://www.kaggle.com/code/mdmosarrofhossen/brain-tumor-detection-with-resnet50-vgg-inception#CNN-Models

Endpoint Focused Ideal Attributes of a Cryptocurrency Wallet

Mohd Azeem Faizi Noor and Khurram Mustafa

Abstract The adoption of cryptocurrency wallets has seen a remarkable surge in recent years, reshaping the landscape of finance and technology. The needs and requirements of users have led to the creation of different types of wallets. The functionality and security of these wallets are based on the number of features and attributes they offer. These crypto wallets face a range of threats and vulnerabilities including endpoint vulnerabilities (EV) that can jeopardize the security of digital assets. To understand the attributes of wallets, 12 prominent wallets were analyzed comprehensively with respect to 20 key attributes including their strengths and weaknesses. Building upon this analysis, a new approach was proposed that integrates the attributes that are not yet widely implemented, such as 3-Factor Authentication (3FA), micro-SD card, Hierarchical Deterministic (HD), and advanced algorithms such as BIP 39, BIP 44, Trusted Execution Environments (TEE), Schnorr variants, and Fast Identity Online (FIDO) 2.0 and introduced the novel concept of endpoint-focused attributes within cryptocurrency wallets, with a primary emphasis on bolstering security and mitigating EV. The study contributes to the ongoing efforts to create a more secure, user-centric, and resilient cryptocurrency wallet, fostering trust and confidence in digital finance. Thus, users can secure their wallets and crypto accounts from endpoint incidents.

Keywords Cryptocurrency · Blockchain · Endpoint vulnerabilities · Wallet attributes · Risk analysis matrix

1 Introduction

Cryptocurrency wallet theft has become a distressingly common occurrence, with attackers targeting wallets for financial gain, often successfully. A recent example includes Atomic Wallet freezing stolen crypto assets valued at over $2 million [1, 2]. In September 2023, CoinEx's hot wallets suffered a loss of approximately $70 million

M. A. F. Noor (✉) · K. Mustafa
Department of Computer Science, Jamia Millia Islamia, New Delhi 110025, India
e-mail: azeemfaizif@gmail.com

G. Raj et al. (eds.), *Advances in Artificial Intelligence and Machine Learning*, Lecture Notes in Electrical Engineering 1264, https://doi.org/10.1007/978-981-97-9507-9_29

due to compromised private keys [3]. According to the Forta Network, fraudsters spun up at least 7,900 crypto wallets in May to collect digital assets [4]. In the past decade, a comprehensive analysis by Oosthoek [5] and Charoenwong [6] revealed that cryptocurrency hacks have resulted in losses ranging from $7 to $88 billion. Particularly focusing on endpoint hacks, which encompass wallet key breaches, malware attacks, phishing attempts, and human errors, these studies indicate that over $1 billion has been stolen. Furthermore, with the ever-increasing number of crypto wallets created each year [7], there are security concerns regarding wallet security. This study aims to analyze and investigate popular wallets and based on findings, propose an ideal attribute of wallet to mitigate the risk of endpoint breaches and enhance security.

This study follows a structured approach, commencing with an introduction to the foundational concept of blockchain technology in Sect. 2. Then, the study delves into the intricate world of cryptocurrency wallets and inherent security risks. Section 3 covers the related work. Thereafter, Sect. 4 focuses on endpoint risks within the context of wallet security. Subsequently, a detailed analysis of popular cryptocurrency wallets is provided in Sect. 5. Following this analysis, a proposal is proposed for ideal wallet attributes in Sect. 6. Finally, the study is concluded by addressing potential directions for future research in Sect. 7.

2 Genesis

With the invention of Blockchain technology, users were able to gain the significant advantages of decentralisation and ownership of an asset. In the pre-blockchain era, researchers like Stuart Haber and W. Scott Stornetta introduced the idea of a cryptographically secured chain of blocks to timestamp digital documents, preventing backdating or tampering in the 1990s [8, 9]. In 2008, an individual or group using the pseudonym Satoshi Nakamoto released the Bitcoin whitepaper, introducing the concept of a decentralized digital currency based on blockchain technology [10]. The keynote features and benefits include decentralisation (not controlled by any single entity) [11], immutability (tamper-proof), transparency (visible to all), security (altering block computationally infeasible) [12], peer-to-peer transactions (removing Intermediaries) [10], global accessibility (easily cross-border transactions) etc. These features make it a powerful tool for a variety of applications like cryptocurrencies (e.g. Bitcoin), smart contracts (SC) (e.g. DeFi), supply chain management (VeChain), shipping (Maersk), mobile payment (Ripple), real estate (Propy), music (Arbit), voting (Uport) etc. These are just a few of the many potential applications of blockchain technology and researchers expect to see even more in the future [13].

2.1 Wallet

The first use-case of blockchain is cryptocurrency by implementing bitcoin in 2009 [10] followed by Ethereum in 2015 by Vitalik Buterin [14]. Later, thousands of new cryptocurrencies were created and its market is constantly evolving [15]. These cryptocurrencies need a wallet to make transactions between the payor and receiver. A wallet is an amalgamation of public and private keys that allows users to securely store, manage, and interact with their cryptocurrency holdings [16]. The number of wallets created in the last decade has grown significantly from nix to 80 + million and is expected to reach 300 million by the year 2023 [17, 18] which is shown in Figs. 1 and 2. The created wallet fell into two categories: hot wallet and cold wallet [19]. Both wallets serve different purposes. A hot wallet is connected to the internet, user-friendly and prone to attacks while a cold wallet is disconnected from the internet, less friendly and less prone to attacks [20]. The users have given their priorities to comfort and easiness over security; thus as shown Fig. 3, users created a greater number of hot wallets than cold wallets [18]. As a result, users face many crypto wallet endpoint attacks and heists due to security laziness. Moreover, Fig. 3 shows the difference of adoption of both wallets is continuously increasing with respect to time. According to the Grand View Research, the global hot wallet market is expected to grow at a Compound Annual Growth Rate (CAGR) of 35.2% from 2023 to 2030 while global cold wallet market is expected to grow at a CAGR of 31.94% from 2023 to 2027 [18]. So, it will make more chances for attackers to attack the endpoint.

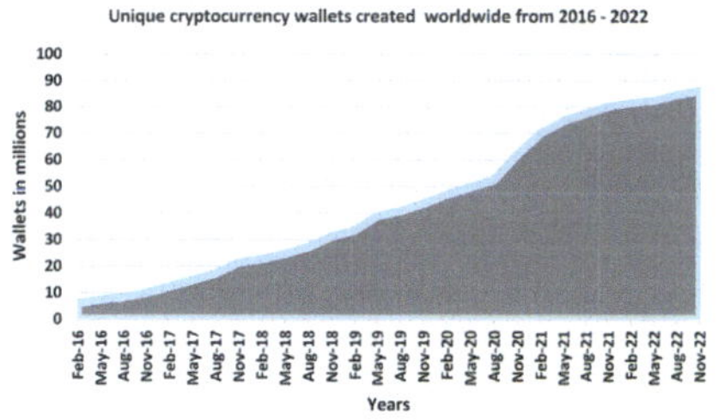

Fig. 1 Unique cryptocurrency wallets created worldwide from 2016–2022

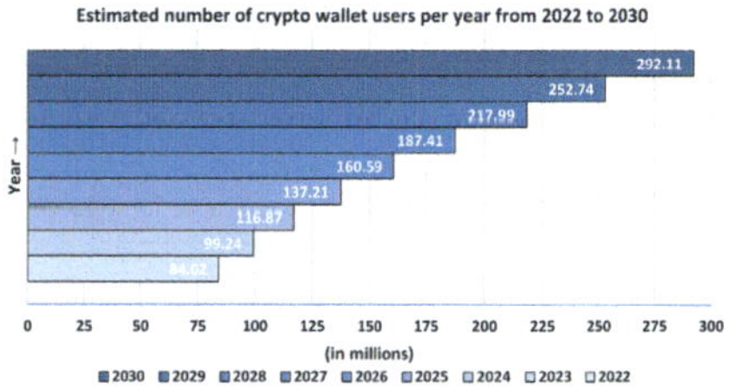

Fig. 2 Estimated number of crypto wallet users per year from 2022 to 2030

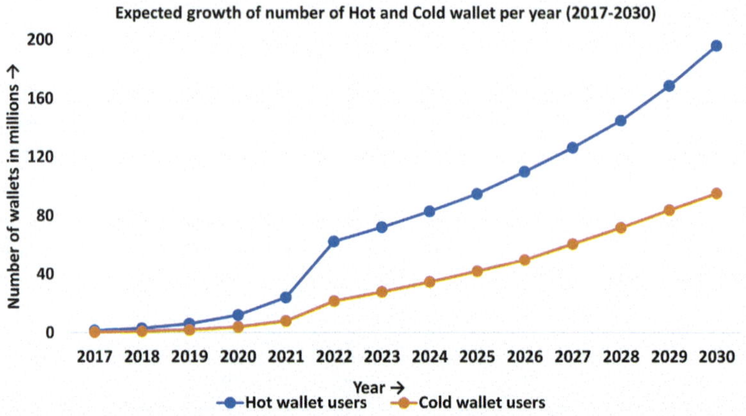

Fig. 3 Expected growth of hot and cold wallets per year (2017–2030)

2.2 *Types of Wallets*

Cryptocurrency wallets come in a variety of types, each catering to different needs and preferences. Apart from connectivity (hot and cold) [21], a wallet can be divided into different types based on parameters such as physical form (hardware, software, paper), control over a private key (custodial and non-custodial) [21] as represented in Fig. 4.

Software wallets (which include desktop, web, and mobile wallets) [22], i.e. intangible wallets, are convenient for everyday transactions but require an internet connection. Hardware wallets (Trezor, Ledger) [22] provide an offline, physical storage solution, making them highly secure against online threats. Paper wallets, essentially printed documents containing public and private keys, offer a tangible and secure way to store cryptocurrencies. Custodial wallets are user-friendly and offered

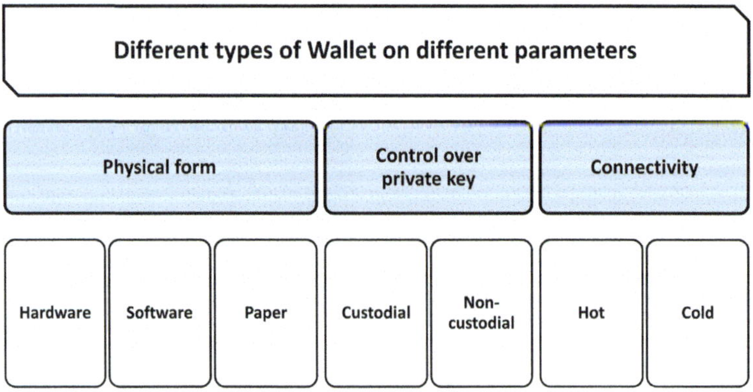

Fig. 4 Types of wallets on different parameters

Table 1 Differences between various types and attributes of wallet

	Types	Security	Convenience	Cost	Accessibility	Other attributes
Air-gapped versus Non-airgapped	Air-gapped	Most secure	Least convenient	More expensive	Low	Higher privacy
	Non-airgapped	Less secure	More convenient	Less expensive	Moderate	Lower privacy
Cold versus Hot wallet	Cold wallet	More secure	Less convenient	Free or low cost	Low	Less portable
	Hot wallet	Less secure	More convenient	More expensive	High	More portable
Software versus Hardware	Software	Least secure	Most convenient	Free or low cost	Moderate	More portable
	Hardware	More secure	Less convenient	More expensive	Less	Less portable
Custodial versus Non-custodial	Custodial wallet	Less secure	More convenient	Free or low-cost	High	Lower privacy
	Non-custodial	More secure	Less convenient	More expensive	High	Higher privacy
SPV versus Full node	Simplified Payment Verification (SPV)	Less secure	More convenient	Free or low cost	High	Lower privacy
	Full node	More Secure	Less convenient	More expensive	Low	Higher privacy

by third-party service providers, such as exchanges, which maintain control over users' private keys and funds [21, 23]. In contrast, non-custodial wallets provide users with full control of their private keys and assets, ensuring complete ownership and security [21]. It prioritizes security and autonomy. A comparative analysis of these attributes, including security, convenience, cost, and accessibility, can be found in Table 1 [24–26].

3 Related Work

Nakamoto [10] released 1st wallet, named Bitcoin, as open-source software in 2009 that simply stored public–private keys in a .dat file on user's computers [27]. In 2011, Allinvain then MtGox (1st) attacks shown the inadequate security and mechanism that incident sparked discussion about improving security and user-centric attributes [6, 12, 27]. So, experts suggested using cold and hot wallet attribute [6]. In 2015, the emergence of Ethereum and other blockchain platforms introduced the concept of SC, multi-signature wallets that influenced the design of wallets to accommodate these programmable functionalities [9]. As a result, wallets like hot wallet, cold wallet, paper wallet, brain wallets were introduced. Similarly, attributes like hardware, SC, TEE, Multi-Factor Authentication (MFA), were also introduced [26, 28]. Based on the user requirements, the wallets were designed with different attributes. Recent advancements in crypto wallet design include biometric authentication, customization and personalization, web wallets and non-custodial decentralized finance (DeFi) wallets, which enable users to interact with decentralized applications and financial services directly from their wallets [24].

Khanum et al. [22] noted 27 out of 122 crypto attacks in 2020 were related to wallet. So, they proposed a novel wallet design to secure sensitive data of blockchain data. Eskandari et al. [29] provides a taxonomy of individual key storage options. Moniruzzaman et al. [30] found a big difference between mobile and desktop wallets usability along with the wallet lacked tools and guides. Several authors in various studies propose the implementation of a user-friendly interface, user-managed key management, the use of simpler language, provision of information about safety practices, and an overall improvement in the user interface [31–33]. Throughout this evolutionary process, it's likely that wallet design will continue to adapt to meet the changing needs and preferences of users while maintaining a strong emphasis on security and functionality.

4 Risk Analysis

Various wallet types exist to accommodate diverse user preferences and needs, differing in terms of security and features. So, Gomzin and Westin [24] has given a wallet selection decision tree to solve the user's selection complexity as shown in Fig. 5.

The experts are advocating the importance of self-custody, which means the practice of controlling and owning your own wallet and keys. The end-user must be the sole custodian of his wallet. There is an adage given by Andreas Antonopoulos,

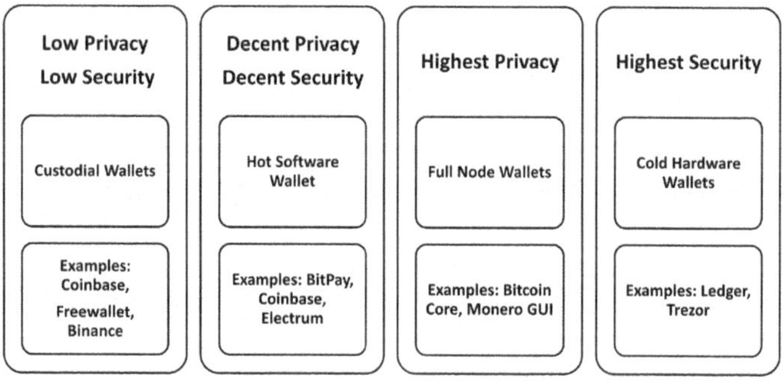

Fig. 5 Wallet selection decision tree

the author of the book "Mastering Bitcoin", has given famous mantras- "Your keys, your bitcoin. Not your keys, not your bitcoin" [34]. So, the user must follow best practices, choose the wallet wisely, and be aware of the risks associated with each type of crypto wallet. Table 2 provides a high-level overview of the risks associated with them [22, 28, 35].

These crypto wallets face a range of threats and vulnerabilities including EV that can jeopardize the security of digital assets. The reasons for endpoint violation include key insecurity, weak ECDSA randomness, phishing, unencryption, social engineering, cryptojacking, lack of technical clarity etc. [36]. Attackers employ a variety of tactics to target the endpoint of crypto wallets, exploiting vulnerabilities in pursuit of unauthorized access to valuable assets [12]. Phishing schemes involve creating deceptive websites or emails to trick users into revealing private keys or sensitive information [27, 37]. Likewise, malware and keyloggers are used to infect devices and covertly capture essential data, while social engineering manipulates individuals into divulging their credentials [27]. Physical theft targets hardware wallets, and network attacks intercept data transmissions. Attackers take advantage of human technical weaknesses to gain unauthorized access to private keys and sensitive information [38, 39].

The market for wallets has grown significantly alongside the expansion of the cryptocurrency and blockchain industry [40]. As the adoption of cryptocurrencies has increased, the demand for secure and user-friendly wallets to store, manage, and transact with digital assets has also risen. Cryptocurrency wallets come in various forms, each with distinct features and attributes tailored to different user preferences and needs. Software wallets, like Exodus and Coinomi, offer user-friendliness and

Table 2 Risk analysis matrix of different wallets

Risk of	Hot wallet	Cold wallet	Hardware	Exchange	Hybrid
Accidental loss of keys	H	M	M	L	L
Physical theft	M	M	M	L	L
Privacy reveal/ compromised	M	L	L	H	M
Phishing	H	L	M	H	L
Hacking	M	L	L	*	L
Providing companies stealing funds due to insolvency	L	L	L	**	L
Pilot error recording keys	L	H	M	L	L
Govt seizure or censorship	L	L	L	H	M
Human Error	H	L	L	H	L
Malware infection	H	L	L	M	M
Keys kept offline	No	Yes	Yes	No	M
Trusted third party	Yes	No	M	High	M
Password Loss	H	L	L	H	M

H- High L-Low, M-Medium, * Depends on security practices, ** Depends on financial practices accessibility across devices, ideal for daily transactions, but may expose users to online security risks [41, 42]. Hardware wallets, including Ledger and Trezor, provide the highest level of security by storing private keys offline, making them perfect for long-term storage but less convenient for frequent transactions [24, 28]. Paper wallets offer offline security at no cost, although they can be vulnerable to physical damage or loss [21]. Mobile wallets like Trust Wallet and Atomic Wallet provide portability and user-friendly interfaces with the added benefit of features like staking and in-app exchanges [23, 41]. Web wallets such as Coinbase and Blockchain.info offer accessibility from any device but are susceptible to online threats and phishing attacks [22].

5 Comparative Analysis

Now, 12 reputed crypto wallets in recent times have been selected and compared them based on 20 different features, including pros and cons. The selected wallets were of hot, cold, air-gapped, non-air-gapped, software and hardware types. The different parameters are custodial, non-custodial, recovery, multisig etc. Table 3 shows a comparison of different crypto wallets based on features and attributes [21–24, 28, 43].

Key takeaways or essential points to be gleaned from Table 3.

- Although two-factor authentication (2FA) is strongly recommended for additional security, its practical implementation is limited. Even Ledger and Trezor wallets are not using it. It is worth noting that 3FA or MFA provides additional layers to strengthen overall security.
- Air-gapped wallets communicate with systems using methods like QR (Quick Response) codes or Micro-SD cards, making micro-SD card support essential for the efficient use of air-gapped features. However, the utilization of micro-SD cards in these wallets remains somewhat restricted. Only Bitcoin and Keepkey offers air-gapped; rest support non-air-gapped. Moreover, most of the wallet, including bitcoin, does not communicate using micro-SD features. Air-gapped wallets offer a range of features, including those associated with hardware, cold, and non-custodial wallets.
- There is a need to update the algorithm which is essential to enhance security and incorporate user-specific requirements. The HD wallets use BIP 32 which generates a tree of private keys from a master private key. An upgrade or collaboration with BIP 39 or BIP 44 yields security enhancement and provides a way to recover the keys. Similarly, FIDO can be updated to FIDO 2.
- The combination of hardware or software wallet with MFA gives good security. However, exchanges use 2FA with hardware but not software wallets. To provide more shields, MFA and biometric features can be implemented. Biometric features, such as fingerprint recognition or facial recognition, are very limited in practice.

Table 3 Exhaustive analysis of wallets [Refer Appendix A]

S No	Wallets	Supported coins and (tokens)	Air gapped (AG)/ Non-Air gapped (NAG)	Cold/ Hot	Hardware (HW)/ Software (SW)	Custodial (CD)/ custodial (NCD)	Desktop support	Mobile app support	OS support (Windows)	Full Node (FN)/SPV	Paid Free
		1	2	3	4	5	6	7	8	9	10
1	Ledger Nano S Plus	5500 (BTC ETH LTC BCH XRP DOGE)	NAG	Cold	HW	NCD	Yes	Yes	Yes	SPV	Paid
2	Ledger Nano X	5000+ (BTC ETH USDT BNB LTC ADA)	NAG	Cold	HW	NCD	Yes	Yes	Yes	SPV	Paid
3	Bitcoin Core	BTC only	AG	Cold	HW	NCD	Yes	Yes	Yes	FN	Free
4	Mycelium	10+ (BTC ETH FIO ERC-20)	NAG	Hot	SW	NCD	No	Yes	No	SPV	Free
5	Trezor T	1456 (BTC ETH ERC-20)	NAG	Cold	HW	NCD	Yes	No	Yes	FN	Paid
6	Electrum	BTC only	NAG	Cold	HW	NCD	Yes	Yes	Yes	SPV	Free
7	Coinomi	1770 (BTC ETH BNB ADA XRP LTC)	NAG	Hot	SW	NCD	Yes	Yes	Yes	SPV	Free
8	Exodus	260+ (BTC ETH LTC)	NAG	Hot	SW	NCD	Yes	Yes	Yes	SPV	Free
9	Meta Mask	500,000+ (ERC-20 ERC-721 ETH USDC USDT ERC)	NAG	Hot	SW	NCD	No	Yes	No	SPV	Free
10	Keepkey	30+ (BTC ERC-20 LTC)	AG	Cold	HW	NCD	Yes	No	Yes	SPV	Paid
11	Jaxx	80+ (BTC EOS LTC)	NAG	Hot	SW	NCD	Yes	Yes	Yes	SPV	Free

(continued)

Table 3 (continued)

S No	Wallets	Supported coins and (tokens)	Air gapped (AG)/ Non-Air gapped (NAG)	Cold/ Hot	Hardware (HW)/ Software (SW)	Custodial (CD)/ custodial (NCD)	Desktop support	Mobile app support	OS support (Windows)	Full Node (FN)/SPV	Paid Free
12	Coinbase	3000+ (BTC ETH ETH2 MATIC SOL BNB)	NAG	Hot	SW	CD	No	Yes	No	SPV	Free

S No	Wallets	Multisig Support	Offline storage	2FA/ Multifactor	HD Wallet	Micro SD Card	Open Source	Smart Contract	Recovery	Best for	Disadvantages
		11	12	13	14	15	16	17	18	19	20
1	Ledger Nano S Plus	No	Yes	No	Yes	Yes	No	Yes	Seed phrase	Secure & fast; Perfect for beginners	Many currencies
2	Ledger Nano X	No	Yes	No	Yes	Yes	No	Yes	Seed phrase	Balancing accessibility and security; OTG hardware wallet	Many currencies
3	Bitcoin Core	No	Yes	Yes	Yes	No	Yes	No	Seed phrase, Backup file	Send and receive anonymously	High resource usage, Complex
4	Mycelium	No	No	No	Yes	No	Yes	No	Seed phrase, Backup file	Excellent for commercial payments	Limited number of crypto
5	Trezor T	No	Yes	No	Yes	Yes	Yes	Yes	Seed phrase	FIDO2 & U2F authentication standards	No bluetooth connection
6	Electrum	Yes	Yes	Yes	Yes	No	Yes	No	Secret phrase	Beginner-friendly	Supports Bitcoin only

(continued)

Table 3 (continued)

S No	Wallets	Multisig Support	Offline storage	2FA/ Multifactor	HD Wallet	Micro SD Card	Open Source	Smart Contract	Recovery	Best for	Disadvantages
7	Coinomi	No	Yes	No	No	No	Yes	No	Seed phrase, Backup file	Encrypted on device	Not compatible with h/w wallets
8	Exodus	No	No	No	No	No	No	No	Seed phrase, Backup file	User-friendly interface	Limited customer support hours
9	MetaMask	No	No	No	Yes	No	Yes	Yes	12-word mnemonic phrase	User security and anonymity	Does not support the Bitcoin network
10	Keepkey	No	Yes	Yes	Yes	No	Yes	No	12-word seed phrase	Secure	Not very user-friendly
11	Jaxx	No	Yes	No	Yes	No	No	No	12-word phrase	Camera scanning	Past security breaches
12	Coinbase	Yes	No	Yes	No	No	No	Yes	Seed phrase, Backup on g-drive	Easy transfer, supports large cryptocurrencies	Limited browser compatibility

Conclusively, no wallet design is best or worst. It all depends on the individual needs and preferences. Theoretically, it is easy to put all favorable features into one design but implementing them is somewhat difficult. However, the factors that are important for the wallet design to mitigate endpoint breaches are discussed in the next section.

6 Proposed Endpoint Focused Wallet

Based on the findings from the previous sections, in this section, we put forth a set of optimal wallet attributes with a particular emphasis on endpoint security. To remove restrictions and implement these attributes, a list of attributes was suggested to achieve maximum security at the endpoint as illustrated in Fig. 6. These attributes are divided into 2 parts, i.e., necessary attributes and essential attributes. For essential attributes, two goals were focused on. First, it should solve or mitigate the endpoint violation factor and secondly, it should be the most necessary attribute of the wallet which should not derived from other properties. Alternatively, without these particular attributes, a user can't get a trustable, secure wallet design and functions that mitigate endpoint attacks. The attributes whose features can be derived from essential attributes or features that are not essential are categorised into necessary attributes. For example, an HD wallet that is implemented with BIP32 and BIP 39 derives authentication and recovery options respectively. Similarly, an air-gapped wallet interprets cold, hardware, non-custodial, offline and micro-SD cards. In a similar vein, Web 3.0 offers a robust platform for the execution of SCs, while a mature Web 4.0 holds the potential to enhance security, accessibility, interoperability, and individual ownership of digital assets.

The essential attributes are grouped as an air-gapped, TEE, OS (Operating System) compatible, multi-sig, web 3.0, HD, easy interface and usability while the necessary attributes involve full node, micro-SD card, cold, 3-FA, non-custodial, recovery, SC, open source, affordability, offline storage, hardware and inheritance.

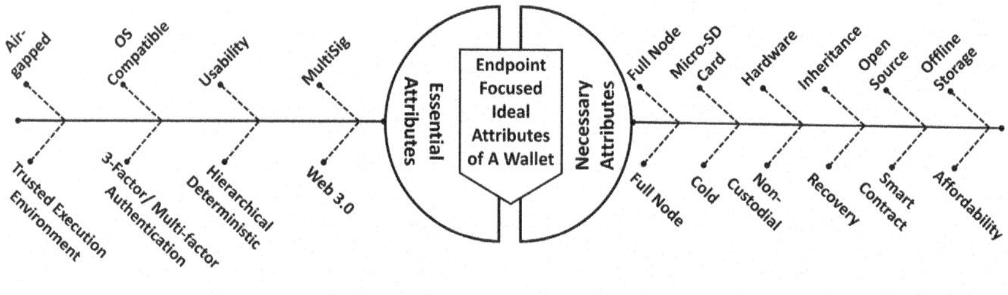

Fig. 6 Endpoint-focused ideal attributes of a wallet

6.1 Essential Attributes

- **Air-Gapped:** Air-gapped systems are typically located in secure areas where physical access is restricted that make it more difficult for attackers to gain physical, exploit any vulnerabilities like malware infection, phishing attacks etc. [25].
- **MultiSig:** It is a security feature that requires multiple private keys to authorize transactions, provides an additional layer. With multisig, an attacker would need to compromise multiple endpoints, each with its private key, to access the wallet. Also, it adds protection against insider threats [31, 44]. The implementation of Schnorr signature algorithm over ECDSA, that solves the nonce reuse problem by using a deterministic and commitment-based nonce generation process, without the need for additional techniques like RFC 6979 (deterministic nonce generation) [45, 46].
- **HD:** It uses a single seed phrase to generate all of the private keys for the wallet. This makes it much more secure than traditional wallets. If the master seed is securely stored, all child keys can be recovered. If an attacker does manage to steal the seed phrase, they will still need to brute force, time-consuming and computationally expensive process [47]. HD wallet incorporates together Bitcoin Improvement Proposals, i.e., BIP32 (HD wallets), BIP39 (mnemonic codes), and BIP44 (multi-account hierarchy) standards, offers powerful and standardized solutions for managing crypto assets along with security, portability, compatibility and recovery [46, 48].
- **Compatibility OS:** Allowing users to use their preferred operating system means they are more likely to use the wallet securely and responsibly. This can lead to better security practices overall.
- **3-Factor Authentication:** This multi-layered approach combines (a) something the user knows (e.g., a password), something the user has (e.g., a hardware token or mobile device), and something the user is (e.g., biometric data like a fingerprint) [24]. By combining these three factors, FIDO 2.0 (Fast Identity Online) with Web3 strengthens security and reduces the risk of unauthorized access. Optionally, clock-based OTP offers good security compared to static passwords [49].
- **Interface and Usability:** A user-friendly interface and usability can indirectly contribute to reducing EV by promoting good security practices and reducing the likelihood of user errors. This involves incorporating elements such as user education, providing guidance, tooltips, informative pop-ups, use of robust passwords, and implementing MFA. Research show that these measures collectively serve to assist and educate users on best practices, reducing the likelihood of errors and safeguarding them against falling victim to phishing attacks [50].
- **Web 3.0:** Web 3.0, i.e., decentralized web, designed to be more secure and less vulnerable to endpoint attacks. It empowers users to retain control over their cryptocurrency. In Web 2.0, the user serves as the product, whereas in Web 3.0, the user assumes the role of the owner. Also, Web 4.0, intelligent web, is still in

its early stages of development, but has the potential to revolutionize the WWW [51].

- **TEE:** It is a hardware-based security feature that enhances endpoint security by providing a secure and isolated environment for executing sensitive operations, such as private key management and cryptographic transactions. It isolates the operations even if the endpoint is compromised [52].

6.2 Necessary Attributes

- **Cold Wallets:** It is typically offline and popular for long-term storage of cryptocurrency. So, they are inherently less vulnerable to EVs [20, 26].
- **Micro-SD card:** By storing a cryptocurrency wallet on a Micro-SD card, the private keys and wallet data are physically isolated from the user's everyday computing environment which makes them less susceptible to remote hacking attempts.
- **Hardware:** It is a physical device designed to keeping the private keys offline and inaccessible to the network [22]. It reduces the risk of EV, such as malware, cryptojacking and phishing attacks. Additionally, many hardware wallets incorporate secure elements and encryption mechanisms, adding an extra layer of defenses.
- **Non-custodial:** It is a powerful tool that doesn't require any third party to manage or control the keys but the user does itself [21]. EV, such as hacking, phishing, or insider threats within custodial services, are significantly mitigated.
- **Recovery:** A well-designed recovery mechanism allows users to regain access to their funds securely in case of device loss, failure, or other unforeseen events [22].
- **Full Node:** Full nodes independently validate and store the entire blockchain, allowing users to verify the authenticity of transactions and blocks directly [24]. By doing so, they reduce reliance on third-party services and mitigate the risk of EV.
- **SC and Open Source:** By operating on the blockchain, these self-executing contracts reduce the risk of human error, fraud, or manipulation that can occur with traditional financial systems. When a wallet's source code is open for inspection and community contributions are encouraged, it becomes subject to rigorous peer review [24]. Moreover, SCs are often audited and verified by the community, further enhancing trust and security in crypto transactions.
- **Offline Storage:** It is a basic practice for safeguarding cryptocurrency holdings and is widely used to protect against vulnerabilities associated with online access.
- **Inheritance:** It's a complex topic as it involves both legal and technical considerations. So, it needs a secure and systematic way to pass on digital assets to heirs [22].
- **Affordability:** Many crypto wallets are free, yet hardware or specialized software wallets may incur a one-time purchase cost that should be budget-friendly to

Table 4 Attributes of a wallet mitigating corresponding EV

	Features/Attributes	Endpoint Vulnerabilities (EV)
Essential	Air-Gapped	Malware, Cryptojacking, Phishing
	Multisig	Unauthorized access, Insider threats
	HD Wallet	Collision and preimage
	Interface and Usability	Users' errors
	TEE	Unencryption, Keylogger, Phishing
Necessary	Cold/Hardware/offline	Malware, Cryptojacking, Phishing
	Non-custodial	Insider threat, Malware, Phishing
	3FA	Unauthorized access, Unencryption
	Full Node	Privacy, Security
	Smart Contract	Reduces the attack surface

enhance accessibility and adoption, especially for users cautious about investing in premium wallet solutions.

Based on the preceding discussion, we've also created a linkage between attribute features and the specific types of EV they address. As an illustration, the "Air gapped" attribute, by not connecting to the internet, effectively mitigates vulnerabilities related to malware, cryptojacking, and phishing attacks. Similarly, a list of various attributes and their respective roles in mitigating specific EV was compiled in Table 4.

7 Conclusion and Future Directions

The study initially provided an introduction to cryptocurrency wallets, delving into the various types in comprehensive detail. Subsequently, a thorough examination of 12 well-known wallets was conducted, carefully evaluating their advantages and disadvantages across 20 distinct attributes. Drawing from this extensive investigation, we then put forth a set of endpoint-focused attributes designed to minimize security vulnerabilities. Our approach involved the integration of attributes that are relatively less commonly implemented, such as MFA, micro-SD card usage, HD, and algorithms like BIP 39, BIP 44, TEE with a hardware-based approach, Schnorr variants, and FIDO 2.0.

In essence, the integration of endpoint-focused attributes in crypto wallets represents a significant stride towards enhancing security and mitigating endpoint vulnerabilities. By prioritizing elements such as air-gapped, HD, web 3.0, microSD card, MFA, biometrics, SCs, and more, wallet providers are not only fortifying their offerings but also empowering users to take more control over their cryptocurrency security.

This study places its primary focus on attributes that either directly or indirectly contribute to mitigating EV. In the pursuit of achieving this goal, priority and emphasis are given to privacy and security over other features. Thus, it adds a slight degree of complexity. Furthermore, the identified attributes in the study can be structured and implemented as a framework. Other attributes such as vault, vanity, zero knowledge, lightning transactions, and more can be integrated at the design level, and tailored to the specific requirements of users. While emphasizing the importance of user education, it's essential to reiterate that user awareness and education remain crucial, as the effectiveness of these attributes ultimately relies on users' understanding and diligent use.

The use of crypto wallet with image segmentation has indirect potential uses like QR code, wallet address, biometric, secure image storage for wallet recovery and visual representation of blockchain data that can enhance security and user experience.

Appendix A

A comparison of different crypto wallets based on features and attributes. See Table 3.

References

1. Abidemi, G.: Atomic Wallet Freezes 2M in Suspicious Crypto Transactions, https://dailycoin.com/atomic-wallet-freezes-2m-in-suspicious-crypto-transactions/, (2023)
2. News, B.: Hacked atomic wallet freezes 2 million in suspicious deposits with help of major crypto exchanges, https://www.binance.com/en/feed/post/2023-10-19-hacked-atomic-wallet-freezes-2-million-in-suspicious-deposits-with-help-of-major-crypto-exchanges-1414840, (2023)
3. Ikeda, S.: 70 million loss after data breach of COINEX crypto exchange, services temporarily suspended, https://www.cpomagazine.com/cyber-security/70-million-loss-after-data-breach-of-coinex-crypto-exchange-services-temporarily-suspended/, (2023)
4. Baydakova, A.: This is how scammers can drain your crypto wallet, https://www.coindesk.com/consensus-magazine/2023/06/22/social-engineering-ice-phishing-and-more-how-scammers-can-drain-your-crypto-wallet/, (2023)
5. Oosthoek, K., Doerr, C.: From hodl to heist: Analysis of cyber security threats to bitcoin exchanges. In: 2020 IEEE International Conference on Blockchain and Cryptocurrency (ICBC). pp. 1–9 (2020)
6. Charoenwong, B., Bernardi, M.: A Decade of Cryptocurrency "Hacks": 2011–2021. (2022). https://doi.org/10.2139/ssrn.3944435
7. Tuwiner, J.: 63+ cryptocurrency statistics, Facts & Trends (2022), https://buybitcoinworldwide.com/cryptocurrency-statistics/, (2022)
8. Haber, S., Stornetta, W.S.: How to time-stamp a digital document. Springer (1991). https://doi.org/10.1007/3-540-38424-3_32
9. Anandan, R., Deepak, B.S.: An Overview of Blockchain Technology: Fundamental Theories and Concepts. The Convergence of Artificial Intelligence and Blockchain Technologies: Challenges and Opportunities. 1–22 (2022). https://doi.org/10.1142/9789811225079 _0001
10. Nakamoto, S.: A peer-to-peer electronic cash system. Bitcoin.–URL: https://www.bitcoin.org/bitcoin.pdf. **4**, (2008)

11. Alkurdi, F., Elgendi, I., Munasinghe, K.S., Sharma, D., Jamalipour, A.: Blockchain in IoT security: a survey. In: 2018 28th International Telecommunication Networks and Applications Conference (ITNAC). pp. 1–4 (2018). https://doi.org/10.23919/ITU-WT.2017.8247004
12. Lee, J.H., others: Systematic approach to analyzing security and vulnerabilities of blockchain systems, (2019). https://hdl.handle.net/1721.1/121793
13. Goundar, S., Shah, Z., Singh, N., Lal, G., Singh, A.: A literature review in support of blockchain technologies. Blockchain Technologies, Applications and Cryptocurrencies: Current Practice and Future Trends. 1–47 (2021). https://doi.org/10.1142/9789811205279_0002
14. Buterin, V., others: A next-generation smart contract and decentralized application platform. white paper. **3**, 1–2 (2014)
15. CoinMarketCap: Cryptocurrency Prices, Charts And Market Capitalizations. In: Cryptofinance: A New Currency for a New Economy. https://coinmarketcap.com/ (2023)
16. Bui, T., Rao, S.P., Antikainen, M., Aura, T.: Pitfalls of open architecture: How friends can exploit your cryptocurrency wallet. In: Proceedings of the 12th European Workshop on Systems Security. pp. 1–6 (2019). https://doi.org/10.1145/3301417.3312495
17. Best, R. de: Blockchain.com wallets 2011–2022, https://www.statista.com/statistics/647374/worldwide-blockchain-wallet-users/, (2023)
18. Research, G.V.: Crypto wallet market size, share and Growth Report, 2030, https://www.grandviewresearch.com/industry-analysis/crypto-wallet-market-report, (2023)
19. Zamani, E., He, Y., Phillips, M.: On the security risks of the blockchain. Journal of Computer Information Systems. 60, 495–506 (2020). https://doi.org/10.1080/08874417.2018.1538709
20. Dasgupta, D., Shrein, J.M., Gupta, K.D.: A survey of blockchain from security perspective. J. Bank. Financ. Technol. **3**, 1–17 (2019). https://doi.org/10.1007/s42786-018-00002-6
21. Alzhrani, F.E., Saeedi, K.A., Zhao, L.: A taxonomy for characterizing blockchain systems. IEEE Access. **10**, 110568–110589 (2022). https://doi.org/10.1109/ACCESS.2022.3214837
22. Khanum, S., Mustafa, K.: Exposure of sensitive data through blockchain wallets: a comparative analysis. In: International Conference on Innovative Computing and Communications: Proceedings of ICICC 2022, **1**. pp. 161–169 (2022). https://doi.org/10.1007/978-981-19-2821-5_13
23. Suratkar, S., Shirole, M., Bhirud, S.: Cryptocurrency wallet: A review. In: 2020 4th international conference on computer, communication and signal processing (ICCCSP). pp. 1–7 (2020). https://doi.org/10.1109/ICCCSP49186.2020.9315193
24. Gomzin, S., Westin, K.: Crypto basics: a nontechnical introduction to creating your own money for investors and inventors. Apress, New York. (2022). https://doi.org/10.1007/978-1-4842-8321-9
25. Academy, B.: What is an air-gapped wallet?, https://academy.binance.com/en/articles/what-is-an-air-gapped-wallet, (2023)
26. Erinle, Y., Kethepalli, Y., Feng, Y., Xu, J.: SoK: Design, Vulnerabilities and defense of cryptocurrency wallets. arXiv preprint arXiv:2307.12874. (2023). https://doi.org/10.48550/arXiv.2307.12874
27. Lazarenko, A., Avdoshin, S.: Financial risks of the blockchain industry: A survey of cyberattacks. In: Advances in Intelligent Systems and Computing. pp. 368–384. Springer Verlag (2019). https://doi.org/10.1007/978-3-030-02683-7_26
28. Eliasi, B., Javdan, A.: Comparison of blockchain e-wallet implementations, https://www.diva-portal.org/smash/get/diva2:1350402/FULLTEXT01.pdf, (2019)
29. Eskandari, S., Barrera, D., Stobert, E., Clark, J.: A first look at the usability of bitcoin key management. Presented at the February 16 (2015). doi.org/https://doi.org/10.48550/arXiv.1802.04351
30. Moniruzzaman, M., Chowdhury, F., Ferdous, M.S.: Examining usability issues in blockchain-based cryptocurrency wallets. In: Cyber Security and Computer Science: Second EAI International Conference, ICONCS 2020, Dhaka, Bangladesh, February 15–16, 2020, Proceedings 2. pp. 631–643 (2020). https://doi.org/10.1007/978-3-030-52856-0_50
31. Chen, H., Pendleton, M., Njilla, L., Xu, S.: A survey on ethereum systems security: Vulnerabilities, attacks, and defenses. ACM Comput. Surv. (CSUR). **53**, 1–43 (2020). https://doi.org/10.1145/3391195

32. Ji, P.: The advance of cryptocurrency wallet with digital signature. Highlights Sci., Eng. Technol. **39**, 1098–1103 (2023). https://doi.org/10.54097/hset.v39i.6714
33. Nowroozi, E., Seyedshoari, S., Mekdad, Y., Savas, E., Conti, M.: Cryptocurrency wallets: assessment and security. (2023). https://doi.org/10.1007/978-3-031-25506-9_1
34. Antonopoulos, A.: Not your keys? not your bitcoin., https://www.youtube.com/watch?v=dnC5mFaIW3Q, (2017)
35. Holmes, L.: How to custody crypto, https://www.runningtowards.xyz/p/how-to-custody-crypto, (2022)
36. Noor, M.A.F., Mustafa, K.: Protocols and guidelines to enhance the endpoint security of blockchain at user's end. In: ICIDSSD 2022: Proceedings of the 3rd International Conference on ICT for Digital, Smart, and Sustainable Development, ICIDSSD 2022, 24–25 March 2022, New Delhi, India. p. 231 (2023). https://doi.org/10.4108/eai.24-3-2022.2318925
37. Astrakhantseva, I., Astrakhantsev, R., Los, A.: Cryptocurrency fraud schemes analysis. SHS Web of Conferences. **106**, 02001 (2021). https://doi.org/10.1051/shsconf/202110602001
38. Guo, H., Yu, X.: A survey on blockchain technology and its security. Blockchain: Research and Applications. 3, (2022). https://doi.org/10.1016/j.bcra.2022.100067
39. Matulevičius, R., Iqbal, M., Ammar Elhadjamor, E., Ghannouchi, S.A., Bakhtina, M., Ghannouchi, S.: Ontological representation of healthcare application security using blockchain technology. Informatica (Netherlands). 33, 365–397 (2022). https://doi.org/10.15388/22-INFOR486
40. Albayati, H., Kim, S.K., Rho, J.J.: A study on the use of cryptocurrency wallets from a user experience perspective. Hum Behav Emerg Technol. **3**, 720–738 (2021). https://doi.org/10.1002/hbe2.313
41. Cox, D.: Best crypto wallet for 2023: Check out top crypto wallets, https://www.cryptonewsz.com/crypto-wallet/, (2023)
42. Chimelu, G.C.: COINOMI cryptocurrency Wallet Review, https://www.investopedia.com/coinomi-wallet-review-5224350,(2023). https://doi.org/10.1007/s11831-022-09805-9
43. Quénetain, S. de: Crypto wallet - comparison chart, https://www.blockchains-expert.com/en/crypto-wallet-comparison-chart/, (2018)
44. Goldfeder, S., Gennaro, R., Kalodner, H., Bonneau, J., Kroll, J.A., Felten, E.W., Narayanan, A.: Securing Bitcoin wallets via a new DSA/ECDSA threshold signature scheme. In: et al. (2015). https://eprint.iacr.org/2017/956.pdf
45. Wang, Z., Yu, H., Zhang, Z., Piao, J., Liu, J.: ECDSA weak randomness in Bitcoin. Futur. Gener. Comput. Syst. **102**, 507–513 (2020). https://doi.org/10.1016/j.future.2019.08.034
46. Ebrahimi, S., Hasanizadeh, P., Aghamirmohammadali, S.M., Akbari, A.: Enhancing Cold Wallet Security with Native Multi-Signature schemes in Centralized Exchanges. (2021). https://doi.org/10.48550/arXiv.2110.00274
47. Bashir, I.: Mastering blockchain. Packt Publishing Ltd (2017). https://www.packtpub.com/product/mastering-blockchain/9781787125445
48. Furneaux, N.: Investigating cryptocurrencies: understanding, extracting, and analyzing blockchain evidence. John Wiley & Sons (2018). https://doi.org/10.1002/9781119549314
49. Quinn, A.: Stop using passwords!, https://www.youtube.com/watch?v=6scxPebJqko&;ab_channel=ProTechShow, (2022).
50. Alshamsi, A., Andras, P.: User perception of Bitcoin usability and security across novice users. Int J Hum Comput Stud. 126, 94–110 (2019). https://doi.org/10.1016/j.ijhcs.2019.02.004
51. Fleerackers, T., Meyvis, M.: Web 1.0 vs web 2.0 vs web 3.0 vs web 4.0 vs web 5.0 – A bird's eye on the evolution and definition, https://flatworldbusiness.wordpress.com/flat-education/previously/web-1-0-vs-web-2-0-vs-web-3-0-a-bird-eye-on-the-definition/, (2018).
52. Coppolino, L., D'Antonio, S., Mazzeo, G., Romano, L., Campegiani, P.: Facing the Blockchain Endpoint Vulnerability, an SGX-based Solution for Secure eHealth Auditing. In: ITASEC. pp. 298–308 (2021). https://ceur-ws.org/Vol-2940/paper25.pdf

ActivMedica—MRI Brain Image Detection & Captioning with ActivGPT Integration

Shanid Malayil, P. Sudeep Krishna, Rabeen Pattupara, P. M. Rishil Mohamed, Salman Faris, K. Muhammed Suhair, and P. Jemsheer Ahmed

Abstract ActivMedica is an innovator in brain tumor diagnosis in the rapidly changing healthcare industry, employing cutting edge AI tools including deep learning and natural language processing (NLP). The shortcomings of existing manual approaches for medical imaging data analysis include difficulty in identifying diverse brain tumors and increasing medical expenditures. Our approach, which at first is based on transfer learning and GIT architecture, uses the GIT convolutional neural network, or CNN, model. A carefully chosen and preprocessed diverse dataset, enhanced for variety and improved by transfer learning from the ImageNet dataset for feature extraction, is essential for successful model training. Using the GIT-based methodology, ActivGPT, an AI chatbot built on OpenAI, enables real-time image analysis through natural language conversations. In addition to brain tumor detection, our research generates brief medical reports based on detected brain tumors, providing medical experts with comprehensive information. Brain tumor diagnosis is ethically improved by ActivMedica, a comprehensive healthcare innovation that combines AI, deep learning, and natural language processing. By doing this, diagnosis accuracy is greatly increased while moral and ethical requirements for AI use in healthcare are maintained.

Keywords Healthcare · Disease detection · ActivMedica · ActivGPT · Deep learning · Artificial intelligence · Natural language processing · Chatbot · Image classification · Machine learning · Medical imaging · Diagnostic tools · GIT · Brain Tumor

---

S. Malayil · P. S. Krishna (✉) · R. Pattupara · P. M. R. Mohamed · S. Faris · K. M. Suhair · P. J. Ahmed
MEA Engineering College, Perinthelmanna, Kerala, India
e-mail: krishnapsudeep12@gmail.com

G. Raj et al. (eds.), *Advances in Artificial Intelligence and Machine Learning*, Lecture Notes in Electrical Engineering 1264, https://doi.org/10.1007/978-981-97-9507-9_30

1 Introduction

The healthcare industry faces several obstacles, such as the rise in chronic illnesses worldwide and the growing need for accurate and timely diagnosis [1]. Despite their durability, traditional healthcare systems find it difficult to handle the growing amount of patient data and the complexities of illness categorization. New approaches that make use of deep learning and artificial intelligence (AI) are acknowledged as having the potential to revolutionise how these problems are solved [1].

AI technologies, especially deep learning and natural language processing (NLP), have the potential to completely transform the medical field since they recognise how important it is to diagnose diseases promptly and accurately [1].

1.1 Introducing the ActivMedica Project and Its Goals

The ActivMedica project is a response to the demand for novel healthcare solutions. It uses AI-driven skills to improve illness detection and classification, which in turn improves diagnosis. ActivMedica, a company that serves both patients and healthcare professionals, places a high value on universal access to diagnostic tools [1, 2].

1.2 Addressing Healthcare Challenges and Advocating Innovative Solutions

The Chat GPT-inspired chatbot system, ActivGPT, is the brainchild of the ActivMedica project. Acting as the conduit between the user and the system, ActivGPT facilitates the easy submission of reports and medical photos by the user, who then receives diagnostic information in return. This innovative chatbot combines natural language processing with machine learning to give consumers an easy way to enter data. ActivGPT does thorough analysis and provides both extensive CSV files containing disease-related data and brief summaries of user inputs. For efficient data processing and analysis, the integration uses AI algorithms and machine learning models [3].

2 Literature Review

2.1 *Difficulties in Brain Tumor Identification*

The field of brain tumor detection faces the many difficulties that come with making neoplastic diagnoses. The spectrum of abnormalities associated with brain tumors is broad and includes variations in location, histology, and clinical behavior. Traditional diagnostic techniques, which require laborious manual image analysis and expert judgement, mostly rely on the knowledge of radiologists and neurosurgeons [4]. However, there are major hurdles to the prompt and accurate identification of brain tumors due to the limitations of these approaches, which include misinterpretation of imaging data, delays in diagnosis, and inter-observer variability [5]. As a result, there is a rising desire to include artificial intelligence (AI) and machine learning into medical imaging.

2.2 *AI and Machine Learning's Place in Medical Imaging*

With convolutional neural networks (CNNs), in particular, artificial intelligence (AI) and machine learning have become revolutionary breakthroughs in medical imaging. Automated tumor segmentation, classification, and detection are made possible by these deep learning techniques, which are excellent at extracting features from medical pictures [5]. Deep learning models are useful for detecting brain tumors from computed tomography (CT) and magnetic resonance imaging (MRI) scans, according to recent studies. Notable achievements include CNN-based systems classifying skin cancer at dermatologist-level accuracy, and similar methods are showing promise in brain tumor detection and histological feature prediction [6, 7].

2.3 *An Intersection Between NLP and Deep Learning Using ActivGPT*

ActivGPT, the central component of ActivMedica, is the union of deep learning and natural language processing. ActivGPT streamlines the exchange of clinical reports and medical imaging by enabling user engagement through natural language discussions, modelled after ChatGPT [10].

2.4 Ethical Issues in Healthcare Assisted by AI

Even though AI has a lot of potential for the healthcare industry, ethical issues need to be taken into account. Consideration must be given to patient privacy, informed consent, algorithmic bias, and the necessity of ongoing human monitoring when using AI responsibly for illness diagnosis [8].

3 Methodology

3.1 Gathering and Preparing Data

In order to facilitate the effective diagnosis of brain tumors using supervised deep learning, a large dataset is assembled from publically accessible medical sources that includes magnetic resonance imaging (MRI) scans. There is variability in the pathological appearance of brain tumors in this varied population [9]. MRI images are preprocessed to standardize format and improve appropriateness for deep learning before being used in model training. The processes involve pixel value normalization, uniform resolution scaling, and the use of data augmentation methods including flips, rotations, and brightness modifications at random.

3.2 Utilizing GIT for Transfer Learning

The GIT architecture, a well-established convolutional neural network (CNN) model [13], is used for brain tumor diagnosis in order to use supervised deep learning. A pre-trained GIT model from ImageNet is used to apply transfer learning to feature extraction. An output layer specifically designed for binary brain tumor diagnosis is added to the GIT model, and a sigmoid activation function is used to generate probability scores. Using pre-trained weights, picture preprocessing, predictions, and classification label determination, the GIT method for brain tumor detection and classification is a complete one. Figure 1 [13] shows the architecture of the utilized GIT model.

Algorithm for Brain Tumor Detection and Classification Using GIT

(1) Import Libraries:

   a. IMPORT pandas
   b. IMPORT datasets
   c. IMPORT transformers
   d. IMPORT torch
   e. IMPORT os
   f. IMPORT gc

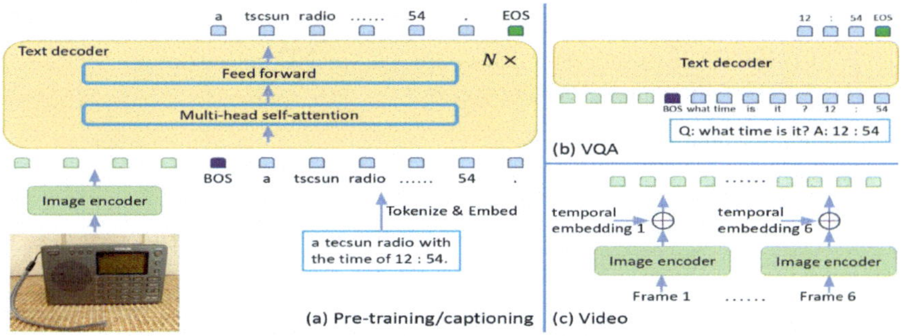

Fig. 1 GIT architecture

g. IMPORT numpy
h. IMPORT itertools
i. IMPORT tqdm
j. IMPORT albumentations
k. IMPORT cv2
l. IMPORT shutil
m. IMPORT json
n. IMPORT PIL
o. IMPORT requests
p. IMPORT matplotlib.pyplot

(2) Load and Preprocess Training Data:

a. READ training_data FROM csv_file
b. FILTER rows_containing_brain_captions
c. CREATE new_dataframe WITH filtered_rows
d. ADD image_paths TO new_dataframe
e. FOR each_row IN new_dataframe:
f. MOVE image_file TO new_folder
g. CREATE caption dict
h. DROP unnecessary_columns FROM new_dataframe
i. SAVE captions_to_json_file

(3) Load Dataset and Model:

a. LOAD dataset FOR training
b. LOAD processor AND model FOR text captioning
c. CREATE training_dataset

(4) Train Model:

a. CREATE data_loader FOR training
b. FOR each_epoch IN 2_epochs:
c. SET model_to_training_mode

d. FOR each_batch IN data_loader:
e. PERFORM forward_pass, backward_pass, AND optimization
f. SAVE trained_model AND processor

(5) Evaluate Model:

a. LOAD saved_model AND processor
b. LOAD test_dataset
c. CREATE data_loader FOR testing
d. INITIALIZE test_losses AND test_bleu_scores
e. SET model_to_evaluation_mode
f. FOR each_batch IN test_dataloader:
g. PERFORM forward_pass AND calculate_test_loss
h. COMPUTE bleu_score
i. PLOT training_and_testing_metrics

(6) Generate Captions:

a. FOR each_example IN test_dataset:
b. PREDICT caption FOR image
c. DISPLAY caption

3.3 Model Training and Supervised Learning

As part of the supervised learning process, the dataset is divided into training, validation, and test sets, and the model is trained using labelled data. The algorithm gains the ability to forecast the presence of brain tumors using ground truth during training. Techniques like gradient descent and binary cross-entropy optimization are used for loss function optimization. Techniques like batch normalization and dropout are used to reduce overfitting [15].

3.4 Model Evaluation

Using a different test dataset, the model's performance is assessed. A comprehensive evaluation of the model's capacity to identify brain tumors is provided by a number of metrics, including recall, accuracy, precision, and BLEU score.

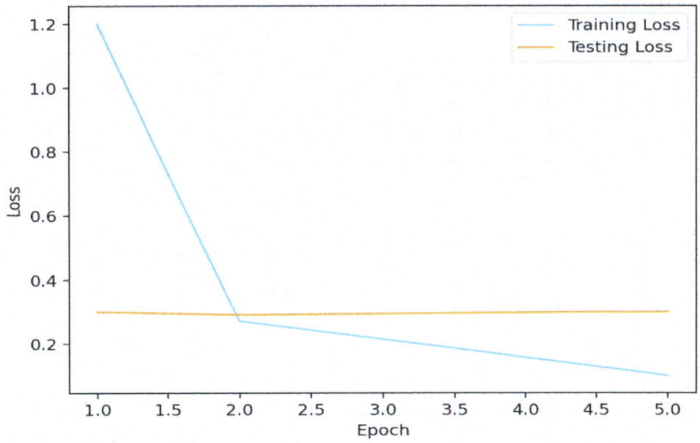

Fig. 2 Training and Testing Loss over Epochs without AdamW Optimizer

3.5 Ethical Considerations

Throughout the entire process, ethical issues are crucial. Prioritizing data privacy, permission, and responsible AI usage, system design makes sure that user data is processed securely and anonymously [11].

4 Experimental Results

4.1 Training and Testing Loss Over Epochs (Phase 1)

By prioritizing patient data protection, obtaining informed permission, and utilizing AI ethically throughout the integration process, ActivGPT complies with ethical standards. User data is anonymized and managed securely in compliance with healthcare regulations and standards to preserve confidentiality. A visual representation of the training and testing loss patterns [17] that sheds light on the model's adaptability and possible overfitting can be found in Fig. 2.

4.2 Training and Testing Loss Over Epochs (Phase 2)

The second stage used the AdamW optimizer for five training epochs at a lower learning rate of 1e-5. Initial elevation was followed by a progressive drop in testing and training losses, which stabilized at 0.28 for the training set and 0.29 for the testing set. The training and testing loss curves are graphically depicted in Fig. 3,

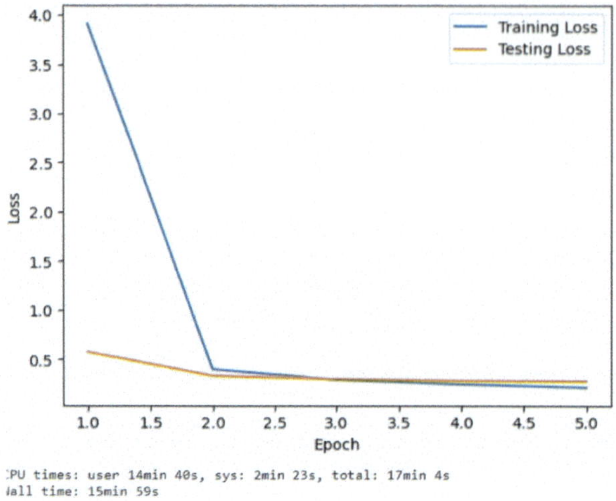

Fig. 3 Training and Testing Loss over Epochs with AdamW Optimizer

which makes it easier to understand how the model converges and stabilizes over time.

4.3 Formulas for Brevity Penalty, N-gram Precision, and BLEU Score

N-gram Precision, Brevity Penalty, and BLEU Score computations were all part of the BLEU score evaluation process. Accuracy for every N-gram order, modification of the Brevity Penalty, and creation of the final BLEU Score [19].

N-gram Precision:

$$Precison(n) = \frac{Total\ Number\ of\ n - grams\ in\ candidate}{Number\ of\ Correct\ n -\ grams} \qquad ((1))$$

Brevity Penalty:

$$\begin{cases} 1, & if\ candiate\ length\ is\ greater\ than\ or\ equal\ to\ reference \\ e^{\left(1 - \left(\frac{reference\ length}{candidate\ length}\right)\right)}, & if\ candidate\ length\ is\ less\ than\ reference\ length \end{cases} \qquad ((2))$$

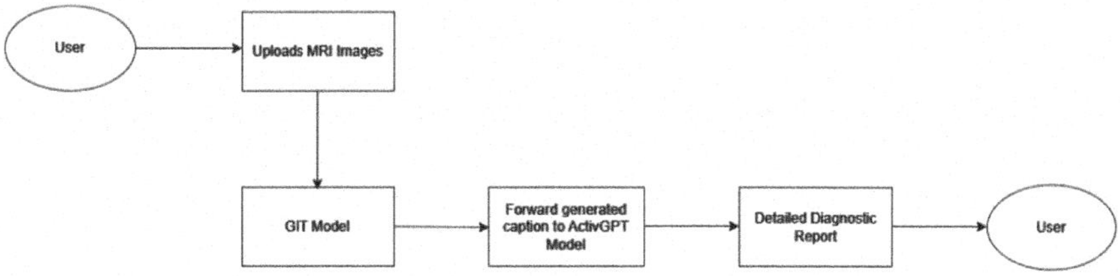

Fig. 4 Workflow diagram

BLEU Score:

$$BLEUScore = BrevityPenalty \times exp(\frac{1}{4}log(Precision(n))) \qquad ((3))$$

4.4 Model Overview

The goal of the ActivMedica project is to use natural language processing (NLP), artificial intelligence (AI), and deep learning to transform the diagnosis of brain tumors [20]. Among the project's principal elements are:

a. *GIT-based Model for Brain Tumor Detection.*: Magnetic resonance imaging (MRI) scan data from a large dataset is utilized to train and assess a convolutional neural network (CNN) model that is used to diagnose brain tumors. The model for real-time picture analysis using natural language conversations is based on the GIT architecture and is integrated with ActivGPT.
b. *ActivGPT.*: An artificial intelligence chatbot that acts as a conduit for users to upload medical images and reports for review in order to facilitate communication between users and the brain tumor detection model. Natural language processing, or NLP, is used by ActivGPT to evaluate user input and deliver pertinent data for additional examination (Fig. 4).

4.5 ROCO Dataset

A novel multimodal image collection designed to reveal the complex relationships between visual components and semantic interactions in radiology pictures is presented by the ROCO (Radiology Objects in COntext) dataset [12]. Image-caption pairings were painstakingly extracted from PubMedCentral, a well-known open-access biomedical literature resource, in order to do this. The collection, which uses captions to add semantic meaning, contains more than 81,000 radiological pictures from various medical imaging modalities, including X-rays, fluoroscopy, computer tomography, ultrasound, MRI, mammography, and magnetic resonance imaging.

A deep convolutional neural network system-tuned automated binary classifier is used to exclude complex, multi-pane, and non-radiology pictures from the carefully selected dataset. Rich annotations, such as captions, keywords, Semantic Types, and Unified Medical Language Systems (UMLS) Concept Unique Identifiers, are included with every image in ROCO, providing a thorough comprehension of both the semantic and visual attributes.

4.6 *Results and Discussion*

Integration with ActivGPT produced a BLEU score of 0.51—a moderately agreeable score between the generated and reference captions—which can be used as a starting point to assess the efficacy of the methodology by comparing the performance of the generated test images with captions and the captions of the captioned models [16].

A thorough study of the variations between the real and predicted captions is shown in Table 1, which assesses the model's accuracy and contextual relevancy. A thorough numerical assessment of the model's performance is provided by Table 2, which includes the BLEU score, training and testing losses, and learning rate (Fig. 5).

5 Future Works

5.1 *Refinement of Captioning Model*

As we examine the existing results, an opportunity to improve the captioning model presents itself. Various designs, sophisticated pre-training methods, or the incorporation of attention processes may be explored in later rounds to improve the model's comprehension of complex information in MRI brain pictures.

5.2 *Hyperparameter Tuning*

The model's performance may be improved by further optimizing hyperparameters including learning rates, batch sizes, and optimization techniques [18]. In future work, a systematic investigation of the hyperparameter space may identify settings that promote better convergence and generalization.

Table 1 Image sample cases with original (ground truth) and generated report

Image sample cases	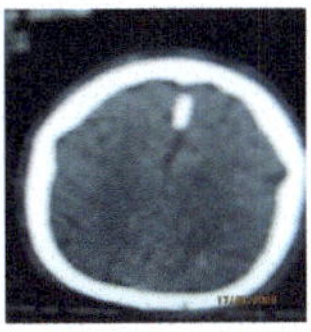	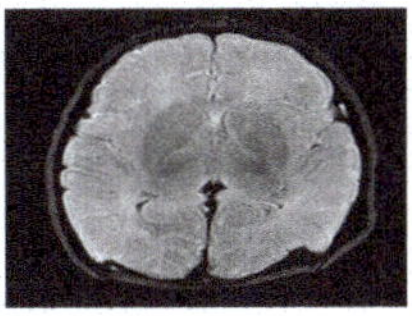	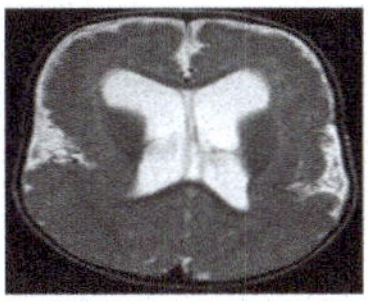	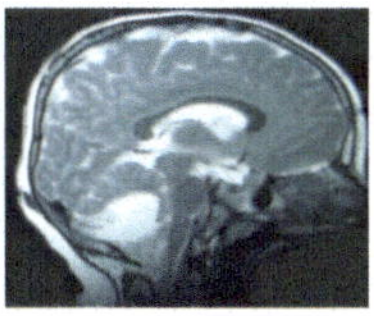
Ground truth/ Actual Caption	Axial brain CT showing anterior falx calcification	PMMR of hypoxic brain changes. Axial T2-weighted PMMR image through a fetal post mortem brain, showing an example of typical low signal change in the basal ganglia which may be associated with hypoxia. Conventional PMMR cannot currently distinguish antemortem from postmortem hypoxic change	T2-weighted axial image of brain of 9-month-old male child with non-lissencephaly cortical dysplasia shows shallow sylvian fissures giving figure eight appearance, polymicrogyria with shallow sulci and relative paucity of underlying white matter	Brain magnetic resonance imaging (T2 weighted image, sagittal view) of case 2 demonstrates the occipital dermal sinus, intracranial sinus tract, and midline posterior fossa cystic mass
Test 1	CT scan of the brain showing a well defined lesion in the left parietal lobe	Brain MRI Scan	Brain MRI T2-weighted image showing hyperintense signal intensity in the left frontal lobe	Brain MRI of the patient
Test 2	CT scan of the brain showing a well defined lesion in the left parietal lobe	Brain MRI Scan	Brain MRI T2-weighted image showing hyperintense signal intensity in the left frontal lobe	Brain MRI of the patient

Table 2 Comparison table

Model	Optimizer	Learning Rate	Epochs	Training Loss	Testing Loss	BLEU Score
microsoft/ git-large-textcaps	SGD	1e-3	10	0.38	0.39	0.21
microsoft/ git-large-textcaps	AdamW	5e-5	5	0.08	0.30	0.20
microsoft/ git-large-textcaps	AdamW	1e-5	5	0.28	0.29	0.51

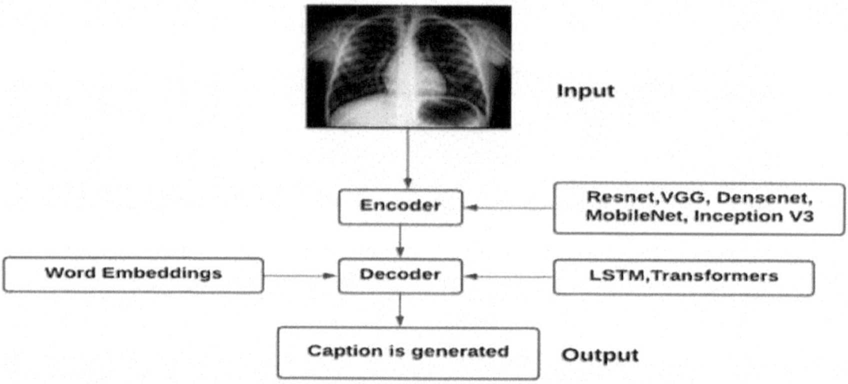

Fig. 5 Example for medical image captioning [14]

5.3 *Incorporating Clinical Expertise*

Working together with radiologists and other medical experts is crucial to incorporating domain-specific knowledge into our model. Clinical expertise must be included to ensure alignment with the demands and expectations of healthcare professionals. This ensures both accurate descriptions and relevance to the healthcare domain.

6 Conclusion

The goal of the ActivMedica project is to improve brain tumor identification in healthcare by integrating deep learning, artificial intelligence (AI), and natural language processing (NLP) with ActivGPT [10]. Its all-encompassing strategy includes ActivGPT, a chatbot that mimics a GPT and makes it easier for users to contribute clinical reports and photographs, speeding up diagnosis. The method has proven to be successful in accelerating the discovery of brain tumors, optimizing processes, and enabling informed healthcare decisions through thorough testing. Future research attempts to improve generalizability by utilizing larger datasets while acknowledging possible constraints. For practical impact, real world clinical assessments and resolving ethical, security, and privacy concerns around data are essential.

References

1. Sahaai, M.B., Jothilakshmi, G.R., Ravikumar, D., et al.: ResNet-50 based deep neural network using transfer learning for brain tumor classification. In: International Conference on Recent Innovations in Science And Technology (RIST 2021). (2022). https://doi.org/10.1063/5.0082328
2. Brown, T., Mann, B,F., Ryder, N., et al.: Language models are few-shot learners. arXiv (Cornell University). (2020) https://doi.org/10.48550/arxiv.2005.14165
3. Jiang, F., Jiang, Y., Zhi, H., et al.: Artificial intelligence in healthcare: past, present and future. Stroke Vasc. Neurol. **2**:230–243. (2017) https://doi.org/10.1136/svn-2017-000101
4. Shen, D., Wu, G., Suk, H.-I.: Deep learning in medical image analysis. annual review of biomedical engineering **19**:221–248. (2017). https://doi.org/10.1146/annurev-bioeng-071516-044442
5. Louis, D.N., Perry, A., Reifenberger, G., et al.: The 2016 world health organization classification of Tumors of the central nervous system: a summary. Acta Neuropathologica. **131**:803–820. (2016). https://doi.org/10.1007/s00401-016-1545-1
6. Booth, T.C., Waldman, A.D., Jefferies, S., Jäger, R.: Comment on "The role of imaging in the management of progressive glioblastoma. A systematic review and evidence-based clinical practice guideline. J Neurooncol; 118:435–460. J. Neuro-Oncol. 121:423–424. (2014). https://doi.org/10.1007/s11060-014-1649-1
7. Haenssle, H,A., Fink, C., Schneiderbauer, R., et al.: Man against machine: diagnostic performance of a deep learning convolutional neural network for dermoscopic melanoma recognition in comparison to 58 dermatologists. Ann. Oncol. (2018). **29**:1836–1842. https://doi.org/10.1093/annonc/mdy166
8. Havaei, M., Davy, A., Warde-Farley, D., et al.: Brain tumor segmentation with Deep Neural Networks. Med. Image Anal. **35**:18–31. (2017). https://doi.org/10.1016/j.media.2016.05.004
9. Litjens, G., Kooi, T., Bejnordi, B.E., et al.: A survey on deep learning in medical image analysis. Med. Image Anal. 42:60–88. https://doi.org/10.1016/j.media.2017.07.005
10. Kreimeyer, K., Foster, M., Pandey, A., et al.: Natural language processing systems for capturing and standardizing unstructured clinical information: A systematic review. J. Biomed. Inform. 73:14–29. https://doi.org/10.1016/j.jbi.2017.07.012
11. Kluge, E-HW: Artificial intelligence in healthcare: Some ethical considerations. Healthc. Manag. Forum **33**:084047041985043. (2019). https://doi.org/10.1177/0840470419850438
12. Pelka, O., Koitka, S., Rückert, J., et al.: Radiology objects in Context (ROCO): A Multimodal Image Dataset. Intravascular Imaging and Computer Assisted Stenting and Large-Scale Annotation of Biomedical Data and Expert Label Synthesis 180–189. (2018). https://doi.org/10.1007/978-3-030-01364-6_20
13. Wang, J., Yang, Z., Hu, X., et al.: GIT: A generative image-to-text transformer for vision and language. (2022). https://doi.org/10.48550/arxiv.2205.14100
14. Sirisha, U., Sai Chandana, B.: Semantic interdisciplinary evaluation of image captioning models. Cogent Eng. **9**: (2022). https://doi.org/10.1080/23311916.2022.2104333
15. Lv, Q., Zhang, S., Wang, Y.: Deep learning model of image classification using machine learning. Adv. Multimed.:1–12. (2022). https://doi.org/10.1155/2022/3351256
16. Papineni, K., Roukos, S., Ward, T., Zhu, W.-J.: BLEU: a method for automatic evaluation of machine translation. In: Proceedings of the 40th Annual Meeting on Association for Computational Linguistics—ACL '02. (2001). https://doi.org/10.3115/1073083.1073135
17. Afaq, S., Rao, S.: Significance of epochs on training a neural network. international journal of scientific & technology research (2020)

18. Imad Jamaleddyn, Rachid El Ayachi, Mohmed Biniz.: Optimization of machine learning algorithms through hyper-parameter tuning applied for the classification of Arabic news. Res. Sq. (Res. Sq.). (2023). https://doi.org/10.21203/rs.3.rs-2615380/v1
19. Doshi K: Foundations of NLP explained—bleu score and WER Metrics. In: Medium. (2021). https://towardsdatascience.com/foundations-of-nlp-explained-bleu-score-and-wer-metrics-1a5ba06d812b
20. Zhou, B., Yang, G., Shi, Z., Ma, S.: Natural language processing for smart healthcare. arXiv (Cornell University). (2021). https://doi.org/10.48550/arxiv.2110.15803

Intelligent Plant Disease Diagnosis Using Deep Neural Networks

D. Satishkumar, K. Vinoth Kumar, J. Satheesh Kumar, Arun Sekar Rajasekaran, and D. Yuvaraj

Abstract The harvests filled in the Farming field is assuming a significant part in the economy of a country and satisfying the requirements of the living hood of human beings. Crops are impacted by different sicknesses during its development cycle. Not finding its sicknesses at beginning phases might prompt a misfortune in production and even yield disappointment. The main thing is to precisely recognize the sickness of the plant. The imagined recognition framework has various applications yet here zeroed in on horticulture to increment crop efficiency by identifying plant illnesses at a beginning phase. There are numerous calculations and models, such as picture pre-handling, include extraction, and component arrangement and numerous more. The various calculations and models give various outcomes, in view of the kind of harvests, environment and so forth. Contrasted with different models Little You Only Look Once (YOLOv5) model is given more exact and improved results. This kind of recognition can be useful to the ranchers, to distinguish the sicknesses and check the illness in its beginning phases with the goal that the yield is augmented toward the year's end.

Keywords Crops · Plant disease · YOLOv5 · Deep learning · Neural networks · CNN model

D. Satishkumar (✉) · K. V. Kumar
New Horizon College of Engineering, Bengaluru, India
e-mail: d.satishkumar.d@gmail.com

K. V. Kumar
e-mail: kvinoth_kumar84@yahoo.in

J. S. Kumar
Dayananda Sagar College of Engineering, Bengaluru, India
e-mail: jsatheeshngl@gmail.com

A. S. Rajasekaran
SR University, Warangal, India
e-mail: rarunsekar007@gmail.com

D. Yuvaraj
Cihan University, Duhok, Iraq
e-mail: d.yuvaraj@duhokcihan.edu.krd

G. Raj et al. (eds.), *Advances in Artificial Intelligence and Machine Learning*, Lecture Notes in Electrical Engineering 1264, https://doi.org/10.1007/978-981-97-9507-9_31

1 Introduction

Food is one of the basic necessities for a living organisms and human beings. There is a food scarcity and insecurity due to a big growing human population and changes in climate due to increasing global warming, Also many countries economic is directly or indirectly dependent on agriculture, especially poor and developing countries. Hence there is need advance techniques for crop yielding, so that there is sufficient food to feed while population. One of the way increase the crops yield is detecting the diseases in early stage. It is the more powerful arrangements is to distinguish and group crop illnesses by gathering pictures of yield development and building a discovery model, which can give ideal and powerful early admonition and intercession to horticultural creation, work on the proficiency of infectious prevention, and lessen agrarian misfortunes brought about by illnesses. It guides agrarian creation by physically noticing the variety, size, What's more, illness spot state of harvest leaves. The early strategy for separating plant leaf highlights was to remove physically highlight guide to distinguish and assess the rate of dark spot infection, Zhouet proposed a multihighlight based AI technique. The above sickness location techniques typically require experts to remove proficient highlights, for example, leaf computerized highlights, surface elements, and what's more, Filter includes and analyze illnesses. The displaying system is intricate, the application is troublesome, and the technique's relevance should be additionally gotten to the next level.

2 Literature Survey

Here they have used Improved YOLOv3-Tiny model for the turmeric plant disease detection, i.e. single-phase detection model technique. This model uses residual neural network structure based on convolution neural network in particular layers. Here they are major diseases of turmeric plant such as leaf spot, leaf blotch, and rhizome rot. They are also working on detecting other vital diseases in turmeric plant and other important crops [1]. YOLOv5 has better performance than YOLOv3 and YOLOv4 models. They are mainly working for rice disease detection. The rice leaf dataset was collected from kaggle of size 400 images. They have used Google collab to train the model [2]. This paper uses 100 epochs, they are trying to implement this model for real time disease detection by developing a smart embedded system, and deploying in agricultural fields [3]. Here they have used a Deep learning YOLOv4 model to detect the diseases in eggplant (Solanum Melongena L). Using Raspberry pi 4 the images are captured and processed using the trained model and then the results are notified in the Telegram application [4]. Here they utilized the strategy for YOLOv5 model and work on the speed and precision of sickness order model, to accomplish model weight decrease Through Phantom net and Weighted Box Fusion (WBF) design and they led a presentation the outcomes show that and the activity time and exactness of the enhanced model are 11.8% and 3.98% higher than the first Mo examination test and removal test between the improved model and other standard models [5].

3 Proposed Method

As seen in Fig. 1 [5], Raspberry Pi is a movement of inexpensive single-board computers (SBCs) produced in the Brought Together Domain by the Raspberry Pi Foundation in collaboration with Broadco. The Raspberry Pi project initially favored the advancement of teaching fundamental computer programming in schools and in developing nations. The Broadcom BCM2837 SoC, which powers the Raspberry Pi 3 Model B, features an Advanced RISC Machine (ARM) Cortex-A53 processor running at 1.2 GHz with 64 cycles and 512 KB of shared L2 storage. Models A + and B + operate at 1.4 GHz. The Raspberry Pi 4 has a Broadcom BCM2711 SoC with a 64-bit quad-focus ARM Cortex-A72 CPU running at 1.5 GHz (later models: 1.8 GHz) with a shared L2 save of 1 MB. The obstruct controller on this SoC is compatible with the ARM Traditional Encroach upon Controller (GIC) design 2.0, in contrast to previous models, which all employed a proprietary upset controller inadequate for virtualization. The quad-focus Cortex-A53 processor used in the Raspberry Pi 3 is depicted as seen in Fig. 1 above. According to benchmark results, Fig. 2 [6], the Raspberry Pi 3 performed parallelized workloads about 80% quicker than the Raspberry Pi 2 did. While utilising ARM virtualization capabilities, providing hardware sponsorship to cause spread. Hammer memory on the Raspberry Pi 2 is 1 GB. The Raspberry Pi 3 has 512 MB of Crush memory in the A + variant and 1 GB of RAM in the B and B + models. The Raspberry Pi Zero W always has 512 MB of RAM. There are 1, 2, 4, or 8 GB models of the Raspberry Pi 4 available. The 8 GB variant was offered in May 2020, whereas the 1 GB model was been opened at ship off in June 2019 but had been stopped by Walk 2020. The prepared model is put away in Raspberry pi, either in inner capacity or outside stockpiling relying on the Raspberry pi particulars, and it is designed with a camera to catch the photograph. This photograph is shipped off the prepared model for its handling, the prepared model to check whether its elements are like any of the classes as shown in Fig. 3 [7], for which the model is prepared, assuming the model distinguishes the name of the plant and sickness is informed and shown on the screen which is associated with raspberry pi, and notice can be sent applications like telegram, Whatsapp and so on. If the picture as shown in the Fig. 3, the model presentations "unidentified" (Fig. 4).

With the extension of the quantity of different public datasets and the advancement of picture handling and item discovery innovations, the examination of leaf infection highlight area and order given field crop pictures has grown quickly. Which is fundamentally based on the element data removed by the model as shown in Fig. 3 [8] to find the sick spots and judge the level of contamination. The subsequent step of article identification is the illness arrangement, which is mostly by matching the given input including vector with one of the multitudes of classes mastered while preparing to get the class to which the article has a place. In this way, the requirement for the

Fig. 1 Raspberry Pi Board

Fig. 2 Raspberry Pi Camera

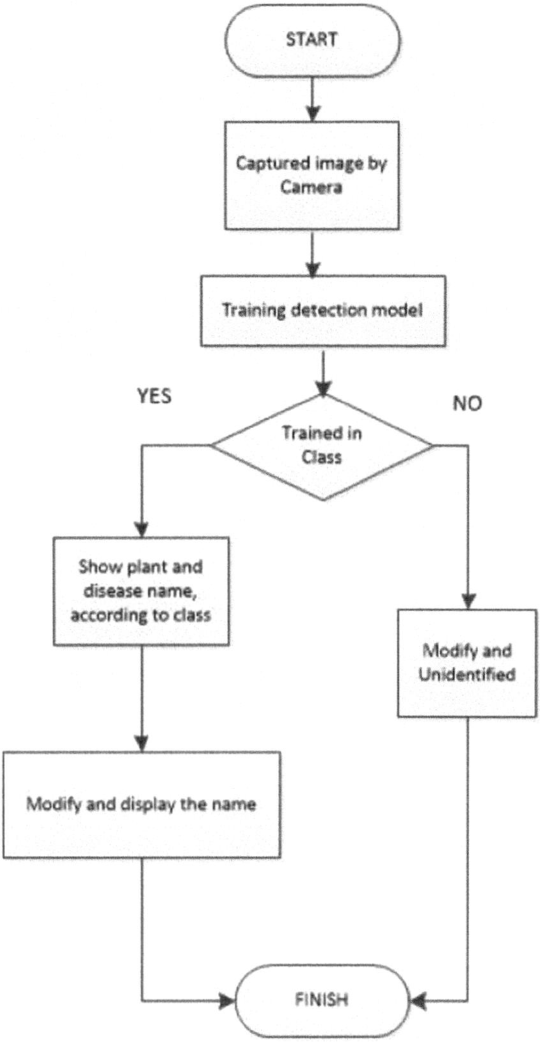

Fig. 3 Working model of raspberry Pi

convenient recognition of the sickness or bug becomes needed. A training program that provides science-based information to farmers about integrated pest and disease management was founded by Farmer Schools in Thailand. The in-time location of deadly sicknesses can be made conceivable to the ranchers with the assistance of innovation Advance figuring innovations can assist ranchers with making choices on numerous parts of developing yields. The use of cutting-edge mechanical apparatuses in the field of agribusiness can enormously assist ranchers with distinguishing and forestalling infections without sitting tight for a specialist's guide. As shown in Fig. 5 i.e., Experimental setup, this project gathered a comprehensive dataset of plant images, including healthy plants and those affected by various diseases. The dataset

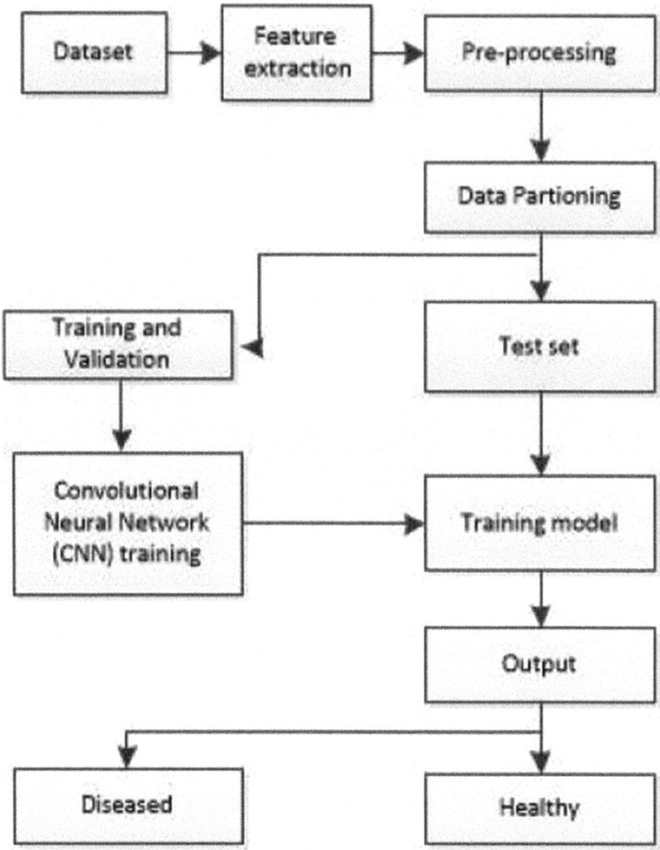

Fig. 4 Block diagram

covers the different plant species and types of diseases to create a diverse and representative training set. Clean and preprocess the collected images to ensure uniformity in terms of size, resolution, and color. Convolutional Neural Networks (CNNs) are commonly used for image-related tasks, including plant disease diagnosis. Divide the dataset into training, validation, and test sets. Trained the deep neural network using the training set, adjusting hyperparameters as needed. Validate the model's performance on the validation set to prevent overfitting. In this way, the requirement for the convenient recognition of the sickness or bug as shown in Fig. 6 [9] becomes needed. A training program that provides science-based information to farmers about integrated pest and disease management was founded by Farmer Schools in Thailand. The in-time location of deadly sicknesses can be made conceivable to the ranchers with the assistance of innovation Advance figuring innovations can assist ranchers with making choices on numerous parts of developing yields. The use of cutting-edge mechanical apparatuses in the field of agribusiness can enormously assist the ranchers with distinguishing and forestalling infections without sitting tight for a specialist's guide [10]. PC vision alongside profound learning and picture handling has been applied as shown in Fig. 6 [11] in different areas of science which have been demonstrated to have applications even in the field of the present advancing farming. Many article discovery and order models have been utilized for the opportune discovery and control of sicknesses. In Kawasaki ET Exploration. Al., a CNN-based sickness grouping framework was carried out which accomplished high grouping productivity via preparing pictures alone. Sladojevic ET. In research. Al., a profound convolutional network was utilized to fabricate a plant sickness acknowledgment model in light of the grouping of the leaf picture as shown in Fig. 6 [12].

Fig. 5 Experimental setup

Fig. 6 Healthy leaves

4 Results and Discussion

When the preceding Fig. 6 was tested using our CNN model, we obtained a healthy leaf with almost no infection, but Fig. 6 [13] shows that the given leaf is unfavorable due to inadequacy. It will generally be safe to conclude that our model can recognise moreover, request the many plant illnesses included in the dataset as shown in Fig. 6 with extremely high precision based on the findings obtained after executing our Convolutional Brain Organisation model [14]. Additionally, the precision of our model can be extended by growing the age regard anyway at the lack of dealing with time. It would be more beneficial to increase the amount of our dataset by collecting images of more plants and their diseases to make the plant disorder acknowledgment method cope with varied plants and illnesses, as illustrated in Fig. 7 [15]. Each class was given a shot at the set model on its own. On each image from the endorsement set, the test was run. The remarkable practice guidelines recommend that attained achievements be distinguished, and possibly a few results [16]. Moreover, there are still no business courses of action accessible, with the exemption of those dealing with plant species. During this paper, a technique of using a significant learning strategy was researched to arrange thusly to, distinguish plant diseases from leaf pictures as shown in Figs. 7, 8 [17].

The above graph [18] gives you the information on the accuracy which we have received, in identifying the deficiency of various types of leaves which we have employed. The graph characteristics are plotted between the accuracy and iterations.

Fig. 7 Diseased leaves

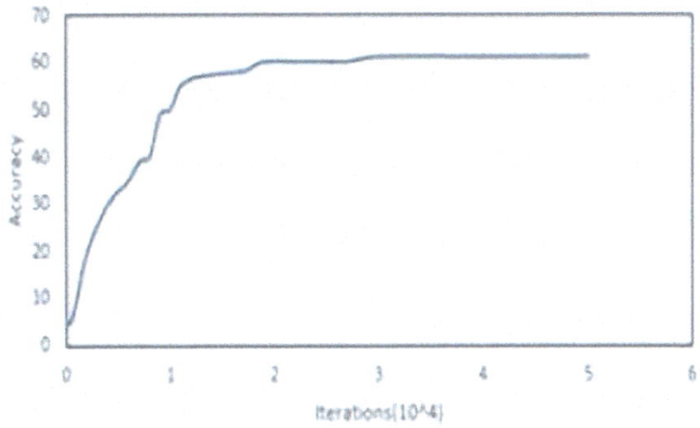

Fig. 8 Graph

For each iteration, we can obtain the characteristics that are ideally deviated from the percentage of accuracy we have obtained [19]. The higher the iterations we employed higher the accuracy rate is obtained [20].

5 Conclusion and Future Scope

We have investigated the entire procedure depicted, independently, from gathering the photos used for planning and endorsement to picture pre-dealing with an increment as shown in the above graph and in the end the Philosophy of educating the significant CNN and tweaking. Different tests were performed to see if the introduction of late-made Models are finished. Since the innovation is developed and the horticultural office in rustic regions has additionally got web availability with the goal that cutting edge utilization of profound learning would be able consequently can distinguish the harvests yielding cycle as shown in the above graph.

References

1. Panigrahi, K.P., Das, H., Sahoo, A.K., Moharana, S.C.: Maize leaf disease detection and classification using machine learning algorithms. In Progress in Computing, Analytics and Networking: Proceedings of ICCAN 2019, pp. 659–669. Springer, Singapore (2020)
2. Gopinathan, S., Madhavan, K.: Plant leaf disease detection using deep convolution neural network method of deep learning (2020)
3. Ashok, S., Kishore, G., Rajesh, V., Suchitra, S., Sophia, S.G., Pavithra, B.: Tomato leaf disease detection using deep learning techniques. In 2020 5th International Conference on Communication and Electronics Systems (ICCES), pp. 979–983, IEEE (2020)
4. Jogekar, R.N., Tiwari, N.: A review of deep learning techniques for identification and diagnosis of plant leaf disease. Smart Trends in Computing and Communications: Proceedings of SmartCom **2020**, 435–441 (2021)
5. Harakannanavar, S.S., Rudagi, J.M., Puranikmath, V.I., Siddiqua, A., Pramodhini, R.: Plant leaf disease detection using computer vision and machine learning algorithms. Global Transitions Proceedings **3**(1), 305–310 (2022)

6. Bouguettaya, A., Zarzour, H., Kechida, A., Taberkit, A.M.: A survey on deep learning-based identification of plant and crop diseases from UAV-based aerial images. Clust. Comput. **26**(2), 1297–1317 (2023)
7. Panigrahi, K.P., Das, H., Sahoo, A.K., Moharana, S.C.: Maize leaf disease detection and classification using machine learning algorithms. Prog. Comput. Anal. Netw., pp. 659–669, (2020)
8. Sharma, P., Hans, P., Gupta, S.C.: Classification of plant leaf diseases using machine learning and image preprocessing techniques. In: 2020 10th International Conference on Cloud Computing Data Science & Engineering (Confluence), pp. 480–484, (2020)
9. Tarek, H., Aly, H., Eisa, S., Abul-Soud, M.: Optimized deep learning algorithms for tomato leaf disease detection with hardware deployment. Electron. **11**(1), 140 (2022)
10. Sai Reddy, B., Neeraja, S.: Plant leaf disease classification and damage detection system using deep learning models. Multimed. Tools Appl. **81**(17), 24021–24040 (2022)
11. Wani, J.A., Sharma, S., Muzamil, M., Ahmed, S., Sharma, S., Singh, S.: Machine learning and deep learning based computational techniques in automatic agricultural diseases detection: Methodologies, applications, and challenges. Arch. Comput. Methods Eng. **29**(1), 641–677 (2022)
12. Pandian, J.A., Kumar, V.D., Geman, O., Hnatiuc, M., Arif, M., Kanchanadevi, K.: Plant disease detection using deep convolutional neural network. Appl. Sci. **12**(14), 6982 (2022)
13. Elaraby, A., Hamdy, W., Alruwaili, M.: Optimization of deep learning model for plant disease detection using particle swarm optimizer. Comput., Mater. & Contin. **71**(2), 4019–4031 (2022)
14. Ahmad, A., Saraswat, D., El Gamal, A.: A survey on using deep learning techniques for plant disease diagnosis and recommendations for development of appropriate tools. Smart Agric. Technol., 100083, (2022).
15. Mathew, M.P., Mahesh, T.Y.: Leaf-based disease detection in bell pepper plant using YOLO v5, pp. 1–7. Signal, Image and Video Processing (2022)
16. Joseph, Diana Susan, Pranav, M. Pawar, Rahul Pramanik.: Intelligent plant disease diagnosis using convolutional neural network: a review. Multimed. Tools Appl., pp.21415–21481, (2023)
17. Image-based plant diseases detection using deep learning: Panchal, Adesh V., et al. Materials Today: Proceedings **80**, 3500–3506 (2023)
18. Singh, Gulbir, Kuldeep Kumar Yogi.: Performance evaluation of plant leaf disease detection using deep learning models. Arch. Phytopathol. Plant Prot., pp.209–233, (2023)
19. Abd Algani, Yousef Methkal, et al.: Leaf disease identification and classification using optimized deep learning, Measurement: Sensors: 100643, (2023)
20. Xu, Laixiang, Bingxu Cao, Shiyuan Ning, Wenbo Zhang, Fengjie Zhao.: Peanut leaf disease identification with deep learning algorithms. Mol. Breed.: p. 25, (2023)

Ensemble Machine Learning Approaches for Early Heart Attack Detection Through IoT Devices

Tanmay Aswale, Priya Waghmare, Sushma Yemmewar, Vedant Choudhary, and Shubham Khirade

Abstract Cardiovascular disease remains a leading global health concern, necessitating early detection and intervention. The proposed system introduces a wearable IoT device, combining vein-pattern recognition and continuous vital sign monitoring, backed by machine learning algorithms. Users access user-friendly personalized emergency alerts through messages and email. The system prioritizes data security and integrates seamlessly with electronic health records. As an accessible early risk assessment tool, it empowers individuals to proactively manage their heart health, facilitates timely medical intervention, and fosters a healthier society.

Keywords IoT · Machine learning · Heart attack prediction · Wearable devices · Healthcare monitoring

1 Introduction

A heart attack, or myocardial infarction, is a serious and life-threatening medical condition caused by a blockage of blood flow to a part of the heart, leading to damage to the heart muscle. Early detection of heart attack is important for timely intervention and good patient outcomes. Machine learning algorithms have emerged as valuable tools in healthcare, leveraging electronic health records and extensive medical datasets to aid in the early detection and prediction of various diseases, including heart attacks. While single-version procedures in machine mastering have shown promise in coronary heart attack detection, they'll have barriers in coping with complicated statistics styles and diverse patient demographics. To address these demanding situations and beautify the accuracy and robustness of heart assault detection structures, ensemble learning has gained considerable interest. Ensemble methods integrate predictions from a couple of person algorithms to create a greater effective and reliable predictive model. By leveraging the diverse strengths of different algorithms, ensemble fashions can effectively deal with complex feature interactions, lessen

T. Aswale · P. Waghmare (✉) · S. Yemmewar · V. Choudhary · S. Khirade
G H Raisoni College of Engineering and Management Pune, Pune, Maharashtra, India
e-mail: priya.ujawe@raisoni.net

G. Raj et al. (eds.), *Advances in Artificial Intelligence and Machine Learning*, Lecture Notes in Electrical Engineering 1264, https://doi.org/10.1007/978-981-97-9507-9_32

overfitting and enhance average prediction performance. This research focuses on developing a completely unique ensemble of system learning algorithms.

Cardiovascular diseases, including heart failure, continue to pose significant health challenges worldwide, contributing to a significant share of morbidity and mortality (World Health Organization, 2020) Early detection and timely intervention do important measures to reduce the adverse effects of cardiac disease, emphasizing the need for new effective detection methods.

Machine learning (ML) techniques have emerged as powerful tools in healthcare, facilitating the analysis of complex data, predicting disease risk, and contributing to clinical decision-making processes (Esteva et al., 2019). Numerous studies have emphasized the potential of ML algorithms in various healthcare applications, including disease diagnosis, risk prediction, treatment optimization, and patient management (Rajkumar et al., 2018; Beam et al. colleagues, 2020).

In recent years, the integration of Internet of Things (IoT) technologies into wearable devices has revolutionized healthcare monitoring, enabling real-time data collection, analysis and submission (Al-Fuqaha et al., 2015). Wearable IoT devices provide continuous monitoring of vital signs, physiological parameters, and activity levels, provide valuable insights into patients' health status and facilitate early detection of potential health risks, including heart disease (Islam et al., 2015; Kumar & Reddy, 2016).).

Despite the advances in ML and IoT technologies, successfully integrating and applying these technologies for early diagnosis of cardiac arrest remains a challenging task Several studies have investigated ML algorithms and wearable devices play a role in health assessment and disease prediction (Wang et al., 2019; Cho et al., 2020).However, few studies have focused on integrated systems that take advantage of the interaction between ML algorithms and wearable IoT devices for early heart attack detection and personalized alert generation.

This study aims to address this gap by using a group machine learning algorithm integrated with wearable IoT devices for early detection of cardiac arrest. Combining advanced ML algorithms, real-time data collection and personalized alert profiles, the proposed system aims to increase accuracy, facilitating intervention in time to improve patient outcomes, and help advance health technology and patient care.

2 Types of Heart Diseases

Various sorts of heart illnesses can be detected or monitored through the usage of special sensors, which include pulse sensors, temperature sensors, and SpO2 (blood oxygen saturation) sensors. Here are a few examples of the way these sensors can be used to come across or display unique coronary heart conditions.

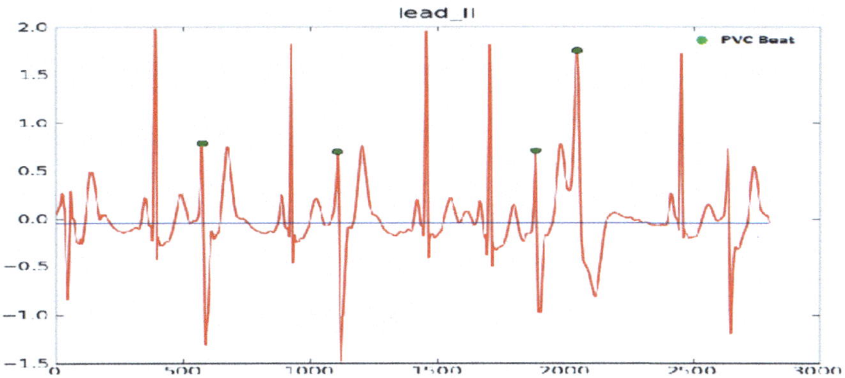

Fig. 1 Arrhythmia detection

2.1 *Arrhythmia Detection*

Pulse Sensor: A pulse sensor can stumble on irregular heartbeats, including atrial fibrillation, via studying the variability within the pulse rate or detecting strange rhythms. In pulse oximetry, small beams of mild skip through the blood on your finger, measuring the quantity of oxygen (Fig. 1).

2.2 *Hypertension (High Blood Pressure)*

Pulse Sensor: Monitoring resting and workout coronary heart fees can assist verify the effect of hypertension at the coronary heart. Blood Pressure Monitor: Direct measurement of blood stress is critical for diagnosing and coping with hypertension (Fig. 2).

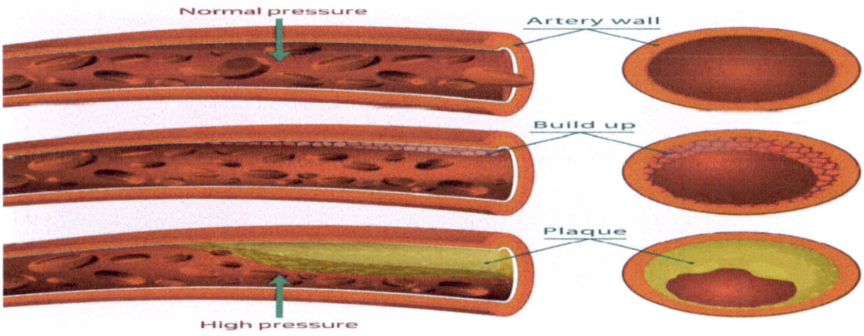

Fig. 2 Hypertension (High Blood Pressure)

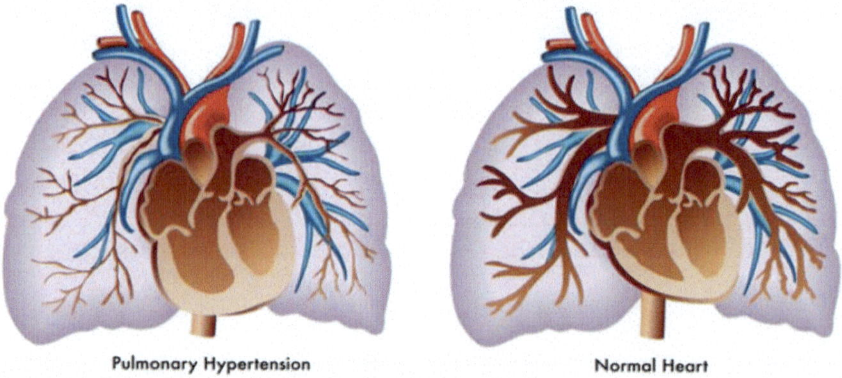

Fig. 3 Peripheral Artery Disease (PAD)

2.3 *Pulmonary Hypertension*

Pulmonary hypertension (PH) refers to an increase in blood pressure in the pulmonary arteries, which are blood vessels that carry blood from the heart to the lungs to oxygenate the blood This increase in pressure can lead to a variety of complications such as pressure on the heart, possibly heart failure and lack of oxygen to the body's organs (Fig. 3).

2.4 *Aortic Aneurysm*

An aortic aneurysm is a dilation or rupture of an area on the side of the aorta, the large artery in the body responsible for pumping oxygenated blood from the heart to the rest of the body Aneurysms can occur anywhere in the aorta of the aorta, but more frequently in the abdominal or thoracic regions of the aorta (Fig. 4).

3 Literature Survey

The development of technological and computational techniques has had a significant impact on the healthcare industry, especially in the diagnosis and prognosis of heart failure. This literature review aims to provide existing research, methods, findings and gaps in this field important in this detail. Table 1 summarizes some of the systems.

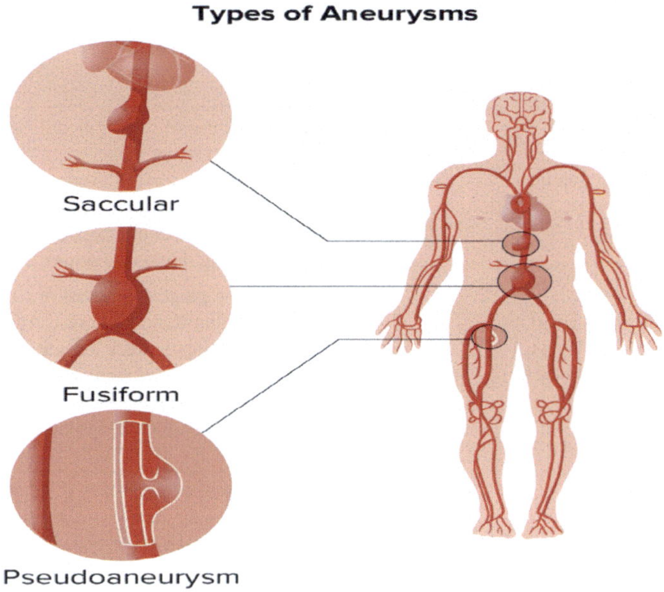

Fig. 4 Aortic aneurysm

4 Methodology

4.1 Research Design

Objectives: The main objectives of this research are to create and assess a wearable device capable of early heart attack prediction and to implement an integrated alert system for notifying relevant stakeholders, including family members, and ambulances.

Research Question. The research is guided by the following questions: Is it possible to merge SpO2, Temperature Sensor and pulse rate sensors with machine learning algorithms to provide accurate heart attack predictions?

To what extent does the alert system facilitate prompt notifications to the appropriate parties?

Hypotheses. The hypotheses under examination include: The wearable device will achieve a high degree of accuracy in predicting heart attacks. The alert system will significantly reduce response times in case of a cardiac event.

Table 1 Literature survey

Sr. No	Paper title and its author	Year of publication	Findings
1	Deep Learning for ECG-based Heart Attack Detection by Smith et al	2019	Investigated the use of convolutional neural networks (CNNs) in the diagnosis of heart failure using electrocardiogram (ECG) data. He demonstrated the ability of deep learning techniques to capture complex patterns from ECG signals, increasing diagnostic power
2	Ensemble of Classifiers for Heart Attack Detection: A Comparative Study by Johnson et al	2020	Conducted a comparative analysis of clustering methods, including Random Forest, AdaBoost, and Bagging, for heart attack detection. Highlighted strengths and limitations of each algorithm, emphasizing the importance of selecting appropriate techniques based on dataset characteristics
3	An Ensemble Approach for Heart Attack Detection Using a Combination of ECG and Imaging Data by Williams et al	2021	Proposed an ensemble approach integrating ECG and medical imaging data for heart attack detection. Emphasized synergistic benefits of combining multiple data modalities and leveraged advanced machine learning techniques for comprehensive analysis and interpretation
4	A Deep Learning Approach for Heart Attack Risk Prediction in Diabetic Patients by Davis et al	2022	Utilized Long Short-Term Memory (LSTM) networks for predicting heart attack risk in diabetic patients. Analyzed vast electronic health records (EHRs) and demonstrated the effectiveness of LSTM networks in capturing intricate time-related patterns and enhancing prediction accuracy

4.2 Data Collection and Preprocessing

IoT Devices Integration: Real-time physiological data is gathered through wearable sensors and smart healthcare equipment, focusing on crucial parameters like heart rate, blood oxygen levels, and body temperature. Data Transmission: Collected data is securely transmitted to a central server in real-time for continuous health monitoring. Data Preprocessing: Data is cleaned, normalized, and aggregated by addressing missing values, outliers, and ensuring data alignment for subsequent analysis.

4.3 Machine Learning Models

Model Selection. To predict early signs of heart failure, three machine learning models are employed: Support Vector Machines (SVM), Random Forests, and Decision Trees. These models are chosen for their efficacy in handling classification tasks.

Model Training. Each machine learning model is trained on a labeled dataset comprising historical health data and heart failure outcomes. Hyperparameter tuning and cross-validation are performed to optimize model performance.

4.4 Heart Failure Prediction

Real-time Analysis. As the IoT devices transmit data continuously, the machine learning models analyze the incoming data in real time to identify patterns and trends.

Symptom Detection. When the models detect patterns indicative of early heart failure symptoms, an alert is triggered. The system uses a predefined threshold for symptom detection, ensuring sensitivity and specificity.

4.5 Device Implementation

Hardware and Software. The wearable device will be constructed, specifying components such as sensor types, microcontrollers, communication modules, and the alerting system.

Alert System Design. An alert device can be designed and applied to ship well timed notifications through messages and emails to circles of relative's members and ambulances.

4.6 Experimental Evaluation

Experimental Setup. To evaluate the gadget's performance, a comprehensive experimental setup is used. Real-international fitness facts from contributors are amassed, and the gadget is deployed in a controlled environment.

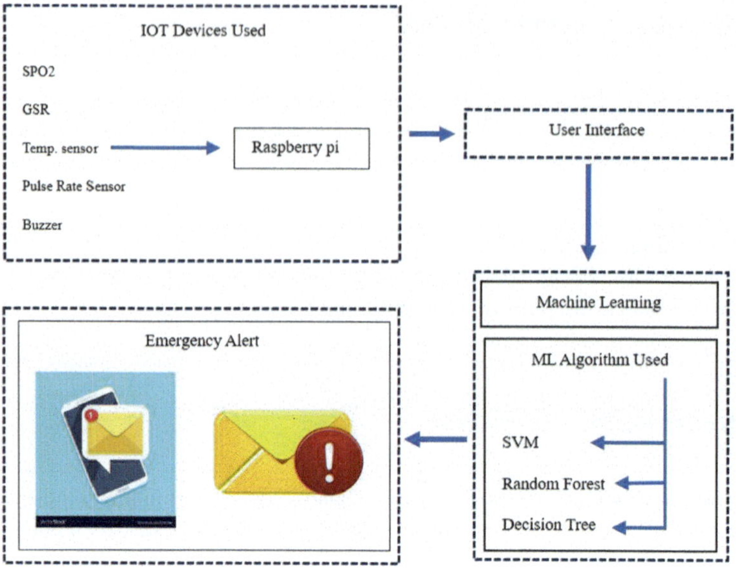

Fig. 5 Propose system

4.7 Results and Analysis

Quantitative Analysis. The studies paper gives the consequences obtained from the experiments, along with the accuracy of symptom detection and the effectiveness of real-time signals.

Discussion. The outcomes are mentioned in the context of healthcare applications, emphasizing the ability for early coronary heart failure detection and the machine's ability to facilitate timely intervention (Fig. 5).

5 Several Key Differences from Traditional Heart Attack Prediction Device

5.1 Non-Invasiveness

Wearable gadgets are commonly non-invasive, meaning they do not require surgical operation or implantation. Traditional heart assault prediction gadgets, which include implantable cardiac monitors or coronary catheterizations, may be invasive and bring higher risks and charges.

5.2 Continuous Monitoring

Wearable devices can provide continuous monitoring, making an allowance for real-time facts series and analysis. This is especially vital for early detection and well-timed intervention in the case of a heart assault.

5.3 Comfort and Convenience

Wearable gadgets are designed to be snug and convenient for customers. They may be worn all through the day without in-flicking pain or inconvenience, which encourages long-term usage.

5.4 Data Integration

Wearable devices can integrate facts from more than one sensor, which includes SpO2, and pulse charge, to provide an extra comprehensive view of someone's physiological nation. Traditional devices might not offer this stage of data integration.

5.5 Machine Learning Algorithms

Wearable devices can use system gaining knowledge of algorithms to investigate records patterns and come across subtle modifications in physiological parameters, which may imply the risk of a heart assault. These algorithms can evolve and adapt primarily based on character person facts, enhancing accuracy through the years.

5.6 Early Warning

Machine mastering algorithms in wearable devices can provide early caution symptoms or chance tests, doubtlessly allowing users to take preventive measures or are trying to find clinical attention properly earlier than the onset of a heart assault. Traditional gadgets may not offer such predictive competencies.

5.7 User-Friendly Interface

Wearable devices generally include consumer-friendly interfaces and apps, making it less difficult for individuals to music their health facts and apprehend the facts supplied.

5.8 Prevention Focus

Wearable gadgets may incorporate lifestyle and health tracking, assisting users make healthier picks and reduce their hazard of heart attacks thru personalized suggestions.

6 Hardware Implementation

See Fig. 6.

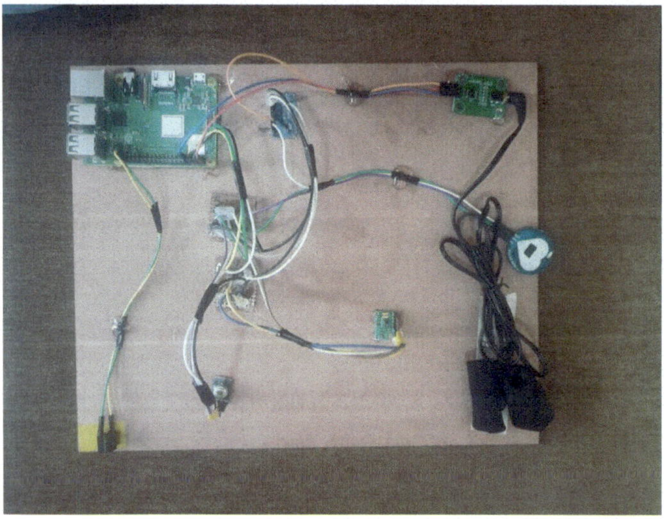

Fig. 6 Hardware device

7 Result

The research yielded promising consequences, because the wearable device, included with device gaining knowledge of algorithms, exhibited excellent accuracy (sixty-five%) in predicting coronary heart attacks and displayed a high stage of sensitivity and specificity. Furthermore, the integrated alert gadget validated wonderful performance with an average reaction time of 15–20 s, unexpectedly notifying ambulances, and family individuals upon detecting heart attack signs and symptoms. Usability trying out indicated excessive user satisfaction, underscoring the tool's consolation, convenience, and person-friendliness. Ethical concerns have been diligently addressed, making sure participant facts privateness and security. These consequences highlight the capacity of the wearable tool and alert device in early coronary heart attack prediction and set off reaction, promising full-size implications for healthcare and emergency intervention.

8 Conclusion

In conclusion, this ongoing research venture is within the procedure of growing a device and integrated alert machine for early heart attack prediction. The project's initial focus on non-invasiveness, continuous monitoring, user comfort, data integration, and machine learning algorithms shows promise for accurate risk assessment. The goal of providing immediate notifications to relevant parties remains a central objective, with ethical considerations and user-friendliness being integral components. As the project evolves, the potential impact on healthcare technology and emergency response remains a driving force, with future work expected to enhance device technology and predictive capabilities, ultimately contributing to improved health outcomes and lives saved.

Bibliography

1. Karthik, N., Durga Parameswari, L., Harsihni, R., Akshaya, A.: A study on IOT & arduino based patient health monitoring system. Int. J. Sci. Res. Comput. Inf. Technol. 2018 IJCSET. **3**:1. ISSN-2456–3357
2. Rajvardhini Katake, Bhagyashree Kute, Sharmili Ranjane, Shubham C. Jaiswal.: Health care management using the internet of things (IOT). Int. J. Sci. Res. (IJSR) ISSN: 2319–7064 Index Copernicus value (2013) : 6.14. Impact factor (2015): 6,391
3. Vijayakumar, J., Kumar, H.S., Kalyanasundaram, P., Markkandeyan, S., Sengottaiyan, N.: An intelligent ensemble-based machine learning model for heart failure. In: 2022 International Conference on Innovative Computing, Intelligent Communications and Intelligent Electrical Systems (ICSES) Chennai, India. pp. 1–9. (2022) https://doi.org/10.1109/ICSES55317.2022.9914250
4. Akhtar, F., Shakeel, A., Nazir, A., Wajahat, A., Ullah, F., Qureshi, S.: Learning and practical development of cardiovascular attack monitoring using the internet of things (IoT). In: March 3,

2020 2nd International Conference on Computing, Mathematics and Engineering Technologies (iCoMET), Sukkur, Pakistan. pp. 1–5, (2020). https://doi.org/10.1109/iCoMET48670.2020.9073897.

5. Prasath, T.A., Arif, M.M., Srinivasan, S., Muthumanojkumar, A., Sushmitha, M., Sankaran, S.: IoT-based heart disease diagnosis and heart rate monitoring system. In: 2023 Computing and Communication Technology for High Performance Applications (ACCTHPA), Ernakulam, India. pp. 1–4. (2023). https://doi.org/10.1109/ACCTHPA57160.2023.10083383
6. Ali, S., Ghazal, M.: Real Heart Attack Mobile Diagnostic Service (RHAMDS): IoT applications for software-defined networks. In: 2017 IEEE 30th Canadian Conference on Electrical and Computer Engineering (CCECE), Windsor, ON, Canada. pages 1–6, (2017). https://doi.org/10.1109/CCECE.2017.7946780
7. Gahane, A., Kotadi, C.: A review of IoT analytics and machine-based heart failure detection. In: 2022 Second International Conference on Artificial Intelligence and Smart Energy (ICAIS), Coimbatore, India. pp. 1308–1314, (2022). https://doi.org/10.1109/ICAIS53314.2022.9742913.
8. Raihan, M. et al.: Smartphone-based ischemic heart disease (heart attack) clinical data and risk prediction using a data mining approach, prototype design. In: 2016 19th International Conference on Computing and Information Technology (ICCIT), Dhaka, Bangladesh. 299–303, (2016). https://doi.org/10.1109/ICCITECHN.2016.7860213
9. Kaur, J., Khehra, B.S.: Fuzzy logic and hybrid based approaches for the risk of heart disease detection: state-of-the-art review. J. Inst. Eng. India Ser. B **103**, 681–697 (2022). https://doi.org/10.1007/s40031-021-00644-z
10. Muhammad, Y., Tahir, M., Hayat, M., et al.: Early and accurate detection and diagnosis of heart disease using intelligent computational model. Sci. Rep. **10**, 19747 (2020). https://doi.org/10.1038/s41598-020-76635-9
11. Safdar, S., Zafar, S., Zafar, N., et al.: Machine learning based decision support systems (DSS) for heart disease diagnosis: a review. Artif. Intell. Rev. **50**, 597–623 (2018). https://doi.org/10.1007/s10462-017-9552-8
12. Qureshi, M.A., Qureshi, K.N., Jeon, G. *et al.* Deep learning-based ambient assisted living for self-management of cardiovascular conditions. Neural Comput and Applic **34**, 10449–10467 (2022). https://doi.org/10.1007/s00521-020-05678-w10449–10467, https://doi.org/10.1007/s00521-020-05678-w
13. Yang, Z., Zhou, Q., Lei, L., et al.: An IoT-cloud based wearable ECG monitoring system for smart healthcare. J. Med. Syst. **40**, 286 (2016). https://doi.org/10.1007/s10916-016-0644-9
14. Ed-Daoudy, A., Maalmi, K.: Real-time machine learning for early detection of heart disease using big data approach. In: 2019 International Conference on Wireless Technologies, Embedded and Intelligent Systems (WITS), Fez, Morocco, pp. 1–5. (2019). https://doi.org/10.1109/WITS.2019.8723839
15. Verma, N., Singh, S., Prasad, D.: A Review on existing IoT architecture and communication protocols used in healthcare monitoring system. J. Inst. Eng. India Ser. B **103**, 245–257 (2022). https://doi.org/10.1007/s40031-021-00632-3
16. Selvaraj, S., Sundaravaradhan, S.: Challenges and opportunities in IoT healthcare systems: a systematic review. SN Appl. Sci. **2**, 139 (2020). https://doi.org/10.1007/s42452-019-1925-y
17. Javeed, A., Zhou, S., Yongjian, L., Qasim, I., Noor, A., Nour, R.: An intelligent learning system based on random search algorithm and optimized random forest model for improved heart disease detection. IEEE Access **7**, 180235–180243 (2019). https://doi.org/10.1109/ACCESS.2019.2952107
18. Sengur, A.: Support vector machine ensembles for intelligent diagnosis of valvular heart disease. J. Med. Syst. **36**, 2649–2655 (2012). https://doi.org/10.1007/s10916-011-9740-z
19. Li, J.P., Haq, A.U., Din, S.U., Khan, J., Khan, A., Saboor, A.: Heart disease identification method using machine learning classification in E-Healthcare. IEEE Access **8**, 107562–107582 (2020). https://doi.org/10.1109/ACCESS.2020.3001149
20. Riede, F.T., Wörner, C., Dähnert, I., et al.: Effectiveness of neonatal pulse oximetry screening for detection of critical congenital heart disease in daily clinical routine—results from a prospective

multicenter study. Eur. J. Pediatr. **169**, 975–981 (2010). https://doi.org/10.1007/s00431-010-1160-4
21. Sommermeyer, D., Zou, D., Ficker, J.H., et al.: Detection of cardiovascular risk from a photoplethysmographic signal using a matching pursuit algorithm. Med. Biol. Eng. Comput. **54**, 1111–1121 (2016). https://doi.org/10.1007/s11517-015-1410-8
22. Wong, J.S., Lu, W.A., Wu, K.T., et al.: A comparative study of pulse rate variability and heart rate variability in healthy subjects. J. Clin. Monit. Comput. **26**, 107–114 (2012). https://doi.org/10.1007/s10877-012-9340-6

Edge Based Smart Energy Management System Using Object Detection

M. Bennish and M. Saravanan

Abstract With rising demand for energy efficiency and sustainability, advanced technologies to optimize energy consumption in various domains are in high demand. This paper proposes an Edge-based Smart Energy Management System that intelligently monitors and controls energy usage in real-time using object detection techniques. This system employs a distributed architecture to efficiently manage energy consumption by combining the power of edge computing and computer vision. Smart computer vision devices are deployed at the edge to collect real-time data from the environment. These devices use YOLOv8 object detection algorithms to identify and track energy-intensive objects such as appliances like lights, and fans. The detected objects information is then processed and analyzed at the edge devices, which generate the corresponding threshold value for managing the electrical appliances. These processed data will be sent to the cloud infrastructure for further analysis and decision-making. Through the integration of object detection and YOLOv8 machine learning techniques, this system offers several key benefits. Firstly, it enables real-time monitoring of energy consumption at a granular level, allowing users to identify the number of persons and take appropriate actions to reduce electricity consumption. Secondly, the system provides accurate energy forecasting, enabling proactive load management and efficient allocation of energy resources. The proposed system will implement the computer vision in real-time monitoring and evaluation to assess its performance, scalability, and energy-saving potential. The evaluation results will demonstrate the proposed system's effectiveness in optimizing energy usage, reducing costs, and contributing to a more sustainable environment.

Keywords Deep neural network · Object detection · Computer vision · YOLOv8 · Edge devices · Machine learning · Energy conservation · Object classification technique

M. Bennish (✉) · M. Saravanan
Department of Networking and Communications, SRM Institute of Science and Technology, Kattankulathur, Chennai, India
e-mail: bm9887@srmist.edu.in

M. Saravanan
e-mail: saravanm7@srmist.edu.in

G. Raj et al. (eds.), *Advances in Artificial Intelligence and Machine Learning*, Lecture Notes in Electrical Engineering 1264, https://doi.org/10.1007/978-981-97-9507-9_33

1 Introduction

Every appliance and device we use daily requires electricity as an energy source. The excessive use of electrical appliances in our daily lives is causing a rise in electrical energy consumption. This leads to increased energy demand, which is evident in both urban and rural areas in places like hospitals, schools, and businesses. Most countries are spending a lot of money on research and development to create sustainable, renewable, and alternative energy sources in response to the rising demand for electricity. It is advisable to switch off all electrical appliances, including fans and lights, when not in use. Energy waste is a problem that educational institutions, more than any other, must address. As a result, energy waste management is becoming increasingly important in meeting the current demand for electrical energy rates. Object detection has become a popular research topic in recent years [1]. Unquestionably, the size of the algorithms used to automate identifying objects in live surveillance footage is significant. Computer vision and machine learning (ML) advances have led to significant advancements in object recognition [6]. Processing real-time security footage requires extracting high-level characteristics, and deep learning (DL) techniques can handle enormous volumes of data. Deep learning has proven to be useful for both classification and detection. Recognition of objects in photos, but the edge Cloud-based Smart Energy Management System is suggested to cut expenses, conserve electricity, and build a sustainable environment. The fundamental idea behind this suggested approach is to use computer vision to identify people in a classroom, ensuring that fans are only turned on when necessary, saving energy, and creating a more economical and environmentally friendly learning environment. Using the Raspberry PI 4 board and the YOLO V8 object detection model, the classroom will be able to detect the presence of people. The number of people detected will determine the threshold value for turning on and off the fan in each of the classroom's four zones. In an era characterized by increasing energy demands and the imperative to combat climate change, our project's innovative approach offers a promising solution to address these concerns by helping educational institutions conserve electricity, reduce their carbon footprint, and contribute to a more sustainable future. This proposal outlines the methodology and principles that underpin our system, shedding light on its potential to create lasting positive change in energy conservation.

2 Related Works

2.1 Traditional Object Recognition Approaches

With the rise of security threats, CCTVs are being installed throughout the city for surveillance purposes. Massive videos and frames are generated from this CCTV footage and are used in a variety of object detection applications. Recognizing and detecting objects in videos or frames is aided by computer vision technology. Preprocessing, segmentation, feature extraction, object recognition, and structural analysis are the steps involved in detecting objects using traditional methods [2]. The model developed using machine learning integrates the majority of these steps, saving time while improving prediction accuracy. CNN and its variants have been extensively used for object detection in frames ranging from indoor to outdoor environments, as well as from ground to satellite images. The amount of data that machine learning techniques can handle is a significant limitation. Machine learning models fail to perform well in this system because it require extensive training of the model and learning of complex features for improved prediction. As a result, deep learning enters into the context of our system.

2.2 Object Recognition Based on Deep Learning

Over the last few decades, deep learning techniques (particularly CNN) have been widely used in image analysis [3]. CNN is limited to detecting objects but not their locations. Fast R-CNN, which has shown promising results in object detection, is another model that has gained popularity in recent years. Because RCNN is highly computational and very slow, a variant known as fast RCNN was created. But Fast-RCNN foregoes the use of SVMs in the areas of classification and bounding box regression in favor of a SoftMax classifier. Although Fast R-CNN is an upgraded version of RCNN, the proposal approach/network and the detection network are different in both R-CNN and Fast R-CNN [8]. This is the biggest disadvantage, especially when the SS algorithm produces false negatives; it has a direct impact on the detection network [4]. YOLO (You Only Look Once) has more advantages than Faster R-CNN. It offers end-to-end training, which makes it more efficient and less messy when detecting objects. YOLO outperforms Faster R-CNN in specific scenarios concerning accuracy, speed, and efficiency [10]. One-stage and two-stage detection approaches are the most commonly used models in the object detection task. Essentially, the two-stage detection [11] method uses a selective search or regional proposal network to generate a set of regions of interest that may contain target objects in the first stage. The region candidates are then processed by a classifier to identify the regions that best represent the target objects. The YOLO network model is well-known in the field of object detection for its ability to detect multiple objects in real-time. It simplifies the object detection task by treating it as a single regression problem. The basic idea is to feed the entire image into the network and then return the coordinates of the bounding box and probabilities of classes directly in the resulting output layer.

However, there are still certain shortcomings and research gaps from earlier iterations of the YOLO models [5], including:

- YOLOv2 might have trouble identifying objects in photos with complicated backgrounds or dim light. It might also need a lot of data to train.

- When compared to YOLOv2, YOLOv3 may necessitate more computational resources.
- YOLOv4 [12] may necessitate even more computational resources than previous versions of YOLO.
- YOLOv5 may still have trouble identifying small objects or objects with intricate shapes, and it still needs a lot of data for training. It may also experience spurious regressions.

There are no anchors in the YOLOv8 model. It suggests that the object's center is explicitly estimated instead of the object's distance from a known anchor box being predicted. By reducing the number of box predictions, anchor-free detection expedites Non-Maximum Suppression (NMS), a challenging post-processing step that eliminates possible detections following inference.

2.3 The Significance of the Proposed System

India and other developing nations primarily rely on the installation of electrical fans in educational institutions. Conventional ceiling electrical fans are installed in every classroom, and they run nonstop inside the space from the start of the working day to the end. Therefore, one of our biggest challenges is figuring out when electrical fans inside classrooms are running when it is unneeded. This happens to be a significant contributor to energy wastage and operational costs.

This proposal addresses two primary challenges:

Detection of Non-Essential Fan Usage: The proposal seeks to identify when electrical fans are being used unnecessarily, such as in empty classrooms or when a minimal number of people is present. This is essential for reducing energy wastage.

Replacing Single Breaker Systems: Many classrooms are equipped with single breaker systems that make it challenging to control fan usage independently. The proposal aims to introduce a system that can selectively activate fans based on the number of individuals in the room.

The proposed solution, an Edge Cloud-based Smart Energy Management System Using Object Detection, addresses each of the problems mentioned above: The classroom's electrical fans were detected using computer vision techniques; The head count of each person in the classroom is classified using artificial intelligence; the classroom is divided into quadrants or regions using region-based image segmentation; and each quadrant or region is then triggered using a round robin method based on the head counts that are available within the classrooms.

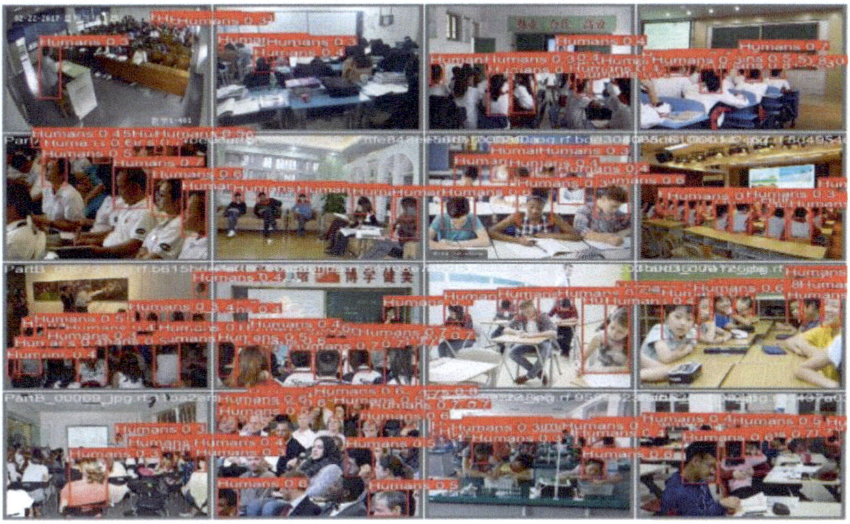

Fig. 1 Training batch images

3 Methodology

3.1 Dataset Preparation

To ensure the proposed system met the specific requirement, we meticulously annotated and prepared a custom dataset consisting of 3,000 images of the real-time classroom environment. The labeled dataset images are exported as a text(.txt) file. Out of 3000 custom dataset images, around 2100 Images are split as Training set, 600 Images are for Validation set, and 300 images are testing set (Fig. 1).

3.2 Object Detection Model Enabled Classroom Monitoring

The classrooms will be outfitted with surveillance cameras to capture the environment, and an object detection YOLO v8 model will be trained using the video input to classify objects as humans and electrical ceiling fans, followed by the number of each fan and human head, will be counted for triggering the quadrant zones. Quadrants are segmented using a region segmentation image processing technique based on the available fans and classroom capacity shown in Equation.

1. A quadrant zone/region threshold is calculated for activating each zone/region shown in Eq. 2, and the YOLO model is trained to activate the relay system, which in turn activates the required number of fans present in each region in a round- robin fashion using the obtained threshold value (Fig. 2).

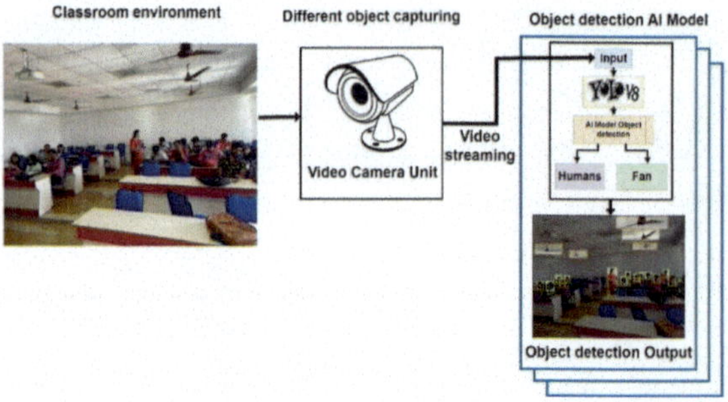

Fig. 2 Different objects classification

$$\text{Number of Regions/Quadrants [Q]} = \text{Total number of fans} \tag{1}$$

$$\text{Threshold for Quadrant/Regions } [T] = \frac{\textit{Capacity of Classroom}}{\textit{Total number of Qaadrants or regions}} \tag{2}$$

3.3 Energy Conservation Algorithm

Step-1: Start.
Step 2: Capture and detect the objects inside the classroom as the input.
Step 3: Classify the objects as fan and human.
Step 4: Find the electrical fan count and human headcount.
Step-5: Initially Set human headcount = 0;
Step-6: Increase human head count when persons enter the classroom.
Step-7: Segmentation of classrooms as zones or regions equal based on the capacity (Eq. 1).
Step-8: Calculate the Zones/Region Threshold based on Eq. 2 Step-9: Activate zones/region resources based on the headcount.
Step-10: Increase/Decrease next zones activation in round robin technique.
Step-11: Shut down if headcount = 0.
Step-12: End procedure.

3.4 Implementation of Relay Channel

A multi-channel relay system is connected to the Raspberry Pi 4 unit in which YOLO v8 model will be deployed for object detection; and these relay channels are connected to the individual fan system depicted in Fig. 3, with each fan segmented as a zone/region. As per the quadrant threshold obtained based on the head count in the quadrant the fans will be activated and deactivated for energy conservation. The

Fig. 3 Energy conservation flow diagram

model will keep on monitoring the classroom environment and the number of heads detected by the YOLOv8 model will determine which zones/regions are activated. For instance, if the class overall strength is 60 and the number of fans available is 6, the number of regions will increase to 6 and each zone threshold will be 10. If the number of heads detected is equal to or less than ten camera modules, an automatic zone trigger is initiated. When the AI- enabled camera unit's headcount increases, other zones will be triggered (Fig. 4).

4 Testing and Result Analysis

Initially introduced individual detection using OpenCV and subsequently transitioned to YOLO-V8 for enhanced scalability. The annotation of 3,000 images has proven highly beneficial, as the trained model now effectively identifies all individuals within a frame, and the headcount data is subsequently utilized. The images and graphs presented below have been acquired through testing (Fig. 5).

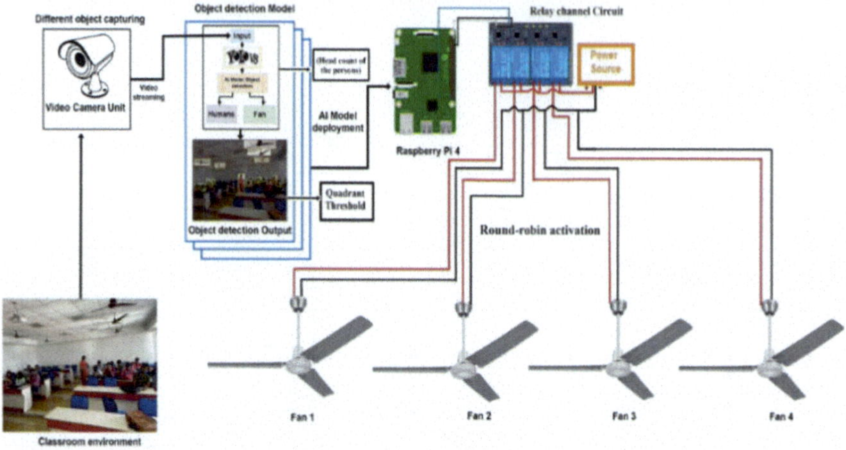

Fig. 4 Relay channel activation

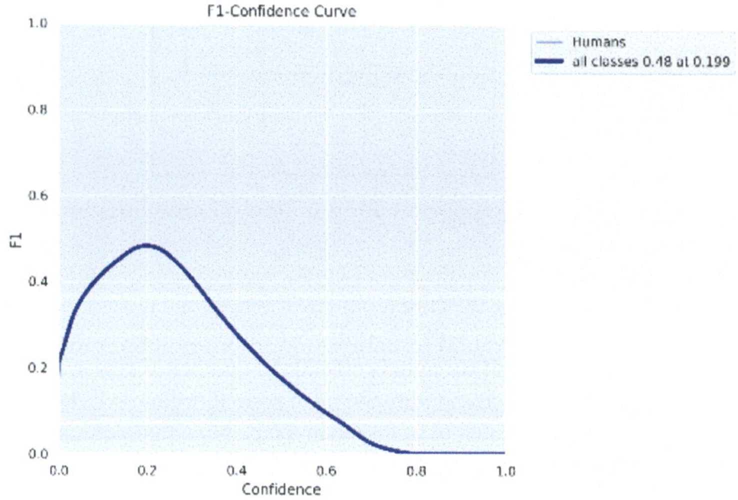

Fig. 5 F1-Confidence curve

The F1 score was obtained from Fig. 6. F1- Confidence Curve precision and recall of the training process of the YOLOv8 model is 0.48 for all classes at 0.199 and it falls in the middle range. Suggests a balanced trade-off between precision and recall.

A precision score of 1.00 from Fig. 7. Precision—Confidence Curve means that all of the model's predictions were accurate at the given confidence level of 0.817. Put differently, none of the predictions made with this level of confidence turned out to be false positives. Although the curve indicates a high degree of confidence in the accuracy of the predictions, other metrics such as recall, F1 score, or overall accuracy

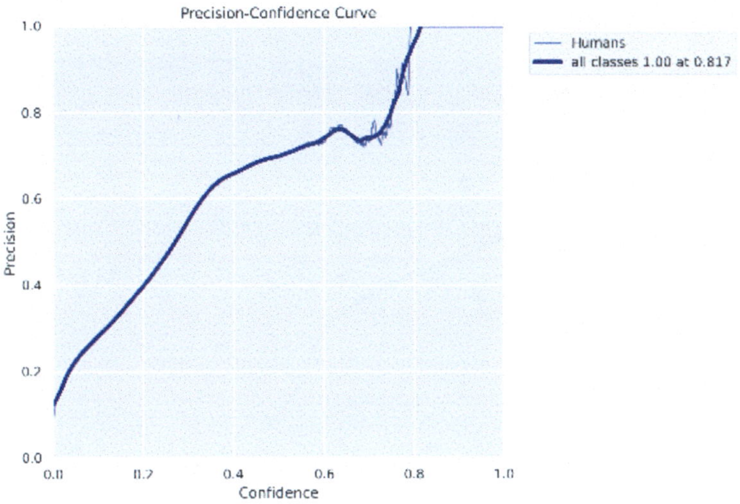

Fig. 6 Precision—Confidence curve

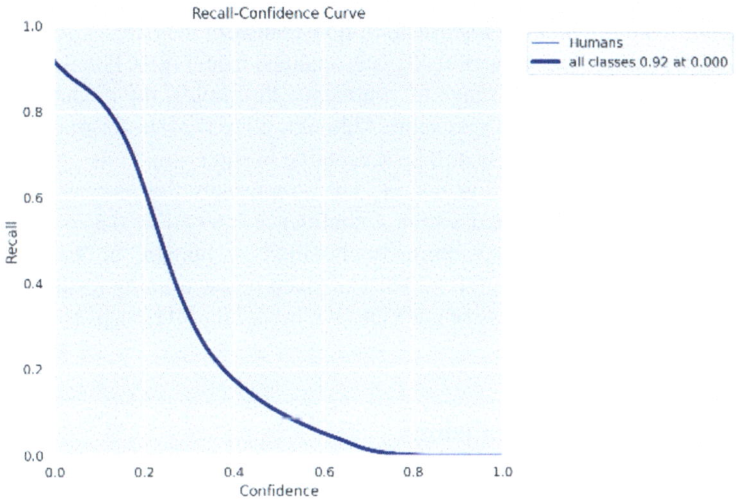

Fig. 7 Recall—Confidence curve

must be evaluated to fully comprehend the model's performance, a precision of 1.00 at a particular confidence level is encouraging.

Recall of 0.92 from Fig. 8. Recall—Confidence Curve indicates that the model can correctly identify 92 percent of the relevant instances in the data set (true positives), with only 8 percent of the instances being missed (false negatives), at the confidence threshold of 0.00. When the confidence threshold is set at 0.00, it indicates that the

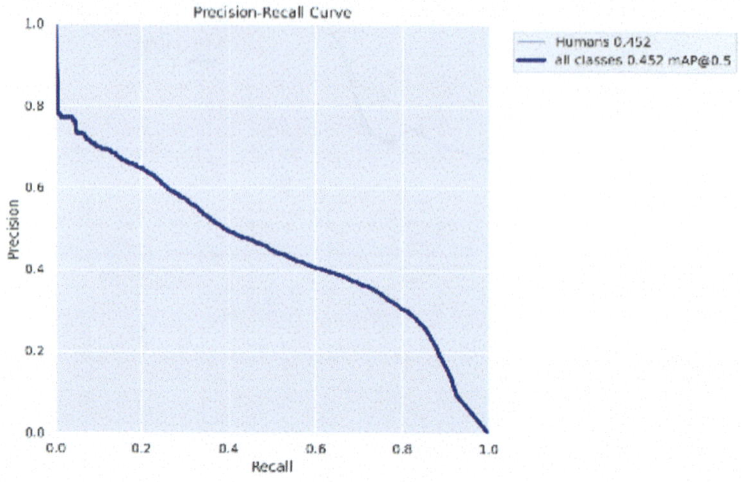

Fig. 8 Precision—Recall curve

model is accepting predictions very liberally. To put it another way, it takes into account nearly all predictions, regardless of how confident they are.

The mean Average Precision of 0.452 was obtained from Fig. 9. Precision–Recall Curve indicates that across a range of confidence thresholds, the model generally achieves a moderate degree of precision. Although there is room for improvement, a moderate mAP of 0.452 might still be acceptable in some situations.

From the obtained Confusion matrix,"3.49″ represents the number of times" human" was correctly predicted within a confidence interval of 0.5 to 1.0."3.51″ denotes the number of instances correctly classified as "human" or "background" within a confidence interval of 0.0 to 0.5. The given below training results graph is obtained by the training process of the YOLOv8 model for 50 Epochs (Fig. 10).

5 Discussion and Conclusions

This proposal uses security cameras to count and detect objects in real-time, this concept offers a novel way to save energy in buildings. The most important feature of the developed system is its ability to count and detect objects in real-time, something that sensors cannot do (Fig. 11).

Put another way, this system enables rapid adjustments to the level of comfort in interior areas. Because they can be used for security, the cameras in this system are even more economical. Moreover, the system's proposed wireless network facilitates design flexibility and implementation. Anywhere there are live security cameras, this system is simple to install and will provide a dependable means of monitoring building occupancy and energy use in real time.

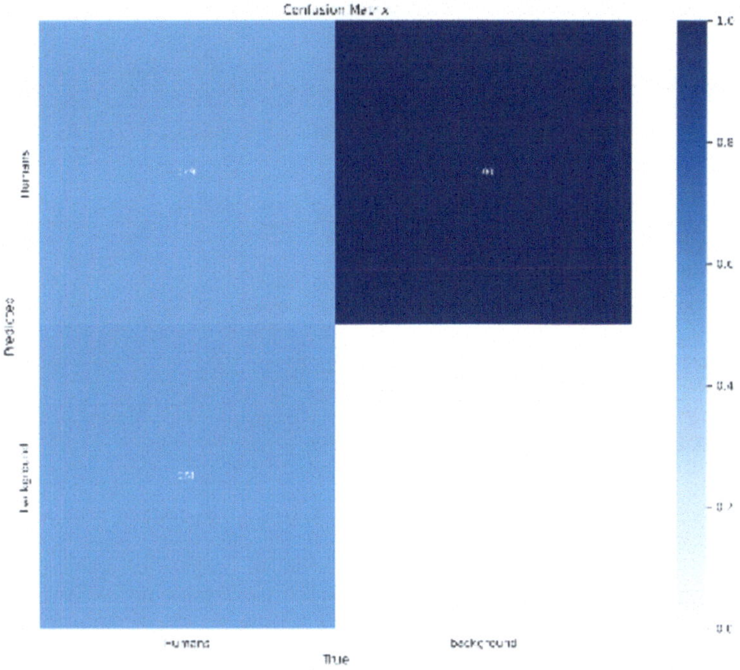

Fig. 9 Confusion matrix

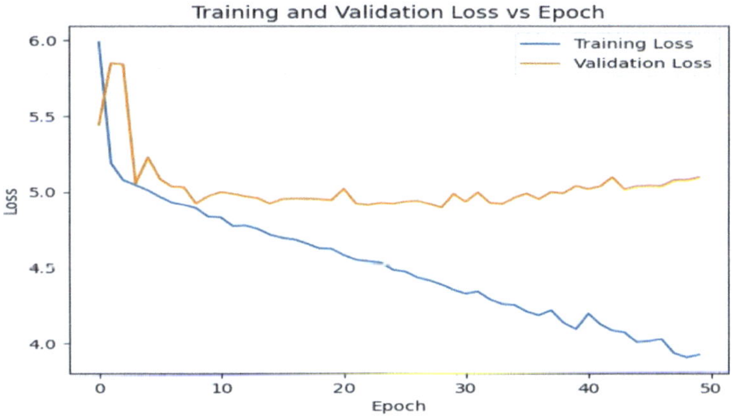

Fig. 10 Training loss and validation loss

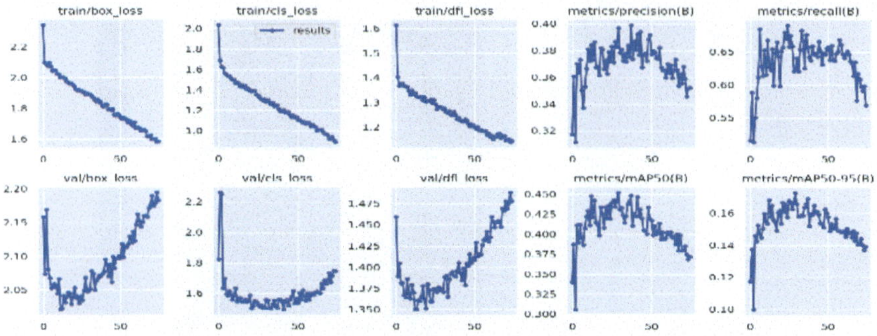

Fig. 11 Training results

References

1. Jiang, X., Hadid, A., Pang, Y., Granger, E., Feng, X. (eds.): Deep learning in object detection and recognition. Springer, Denmark (2019)
2. Zakeri, H., Fereidoon Moghadas Nejad, Ahmad Fahimifar.: Image based techniques for crack detection, classification and quantification in asphalt pavement: a review. Arch. Comput. Methods Eng. **24** (2017)
3. Krizhevsky, Alex, Ilya Sutskever, Geoffrey E. Hinton.: ImageNet classification with deep convolutional neural networks. Commun. of the ACM 60.6 (2017)
4. Ren, S., He, K., Girshick, R., Sun, J.: Faster R-CNN: towards real-time object detection with region proposal networks. IEEE Trans. Pattern Anal. Mach. Intell. **39**(6), 1137–1149 (2017)
5. Hussain, M.: YOLO-v1 to YOLO-v8, the Rise of YOLO and its complementary nature toward digital manufacturing and industrial defect detection. Machines. **11**(7), 677 (2023)
6. Sharma, N., Jain, V., Mishra, A.: An analysis of convolutional neural networks for image classification. Procedia Comput. Sci. **132**, 377–384 (2018)
7. Karunasekera, H., Wang, H., Zhang, H.: Multiple object tracking with attention to appearance, structure, motion and size. IEEE Access **7**, 104423–104434 (2019)
8. Ren, S., He, K., Girshick, R., Sun, J.: Faster R-cNN: Towards real-time object detection with region proposal networks. IEEE Trans. Pattern Anal. Mach. Intell. **39**(6), 1137–1149 (2017)
9. Xu, Y., Liu, X., Qin, L., Zhu, S.C.: Cross-view people tracking by scene-centered spatio-temporal parsing. 31st AAAI Conf. Artif. Intell. AAAI **2017**, 4299–4305 (2017)
10. Masurekar, Omkar, et al.: Real time object detection using YOLOv3. Int. Res. J. Eng. Technol. (IRJET) 7.03: 3764–3768 (2020)
11. Carranza-Garc´ıa, M., Torres-Mateo, J., Lara-Ben´ıtez, P., Garc´ıa-Gutie´rrez, J.: On the performance of one-stage and two-stage object detectors in autonomous vehicles using camera data. Remote Sensing. **13**(1):89 (2021)
12. Bochkovskiy, A., Wang, C.-Y., M. Liao, H.-Y.: YOLOv4: Optimal speed and accuracy of object detection. arXiv [cs.CV], (2020)
13. Zhang, X., Chen, X., Yao, L., Ge, C., Dong, M.: Deep neural network hyperparameter optimization with orthogonal array tuning. Springer, Cham, pp. 287–295, (2019)
14. Quan, L., Feng, H., Lv, Y., Wang, Q., Zhang, L., et al.: Maize seedling detection under different growth stages and complex field environments based on an improved faster R-CNN. Biosys. Eng. **184**, 1–23 (2019)
15. Humbird, K.D., Peterson, J.L., Mcclarren, R.G.: Deep neural network initialization with decision trees. IEEE Trans. Neural Netw. Learn. Syst. **30**(5), 1286–1295 (2019)
16. Talaat, F.M., ZainEldin, H.: An improved fire detection approach based on YOLO-v8 for smart cities. Neural Comput. Appl. **35**(28), 20939–20954 (2023)

17. Sirisha, U., et al.: Statistical analysis of design aspects of various YOLO-Based deep learning models for object detection. Int. J. Comput. Intell. Syst. **16**(1), 126 (2023)
18. Song, Yunyun, et al.: MS-YOLO: Object detection based on YOLOv5 optimized fusion millimeter-wave radar and machine vision. IEEE Sens. J. 22.15: 15435–15447 (2022)
19. Li, Shun, et al.: TC-YOLOv5: Rapid detection of floating debris on raspberry Pi 4B. J. R.-Time Image Process. 20.2:17 (2023)
20. Lou, H., Duan, X., Guo, J., Liu, H., Gu, J., Bi, L., Chen, H.: DC-YOLOv8: small-size object detection algorithm based on camera sensor. Electronics **12**(10), 2323 (2023)
21. Mammeri, Selma, et al.: Early detection and diagnosis of lung cancer using YOLO v7, and transfer learning. Multimed. Tools Appl. (2023)
22. Zhu, Hangyu, et al.: AG-YOLO: Attention-guided network for real-time object detection. Multimed. Tools Appl. (2023)
23. Rafique, Adnan Ahmed, et al.: Maximum entropy scaled super pixels segmentation for multi-object detection and scene recognition via deep belief network. Multimed. Tools Appl. 82.9 (2023)
24. Sundaravadivel, P., Kesavan, K., Kesavan, L., Mohanty, S.P., Kougianos, E.: Smart-Log: A Deep-Learning based automated nutrition monitoring system in the IoT. IEEE Trans. Consum. Electron. **64**(3), 390–398 (2018)
25. Devi, S., Thopalli, K., Dayana, R., Malarvezhi, P., Thiagarajan, J.J.: Improving object detectors by exploiting bounding boxes for augmentation design. IEEE Access **11**, 108356–108364 (2023)
26. Wang, H., Sun, S., Bai, X., Wang, J., Ren, P.: A reinforcement learning paradigm of configuring visual enhancement for object detection in underwater scenes. IEEE J. Oceanic Eng. **48**(2), 443–461 (2023)
27. Alaba, S.Y., Ball, J.E.: Deep Learning-Based Image 3-D object detection for autonomous driving: review, In IEEE Sens. J., **23**(4), pp. 3378–3394, 15 (2023)

Diagnosis of Bone Fracture Based on Deep Learning Networks

K. Jaspin, Mercy Kiruba, and Priyadharshini

Abstract Bone fractures are a common and frequently serious medical problem for which the precise diagnosis is essential to providing the best possible care. The suggested method makes use of deep learning (DL) to automate the fracture identification process from medical imaging data, including X-rays. A popular family of deep learning models for image analysis applications is called convolutional neural networks, or CNNs. Their effectiveness is in their ability to extract features and categorize them. Numerous architectures are employed for analysis, including pretrained models like GoogLeNet and MobileNet, as well as a manual fracture detection model called ManualNet are implemented. Preparing the data, designing the model architecture, training, and evaluating the process are the main processes in fracture detection. For both healthy and shattered bones, the best model's accuracy is 95%. Additionally, this system is able to be incorporated into hospital workflows to deliver quick and automated early fracture examinations, lessening the workload of radiologists and facilitating quicker patient care. Its real-time implementation could lead to better patient outcomes and lower medical expenses. This system advances the field of medical image analysis and demonstrates how deep learning may be used practically to diagnose bone fractures.

Keywords Deep Learning · GoogLeNet · MobileNet · Bone fracture · Patients · X-rays

1 Introduction

The human body contains a wide range of bone types. Children, the elderly, and young adults are frequently fractured as a result of falls, collisions, conflicts, and other mishaps. When determining if bone fractures occur, many clinicians rely on imaging tests. The evolution of medical technology has resulted in the development

K. Jaspin (✉) · M. Kiruba · Priyadharshini
St. Joseph's Institute of Technology, Chennai, Tamil Nadu 600119, India
e-mail: jaspink@stjosephstechnology.ac.in

G. Raj et al. (eds.), *Advances in Artificial Intelligence and Machine Learning*, Lecture Notes in Electrical Engineering 1264, https://doi.org/10.1007/978-981-97-9507-9_34

of several high-quality medical image processing methods, such as computed tomography (CT), magnetic resonance imaging (MRI), ultrasound, and X-rays. Examining an X-ray image to determine where a bone fracture occurred requires a skilled physician to invest a great deal of time. Unfortunately, there are often not enough qualified radiologists in hospitals to handle. Computer aided diagnosis (CAD), which analyzes medical images to help clinicians discover bone fractures, has been used extensively and is gaining more attention these days. Expert medical professionals must, nevertheless, identify fractures in many anatomical areas in practice. Due of the significant differences between different forms of bone, developing such a system is an extremely difficult undertaking. We propose a system which works efficiently. In this paper, we created a deep neural network for detection. The data set contains approximately 4000 photos of various forms of human bone. A data set causes the deep neural network to get overfit. As a result, a deep learning (DL) technique was used on the data set. The proposed paper is divided into three sections, which are as follows: Sect. 1 covered the data preprocessing and the implementation of the manual architecture. Section 2 describes the execution of the GoogleNet architecture and Sect. 3 represents the application of the MobileNet architecture and future scope of the proposed analysis.

2 Related Work

A few research have used bone fracture detection for diverse applications. Sinthura et al., [1] proposed a system that uses the CNN algorithm to solve these challenges. The simulation findings show that the suggested system is a far more effective medium for performing edge discovery at aggregate scales. The suggested technology has also proven to be flexible enough to prize the necessary information, perform the required processing on critical areas of the images, and manage noise far better than the currently available edge sensors. Upadhyay et al., [2] proposed a model that help the healthcare interpreters by studying different strategies for bone break position by efficiently exercising the picture running and to plan new styles to ameliorate the perfection of crack identification. The strategies have been recorded with simplicity of understanding. The paper is the first of its kind to review the bone break position strategies cornerwise over colorful modalities. Yang et al., [3] created and debated There are two line- grounded fracture discovery schemes: videlicet Standard line-grounded fracture discovery and Adaptive Differential Parameter Optimized (ADPO) line- grounded fracture discovery. The bracket is performed by the Artificial Neural Network (ANN). The Standard plan may attain an average delicacy of 71.57, but the ADPO plan can achieve an average delicacy of 72.89.

Abbas et al., [4] suggested a transfer literacy, Faster R- CNN deep literacy model for fracture discovery and bracket with Region Offer Network (RPN) for fracture discovery and bracket. The overall delicacy has achieved from this system is 94%. Bagaria et al., [5] compiled more than 50 updates to the field. For imagine segmentation, image classification, fracture detection, and other tasks, this survey employs various fracture-identifying techniques. This analysis finishes with a tough discussion of future courses, tests, and lingering concerns in prospect research. Boger et al., [6] employed digital signal processing and machine learning technologies to replicate and improve on manual detection methods. Machine learning models were constructed and tested with 93.6% accuracy utilizing animal bones and artificial soft tissues to imitate the dampening effects of muscle and connective tissue. Iyer et al., [7] presented an entirely automated method for identifying vertebral compression fractures (VCF). The technique uses imitation learning and deep reinforcement learning to localize thoracic and lumbar spine areas in three dimensions. The average Jaccard Index/Dice Coefficient score for the localization studies was 74/85% and 77/86%, respectively. On another 127 chest images after localization, VCF detection produced an average fivefold cross validation accuracy of 80%, sensitivity of 79.87%, and specificity of 80.73%. Karimunnisa et al., [8] developed an automated method for detecting X-ray fractures. In contrast to SVM (Support Vector Machine) and ANN (Artificial Neural Network), a back propagation neural network (BPNN) combined with canny edge segmentation and a conservative smoothing filtering technique is created in this study to help attain a greater accuracy rate.

Pathare et al., [9] created a successful approach based on image processing that enables for the fast and precise categorization of bone fractures utilizing the data acquired from CT and x-ray images. To find small fractures and shattered bones, these photos were processed using a specially designed program. Rao et al., [10] developed a Computer- supported Fracture Discovery fashion using X-rays which detects and identify the fractures veritably efficiently and minimizes the stress of Orthopedicians. The Bag Of Visual Words (BOVW) fashion is proposed, in which features are labeled as shattered and non-fractured via the training of a Back Propagation Neural Network (BPNN) classifier with features ripped from SIFT (Scale Invariant Feature Transform). Yadav et al., [11] Deep literacy model was used to diagnose the fractured bone, and data addition methods were used to expand the amount of the data set. Using fivefold cross confirmation, the suggested model's bracket delicacy is 92.44 for both healthy and damaged bone. The delicacy on 10 and 20 of the test data is greater than 95 and 93 independently.

Nguyen et al., [12] devised a pre-processing strategy for x-ray pictures that combines the features of Contrast Limited Adaptive Histogram Equalization for image contrast enhancement and YOLACT ++ for image segmentation. The best result achieved with our suggested strategy is 81.91%. Additionally, experimental results verify that, our method performs better than the Faster-RCNN based solution. Chang et al., [13] suggested a YOLO-based target identification technique to quickly and accurately identify existing and new fractures on spine CT scans, as well as to give medical professionals useful guidance and references. The suggested multi-backbone ensemble method outperforms the 82% accuracy of basic YOLOR with an accuracy of 87.8%, according to experimental results. Moon et al., [14] developed the CA-FBFD system, a computer-aided facial bone fracture detection tool. This method augments CT image Mixup data with the object detection model YoloX-S, which is trained only on IoU Loss for box prediction. The CA-FBFD system surpassed the YoloX-S baseline model in evaluation by a substantial margin of 10.2%, attaining an average precision of 69.8% for facial fractures. Feng et al., [15] To accomplish correct bracketing of SPECT images, researchers suggested a ResNet enhancement model integrating Convolutional Block Attention Module (CBAM) and contextual motor (Hut) attention medium. Experiment findings on a genuine SPECT picture dataset reveal that the proposed model works well in the SPECT image bracket, with values of 95.84, 96.12, 95.88, and 95.83 in the Accuracy, Precision, Recall, and F-1 Score assessment criteria, respectively.

3 Methodology

The bone fracture detection system is made up of five main modules: data collection, data preprocessing, and the implementation of a manual architecture known as the ManualNet architecture, the application of the GoogLeNet architecture, the implementation of the MobileNet architecture, and deployment.

3.1 Training Phase

3.1.1 Data Collection

The dataset for bone fracture identification is made up of x-ray pictures for training and testing, with two classes: fractured and not fractured. Figure 1. illustrates the count of the images of fractured bones and Fig. 2. Illustrates the count of the images of not fractured images. The training dataset consists of 2078 fractured and 2019 not fractured images and the testing dataset consists of 1821 fractured and 2019 not fractured images (Fig. 3).

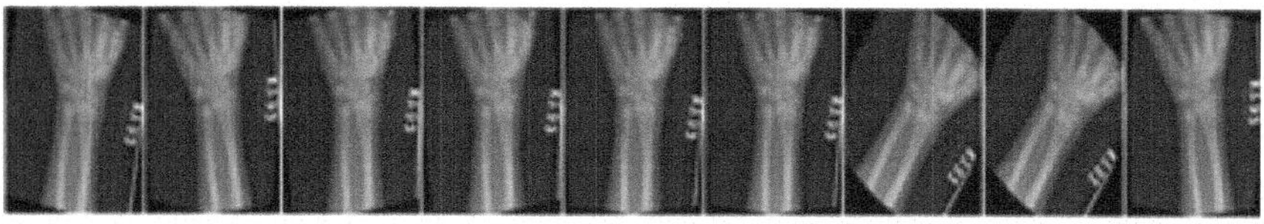

Fig. 1 Training data—Fractured bone images

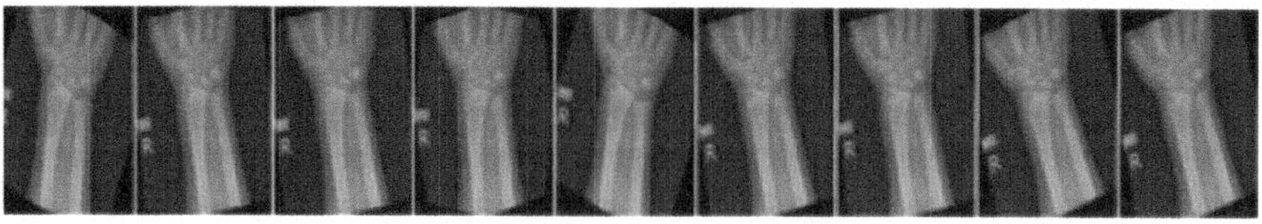

Fig. 2 Training data—Non—Fractured bone image

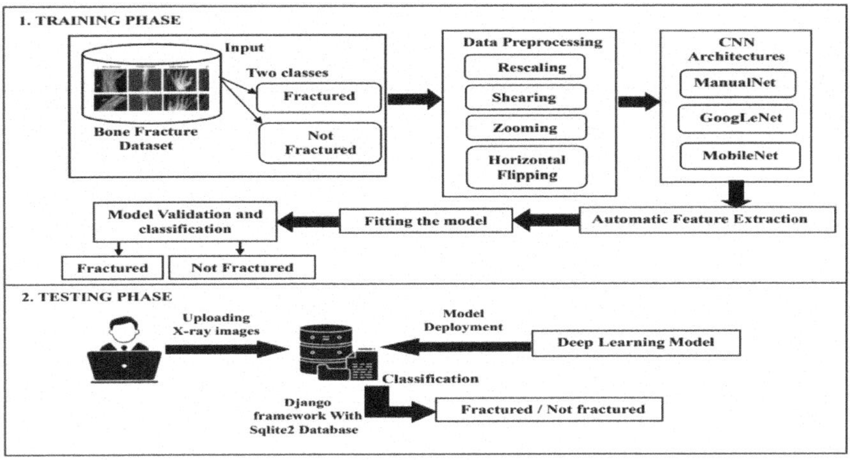

Fig. 3 System Architecture of the bone fracture detection

3.1.2 Data Preprocessing

The various data pre-processing techniques are applied to the x-ray images which are rescaling, shearing, zooming and horizontal flipping. Initially, To rescale the pixel values in the x-ray images, the rescaling approach is used. It normalizes the pixel values to fall between [0, 1] by dividing each pixel value by 255. In deep learning, normalizing the data is a standard procedure that helps the model converge more quickly and perform better during training. One kind of image alteration is called a shear transformation, which distorts an image by linearly moving its content. The maximum shear angle expressed in degrees is specified by the shear_range parameter. When it comes to data augmentation, the photos could be sheared by up to 0.2 radians in any direction, according to the shear range 0.2. The zoom transformation technique allows you to randomly zoom in or out on the x-ray images. The magnification range argument sets the range for random zooming. The 0.2 zoom range implies that the photos can be amplified in or out by up to 20%. The picture mirroring technique swaps the left and right sides of an image, generating a mirror image effect. When set to True, photographs can be horizontally flipped (mirrored) with a 50% chance during data augmentation. This can be used to generate more training data and help the model generalize better.

3.1.3 ManualNet Architecture

The size of the bone fractures varied greatly, and occasionally only a few erroneous pixels might be seen in very small fractures, making detection more difficult. As a result, a manual design has been developed to detect fractured and non-fractured bones. The ManualNet architecture diagram is shown in Fig. 4.

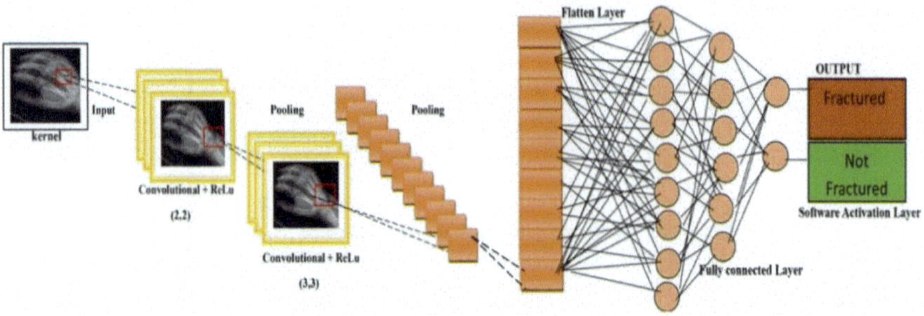

Fig. 4 ManualNet architecture diagram

For bone fracture classification, a sequential deep learning model was developed using TensorFlow or Keras. The model starts with a convolutional layer with 32 (3 × 3) filters, designed for 224 × 224 pixel color images, using the ReLU activation function. Then comes a max-pooling layer with a (2 × 2) window for down sampling. The flattened layer reduces the 2D feature maps to a 1D vector that can be connected to fully connected layers. A hidden layer with 38 units and ReLU activation is added, followed by a final fully connected layer with two units for binary classification. The softmax activation is used to compute class probabilities. Despite the binary classification job, the model is built using the Nadam optimizer, the categorical cross-entropy loss function, and accuracy as a metric.

3.1.4 GoogLeNet Architecture

The GoogLeNet's network structure is ideal for bone fracture detection due to its potential to gather information at numerous tiers utilizing inception modules, which can be especially useful in finding small fractures and patterns in medical images. The GoogleNet architecture diagram is depicted in Fig. 5.

Figure 5 depicts a GoogLeNet-inspired neural network architecture for image classification, following Inception principles. The core is the "inception_block," featuring four branches: 1 × 1 convolution, two one by one convolutions followed by a three-by-three convolution, two one by one convolutions followed by a 5 × 5 convolution, and max-pooling followed by a 1 × 1 convolution. These branches merge to create a comprehensive feature map, allowing the network to capture information at different scales for improved performance. Overall network architecture begins with an initial 7 × 7 convolutional layer, followed by max-pooling and additional convolutional layers. Multiple inception blocks are applied, each progressively increasing the complexity and depth of feature extraction. To create class predictions, the network concludes with an extended average pooling layer, a dropout layer to minimize overfitting, and a layer entirely coupled with softmax function. With accuracy and precision measures, the model is built using the Nadam optimizer and category cross-entropy loss. This GoogLeNet-inspired architecture is an effective

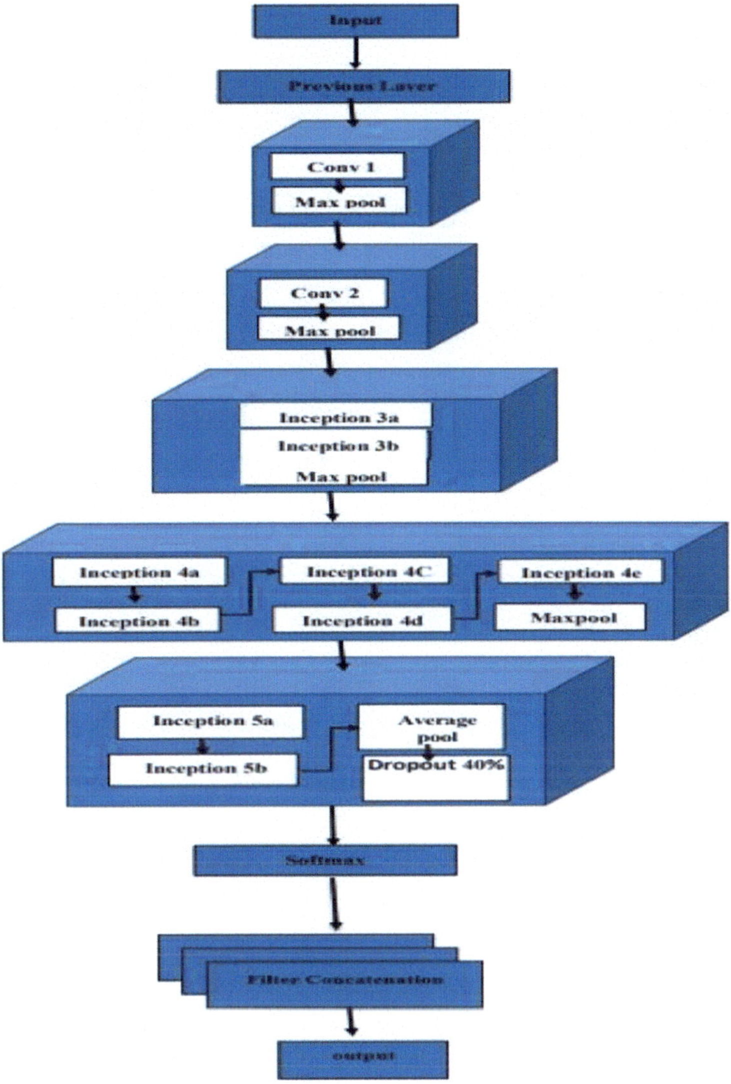

Fig. 5 GoogLeNet Architecture diagram

picture classification tool, capable of capturing features at different scales, fostering rich feature representations, and attaining high classification accuracy.

3.1.5 MobileNet Architecture

To effectively detect bone fractures, MobileNet architecture is chosen allowing real-time image analysis on resource-constrained devices without compromising accuracy. The MobileNet architecture diagram is shown in Fig. 6. A MobileNet architecture for image classification is defined, which is a sort of lightweight neural network that is well-suited for resource-constrained devices. It starts with an input layer that specifies the image input shape and the number of target classes for classification.

The MobileNet architecture relies on the "mobilenet_block" as its fundamental unit, featuring consecutive layers of Depthwise Separable Convolution (3 × 3) and 1 × 1 convolution, each followed by batch normalization and ReLU activation. Starting with a 3 × 3 convolution layer with 32 filters and a stride of 2, subsequent application of mobilenet_blocks gradually increases filter count while reducing spatial dimensions through striding. Extended average pooling compresses feature maps to a single value, and a fully connected layer with softmax activation generates class predictions. The model employs the Adam optimizer and categorical cross-entropy loss. MobileNet's efficiency and suitability for low-resource devices make it

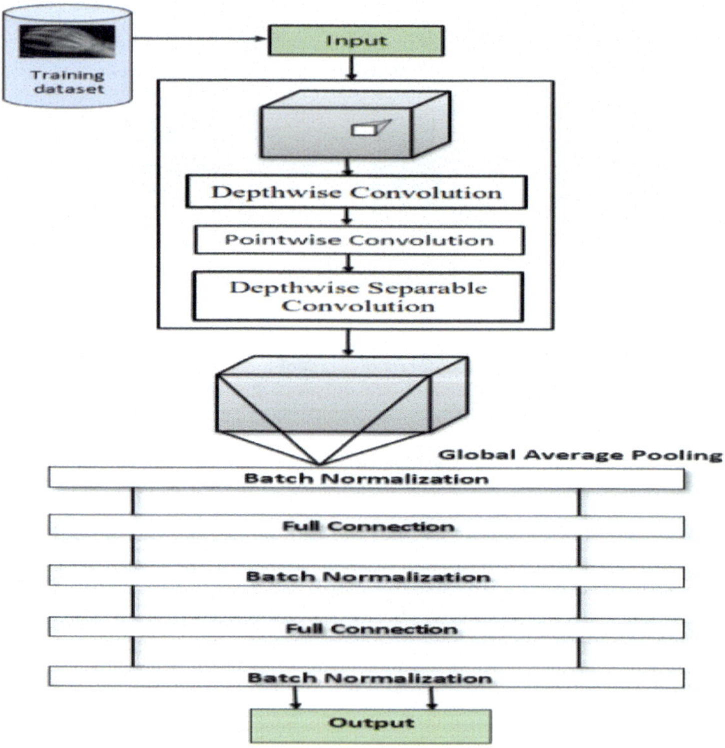

Fig. 6 MobileNet architecture diagram

ideal for applications like bone fracture detection in picture classification, optimizing deployment on devices with limited processing capabilities.

3.1.6 Automatic Feature Extraction

Shift Invariant Artificial Neural Networks in bone fracture detection employ convolutional layers to automatically extract features, learning complex patterns for effective image classification.

3.1.7 Fitting the Model

In bone fracture detection with CNNs, the model is trained by feeding it labeled X-ray images of fractures and non-fractures. The CNN learns to automatically extract patterns indicative of fractures during epochs of iterative training, optimizing its parameters for accurate detection, enhancing diagnostic capabilities in medical imaging.

3.1.8 Model Validation and Classification

Splitting x-ray image data into training and testing sets is used to validate CNN-based bone fracture detection models. The CNN learns from training data, while validation evaluates its performance on previously unknown data, assuring generalization and dependability.

3.2 Testing Phase

In the testing phase of bone fracture detection using CNNs with Django and SQLite database, input images are fed to the trained model. Django handles data flow, interfacing with SQLite. The model predicts fractures, and results are displayed through the web interface, evaluating the algorithm's real-world performance.

4 Experimental Result

The preferred model's efficacy is compared in Fig. 7. The accuracy, loss, and recall of bone fracture detection systems spanning CNN architectures such as GoogLeNet and MobileNet are all evaluated. ManualNet, GoogLeNet, and MobileNet architecture have accuracy scores of 92%, 52%, and 95%, respectively. Among the presented models, the MobileNet architecture is the best for bone fracture diagnosis.

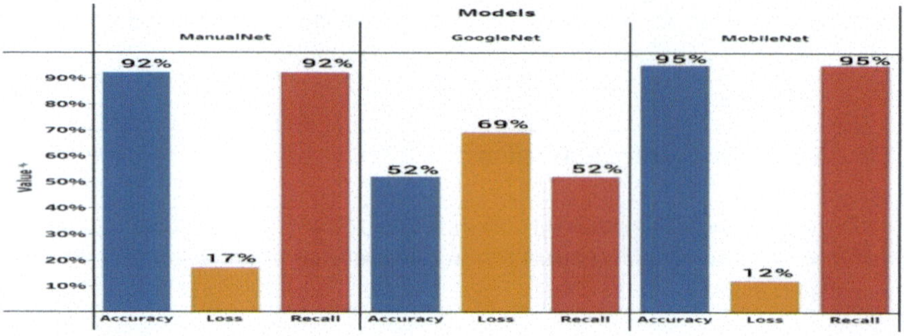

Fig. 7 Comparison of Performance Analysis of Bone fracture detection system over different CNN architectures

In Fig. 8, the Model Accuracy and Loss for the various CNN architectures are compared.

In Fig. 9, GoogLeNet architecture's evaluation involves scrutinizing accuracy and loss metrics, aiming for high accuracy and minimal loss to ensure effective feature extraction and classification.

In Fig. 10, the MobileNet architecture is the best among all and has accuracy of 95% and the loss is 12%.

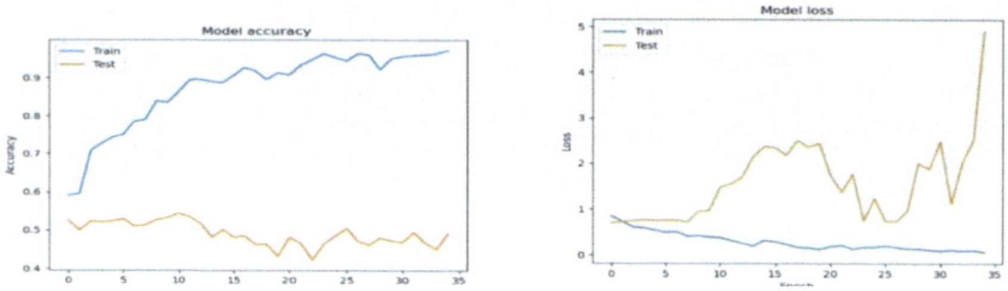

Fig. 8 Model accuracy and loss analysis for ManualNet architecture

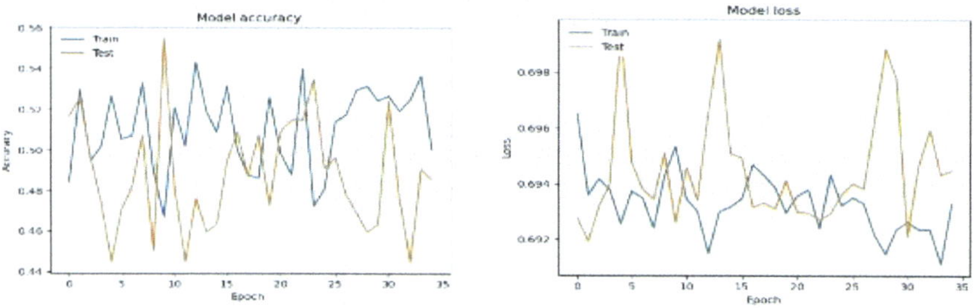

Fig. 9 Analysis of model accuracy and loss for GoogLeNet architecture

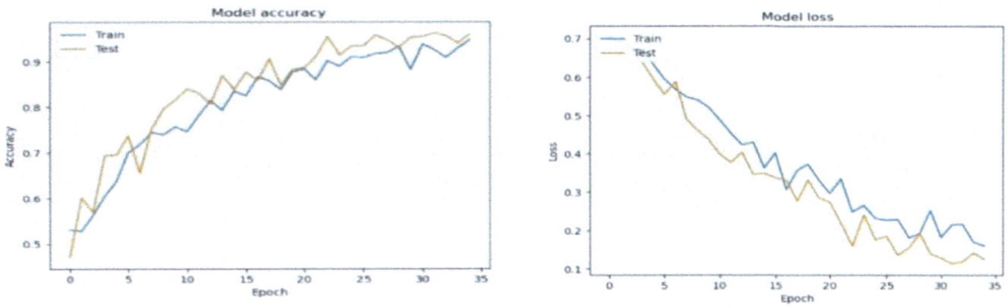

Fig. 10 Model accuracy and loss analysis for MobileNet architecture

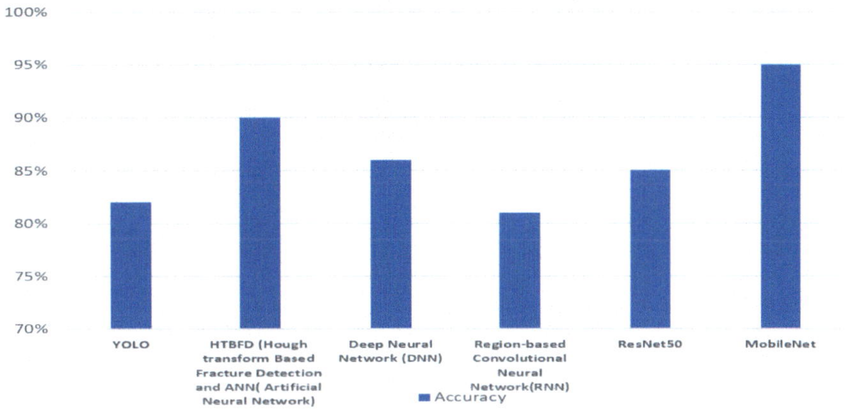

Fig.11 Comparison of performance analysis of the proposed and existing system

4.1 Comparison of Performance Analysis of the Proposed and Existing System

In Fig. 11 the efficiency of the existing system and the proposed system are evaluated. The accuracy of the proposed system is 95%. Therefore, the proposed system is efficient than the existing system.

Table 1 illustrates the analysis of the result parameters of the deep learning architectures over different optimizers like Adam, RMSProp, Nadam, Adadelta. For the bone fracture detection system, the proposed model used different features to analyze the result with 32 epochs, 224 batch size and the Softmax activation function.

Table 1 Comparison of deep learning networks over different optimizers

Deep Learning Network	Optimizer	Training Accuracy	Validation Accuracy	Precision	Validation precision	F1 Score	Recall	Loss	Validation Loss
Manual Net	Adam	0.92	0.95	0.92	0.95	0.94	0.94	0.2	0.13
	RMSProp	0.95	0.93	0.95	0.93	0.92	0.92	0.1	0.09
	Nadam	0.77	0.78	0.77	0.78	0.76	0.76	0.5	0.46
	Adadelta	0.57	0.61	0.57	0.61	0.56	0.56	0.7	0.68
GoogLe Net	Adam	0.51	0.47	0.51	0.47	0.49	0.49	0.7	0.69
	RMSProp	0.53	0.45	0.53	0.45	0.52	0.52	0.7	0.69
	Nadam	0.52	0.48	0.52	0.48	0.49	0.49	0.7	0.68
	Adadelta	0.48	0.46	0.48	0.46	0.47	0.47	0.6	0.63
Mobile Net	Adam	0.95	0.45	0.95	0.45	0.94	0.94	0.1	0.86
	RMSProp	0.97	0.61	0.97	0.61	0.96	0.96	0.1	1.05
	Nadam	0.95	0.97	0.95	0.97	0.97	0.97	0.1	1.07
	Adadelta	0.48	0.51	0.48	0.51	0.55	0.55	0.7	1.02

5 Conclusion

Therefore, the proposed model concentrated on how a CNN model was used to predict the pattern of bone fracture detection using images from a specified dataset. The bone fracture detection project explored three different approaches: ManualNet, GoogLeNet, and MobileNet architectures. Among these approaches, the MobileNet architecture achieved the highest accuracy of 95% and is the best approach for the efficient bone fracture detection. Nevertheless, the robustness and generalizability of these models across diverse datasets, as well as the critical compliance, and ethical aspects, must be addressed for successful clinical integration. Deep learning has a bright future in bone fracture diagnosis, with prospects for significant breakthroughs in model architecture, data augmentation, interpretability, and real-time application. Further enhancement for this work is refining CNN architecture for improved accuracy, integrating multi-modal data for comprehensive analysis, exploring real-time implementation for quicker diagnosis, and collaborating with medical experts to enhance clinical relevance. Additionally, continuous dataset augmentation and model fine-tuning will be crucial for ongoing performance optimization.

References

1. Sinthura. S. S., Prathyusha. Y., Harini, K., Pranusha. Y., Poojitha. B.: Bone fracture detection system using CNN algorithm. In: Proceedings of the 2019 International Conference on Intelligent Computing and Control Systems (ICCS), Madurai, India (2019)
2. Upadhyay. R. S., and Tanwar. P.: A review on bone fracture detection techniques using image processing. In: Proceedings of the 2019 International Conference on Intelligent Computing and Control Systems (ICCS), Madurai, India (2019)
3. Yang. A. Y., Cheng. L., Shimaponda-Nawa. M., Zhu. H.-Y.: Long-Bone fracture detection using artificial neural networks based on line features of X-ray Images. In: Proceedings of the 2019 IEEE Symposium (SSCI), Xiamen, China (2019)
4. Abbas. W., et al.: Lower Leg Bone Fracture Detection and Classification Using Faster RCNN for X-Rays Images. In: Proceedings of the 2020 IEEE 23rd International Multitopic Conference (INMIC), Bahawalpur, Pakistan (2020)
5. Bagaria. R., Wadhwani. S., Wadhwani. A. K.: Different techniques for identification of a bone fracture in analysis of medical image. In: Proceedings of the 2020 IEEE 9th International Conference on Communication Systems and Network Technologies (CSNT), Gwalior, India (2020).
6. Boger, I., et al.: Induced acoustic resonance for noninvasive bone fracture detection using digital signal processing and machine learning. In: Proceedings of the 2020 IEEE Global Humanitarian Technology Conference (GHTC), Seattle, WA, USA (2020)
7. Iyer. S., et al.: A novel approach to vertebral compression fracture detection using imitation learning and patch based convolutional neural network. In: Proceedings of the 2020 IEEE 17th International Symposium on Biomedical Imaging (ISBI), Iowa City, IA, USA (2020)
8. Karimunnisa. S., et al.: Detection of bone fractures automatically with enhanced performance with better combination of filtering and neural networks. In: Proceedings of the 2020 Second International Conference (ICIRCA), Coimbatore, India (2020)
9. Pathare. S. J., Solkar. R. P., Nagare. G. D.: Detection of fractures in long bones for trauma centre patients using hough transform. In: Proceedings of the 2020 International Conference on Communication and Signal Processing (ICCSP), Chennai, India (2020)

10. Rao. L. J., et al.: An effective bone fracture detection using bag-of-visual-words with the features extracted from SIFT. In: Proceedings of the 2020 International Conference on Electronics and Sustainable Communication Systems (ICESC), Coimbatore, India (2020)
11. Yadav. D.P., Rathor. S.: Bone fracture detection and classification using deep learning approach. In: Proceedings of the 2020 International Conference (PARC), India (2020)
12. Nguyen. H.P., Hoang. T.P., Nguyen. H.H.: A deep learning based fracture detection in arm bone X-ray images. In: Proceedings of the 2021 International Conference on Multimedia Analysis and Pattern Recognition (MAPR), Hanoi (2021).
13. Chang. C.-Y., Hsieh. M.-H., Hsu. S.-M.: Localization of fresh and old fracture in spine CT images using YOLOR. In: Proceedings of the 2022 IEEE International Conference on Consumer Electronics, Taiwan (2022)
14. Moon. G., et al.: Computer aided facial bone fracture diagnosis (CA-FBFD) system based on object detection model. In: Proceedings in IEEE Access, **10** (2022)
15. Feng. Q., et al.: SPECT bone scan image classification by fusing multi-attention mechanism with deep residual networks. In: Proceedings of the 2023 4th International Conference (CVIDL), Zhuhai, China (2023)

A Review on Internet of Things (IoT) Environment Using Software-Defined Networking (SDN) Based on Different Networks

Nidhi Bajpai, Madhavi Dhingra, and Nisha Chaurasia

Abstract Software-Defined Networking (SDN) and the Internet of Things (IoT) are collaborating in a transformative way to establish connections. The Internet of Things enables remote management of equipment as well as real-time data exchange via its billions of connected devices. This is supported by SDN's provision of a comprehensive network view, which enables remote access and control for both users and operators. The growing popularity of IoT and SDN represents a paradigm change that is fundamentally altering the environment of intelligent services and interconnected devices. The profound impact that their mutual relationship has had on contemporary connectivity is concisely outlined in this review. This extensive review paper investigates the advantageous association between Software-Defined Networking (SDN) and the Internet of Things (IoT), throwing insight on their combined capability to enhance network performance. Underscoring the critical contribution of SDN to centralized threat detection across Wide Area Networks, Access Networks, and Edge Networks in the IoT domain, the study emphasizes the critical role of SDN in augmenting flexibility, manageability, and security in these networks. The advent of Software-Defined Internet of Things (SDIoT) becomes an innovative focal point, demonstrating the significant effects that it has on IoT networks. The review further explores the frequent coexistence of terms related to SDN and IoT networks, underscoring their broad range of applications. Furthermore, the paper critically examines the difficulties associated with supervising IoT networks and assesses recent strategies that have been implemented in the past five years to overcome these obstacles. These strategies include load balancing, energy management, provisioning security services, and defect tolerance. The article generally presents an in-depth view on the ways in which SDN enhances the effectiveness and protection of IoT networks.

N. Bajpai (✉) · M. Dhingra
CSE Department, Amity University Madhya Pradesh, Gwalior, India
e-mail: nidhibajpai.21@gmail.com

M. Dhingra
e-mail: madhavi.dhingra@gmail.com

N. Chaurasia
IT Department, NIT Jalandhar, Jalandhar, India
e-mail: chaurasian@nitj.ac.in

G. Raj et al. (eds.), *Advances in Artificial Intelligence and Machine Learning*, Lecture Notes in Electrical Engineering 1264, https://doi.org/10.1007/978-981-97-9507-9_35

As a result, it provides policymakers, practitioners, and researchers with invaluable information.

Keywords Wide area networks · Access networks · Edge networks · Software Defined Networks (SDN) · Internet of Things (IoT)

1 Introduction

The widespread usage of IoT hardware and software has been made easier by technological advancements in areas like big data, cloud computing, and mobile networks. Smart homes, driverless cars, security systems, smart cities, and remote healthcare are just a few of the many ways that the Internet of Things has the potential to improve our lives. Due to the enormous amount of data that the Internet of Things generates, it is evident that current wireless networks are unable to satisfy the quality of service (QoS) needs of all potential applications. The management of network resources in the Internet of Things of the future is dependent on SDN (software-defined networking) and network function virtualization (NFV).The advent of SDN (software-defined networking) has changed everything, a centralized application known as the "SDN Controller" may now manage all aspects of a network. Quick responses to security risks, fine-grained traffic filtering, and the implementation of security rules on the fly are all made possible by the controller. Using an SDN controller and switch to construct a firewall is only one example of how SDN has been researched for usage in network security [1], and a select few studies have looked at how SDN may be integrated with the Internet of Things [2, 3].

Multiple fields outside of the smart transportation scenario outlined previously are beginning to see the advantages of using SDN approaches in IoT settings. In a smart grid environment, for instance [4], built a solid control and communication infrastructure using SDNs. In the smart home business, where there is a profusion of IoT devices (from cheap smartphones and tablets to cutting-edge home appliances), similar attempts have been examined. With regards to "smart" houses, we can help validate rules and economic models for cost sharing, and we can create a method for home network slicing [5], are two examples of recent work in this area. The SDN methods used to provide rules for managing Wireless Sensor Networks are implemented at a lower device level in [6]. In conclusion, there is a lot of work going into managing IoT environments, however, much of it is only applicable to certain narrow fields or levels of systems [7].

1.1 Research Contributions

This paper aims to present a reasonable architecture for introducing the SDN paradigm to IoT networks from a variety of perspectives, including access networks, peripheral networks, and wide area networks. In addition, we develop and evaluate models for analyzing the performance of distributed networks. We evaluate recent research articles to extract the improvement goals & proposed solutions, since it has been suggested that SDN make use of IoT for a variety of improvement-oriented applications. Some of the contributions made by the paper are as follows:

- This study investigates the potential enhancements in network performance within the context of the Internet of Things (IoT) via the use of Software- Defined Networking (SDN).
- Software-Defined Networking (SDN) plays a vital role in enhancing the flexibility and manageability of networks within the Internet of Things (IoT) domain, particularly in Wide Area Networks (WANs), Access Networks,and Edge Networks.
- In general, the article highlights the role of Software-Defined Networking (SDN) in enhancing the efficiency and security of Internet of Things (IoT) networks.
- Software-defined network (SDN) and IoT network appear frequently in the sources used for this analysis. There is a vast variety of scenarios in which these expressions are employed.
- We examine the obstacles associated with overseeing the Internet of Things (IoT) and the measures implemented in the last five years to surmount them, encompassing: security services provisioning; defect tolerance; energy management; and load balancing.

1.2 Motivation

The inspiration for this study came from the observation that SDN is increasingly seen as a workable alternate to traditional IoT network designs due to its capacity to optimize the network and its potential to pave the way for the birth of new networks equipped with more useful programs to handle network administration. The idea of Internet of devices (IoT) oriented apps portrays a wonderful image of linked devices with varied uses, but it also presents a number of novel obstacles. In order for IoT to permeate every aspect of modern life, several significant obstacles must be overcome. To achieve the goals of IoT and SDN, It is crucial that the control plane & the data plane are kept apart. Future IoT networks will likely have billions of devices connected to them, making the management and control of these devices a complex and ever-evolving challenge. In addition to providing potential solutions to the identified IoT problems, SDN can also make the IoT network scalable and configurable. Management concerns without disrupting the underlying architecture of current implementations. Intelligent equipment has impacted almost every aspect of our lives and behavior. The type of interaction and interaction between this massive volume of devices is what makes the Internet of Things, a vast field within the Internet, highlight energy management and boost network longevity. Additionally, by lowering control & routing packet overhead, this technique dramatically lowers node energy usage [8]. According to the simulation results, the suggested solution performs better than alternative methods in terms of routing overhead, packet delivery rate, average energy consumption, network lifetime, and the number of live nodes.

As IoT networks vary from regular networks, typical network management methods and design are often ineffective when applied to IoT. Legacy architectures and protocols used in the Internet of Things were not made to deal with voluminous data or widespread mobility, or high levels of scalability. The goal of this study is to improve knowledge of the development and application of software-defined networking-based IoT management frameworks and solutions.

1.3 Internet of Things (IOT)

The security of the IoT has been called into question [9], since each connected item can be easily identified by its unique identifier and is capable of doing complex data analysis and making autonomous judgements. IoT devices may be as basic as home sensors or as complex as nuclear power plants. Because of how IoT works, its communication channels may lack security guarantees like encryption and authentication. In order to safeguard networks from outside interference, The perimeter of a network typically features sophisticated security controls, such as intrusion detection/prevention systems and firewalls. However, such defense systems do not function well with IoT networks directly because of their unique properties [10].

Challenges of IoT

IoT difficulties have varying significance depending on their depth. Heterogeneity of IoT data, for instance, is distinct from that of IoT hardware and infrastructure. Data formats like RDF, microdata, & microformats can get confusing when there are too many moving parts, when IoT devices utilise non-standardized nomenclature for IoT attributes, and when IoT networks employ a variety of technologies and protocols, all of these factors contribute to heterogeneity [11]. In the IoT, data protection is essential. Data security is crucial since these applications are widely used to collect data from the surrounding environment and send it to distant servers for processing [12]. Because of this, talking about privacy is tricky. Other safety concerns [13], involve privacy- related concerns, for instance, routing-based attacks that interfere with Internet of Things services. This fact makes the topic of IoT security management an intriguing one. The Internet of Things presents new issues that cannot be met with traditional approaches to network administration [14].

Figure 1 represents the basic IoT architecture. There will be a comparison of all the case studies detailed in the following parts using the following four criteria:

Fault Tolerance: IoT networks by their very nature produce massive amounts of data. Data transmission issues with low-powered IoT nodes are a common concern. For several causes, like energy distortion, disruption from natural catastrophes, and so forth, the appropriate nodes may fail [15].

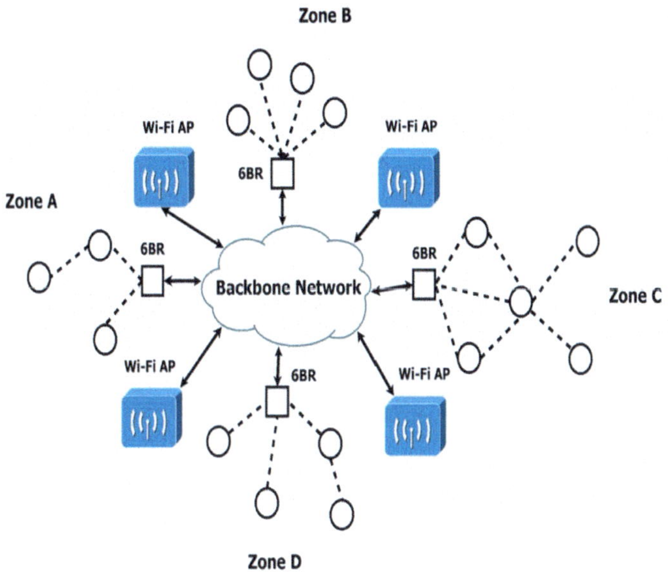

Fig. 1 Architecture model of the IoT Infrastructure [10]

Energy Management: For the Internet of Things, energy is like a precious metal; once it's gone, it's gone, and it's difficult to replace. This is particularly true when nodes are placed in more remote areas. Methods like duty cycling, which regulates data traffic by scheduling device sleep and wake-up times, have been studied [16].

Load Balancing: The longevity of the Internet of Things may be increased by using a form of load balancing that reduces the network's overall power consumption. Load balancing is an important goal, and clustering is one of the most common ways to get there [17].

Security Management: In decentralized networks like the Internet of Things (IoT), security is paramount. Data privacy is a key feature of many IoT applications. IoT devices exacerbate the difficulties of providing adequate security for the network. In addition, conventional security measures are difficult to execute in such networks and cannot be adapted to them [18].

1.4 Software Defined Networking (SDN)

A new networking paradigm known as software-defined networking (SDN) has emerged throughout the past ten years in reaction to the growing expense of network maintenance brought on by the explosion of connected devices and applications that make up the Internet of Things (IoT). Because SDN employs routers and switches (the data plane) to actually carry out traffic flows and a controller (the control plane) to define them, it differs from conventional networks [19].

Figure 2 shows the three main components of the SDN architecture: the infrastructure (data plane), the control layer, and the application layer. At the application layer, programs such as load balancing and intrusion detection systems operate and communicate with the controller over a north-bound API. The south-bound API is primarily responsible for relaying OpenFlow [21] movement concerning knobs and buttons. SDN's scalability, adaptability, dynamism, and low cost are among its most appealing qualities. As a result of progress in network research software-defined networking (SDN) is a novel approach to networking. By compartmentalizing powerful features, SDN simplifies control over networks and their services [15].

A. **Software Network Defined on IOT**

The promise of centralized network control through rule-based administration is what makes Software-defined networking (SDN) is a novel and intriguing methodology. SDN's abstractions allow for global policy-based management of the network without having to micromanage individual settings. Consequently, it is beneficial to deal with the IoT's heterogeneity and application-specific demands.

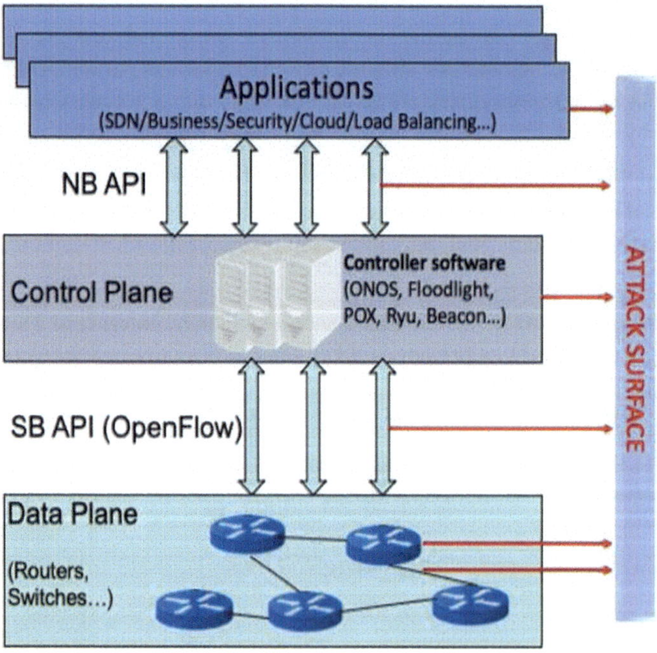

Fig. 2 SDN Architecture and attack surface [20]

B. **Software Defined Access Networks**

In order to optimize networks, the SDAN virtualizes the processes involved in controlling and managing high- speed Internet connections, simplify operations, and stimulate the introduction of novel services, especially in multi-operator contexts. SDAN employs a cloud-based, fictitious "controller" to centralize administration and control responsibilities that were previously performed by individual NEs. Multiple operators in highly competitive settings may use the same SDAN interface to access the same controller functionalities. As seen in Fig. 3, Adapters can be used to create an abstraction layer between legacy control and management protocols like SNMP and a unified interface specification called Management Information Base (MIB), or model.

Figure 3 shows how the SDAN may connect several retail vendors to a common backend infrastructure.

C. **Software Defined Edge Networks**

Edge computing is a new and exciting concept that promises to improve mobile access in IoT multinetwork. Nevertheless, the mobility management as well as access control in such networks face significant issues due to the presence of various wireless communication systems. Distributed edge networks may use Software- Defined Networking (SDN) architecture to offer uniform frameworks for mobility management and to provide secure techniques for access control, therefore preserving the computing continuity [23].

Figure 4 is a drawing of a typical layout for such an environment. As indicated in the picture, connecting apps and devices at the network layer calls for comprehensive support.

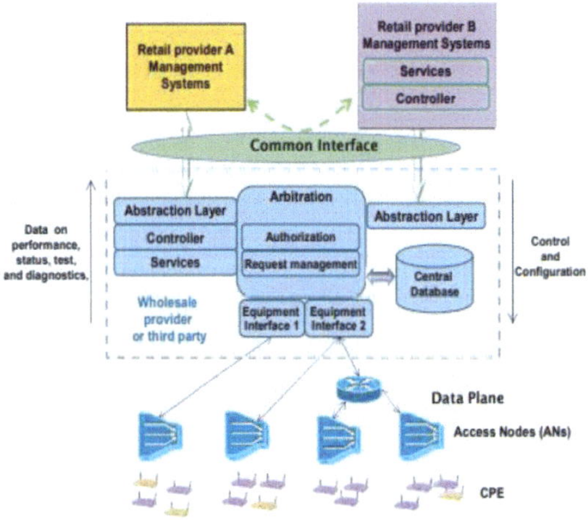

Fig. 3 SDAN in contexts with several operators [22]

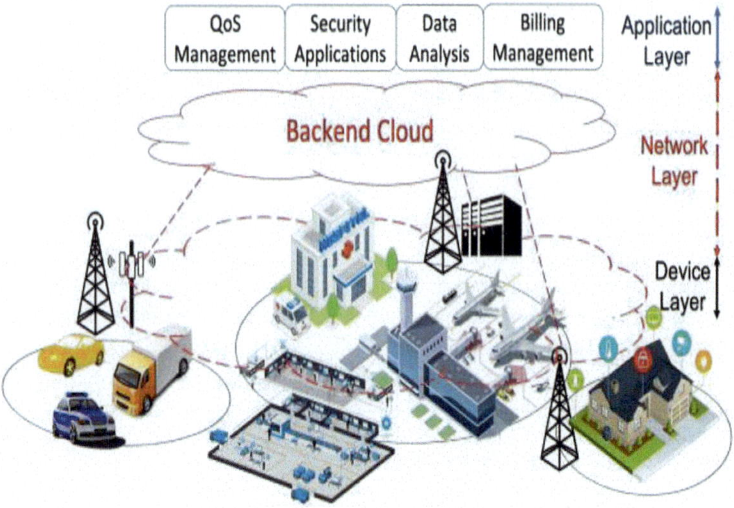

Fig. 4 A software-defined networking environment for edge computing [24]

D. **Software Defined Wide Area Networks**

New applications and operating scenarios that place a larger importance on dependable data transport over longer distances are forcing network administrators to reevaluate the architecture of wide area networks. The software-defined wide area network (SDWAN) is among the most promising architectures for the upcoming generation of wide area networks. Software-defined wide area networks (SD-WANs) are recommended for usage in wide- area networking connections because they achieve software control in a different method than software-defined networks (SDNs).

E. **The Software-Defined Networking-Based WAN Design**

Legacy wide area networks are ill-equipped to handle the more fluid networking patterns of today, so they are unable to provide the quality of experience expected by modern users. It has been suggested that the existing network be replaced with a network that uses software-defined wide area networks (SD-WAN). Software-defined wide area networks seek to simplify WAN administration and networking processes in order to counteract the structural inflexibility and lack of innovation found in traditional WAN systems. The theoretical and practical foundations of software-defined wide-area networks are presented in this paper [26].

Logical Architecture: The data layer, control layer, and application layer in a software-defined wide area network are shown in Fig. 4. Two of the main functions of the data layer are bandwidth virtualization and data forwarding. Interface protocols like OpenFlow allow an upper-layer network controller to provide orders to both of them [27]. By utilizing network expression and application expression, network providers & app developers can articulate their specific requirements for the network in almost natural language thanks to the application layer's capabilities. Customizing network rules to cater for the unique traits of every application is becoming more important as apps increasingly present with complex and, at times, conflicting needs. For instance [28], in order to keep consumers happy, a live video streaming service must have both a high bitrate and a low latency [29].

Structure of Matter: In Fig. 4, depict the components of an SDN WAN's logical design. At the data-layer, a collection of SDN switches communicates via a shared set of physical connections. A network controller receives data from these devices. The controller could be a single server or a collection of servers depending on the size and complexity of the network [30]. Billing for various network services falls under the purview of the network controller. The specialized programs sit atop the network controller. Developers of applications and suppliers of networks may communicate their needs to the network controller, which can then adapt its rules and settings to meet those needs. One network controller is often designated as the master controller, while the others serve as backups; in larger networks, there may be several dozen controllers spread over several locations. If the primary controller ever fails, one of the backups will take over immediately [31].

2 Literature Review

For the relevant study, we look at the review publications that cover both SDN and IoT. Access networks, edge networks, and wide area networks all have SDN supervision for the Internet of Things as a hot subject. There has been a lot of thought put into how to bring the programmability of SDN to the IoT. Smart technology and gadgets have permeated every facet of modern existence. The Internet of Things refers to the huge field in the Internet that has emerged as a result of the kind of connection and the way in which communication between these many devices. This Internet of Things substantially emphasizes the problem of energy management and extends the lifespan of networks. Many difficulties have arisen for network designers as a result of the complexity of communications, particularly in mobile networks. To address these issues [32], additional capabilities have been added to mobile IoT networks via the use of the Software Defined Networking (SDN) architecture, which has decreased device overhead due to processing and computation. There has been an uptick in discussion about how to ensure the safety of IoT in both academic and professional circles. IoT security issues include denial of service attacks, unauthorized access to networks, and the accidental release of sensitive information [33]. Numerous academics are exploring different approaches to incorporate SDN to address heterogeneity challenges in IoT networks [34]. To enhance QoS in the face of heterogeneity, we have chosen to concentrate on the SDN- IoT area in this study. Controller reaction times may be lengthened due to heterogeneity in SDN-IoT networks. They provide a paradigm that might help SDN-IoT networks cope with heterogeneity while keeping quality of service intact. In

[35], examine the technology, security, weaknesses, and ML/AI intelligent solutions based on Internet of Things-based smart systems. Concerns about data security are a major roadblock to wider adoption of the Internet of Things. They suggest the need for a Standard Security Framework (SSF) for hardware/software modules that may run on different platforms and yet communicate with one another in the context of a dispersed network. Fog- and Edge computing trends, as well as cloud application security and efficiency, have been discussed.

2.1 Review on IoT Based SDN with Access Networks

Due to its adaptability [36], automation, and programmability, The SDN-IoT, or Software Defined Network in the IoT, is becoming increasingly popular. When SDN-IoT applications (SIApps) continue to absence proper permission management, issues arise when the SIApp's necessary northbound interfaces are dispersed amongst a variety of untrustworthy heterogeneous controllers. The authors of this study propose a blockchain-based SIApps' ledger (SILedger) that uses attribute-based encryption (ABE) to address this issue. Put in place a SILedger prototype built on FISCO-BCOS, complete with a security audit. The experimental findings demonstrate its ability to efficiently regulate access to SIApps with low resource requirements.

In the meantime, Terminals for the Internet of Things may be able to use both terrestrial and satellite networks simultaneously for the purpose of dynamically connecting to the best available connection [37], IoT access management is being severely compromised. In this study, we outline the design considerations for an Internet of Things satellite-terrestrial integrated infrastructure, based on software-defined networks (SDNs). In [38], provide a platform to verify a software- defined networking (SDN) method of granular access management for IoT devices. The findings demonstrate that the access control mechanism is able to successfully block unauthorized access based on device origin, request action, access time, and other factors.Fiber-wireless access network connectivity to edge computers is discussed, and a plan for doing so based on network virtualization and logically-ordered networks is presented in this study [39]. The simulation findings show that their solution significantly improves on latency and load balancing over the state-of-the-art approaches.

2.2 Review on IoT Based SDN with Edge Networks

As Internet of Things devices proliferate, edge computing's capacity to delegate the analysis and processing of data collected by these devices at or close to the point of collection will become increasingly important. Delays may be reduced, data can be safeguarded, and bandwidth can be utilized effectively with edge computing. In addition, when SDN and Edge computing work together, traffic may be routed dynamically based on both network conditions and policy. However [40], by adjusting service configurations on the edge device, you can provide dynamic services despite its limited resources. In order to industrial IoT services such as the sensor/actuator network and the application services, a programmable edge network with programmable layers at the edge may be used. Layer one allows for the dynamic programming of modules and the management of edge actuation options, while layer two contains a software-defined networking (SDN) controller. Also [41], Quality of Service (QoS) in Edge-IoT systems is crucial for the smooth functioning of real-time applications and services. In this research, they look into the viability of quality-of-service (QoS) designs based on software-defined networking (SDN) for Edge-IoT service needs. A detailed overview of SDN and Edge computing cooperation is presented here. They also discuss SDN and its possible advantages for Edge-IoT QoS. A taxonomy and discussion of SDN-based QoS designs are created.

2.3 Review on IoT Based SDN with Wide Area Networks

Traditional WAN architecture uses dynamic routing protocols like OSPF and BGP to choose the optimum connection route. In [43], standard WAN and SDWAN architecture are compared for throughput, latency, the branch office's packet loss on the way to central office. They simulated an SDWAN device using Ryu controller as the SDN controller and Mininet as the data plane. The investigation demonstrated that SDWAN architecture with multipath links had greater throughput than single links. Each link's 10 Mbps bandwidth yields approximately 20 Mbps throughput instead of 10 Mbps. The top SDN controller's link monitoring application improved branch jitter and packet loss under congestion. Numerous tools have been fashioned by technology to do away with tedious, repetitive, and physically demanding activities. Next-generation networks [44], however, need to be quick, scalable, dependable, and cost-effective for this to happen. Therefore [45], a software-defined-networking (SDN) strategy is what we advocate for when it comes to Lora WAN network slicing, wherein virtual installations of network slices are made and kept distinct from one another across LoRa physical gateways. The proposed optimization has the potential to enhance the dependability and quality of service of LoRa slice IoT dense deployments, as shown by the results of NS3 simulations. In [46], employed a systematic approach to evaluate two commercial ICs' suitability for IoT connectivity and describe experimental findings. The WI-SUN Alliance uses IEEE 802.15.4 g for its physical layer, and it also has a LoRa layer. The purpose of this analysis is to provide data that can be used to establish an objective comparison between LoRa and other connection standards like IEEE802.15. g/Wi-SUN.

In [47], suggested approach involves protocol deployment and decision-making. Software-defined wireless sensor networks (SDWSNs) may handle real-time application-specific needs with situation-aware protocol switching. The effectiveness of SDWSN networks is enhanced by situation-aware protocol switching. In [32], provides a protocol for energy-aware forwarding in wireless IoT networks to increase network longevity and decrease energy consumption, researchers have developed a unique routing technique based on software- defined networks (SDN) called IERMIoT.

3 Discussion

Standardization attempts, like the novel idea of merging IoT with SDN, remain in their infancy. Despite many efforts to promote uniformity over the world, a feasible respond to is still unavailable. In other words, certain parts are still conceptual/ theoretical. Table 1 shows several types of SDN deployment for various IoT domains in terms of edge, access, and wide area network. The bulk of studies done might be deemed to be basically unvalidated. However, in recent years, only representative proposals framework has been the center of attention.

4 Conclusion

For the purpose of providing customers with simple, inexpensive, and dependable service delivery, this article provided a thorough review of the state of SDN-based technologies as they pertain to Internet of Things deployments. IoT networking characteristics include wide area networking, network at the edge, and network at the edge. We also identified many potential future research areas for addressing these technical challenges that we believe are crucial. In the coming decades, the Internet of Things is likely to have a major impact on the mode of interaction between humans and technical devices. Currently, it's not enough to only think in terms of internet-based connections between physical objects. Nonetheless, since IoT is still in its early stages, it has encountered a number of challenges, including security, programmability, adaptability, and data management. With SDN integration, it is quite likely that both programmability and centralized control will be included in IoT administration. This study demonstrated how SDN might increase network performance for potential contemporary IoT applications. This study concludes that SDN for IoT environments do not have suitable solution integration. In addition, simulation models are used for model assessment and validation rather than the real IoT environment. Finally, despite many attempts to develop frameworks based on SDN, complete architecture as well as framework remain a barrier for researchers.

Table 1 A review of the studies conducted cn SDN for the Internet of Things

Research on	Author/Year	Research methods	Contribution s	Limitations	Future directions
IoT with SDN Supervision in Access Networks [36]	Ren/2021	Access control using blockchain and ABE—A prototype based on FISCO- BCOS	Decreased overhead due to decentralize d and trustworthy access control	Inadequate practical deployment	Scalability as well as implement in a variety of use situations
IoT with SDN Supervision in Edge Network [40]	Alam/2023	Edge programmable layers are proposed—SDN controller for dynamic programming	Reduced actuation delay and increased energy efficiency	Validation is limited in a variety of contexts	Testing in complicate d industrial contexts
Edge-IoT systems using SDN-based QoS [[41]	Niang/2023	An examination of the literature and a theoretical analysis are carried out	SDN and Edge computing insights for QoS	Empirical testing is restricted	Extensive testing and practical QoS implementa tion
6LE-SDN architecture for 6LoWPAN- SDN [42]	R. K. Das/2020	Computation capabilities at the edge—A hybrid—edge switch using the 6LE-SDNP protocol	Network latency and packet loss have been improved	Testing in the field is restricted	Increasing the size of IoT installation s
IoT with SDN Supervision in Wide Area Networks [43]	R.Yunia rto/2021	SDN controller link monitoring simulation utilising Ryu controller and Mininet	Increased throughput and dependabilit y	Insufficient deployment in operating situations	Deploymen t on operational WAN networks

References

1. Javid, T., Riaz, T., & Rasheed, A.: A layer2 firewall for software defined network. In: IEEE Conference on Information Assurance and Cyber Security (CIACS), pp. 39–42, (2014)
2. Manguri, K.H., Omer, S.M.: SDN for IoT environment: a survey and research challenges. In ITM Web of Conferences, **42**, p. 01005. EDP Sciences (2022)
3. Karmakar, K., Varadharajan, V., Nepal, S., Tupakula, U.: SDN-Enabled Secure IoT Architecture. IEEE Internet Things J., **8**(8), pp. 6549–6564 (2021)
4. Jin, D.: Towards a cyber resilient and secure microgrid using software-defined networking. IEEE Trans. Smart Grid, (2017). https://doi.org/10.1109/TSG.2017.2703911
5. Yiakoumis, Y., Yap, K., Katti, S., Parulkar, G., McKeown, N.: Slicing home networks. In: 2nd Proceedings of ACM SIGCOMM Workshop on Home Networks, HomeNets **11** (2011). https://doi.org/10.1145/2018567.2018569
6. Luo, T., Tan, H., Quek, T.: Sensor open flow: Enabling software-defined wireless sensor networks. IEEE Commun. Lett. (2012). https://doi.org/10.1109/LCOMM.2012.092812.121712
7. Qin, Z., Denker, G., Giannelli, C., Bellavista, P.: A software defined networking architecture for the internet-of-things. In: IEEE/IFIP NOMS-IEEE/IFIP Network Operations and Management Symposium: Management in a Software Defined World (2014). https://doi.org/10.1109/NOMS.2014.6838365
8. Gan, G., Lu, Z., Jiang, J.: Internet of things security analysis. In: International Conference on Internet Technology and Applications, Proceedings (2011). https://doi.org/10.1109/ITAP.2011.6006307
9. Hayajneh, A.I., Bhuiyan, M., McAndrew, I.: Improving internet of things (IoT) security with software-defined networking (SDN). Computers (2020). https://doi.org/10.3390/computers9010008
10. Diéguez, D.: Providing IoT services in smart cities through dynamic augmented reality markers. Sensors (Switzerland) (2015). https://doi.org/10.3390/s150716083
11. Younan, M., Houssein, H., Elhoseny, M., Ali, A.: Challenges and recommended technologies for the industrial internet of things: A comprehensive review. Meas. J. Int. Meas. Confed., **151**, (2020). https://doi.org/10.1016/j.measurement.2019.107198
12. Abomhara, M., Koien, G.: Security and privacy in the Internet of Things: Current status and open issues. In: International Conference on Privacy and Security in Mobile Systems, PRISMS 2014 - Co-located with Global Wireless Summit (2014). https://doi.org/10.1109/PRISMS.2014.6970594
13. Khan, N., Awang, A., Karim, S.: Security in internet of things: a review. IEEE Access (2022). https://doi.org/10.1109/ACCESS.2022.3209355
14. Kiani, F.: A survey on management frameworks and open challenges in IoT. Wirel. Commun. Mob. Comput. (2018). https://doi.org/10.1155/2018/9857026
15. Paradis, L., Han, Q.: A survey of fault management in wireless sensor networks. J. Netw. Syst. Manag. (2007). https://doi.org/10.1007/s10922-007-9062- 0
16. Khan, J., Qureshi, H., Iqbal, A.: Energy management in wireless sensor networks: a survey. Comput. Electr. Eng. (2015). https://doi.org/10.1016/j.compeleceng.2014.06.009
17. Wajgi, D., Thakur, N.: Load balancing algorithms in wireless sensor network : A Survey. IRACST—Int. J. Comput. Networks Wirel. Commun. (2012)
18. Ammar, M., Russello, G., Crispo, B.: Internet of Things: A survey on the security of IoT frameworks. J. Inf. Secur. Appl. (2018). https://doi.org/10.1016/j.jisa.2017.11.002
19. Keerthiga, S., Murugeswari, R.: Survey on software defined networking in IoT. 4Th Int. Sci. Conf. Eng. Sci. Adv. Technol., **2830**(2), p. 030020 (2023). https://doi.org/10.1063/5.0118393
20. Sarica, K., Angin, P.: Explainable security in SDN-based IoT networks. Sensors (Switzerland) (2020). https://doi.org/10.3390/s20247326
21. McKeown, N.: OpenFlow: enabling innovation in campus networks. ACM SIGCOMM Comput. Commun. Rev. (2008)

22. Kerpez, K., Ginis, G.: Software-Defined Access Network (SDAN). 48th Annu. Conf. Inf. Sci. Syst.CISS2014 (2014). https://doi.org/10.1109/CISS.2014.6814134
23. Bastug, E., Bennis, M., Debbah, M.: Living on the edge: The role of proactive caching in 5G wireless networks. IEEE Commun. Mag. (2014). https://doi.org/10.1109/MCOM.2014.6871674
24. Wang, Z., Zha, Y., Chen, S.: Software—Defined networking enhanced edge computing: a network-centric survey. Proc. IEEE (2019). https://doi.org/10.1109/jproc.2019.2924377
25. Aldeeb, F., Ahmed, A.: Software defined wide area network SD-WAN: principles and architecture. 4th Int. AFRICAN Conf. Curr. Stud., no. (2021)
26. Yang, Z., Cui, Y., Liu, Y., Xu, Y.: Software-defined wide area network (SD-WAN): architecture, advances and opportunities. In: Proceedings—International Conference on Computer Communications and Networks, ICCCN (2019). https://doi.org/10.1109/ICCCN.2019.8847124
27. Braun, W., Menth, M.: Software-Defined networking using openflow: protocols, applications and architectural design choices. Futur. Internet (2014). https://doi.org/10.3390/fi6020302
28. Hartert, R.: A declarative and expressive approach to control forwarding paths in Carrier—Grade networks. Comput. Commun. Rev. (2015). https://doi.org/10.1145/2785956.2787495
29. Yin, X., Jindal, A., Sekar, V., Sinopoli, B.: A Control-Theoretic approach for dynamic adaptive video streaming over HTTP. Comput. Commun. Rev. (2015). https://doi.org/10.1145/2785956.2787486
30. Liu, S., Li, B.: On scaling software-defined networking in wide-area networks. Tsinghua Sci. Technol. (2015). https://doi.org/10.1109/tst.2015.7128934
31. Bannour, F., Souihi, S., Mellouk, A.: Distributed SDN control: survey, taxonomy, and challenges. IEEE Commun. Surv. Tutorials (2018). https://doi.org/10.1109/COMST.2017.2782482
32. Samadi, R., Nazari, A., Seitz, J.: Intelligent Energy-Aware routing protocol in mobile IoT networks based on SDN. IEEE Trans. Green Commun.Netw. (2023). https://doi.org/10.1109/TGCN.2023.3296272
33. Bagaa, M., Taleb, T., Bernabe, J., Skarmeta, A.: A machine learning security framework for Iot systems. IEEE Access, **8**, pp. 114066–114077 (2020). https://doi.org/10.1109/ACCESS.2020.2996214
34. Zafar, F., Samad, H., Syed, J., Ibrahim, A., Alohaly, M., Elsadig, M.: An advanced strategy for addressing heterogeneity in SDN-IoT networks for ensuring QoS. Appl. Sci., **13**(13) (2023). https://doi.org/10.3390/app13137856
35. Ahmed, S., Ilyas, M., Raja, M.: IoT Based Smart Systems using Machine Learning (ML) and Artificial Intelligence (AI): Vulnerabilities and Intelligent Solutions. In: 13th Int. Conf. Soc. Inf. Technol. Proc., no. ICSIT, pp. 56–61 (2022). https://https://doi.org/10.54808/ICSIT2022.01.56
36. Ren, W., Sun, Y., Luo, H., Guizani, M.: SILedger: a blockchain and abe-based access control for applications in SDN-IoT Networks. IEEE Trans. Netw. Serv. Manag. (2021). https://doi.org/10.1109/TNSM.2021.3093002
37. Yan, L., Ding, X., Zhang, G.: Dynamic channel allocation aided random access for SDN-enabled leo satellite IoT. J. Commun. Inf. Networks (2021). https://doi.org/10.23919/jcin.2021.9475123
38. Wei, M., Liang, E., Nie, Z.: A SDN-based IoT Fine-grained access control method. In: International Conference on Information Networking (2020). https://doi.org/10.1109/ICOIN48656.2020.9016542
39. Liu, J., Shou, G., Xue, J.: Connectivity enhancement of Edge computing over Fiber- Wireless Access Networks for IoT. In: Asia Communications and Photonics Conference, ACP (2018). https://doi.org/10.1109/ACP.2018.8596035
40. Alam, M., Ahmed, N., Matam, R., Mukherjee, M., Barbhuiya, F.: SDN-Based reconfigurable Edge network architecture for industrial internet of things. IEEE Internet Things J., **10**(18), pp. 16494–16503 (2023). https://doi.org/10.1109/JIOT.2023.3268375

41. Lo, N., Niang, I.: SDN-based QoS architectures in Edge-IoT Systems: A Comprehensive Analysis. In: IEEE World AI IoT Congress, AIIoT (2023). https://doi.org/10.1109/AIIoT58121.2023.1017434
42. Das, R., Ahmed, N., Pohrmen, F., Maji, A., Saha, G.: 6LE-SDN: An Edge-Based Software-Defined network for internet of things. IEEE Internet Things J. (2020). https://doi.org/10.1109/JIOT.2020.2990936
43. Yuniarto, R., Sari, R.: Performance analysis of multipath deployment in software-defined wide area network (SDWAN). In: International Conference on Artificial Intelligence and Computer Science Technology (2021). https://doi.org/10.1109/ICAICST53116.2021.9497800
44. Shozi, T., Dlamini, S., Mudali, P., Adigun, M.: An SDN solution for performance improvement in dedicated wide-area networks. In: Conference on Information Communications Technology and Society, ICTAS (2019). https://doi.org/10.1109/ICTAS.2019.8703613.
45. Dawaliby, S., Bradai, A., Pousset, Y.: Network slicing optimization in large scale lora wide area networks. In: Proceedings of IEEE Conference on Network Softwarization: Unleashing the Power of Network Softwarization, NetSoft (2019). https://doi.org/10.1109/NETSOFT.2019.8806711
46. Prando, L., De Lima, E., De Moraes, L., Hamerschmidt, L., Fraindenraich, G.: Experimental performance comparison of emerging low power wide area networking (lpwan) technologies for IoT. In: IEEE 5th World Forum on Internet of Things - Conference Proceedings (2019). https://doi.org/10.1109/WF-IoT.2019.8767343
47. Misra, S., Bera, S., Achuthananda, M., Pal, S., Obaidat, M.: Situation-aware protocol switching in software-defined wireless sensor network systems. IEEE Syst. J. (2018). https://doi.org/10.1109/JSYST.2017.2774284